19th International Summer School on Vacuum, Electron and Ion Technologies (VEIT2015)

Journal of Physics: Conference Series Volume 700

Sozopol, Bulgaria
21 – 25 September 2015

Editors:

M.C.M. van de Sanden
Miglena Dimitrova
Chavdar Ghelev

ISBN: 978-1-5108-2283-2
ISSN: 1742-6588

Printed from e-media with permission by:

Curran Associates, Inc.
57 Morehouse Lane
Red Hook, NY 12571

Some format issues inherent in the e-media version may also appear in this print version.

Printed by Curran Associates, Inc. (2016)

For permission requests, please contact the Institute of Physics
at the address below.

Institute of Physics
Dirac House, Temple Back
Bristol BS1 6BE UK

Phone: 44 1 17 929 7481
Fax: 44 1 17 920 0979

techtracking@iop.org

Additional copies of this publication are available from:

Curran Associates, Inc.
57 Morehouse Lane
Red Hook, NY 12571 USA
Phone: 845-758-0400
Fax: 845-758-2633
Email: curran@proceedings.com
Web: www.proceedings.com

Table of contents

Volume 700

19th International Summer School on Vacuum, Electron and Ion Technologies (VEIT2015)21–25 September 2015, Sozopol, Bulgaria

Accepted papers received: 29 February 2016Published online: 7 April 2016

Preface

Papers

PLASMA-SURFACE INTERACTION AND PLASMA DIAGNOSTICS. MODELING AND COMPUTER SIMULATION

SURFACES AND THIN FILMS PROCESSING AND ANALYSIS

THIN FILMS DEPOSITION. COATINGS FOR ADVANCED APPLICATIONS

Journal of Physics: Conference Series **700** (2016) 011001 doi:10.1088/1742-6596/700/1/011001

VEIT 2015

The International Summer School on Vacuum, Electron and Ion Technologies (VEIT) has been organized biennially since 1978, when the series of VEIT Schools was launched by the Institute of Electronics, Bulgarian Academy of Sciences with the aim to act as a forum for exchange and dissemination of knowledge and ideas on the latest developments in electron-, ion-, and plasma-assisted technologies. The organizers of the 2015 edition of the event were the Institute of Electronics, Bulgarian Academy of Sciences, Sofia, Bulgaria and the Dutch Institute for Fundamental Energy Research, Eindhoven, The Netherlands. While the school has initially been providing a meeting place for researchers mainly from Eastern and Central European countries, its importance has grown issue by issue. The school is now a major scientific event and a meeting place for young scientists from Eastern and Western Europe involved in research and development associated with high-tech industries. Many former school participants have gone on to become leading scientists in research establishments and companies throughout the world. Leading international companies, such as High Voltage Engineering, Balzers, Varian, and Hauzer have used the VEIT forum to present their products through oral presentations, poster contributions or exhibits. The School Proceedings have been published in special issues of the international journals Vacuum, Plasma Processes and Polymers, Journal of Physics: Conference Series.

The Nineteenth edition of VEIT was held in the Black Sea resort Sozopol, Bulgaria, on 21–25 September 2015. It was attended by 101 participants from 16 countries: Bulgaria, Czech Republic, France, Germany, Greece, The Netherlands, Portugal, Romania, Russia, Serbia, Slovakia, Slovenia, Sweden, Turkey, Ukraine and UK.

Following the tradition of publishing the VEIT Proceedings, a selection of papers presented at the event is published in this special issue of Journal of Physics: Conference Series, under the originality and quality criteria of acceptance by the journal, including peer reviewing.

The school comprised fifteen plenary and three poster sessions. At the plenary sessions, twenty-one invited talks of general interest were presented by highly renowned international experts in the field (5 invited talks and 16 progress reports), as well as ten progress reports by young scientists. In total, 74 contributed papers were presented during the poster sessions. There were several scientific highlights covering fundamentals of interaction of fast particles with solids and challenging practical applications ranging from novel techniques for creating hard coatings, optical/protective layers, biocompatible materials to nanosized structures produced by evaporation, sputtering or external irradiation. Latest results were presented on ion-beam synthesis and modification in both low-energy (deposition and film growth) and high-energy (sputtering, implantation) regimes, and processing of solid materials aiming at patterning the surface or at creating nanophase systems for electronic or tribological/wear resistant applications.

Despite the busy scientific program, the atmosphere was relaxed and informal. The early afternoons of most conference days were free to stimulate both scientific and social interaction between participants, which often took place on the beach. The social program included a welcome party, an official dinner, and an outing to a winery in the town of Pomorie.

VEIT 2015 owes its success to many people. The International Advisory Committee shaped the scientific program and ensured high-quality plenary presentations by careful selection of invited speakers. The Local Committee bore the brunt of the organization both at the conference site and in dealing with correspondence, abstracts, and manuscripts for these proceedings. We are grateful to our sponsors, the Dutch Institute for Fundamental Energy Research, Eindhoven, The Netherlands, IUVSTA, and the Ministry of Education and Science of Republic of Bulgaria for there generosity that enabled us to support the attendance of students and provided support to deal with mailing, printing, renting the conference site, etc. We are also grateful to the members of the International Advisory Committee - Th. Czerwiec, M. Dimitrova, G. Dinescu, A. Ehiasarian, N. Guerassimov, V. Guerra, I. Katardjiev, H. Kersten, K. Larsson, D. Mataras, W. Möller, M. Mozetic, R. Panek, I. Petrov, Z. Petrovic, B. Rauschenbach, M.C.M. (Richard) van de Sanden, M. Ürgen, for their continuous concern about the high scientific level and the future of the event, as well as for their contribution to the formation of a new generation of scientists and experts in the field. And, last but not least, we would also like to thank all authors for their valuable contributions to these proceedings and to the school, and all reviewers for their hard and tedious, but very important, work.

The next conference in the series will be held in September 2017.

M.C.M. van de Sanden, Miglena Dimitrova, Chavdar Ghelev,
Guest Editors

Peer review statement

All papers published in this volume of *Journal of Physics: Conference Series* have been peer reviewed through processes administered by the proceedings Editors. Reviews were conducted by expert referees to the professional and scientific standards expected of a proceedings journal published by IOP Publishing.

Optical emission spectroscopy of plasma produced by laser ablation of iron sulfide

S Karatodorov[1,3], V Mihailov[1], T Křenek[2] and M Grozeva[1]

[1]Georgi Nadjakov Institute of Solid State Physics, Bulgarian Academy of Sciences, 72 Tzarigradsko Chaussee, 1784 Sofia, Bulgaria
[2]Research Center of New Technologies, University of West Bohemia, 30614 Plzeň, Czech Republic

E-mail: skarat@issp.bas.bg

Abstract. Optical emission spectroscopy studies are reported of laser ablation plasma from an iron sulphide (FeS) target in vacuum. The plasma emission is characterized with spatial and tempoal resolution. The spectral observations are used to calculate the electron plasma parameters. The electron temperature is calculated by the Boltzmann plot method and the electron number density, by the Stark broadening of the emission lines.

1. Introduction

One of the key applications of laser ablation is pulsed laser deposition. The pulsed laser deposition (PLD) is a widespread technique for preparing micro- and nano-structures, e.g. thin films, nano-particles, etc. One of the key advantages of this technique is the possibility to evaporate various kind of materials, i.e. solid, liquid, non-conductive [1], and to transfer the stoichiometry of the target materials to the deposit on the substrate [2]. Recently, interest has been shown in pulsed laser deposition of iron sulfide (FeS) thin films [3]. The FeS semiconducting and magnetic properties and the abundance of a low-price non-toxic raw material [4], make FeS thin films produced by pulsed laser deposition very attractive for solar cell fabrication.

Under PLD conditions, the laser-generated plasma is the carrier of evaporated material from the target onto the substrate surface. Thus, the deposition results depend on the plasma dynamics. This stresses the importance of investigating the plasma parameters and their evolution in time and space. The plasma parameters can be correlated to the thin-film structure deposited. This would give one the opportunity to tune the resulting deposition by tuning the plasma parameters. Thus, knowledge of the plasma evolution would enable one to control the deposits produced.

This work was aimed at obtaining the plasma parameters under conditions similar to those for actual pulsed laser deposition of iron sulfide. The laser ablation was investigated of an iron sulfide (FeS) target in vacuum by a pulsed 1064 nm Nd:YAG. The plasma evolution was studied by optical emission spectroscopy with spatial and temporal resolution. The electron temperature and number density were extracted from the time-resolved spectral observations. The electron temperature was calculated by the Boltzmann plot method and the electron number density was evaluated by the Stark broadening of the excited emission lines.

[3] To whom any correspondence should be addressed.

2. Experimental setup and conditions

The experimental setup for plasma emission investigation of laser ablation from an iron sulphide target consists of three parts – an excitation laser with delivery optics, a vacuum chamber with the target and a spectroscopic registration system. The setup is shown schematically in figure 1.

The laser is a pulsed Nd:YAG Quanta Ray GCR 3 emitting at the fundamental wavelength of 1064 nm. The pulse energy is fixed at 20 mJ and the pulse duration is 8 ns. The laser is directed to the target by a dielectric mirror at 45 degrees. The laser beam is focused on the target by a glass lens with a 25-cm focal distance.

The FeS pellet target is situated in a rectangular metal vacuum chamber. The chamber has quartz windows on three of the sides enabling laser excitation and spectra collection from orthogonal directions. The vacuum chamber is evacuated by an AV 63 Lavat a.s. diffusion vacuum pump ($< 10^{-6}$ Torr).

The registration system consists of a Mechelle 5000 monochromator with Andor iStar DH734-18F-03 ICCD camera and optical fiber (50 μm). The Mechelle 5000 system has the advantage of recording the whole spectral range 240÷800 nm in one camera shot.

In performing each measurement, the iron sulfide pellet was placed in the vacuum chamber. The chamber was evacuated to $< 10^{-5}$ Torr. The optical fiber was placed parallel to the target surface at around 1 mm (close to the ablation volume) or 3 mm above it. The chosen laser pulse-registration delay was set in the range 20÷1000 ns. The laser pulse was fired to generate the plasma from the target material. The laser pulse triggered the optical diode and the registration system. The spectrum was recorded at the set delay time with a camera gate of 40 ns.

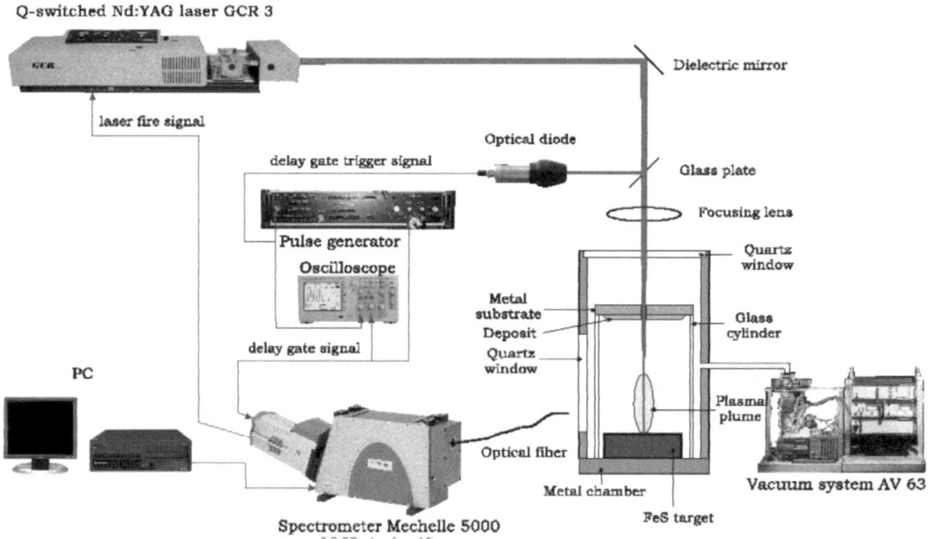

Figure 1. Experimental setup.

3. Experimental results

The laser-induced plasma exhibits a fast expansion in vacuum. It has a weaker spectral emission compared to the emission at higher pressures because of the predominant dissipation of the absorbed laser energy in the directed kinetic motion of the plasma species [5, 6].

Typical emission spectra obtained for 1-mm distance and time-delays between 20÷600 ns are presented in figure 2. The spectra shown are limited to the wavelength region 400 ÷425 nm. As can be seen, there is a decay of the plasma emission with a lifetime of around 500 ns. These characteristics are valid for the full wavelength interval 240÷800 nm recorded in the experiments.

The plasma emission spectra recorded contain a large number of Fe lines and almost completely lack sulfur lines. This is due to the fact that the strong emission lines of neutral sulfur are in the UV and near IR and only weak singly ionized lines are present in the visible region ([7] and references therein). This fact limited the investigation presented here to using only Fe lines for the calculation of the plasma parameters.

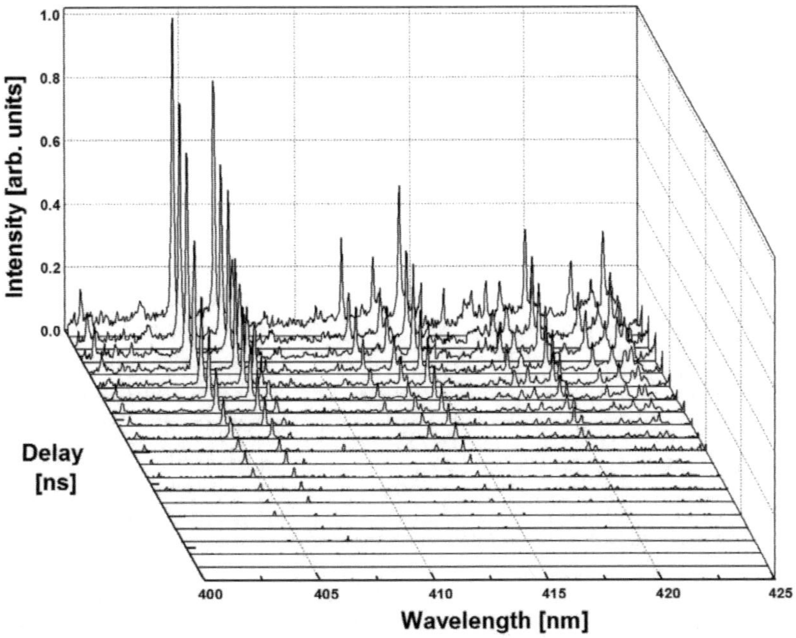

Figure 2. Emission spectra obtained from a FeS sample for 1-mm distance and time delays between 20÷600 ns.

3.1. Electron temperature

The electron temperature is an essential plasma parameter. In the presence of a local thermodynamic equilibrium (LTE), the electron temperature governs the excited levels distribution as described by the Boltzmann equation. In order for the LTE to be satisfied, the electron number density should be high enough so that the excitation rates are higher than the radiative rates. Under these conditions, the excitation electron temperature from the Boltzmann equation coincides with the electronic temperature from the Maxwellian velocity distribution.

The most widely used way of obtaining the electron temperature of laser-induced plasma is the Boltzmann plot method, which is applicable when the local thermodynamic equilibrium is satisfied. The method calculates the electron excitation temperature by measuring the lines intensities from single species. If multiple lines are included, then the good correlation of the experimental data to its linear fit is a reasonable validation of the use of the Boltzmann plot method.

The simplest form of the electron excitation temperature dependence on the species line intensities [8] is given by:

$$\ln\frac{I\lambda}{Ag} = -\frac{E}{kT},$$

where I is the measured line intensity, λ is the line wavelength, A is the transition probability, g is the upper level degeneracy, E is the upper level energy, k is the Boltzmann constant and T is the electron excitation temperature.

A set of 14 Fe I emission lines was used to construct the Boltzmann plot. The lines characteristics are shown in table 1. The set of lines was adapted from [8].

Table 1. Wavelength, upper-level energy, lower-level energy, upper-level degeneracy and transition probability of the Fe I lines used for the Boltzmann plot.

Fe I lines parameters				
Wavelength [nm]	E_j [eV]	E_i [eV]	g_j	A_{ij} [$10^8 s^{-1}$]
368.22	6.91	3.55	5	1.7
368.41	6.09	2.73	7	0.34
368.74	4.22	0.86	9	0.0801
370.56	3.4	0.052	7	0.0322
370.92	4.26	0.91	7	0.156
372.76	4.28	0.96	5	0.225
373.24	5.52	2.2	5	0.28
382.78	4.8	1.56	5	1.05
385.64	3.27	0.051	5	0.0464
386.55	4.22	1.01	3	0.155
387.25	4.19	0.99	5	0.105
389.57	3.29	0.11	1	0.094
389.97	3.27	0.087	5	0.0258
390.29	4.73	1.56	7	0.214

Figure 3 shows the Boltzmann plot obtained for the lines in Table 1 for a delay of 40 ns and a distance of 1 mm. From the plot, the electron excitation temperature is calculated. It is seen in the figure that the linear fit lies well within the error margins of the experimental data, which justifies the assumption of LTE. The value obtained for the excitation electron temperature is 9800 K (0.84 eV).

3.2. Electron number density

The high concentration of electrons and ions in the laser-induced plasmas make them appropriate sources for the measurement of Stark broadening of the emission lines. The classical Stark broadening theory [9] states that isolated lines of neutral elements, with the exception of hydrogen, are primarily broadened by electrons. The collision with electron produces a broadened Lorentzian profile of the emission line.

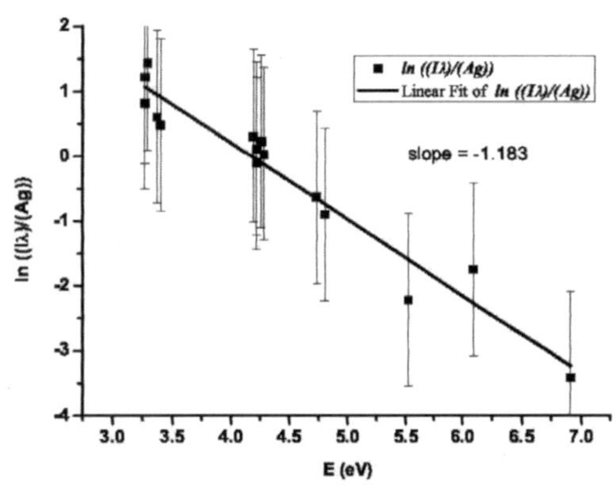

Figure 3. Boltzmann plot of the chosen set of 15 Fe I emission lines.

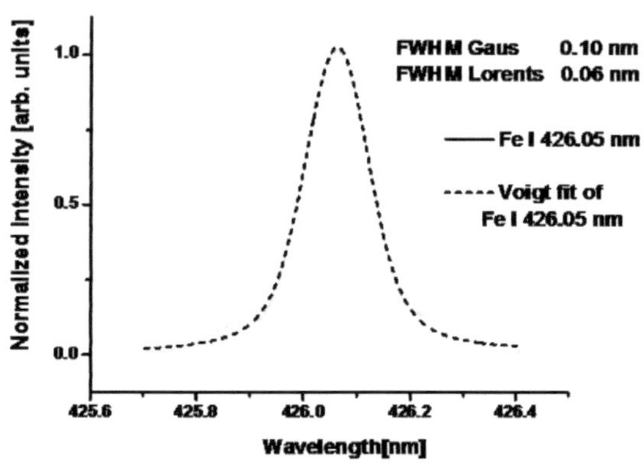

Figure 4. Voigt fit of the Fe I 426.05 nm line.

For calculation of the electron number density, the dependence of the Stark broadening on the electron number density is used:

$$\Delta\lambda_{1/2} = 2w\frac{N_e}{10^{16}},$$

where $\Delta\lambda_{1/2}$ is the measured FWHM Lorentzian profile broadening, w is the natural width of the line and N_e is the electron number density.

In order to obtain the Lorentzian broadening profile, the experimentally observed line Fe I 426.05 nm was deconvoluted. For the deconvolution, the apparatus function of the spectrometer was determined by measuring the linewidth of the Nd:YAG laser second harmonic line (532 nm). In figure 4, the Fe I 426.05 nm spectral line is fitted with a Voigt profile having a Gauss component of 0.10 nm and a Lorentzian component of 0.06 nm. The Lorentzian profile FWHM obtained in this way was inserted in the Stark broadening formula. The data for the *w* Stark parameter of the Fe I line 426.05 nm (0.11 Å) was taken from [10]. The electron number density was calculated to be 2.7×10^{16} cm^{-3} at 60 ns after the laser pulse and was decreasing by one order of magnitude in the next 200 ns.

4. Conclusions

The laser-induced plasma emission from iron sulfide target in vacuum was investigated with spatial and temporal resolution. The electron plasma parameters were calculated from the plasma emission spectra obtained. The Boltzmann plot of the line emission intensities showed an electron excitation temperature of 0.84 eV in the first few tenths of ns after the ablation pulse. For the same time window, the Stark broadening of the Fe I emission line indicated an electron number density of 2.7×10^{16} cm^{-3} with a fast decay in the next 200 ns.

In a future work, the plasma parameters measured here will be used to obtain a correlation with the actual structure of the deposited FeS thin films.

Acknowledgments

This work was performed with financial support under projects INERA (REGPOT-2012-2013-1) and CENTEM, reg. no. CZ.1.05/2.1.00/03.0088, co-funded by the ERDF as part of the Ministry of Education, Youth and Sports' OP RDI program.

References

[1] Cremers D and Radziemski L 2013 *Handbook of Laser-Induced Breakdown Spectroscopy* 2-nd ed (London: Wiley)
[2] Cheung J 1994 *Pulsed Laser Deposition of Thin Films* eds D B Chrisey and G K Hubler (New York:Wiley)
[3] Pola J, Urbanová M, Pokorná D, Bezdička P, Kupčık J and Křenek T 2014 *RSC Advances* **4** 11543
[4] Hofmann W K and Birkholz M 1990 *Solar Energy Mater.* **20** 149
[5] Anabitarte F, Cobo A and Lopez-Higuera J M 2012 *ISRN Spectroscopy* **1** 2012
[6] Karatodorov S, Mihailov V and Grozeva M 2015 *Open Chem.* **13** 187
[7] Gondal M A, Dastageer A, Maslehuddin M, Alnehmi A J, Al-Amoudi O S B 2012 *Optics & Laser Technol.* **44** 566
[8] Aguilera J A and Aragón C 2004 *Spectrochimica Acta* B **59** 1861
[9] Griem H R 1997 *Principles of Plasma Spectroscopy* (Cambridge: Cambridge University Press)
[10] Konjević N, Lesage A, Fuhra J R and Wiese W L 2002 *J. Phys. Chem. Ref. Data* **31** 819

Investigation of the hydrogen neutrals in a discharge source used for production of metal hydrides

I Bozhinova[1,2,3], S Iordanova[1] and A Pashov[1]

[1] Faculty of Physics, St. Kliment Ohridski University of Sofia, 5 J. Bourchier blvd. 1164 Sofia, Bulgaria
[2] Emil Djakov Institute of Electronics, Bulgarian Academy of Sciences, 72 Tsarigradsko Chaussee 1784 Sofia, Bulgaria

E-mail: i.bozhinova@phys.uni-sofia.bg

Abstract. The paper discusses the possible mechanisms for production of metal hydrides (MH) in a DC discharge source. The results of different experiments suggest that the molecules are sputtered directly from the surface of the cathode, where they are formed after adsorption of atomic hydrogen. This hypothesis allows one to understand the operation of the source studied and to optimize its working conditions.

1. Introduction

The high-resolution studies of transition MH (e.g. NiH, FeH, CrH) is motivated mainly by their importance in astrophysics [1, 2]. These molecules have a significant permanent magnetic moment and their presence in the spectra of cold stars allows for the assessment of magnetic fields by the Zeeman splitting and/or broadening. The interpretation of the astronomical observations needs reliable data on the electronic structure of the molecules and on the Landé factors, which can be obtained by high-resolution laboratory measurements [3, 4]. Along with the analysis of a complicated electronic structure, a challenge for the laboratory studies is the effective production of MH. Several sources can be found in the literature, most of them utilizing gas discharges, e.g. [2, 5, 6]. Their performance is comparable, each of them having advantages and disadvantages regarding the densities of the molecules, absorption length, suitability to perform measurements in magnetic fields etc. Some of the sources work effectively in pure hydrogen [5], others, in Ar:H_2 mixtures [2, 6], which indicates different mechanisms for production of MH. This paper presents a study on the performance of the recently reported discharge tube with coaxial geometry [7] in order to shed more light on the pathways for production of MH. Understanding these processes is important for finding the optimal working conditions and, generally, for future construction of sources for MH.

The experimental results on NiH, reported in [7], were obtained by applying laser absorption spectroscopy in pure H_2. It was established that during the operation of the tube, the molecular signal decreases, presumably due to the heating of the cathode. Therefore it was suggested that the formation of NiH is connected with the adsorbed hydrogen on the cathode surface. In this contribution we study this hypothesis and give further evidence that supports it.

[3] To whom any correspondence should be addressed,

2. Operation of the NiH source

A detailed description of the coaxial discharge tube is given in [7]. The cathode consists of eight separate nickel rings with an inner diameter of 65.5 mm, forming a cylinder with a total length of 250 mm. The anode is formed by six stainless steel rods (1 mm in diameter, 300 mm long) laid equidistantly on a cylindrical surface (coaxial with the cathode) with a diameter of 31 mm. The electrodes are mounted in a glass tube equipped with separate electrical feedthroughs for each of them. The discharge is investigated at gas pressure p = 200, 500 and 1600 mTorr and applied DC current I = (20–200) mA. The voltage drop between the cathode and the anode amounts to (350–650) V.

The pressure in the tube changes significantly in a regime of operation without a gas flow (figure 1(a)). After the ignition of the discharge, the pressure initially drops, than it gradually increases and the higher the current the sooner it happens. Depending on the initial conditions (p, I), within 5–6 minutes the gas pressure may triple. The possible reason for this behavior could be the competition between the adsorption of hydrogen on the cathode surface and its desorption. At the present range of gas pressures p = (200–1600) mTorr, and when the discharge is turned off, the adsorption rate is bellow the sensitivity of our detectors (0.1 mTorr/s) for cathode temperatures in the estimated range of 20–70 °C. This means that the adsorption in the current study is mainly due to hydrogen atoms or hydrogen ions – particles produced in the discharge. On reaching the cathode, the hydrogen atoms may be reflected back as atoms or molecules or they may stick to the surface, the latter process contributing to the adsorption. In contrast with the atoms, the ions are accelerated in the constant electric field and they reach the cathode with a significant energy – tens to hundreds of eV depending on the applied voltage and the collision rates with other particles. At the cathode, the ions may be neutralized and back reflected as neutral atoms or they may sputter particles from the cathode, including H, H_2 [8, 9] and NiH. The later process will influence the desorption. The ions may contribute also to the adsorption if they stick to the cathode. Most of the processes associated with ions lead to energy transfer to the cathode and its heating, which in turn reflects on the probabilities of some of these reactions.

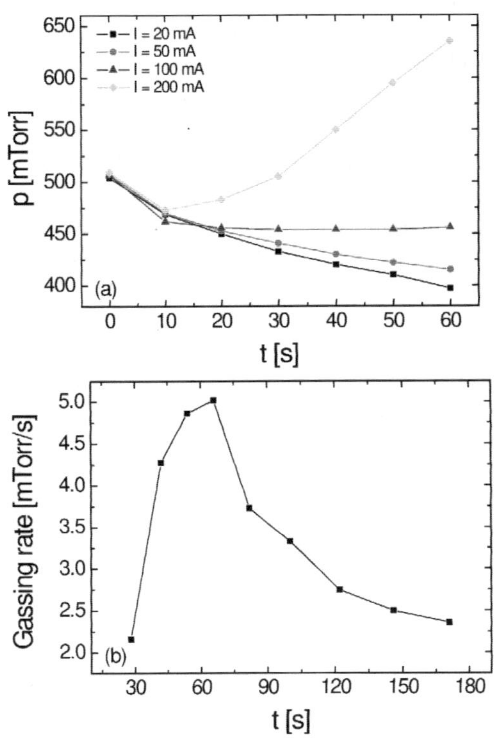

Figure 1. Time dependence of the pressure in the tube at different applied currents (a). Time dependence of the gassing rate at p = 500 mTorr and I = 200 mA (b).

At low currents, respectively low voltages, the processes of adsorption prevail over the desorption, resulting in a pressure reduction (see figure 1(a)). The ions density is low and the cathode remains at a low temperature. At higher currents, the density and the energy of the ions also grow, which leads to heating and intensive sputtering of particles from the cathode. These processes assist the desorption of the hydrogen back to the volume, raising the gas pressure. The interplay between adsorption and desorption in time can be traced out in figure 1(b), where the difference between the desorption and the adsorption rate (called here gassing rate) is illustrated. Before the measurement, the cathode is "filled" with hydrogen by running the discharge for few minutes at a low current (I = 20 mA) to prevent heating. When the current is set to 200 mA, the temperature of the cathode rises and, initially, the rate grows because of the increasing desorption. After about 60 seconds, however, a substantial amount of hydrogen is sputtered from the cathode and the desorption rate starts to decrease. Without "filling" the cathode, the gassing rate decreases immediately after the ignition of the discharge.

It is well known that hydrogen sticks to the metal cathode, forming a molecular bound with atoms from the crystal structure [8, 10]. Therefore, the process of sputtering of NiH molecules seems to be plausible – it was considered as the main process for formation of NiH in the current source [7]. The experimental observations in [7] show that (i) the density of the molecules decreases with the time, (ii) increases with the current and (iii) has an optimum at about p = (300–500) mTorr. The time dependence (i) of the molecular signal can be explained with the gradual heating of the cathode, which increases the hydrogen desorption from the surface and leads eventually to a reduction of the adsorbed hydrogen, i.e. reaching conditions for sputtering of less NiH molecules. At a given gas pressure (ii), the increase of the current results in higher density and energy of the ions, higher sputtering rates and, consequently, to an increase of the NiH signal. The main reasons for the pressure dependence (iii) of the molecular signal at a fixed current could be explained as follows. The lower pressures presume ions with higher energies, resulting in significant sputtering of NiH. This happens for a very short time, because of the fast heating of the cathode and the desorption of the hydrogen. At higher pressures, the density of ions is higher but their energies are lower and provide lower sputtering rates.

3. Spectroscopic observations

The spectroscopic observations are carried out at a current I = 200 mA and at stable gas pressures p = 200, 500 and 1600 mTorr, ensured by the flow regime of operation (gas flow rates between 10 and 40 sccm). The light emitted from the discharge along the axis of the cathode, i.e. perpendicularly to the direction of the electric field, is delivered simultaneously to a spectrometer for detection of emission spectra (Hamamatsu TM-VIS/NIR, 320–1000 nm) and to an interferometric system for analyzing the profile of the H_α spectral line. Detailed description of profile investigation, including the experimental apparatus, can be found in [11]. Briefly, the light collected from the discharge by a light guide is directed to a Fabry-Pérot interferometer through a band pass interference filter for separation of the H_α line. The corresponding interferogram is recorded with a digital camera and analyzed in order to extract the profile of the spectral line. During the discharge operation (3–5 minutes) a series of several spectroscopic measurements (emission spectrum and interferogram) is performed.

The recorded emission spectra show that the ratio of the atomic to molecular line intensities remains almost constant in time (at a fixed gas pressure) within the experimental uncertainty. Contrary to the observations in the RF discharge [11], in the investigated DC discharge the atomic to molecular line intensity ratio decreases with the pressure, being the highest at p = 200 mTorr (figure 2(a)). It may be speculated that the difference comes from the different discharge conditions, determining the possible channels for production of hydrogen atoms. In the RF discharge, H atoms originate mainly from dissociation of H_2. In the DC discharge, the nickel cathode acts as an additional source of H atoms when the desorption processes dominate over the adsorption. A significant contribution to the production of hydrogen atoms may be also expected from reactions between H_2 and ions accelerated in the electric field (see the discussion below). Favorable conditions for these channels are achieved at low pressures and high currents, i.e. high energy of the ions.

Significant differences in the processes leading to hydrogen atoms in both types of discharges are also detected through the interferometric measurements. Figure 2(b) presents a comparison between the fitted profiles of the H_2 line in the RF [11] and in the current DC source. As it could be expected, the narrow central peak, due to emission of thermal hydrogen atoms in both types of discharges, has comparable widths, corresponding to atomic temperatures of about 1000–2000 K (results from this study are shown in figure 2(g)). This emission follows the electron-impact excitation of thermalized ground state atoms arising from dissociation of H_2. The difference between the profiles emerges in the shape of the broad pedestal, although the widths are again comparable (about 10 eV). In the case of the RF discharge the pedestal is flat, which corresponds to emission from almost monoenergetic fast atoms [11]. The latter are products of various reactions passing through excited molecular states (see [11] and references therein). A specific feature of the DC discharges is the existence of a strong electric field leading to acceleration of charged particles towards the cathode – this should be one of the important factors leading to the different appearance of the pedestal shape. The channels of great importance for production of fast excited hydrogen atoms in DC discharges are these with accelerated hydrogen ions taking part in a charge-exchange reaction with H_2 [12,13] and reflection of fast neutralized ions from the cathode [9]. Some of the ions on their way to the cathode collide with H_2 and produce fast excited hydrogen neutrals moving in the same direction. The rest of them reach the cathode, where they may neutralize or neutralize and fragmentize and, as a result of all interactions

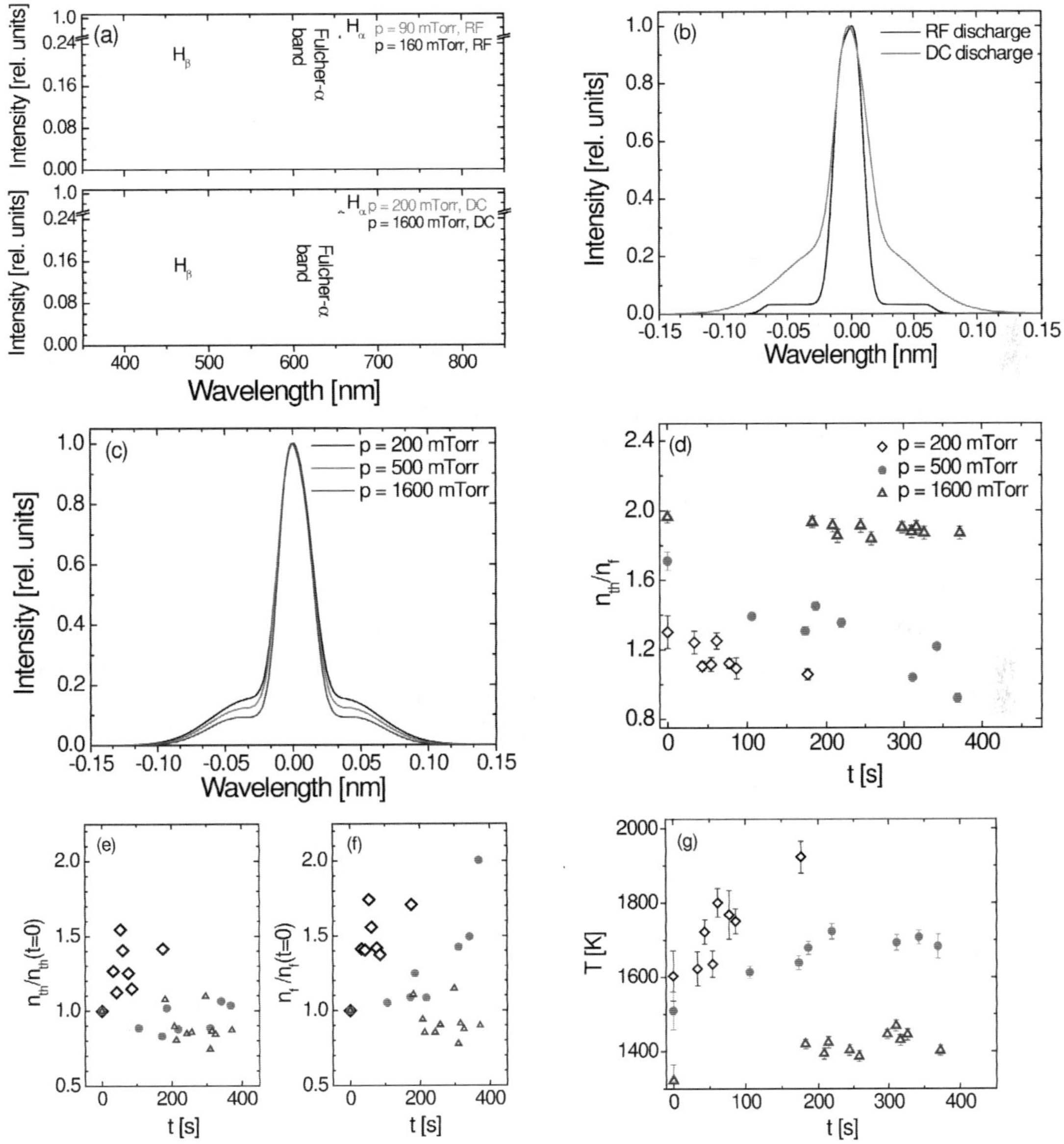

Figure 2. Comparison between emission spectra in the RF and DC discharges (a). Typical H_2 spectral line profile in the RF [11] and the current DC source (b). Change of the H_2 profile with the pressure (c). Time dependence of the ratio between thermal and fast atom densities (d). Time dependence of the relative densities n_{th} (e) and n_f (f). Time dependence of the temperature of thermal hydrogen atoms (g).

with the cathode, fast hydrogen atoms are directed back towards the anode. The ions may also sputter adsorbed hydrogen atoms with significant energies [9]. The fact that the pedestal width in this study stays almost constant (about 10 eV) suggests that the most important source of fast atoms is dissociation of molecular hydrogen (H_2 or H_2^+) which leads to nearly monoenergetic fast atoms, while the change of the wings of the profile may be explained by the significant initial energy of the H_2^+ ions accelerated in the DC electric field. These observations differ from those in the hollow cathode discharge in [13] operated at somewhat higher pressures (2.25–3.75 Torr), where the pedestal width is significantly broader (36–59 eV) and its FWHM changes with the experimental conditions.

Let us now look in more detail at the fitted H_α profiles from the DC discharge at the investigated gas pressures (figure 2(c)). The variation of the experimental conditions leads to changes in the ratio between the areas under the curves describing the central peak and the pedestal. This ratio reflects the ratio between the densities of the thermalized (n_{th}) and the fast (n_f) excited atoms. At higher pressures, the density of all types of particles is higher, leading to significant collision rates and a shorter mean free path. The energy of the accelerated ions is lower and the density of the fast atoms originating from reactions specific for the DC discharges is relatively low. Therefore, at high pressures the pedestal shape is flatter and is more similar to the shape at the RF discharge.

In order to clarify the importance of the interactions between the accelerated ions and the cathode one may follow the evolution of the H_α profile with the time. In figure 2(d), the ratio between the thermal and the fast excited atoms n_{th}/n_f is plotted as a function of the time. For the three working pressures the ratio drops, which means that the relative density of the fast atoms increases. This behavior cannot be explained by the charge-exchange reactions between ions and H_2 because the discharge conditions remain unchanged. The only parameters that change with the time are the temperature of the cathode and the amount of adsorbed hydrogen. As it was shown in [9], the reflection coefficient for hydrogen atoms from nickel depends on the amount of the adsorbed hydrogen on the surface, which is higher for a clean Ni surface. Due to the intensive desorption of hydrogen (see figure 1(b)) during the discharge, the density of the backscattered fast neutralized ions increases, which agrees with the experimental observations. The difference in the slopes of the three n_{th}/n_f dependences in figure 2(d) probably comes from the different desorption rates, which in turn should be associated with the faster increase of the cathode temperature at lower pressures.

From our experimental data it is also possible to asses the time dependence of the densities of the thermalized (n_{th}) and the fast (n_f) excited atoms. They are proportional to the areas of the central peak and the pedestal (figures 2(e) and 2(f)). At 200 mTorr, both densities grow with the time, although (n_f) grows somewhat faster. At 500 mTorr, the density of the fast atoms grows, while the contribution of the thermalized atoms remains almost constant. At the higher pressure (p = 1600 mTorr), both densities are almost constant within the time interval observation. These observations again support the hypothesis that during the discharge operation the density of the fast atoms increases. Nevertheless, more work is needed in order to analyze the contribution of all possible reactions leading to production of fast atoms.

4. Conclusions

In this paper, experimental evidence is discussed supporting the hypothesis that in the coaxial DC source considered the NiH molecules are sputtered from the cathode surface. This hypothesis may help to explain the differences between existing sources and to optimize their working conditions. It turns out that it is important to keep the temperature of the cathode low to prevent the predominance of the desorption processes which reduce the density of NiH molecules on the surface.

The spectroscopic part of this study allows for additional characterization of the hydrogen neutrals in the present discharge source and for evaluation of the relative contribution of reactions leading to formation of excited H atoms. The analysis is in progress and the results will be reported in a separate study.

Acknowledgements
Partial support from the Sofia University Grant 58/2015 and Rila 01/7/2015 Bulgarian Science Found Grant is acknowledged.

References

[1] Kirkpatrick J D 2005 *Annu. Rev. Astron. Astrophys.* **43** 195-245
[2] Vallon R, Ashworth S H, Crozet P, Field R W, Forthomme D, Harker H, Richard C and Ross A 2009 *J. Phys. Chem.* A **113** 13159-66
[3] Vallon R, Richard C, Crozet P, Wannous G and Ross A J 2009 *Astrophys. J.* **696** 172-5
[4] Valenti J A 2001 *ASP Conf. Series* **248** 179-87
[5] O'Brien L C and O'Brien J 2005 *Astrophys. J.* **621** 554-6
[6] Li M, Gray J A and Field R W 1989 *Chem. Phys.* **117** 171-6
[7] Bozhinova I, Kolev St, Dimitrova M, Popov Tsv and Pashov A 2013 *Rev. Sci. Instrum.* **84** 093107
[8] Ehrlich G 1959 *J. Chem. Phys.* **31** 1111-26
[9] Aratari R and Eckstein W 1989 *Nucl. Instrum. and Methods in Phys. Res.* B **42** 11-8
[10] Bartholomew C H 1990 *Catalysis Lett.* **7** 27-52
[11] Iordanova S and Pashov A 2015 *AIP Conf. Proc.* **1655** 004012
[12] Gemišić M R Adamov, Obradović B M, Kuraica M M and Konjević N 2003 *IEEE transactions on plasma science* **31** 44-454
[13] Šišovic N M, Majstorović G Lj and Konjević 2007 *Eur. Phys. J.* D **41** 143–50

Journal of Physics: Conference Series **700** (2016) 012003 doi:10.1088/1742-6596/700/1/012003

Qualitative and quantitative laser-induced breakdown spectroscopy of bronze objects

V Tankova[1,3], K Blagoev[1], M Grozeva[1], G Malcheva[1] and P Penkova[2]

[1]G. Nadjakov Institute of Solid State Physics, Bulgarian Academy of Sciences, 72 Tzarigradsko Chaussee, 1784 Sofia, Bulgaria
[2]National Institute of Archaeology with Museum, Bulgarian Academy of Sciences, 2 Saborna Str., 1000 Sofia, Bulgaria

E-mail: vtankova@issp.bas.bg

Abstract. Laser-induced breakdown spectroscopy (LIBS) is an analytical technique for qualitative and quantitative elemental analysis of solids, liquids and gases. In this work, the method was applied for investigation of archaeological bronze objects. The analytical information obtained by LIBS was used for qualitative determination of the elements in the material used for manufacturing of the objects under study. Quantitative chemical analysis was also performed after generating calibration curves with standard samples of similar matrix composition. Quantitative estimation of the elemental concentration of the bulk of the samples was performed, together with investigation of the surface layer of the objects. The results of the quantitative analyses gave indications about the manufacturing process of the investigated objects.

1. Introduction

During the last twenty years, the increasing popularity of the LIBS method has been due mainly to the relative simplicity and versatility of the experimental setup for investigating different types of objects – inorganic and organic, solids and fluids, located in various environment – air, gas or liquid. Compared to the traditional techniques for analysis, LIBS has a number of advantages: this technique is almost non-destructive and the study can be performed over the whole piece in air at room temperature without preparation or fragmentation requirements. There are no requirements about the shape and the size of the investigated object. The possibility of focusing the laser beam on a microscopic spot allows spatial characterization of the object on the surface and in depth as well. These advantages are of decisive importance when the investigated objects are archaeological artefacts [1, 2, 8].

During the last years, the laser-induced breakdown spectroscopy has demonstrated its possibilities as a promising instrument for spectral analysis of the elemental content of different materials [1, 2]. Since then LIBS has found application in various areas of industry and science. Recently, this analytical method demonstrated the prospects of the technique to become a useful analytical tool in art and archaeology [3-7].

In this work, a set of ancient bronze objects discovered in the North-West part of Bulgaria was

[3] To whom any correspondence should be addressed.

investigated by LIBS. Analytical information obtained from the LIBS measurements was used in order to investigate the surface layer of the objects and to determine the elemental composition of the alloy. A comparison of the results obtained by LIBS and X-ray fluorescence was performed.

2. Experimental

2.1. Samples
In this work we present results from a study of ten bronze finds from the Late Bronze Age (2000 years B.C.). The archaeological samples investigated are part of a bigger collection of various bronze objects that has been discovered in a prehistoric settlement situated in the North-West part of Bulgaria. In figure 1, some of the bronze archaeological samples under study are presented.

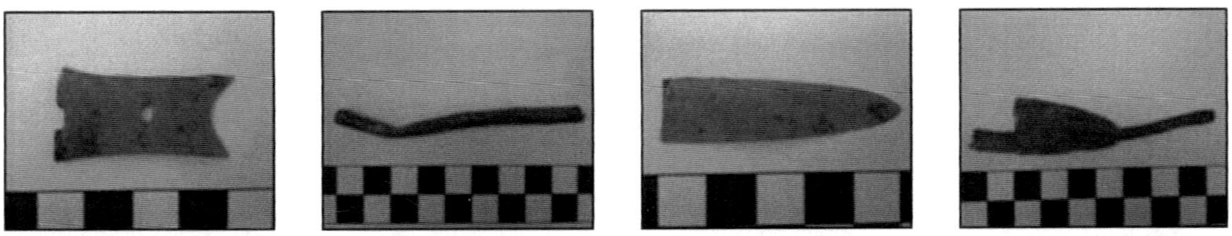

Sample 237 Sample 452 Sample 285 Sample 748

Figure1. Photographs of some of the analyzed archaeological bronze objects.

2.2. Experimental set-up
In the present study, a nanosecond Q-switched Nd:YAG laser (Quanta Ray GC3) was used operating at the fundamental wavelength 1064 nm, with pulse duration 10 ns, pulse repetition rate 10 Hz and energy 750 mJ per pulse. The energy of the laser was reduced to 5-10 mJ using a variable attenuator. The laser beam was focused by a lens (f = 170 mm) on the surface of the sample thus creating a plasma plume. The emitted light was collected by an optical fiber (with diameter 10 μm) and analyzed by an Eschelle spectrometer (type Mechelle 5000, with a grating of 52 grooves/mm and a reverse linear dispersion of 1-4 nm/mm); the spectra were recorded via an intensified charge coupled device (ICCD) detector (DH734-18F-03, Andor Technology) having 1024 × 1024 pixels. The detector was gated by means of a delay/pulse generator type Г5-56. In order to discriminate the atomic and ionic emission from the continuum background of the plasma emission, the delay time of 1μs and gate 1μs were used in the measurements. The spectrometer covers the spectral range 220-850 nm and was calibrated by W/De and Hg/Ar standard lamps for intensity and wavelength, respectively. During the experiments, every sample was placed on a translatable stage and the measurements were carried out within 1 mm from the focal point of the lens, so that a high spatial resolution was achieved. The sample surface was held in a vertical position and the laser beam was focused horizontally at a right angle onto the surface. The experimental parameters were optimized in order to obtain a good signal-to-noise ratio, ensure the best measurements reproducibility and, at the same time, to avoid damaging of the samples.

For surface investigations, emission spectra were recorded for a single laser pulse while for depth profiling measurements spectra were collected and averaged for several successive laser shots.

X-ray fluorescence analysis was performed in order to verify the qualitative and quantitative results obtained by LIBS. For this purpose, a Rayny – EDX-720 spectrometer (Simadzu) was used.

3. Results and discussion

3.1. Qualitative analysis
The main materials used in the Bronze Age have been copper and bronze (copper-tin alloys). Addition of tin to copper at a level of 5-10 % by weight was found to produce a slightly harder alloy, which was easier to cast. Other metals used included lead (Pb) and tin (Sn). The first important analytical

question is to identify the type of alloy and then to determine the quantity of the metals and trace elements contained in order to make a complete characterization of the investigated objects [9].

The typical LIBS spectrum from the archaeological finds analyzed is shown in figure 2. During the experiments, the signal from 10 laser shots on a single position of the sample was accumulated and averaged. All of the samples were found to contain copper (Cu), tin (Sn) and lead (Pb) as main elements. Other elements, such as iron (Fe), nickel (Ni), manganese (Mn), arsenic (As), silicon (Si), calcium (Ca) and phosphorus (P) were also detected as minor and trace elements. The elements like silicon and calcium originate from the soil, where the objects have remained for 4000 years.

The results from the qualitative LIBS measurements were confirmed by the XRF analysis.

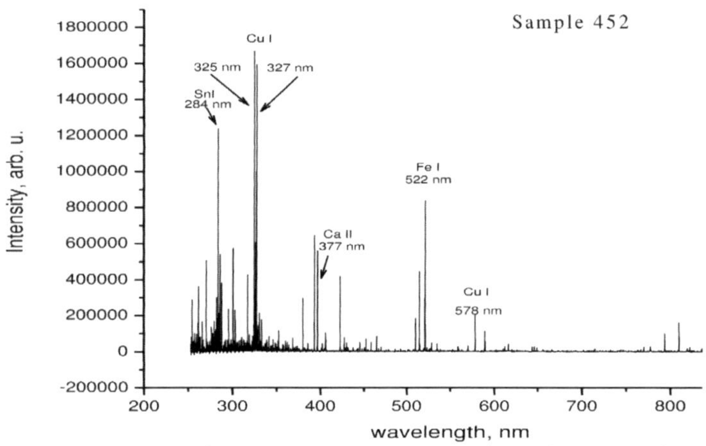

Figure 2. Typical LIBS spectrum obtained for one of the investigated bronze archaeological samples.

3.2. Quantitative analysis

The plasma composition depends not only on the composition of the sample, but also on the laser parameters, sample surface conditions, as well as on the optical and thermal properties of the investigated object. These are known as matrix effects and affect the intensity of the spectral lines of the elements contained in a particular matrix.

In our work, for quantitative determination of the elements in the investigated artefacts, 4 copper alloy reference samples of similar elemental composition with a suitable dominant element were used. In this way, the plasma emission is almost the same for all the samples.

The investigation of the composition of the surface layer was performed using the first several laser pulses applied on the same point on the surface of the sample. In our experiments, in order to reach the pure metal, the surface of the artefacts was mechanically cleaned and after that 100 laser shots were applied before starting the measurements. The concentrations of Sn and Pb in the archaeological objects were obtained adopting the plotted calibration curves for certified bronze alloys (figures 3 and 4).

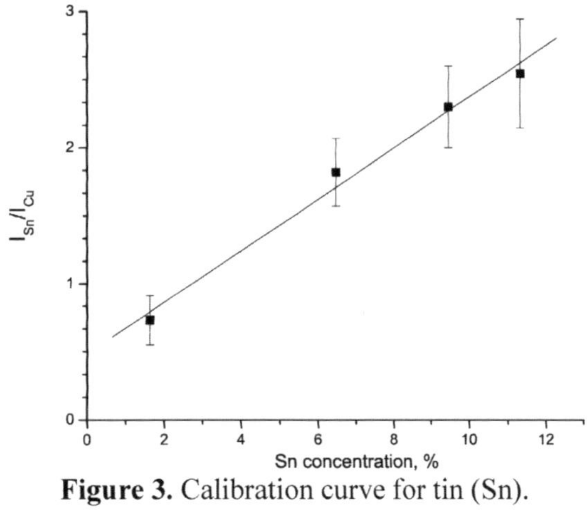

Figure 3. Calibration curve for tin (Sn).

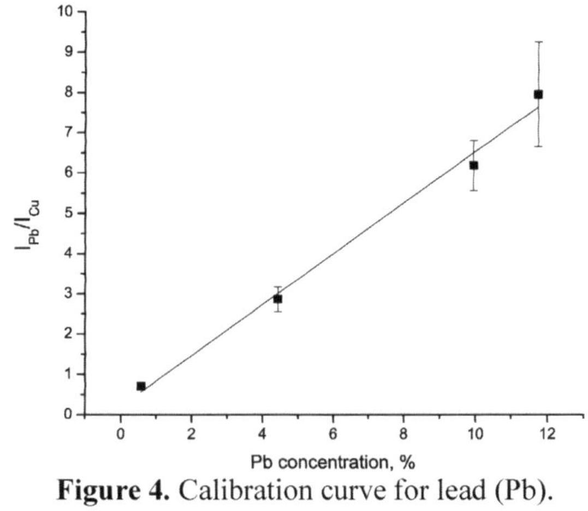

Figure 4. Calibration curve for lead (Pb).

For good statistics, a kinetic series of 10 spectra accumulated for 10 laser pulses were made for bronze archaeological objects and the standard samples.

The results obtained by quantitative LIBS and XRF analyses in the bulk of the samples are shown in table 1. For two of the samples (748, 404), Pb was not observed by XRF because the concentration is close to the detection limit of the experimental system. For some of the results, obtained by LIBS analysis, the error bars are bigger, which could be explained by the sensitivity of this method to the inhomogeneity of the material.

Table 1. Concentrations of tin and lead in the bulk of the samples obtained by LIBS and XRF.

Sample	Concentration of Sn, %		Concentration of Pb, %	
	LIBS	XRF	LIBS	XRF
748	7.8±3.1	10.5±0.1	0.06±0.01	-
404	8.2±0.8	11.5±0.1	0.17±0.03	-
285	8.5±4.7	13±0.1	0.9±0.1	0.9±0.2
409	9.5±0.5	-	0.05±0.01	-
779	10.8±1.7	9.8±0.1	3.6±0.4	3.3±0.2
237	12.1±9.3	14.5±0.1	0.05±0.01	0.6±0.2
318	12.7±1.3	11.5±0.1	0.5±0.1	1.5±0.2
520	13.1±3.4	16±0.1	0.11±0.02	0.7±0.2
492	15.3±6.0	-	1.54±0.15	-
452	17±3.0	19.5±0.1	2.3±0.2	1.5±0.2

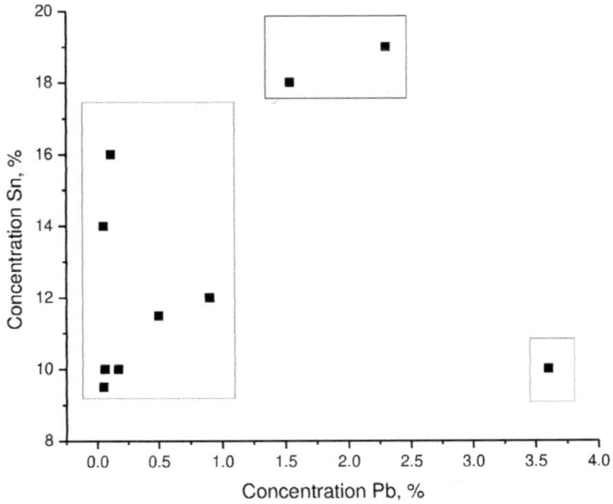

Figure 5. Binary diagram showing compositional differences between the bronze archaeological artefacts investigated.

The quantitative LIBS analysis of the surface content of tin in the bronze samples investigated shows high tin concentration for some of the objects (around 80% tin). This could be an indication of surface treatment of these artefacts during the manufacturing process. For these samples (404, 285, 237, 520, 748), the XRF analysis shows a higher concentration of Sn in the bulk, compared to the LIBS results. The probable reason for this discrepancy is that for the LIBS analysis the surface of the samples was cleaned by 100 laser pulses before the measurements.

The samples 409 and 492 are not analyzed by XRF, and for samples 748 and 404 lead is not detected by XRF.

The results obtained for the Sn and Pb concentrations show that most of the archaeological bronze samples under study contain these elements in concentrations that are typical for the Late Bronze Age [9]. Two of the objects contain more Sn (up to 20%) and Pb and one is with the highest Pb content (about 4 %) (figure 5).

The higher lead and tin concentration might be explained as additionally added for improving the casting properties of the alloy, or might be a result of the process of recycling other metal objects.

4. Conclusions

The results from the analyses show that the measured tin concentration is between 9 % and 20 % in the analyzed samples, which is typical for the Late Bronze Age. A high tin content on the surface, compared to the main body of the objects, was measured for some of the samples. The higher lead and tin content of some of the finds might be an indication of different manufacturing processes. The LIBS results were checked and confirmed by XRF measurements.

Acknowledgements

This work was financially supported by FP7 project REGPOT 316309 INERA. The authors would like to thank the colleagues from the Institute of Archaeology with Museum, BAS, providing the artefacts to be analyzed.

References

[1] Cremers D A and Radziemski L J 2006 *Handbook of Laser-Induced Breakdown Spectroscopy* (John Wiley & Sons Ltd, West Sussex, England)

[2] Miziolek W, Palleschi V and Schechter I 2006 *Laser Induced Breakdown Spectroscopy (LIBS): Fundamental and Applications* (Cambridge University Press, Cambridge, UK)

[3] Castillejo M, Martin M, Silva D, Stratoudaki T, Anglos D, Burgio L and Clark P J 2000 *J. Mol. Struct.* **550-551** 191-8

[4] Maravelaki-Kalaitzaki P, Anglos D, Kilikoglou V and Zafiropulos V 2001 *Spectrochim. Acta* B **56** 887-903

[5] Caneve L, Colao F, Fantoni R and Spizzichino V 2006 *Appl. Phys.* A **85** 151-7

[6] Anzano J M, Villoria M A, Gomushkin I B, Smith B W and Winefordner J D 2002 *Can. J. Anal. Sci. Spectrosc.* **47** 134-40

[7] Lazic V, Colao F, Fantoni R, Palucci A, Spizzichino V, Borgia I, Brunetti B and Sgamellotti A 2003 *J. Cult. Herit.* **4** 303s-8s

[8] Anglos D 2001 *Appl. Spectrosc.* **55** 186A

Tylecote R F 1992 *A History of Metallurgy* Second ed. (London: Maney Publishing, for the Institute of Materials)

Determination of space-time resolved electron temperature in nanosecond pulsed longitudinal discharge in various noble gases and discharge tube constructions

K A Temelkov[1] , S I Slaveeva[1,3] and T P Chernogorova[2]

[1]Metal Vapor Lasers Laboratory, Georgi Nadjakov Institute of Solid State Physics, Bulgarian Academy of Sciences, 72 Tzarigradsko Chaussee, 1784 Sofia, Bulgaria
[2]Faculty of Mathematics and Informatics, St. Kl. Ohridski University of Sofia, 5 James Bourchier Blvd., 1164 Sofia, Bulgaria

E-mail: stisl80@abv.bg

Abstract. Using our results obtained by the analytical solution of the steady-state heat conduction equation for electrons and deriving a new thermal conductivity, 2D (r, t) numerical solution of nonstationary heat conduction, an equation for electrons is found for nanosecond pulsed longitudinal discharge in helium for two different pressures and in neon.

1. Introduction

One of the main fundamental and extremely relevant problems in plasma physics is the determination of the main plasma parameters, namely gas and electron temperatures, electron density, etc. It is very important for gas-discharge laser physics, gaseous discharges, plasma technologies, gas-discharge mass spectroscopy, absorption and emission spectroscopy and plasma in general. The electron temperature determines thoroughly the characteristic constants for elastic and inelastic electron-atom and electron-ion collisions, as well as the three-body electron-ion recombination [1, 2]. In particular for gas discharge lasers, the electron temperature directly influences the formation of the population inversion with the abovementioned processes or indirectly through heavy particles' interaction, such as asymmetric charge transfer, Penning ionization, etc. Determining the electron temperature will be of benefit for theoretical optimization of gas-discharge lasers by self-consistent or simplified kinetic models. The widely applied methods for the electron temperature measurement include the use of a Langmuir probe and laser Thomson scattering, which are inapplicable to the nanosecond pulsed longitudinal discharges (NPLDs) used in our experiments. There exist in the literature several years-consuming models, which predict, among the other parameters, the electron temperature values with considerable variation, and, furthermore, there is no overlap.

Under the conditions of a local thermodynamic equilibrium (LTE), measurement of the relative intensities of some He and Ne spectral lines, originating from different upper levels, enabled us to determine experimentally the electron temperature in a NPLD in He, Ne and Ne-He mixtures, using the relative intensity ratio, Boltzmann relation for the relative equilibrium populations of the two upper levels and the expression derived from them. This electron temperature was space-time

[3] To whom any correspondence should be addressed,

averaged, because space-time averaged relative intensities were measured [3-5]. In spite of the fact that the previous results were space-time averaged, the technique of line-ratio measurement of the temperatures proved to be applicable to NPLDs and to provide a wealth of information on those discharges with further refinement.

Unfortunately, time-resolved spectral investigation using devices equipped by a photomultiplier are prohibited for the discharge period of NPLDs, because of the noise from the high-power electrical pulses and the steep rise of voltage and current pulses of $TV.s^{-1}$ and $GA.s^{-1}$. For the discharge period, i.e. the electrical power pulse period, the spectral signal and the noise were definitely equal, i.e. the signal remained the same using a shutter in front of the spectrometer slit. That is why the time-resolved electron temperature was determined in a NPLD for the afterglow period only [5]. Moreover, despite the tremendous efforts, determination of the radial electron temperature distribution, i.e. the space-resolved electron temperature, also encounters some insuperable obstacles for NPLDs, because of the low signal and the unavoidable excessive noise from the photomultiplier due to the high-power electrical pulses in the kHz range, the secondary electrical pulses in the MHz region and the steep rise in the voltage and discharge current of $TV.s^{-1}$ and $GA.s^{-1}$ (white noise), i.e. a low signal-to-noise ratio.

Assuming that the electron temperature varied only in the radial direction and using the experimental results obtained for the space-time averaged electron temperature, a simple method based on the analytical solution of the steady-state heat conduction equation for electrons was developed for uniform and non-uniform power input in NPLD in He, Ne and Ne-He mixtures [6]. The electronic thermal conductivity was determined from the Wiedemann-Franz law $\frac{k_e}{\sigma} = c.T_e$ for plasma physics, where σ is the specific electrical conductivity, $c = \frac{\pi^2}{3}.\left(\frac{k_b}{e}\right)^2$ is a constant, k_b is the Boltzmann constant, and e is the electron charge, and assuming that σ is independent of the spatial coordinates and had a value of $\bar{\sigma}$. The discharge conditions, such as discharge tubes with special construction and corresponding sizes, buffer gas or buffer-gas mixture at the corresponding pressures, were optimal for laser oscillation in the deep ultraviolet (DUV), and in the near and middle infrared (NIR and MIR) spectral ranges. A comparison was made between the radial electron temperature distributions for an electrical power density independent of the radial coordinate and electrical power density that depends on the radius with a polynomial profile of the second degree. The discrepancy was less than 1×10^{-5} eV and was impossible to be presented in a figure. These results, although satisfactory, enabled us to obtain the electric power density deposited for electrons heating and the initial and boundary conditions for the next advance in our research presented in this paper, namely numerical solution of the nonstationary heat conduction equation for electrons in axial symmetry, i.e. determination of the radial and time dependence of the electron temperature in the NPLDs investigated.

2. Experimental setup

A schematic diagram of the first discharge tube, which is typical for the powerful DUV Cu^+ Ne-CuBr and the NIR Cu^+ He-CuBr lasers, is shown in figure 1 (a). The basic tube with an 11.8-mm inside diameter and a 15-mm outside diameter is made of fused quartz. A ceramic insert, confining the active zone, with an inside diameter of 7.1 mm, an 11.2-mm outside diameter and a length of 100 cm is coaxially sleeved in the basic tube.

Schematic diagram of the second discharge tube, which is typical for the high-power MIR He-$SrBr_2$ laser, is presented in figure 1 (b). The basic tube with a 71.5-mm inside diameter and a 76-mm outside diameter is made of fused quartz. A ceramic insert, confining the active zone, with a 30.5- mm inside diameter, a 38.5-mm outside diameter and a length of 98 cm is coaxially sleeved in the basic tube. The temperature at the quartz tube surface is measured by a thermocouple. The investigated NPLDs are excited by an electrical scheme with interacting circuits (IC scheme) described in detail in [2]. Electric power density ξq_V deposited for electrons heating was 2,0 W cm^{-3} for the first discharge tube (DUV Cu^+ Ne-CuBr and NIR Cu^+ He-CuBr lasers) and 0,2 W cm^{-3} for the second discharge tube (MIR He-$SrBr_2$ laser), as was found in [6].

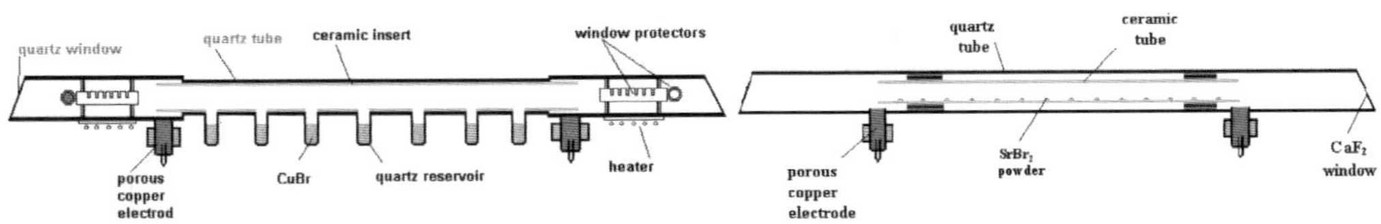

Figure 1. Discharge tube construction used for: **(a)** DUV Cu^+ Ne-CuBr and NIR Cu^+ He-CuBr lasers; **(b)** MIR Sr He-$SrBr_2$ laser.

3. Theoretical results and discussion

The following nonstationary heat conduction equation is solved:

$$c_p\rho\frac{\partial T_e}{\partial t}=\frac{1}{r}\frac{\partial}{\partial r}\left(rk_e(T_e)\frac{\partial T_e}{\partial r}\right)+\xi q_V,\ 0\le r\le R,\ 0\le t\le T \tag{1}$$

under the initial conditions

$$T_e(r,0)=T_e^{\,0}(r)=\left[T_e(R,t)^{1+a}+\frac{1+a}{4B}q_V(R^2-r^2)\right]^{\frac{1}{1+a}} \tag{2}$$

and boundary condition

$$\lim_{r\to 0} rk_e(T_e)\frac{\partial T_e}{\partial r}=0\,. \tag{3}$$

The second boundary condition $T_e(R,0)$, together with the discharge zone radii R and some required thermodynamic properties are presented in table 1. The excitation pulse duration T_1 is set at 140 ns for the NPLDs studied. The period between the excitation pulses T is 50 μs at a pulse repetition rate of 20 kHz.

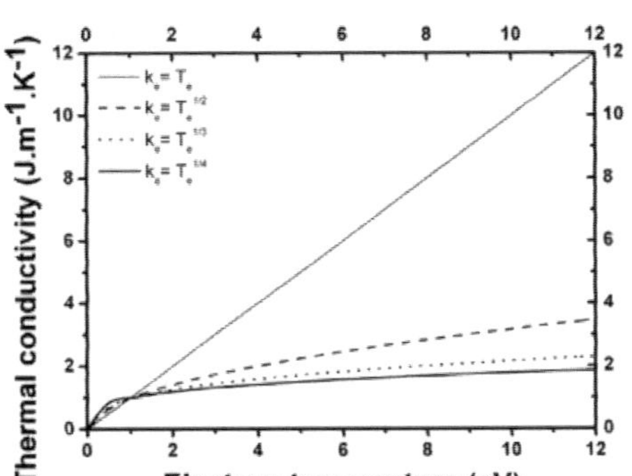

Figure 2. Electronic thermal conductivity as a function of the electron temperature.

Our preliminary results from the computer simulation show that the electronic thermal conductivity derived from the Wiedemann-Franz law does not reflect the physical reality for the nonstationary electron temperature. This is why, a power index scanning was undertaken for the thermal conductivity in the allometric form, i. e. $k_e = B.T_e^{\,a}$, which was widely used for the analytical solution of the stationary heat conduction equation. The results are presented in figure 2.

Using the new thermal conductivity obtained through the power index scanning, theoretical determination of radius- and time-resolved electron temperature is made for the three cases shown in table 1.

Table 1. c_p – specific heat capacity, ρ – gas density, R – discharge radius, $T_e(R,0)$ – second boundary condition.

Lasers	**Buffer gas**	c_p **(J.kg^{-1}K^{-1})**	ρ **(kg.m^{-3})**	***R*** **(mm)**	$T_e(R,0)$ **(K)**
DUV Cu^+ Ne-CuBr laser	Ne (16,7 Torr)	1300	9,525.10^{-3}	3,55	880
NIR Cu^+ He-CuBr laser	He (10 Torr)	5200	1,131.10^{-3}	3,55	880
MIR Sr He-$SrBr_2$ laser	He (45 Torr)	5200	3,395.10^{-3}	15,25	1240

The time-resolved electron temperature for different radial distances is shown in figure 3 in NPLD in Ne (a) and He (b) buffer gases at pressures of 16.7 Torr and 10 Torr, respectively.

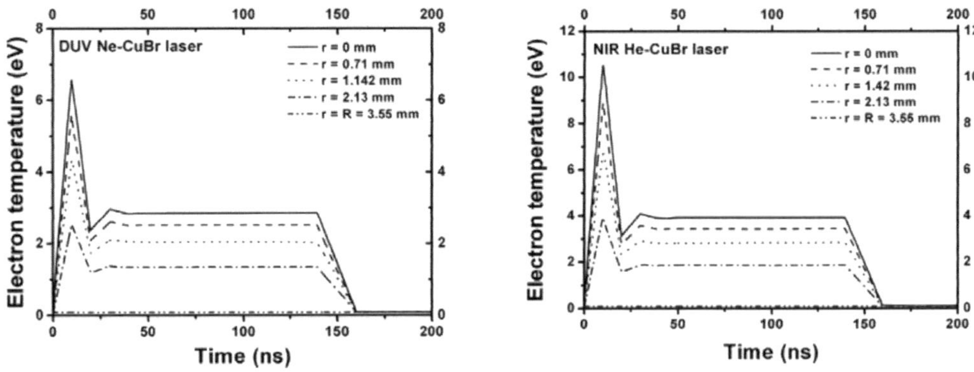

(a) (b)

Figure 3. Time dependence of the electron temperature in NPLD in Ne **(a)** and He **(b)** used for excitation of DUV and NIR Cu^+ lasers, respectively, for different radial distances.

The time-resolved electron temperature for different radial distances is shown in figure 4 in NPLD in He (45 Torr) buffer gas.

The results obtained are in a fair agreement with the self-consistent models developed for NPLD in Ne-Cu and Ne-CuBr lasers [1, 2], as well as to a certain extent with the time-resolved space-averaged electron temperature presented in figure 5, which was obtained by time-resolved measurement of the electrical discharge parameters, such as tube voltage and discharge current [7].

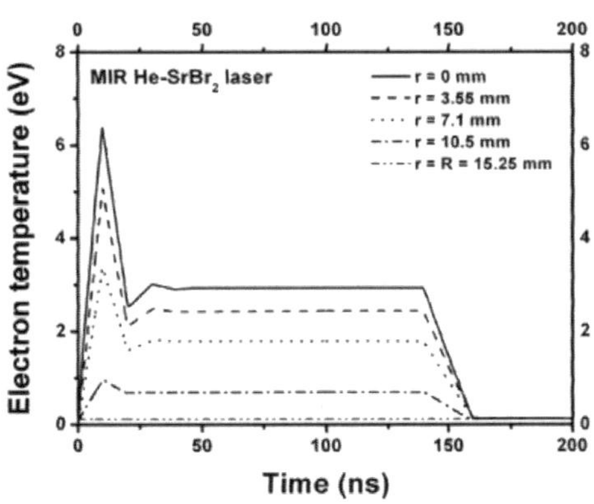

Figure 4. Time dependence of the electron temperature in NPLD in He used for excitation of MIR Sr He-$SrBr_2$ laser for different radial distances.

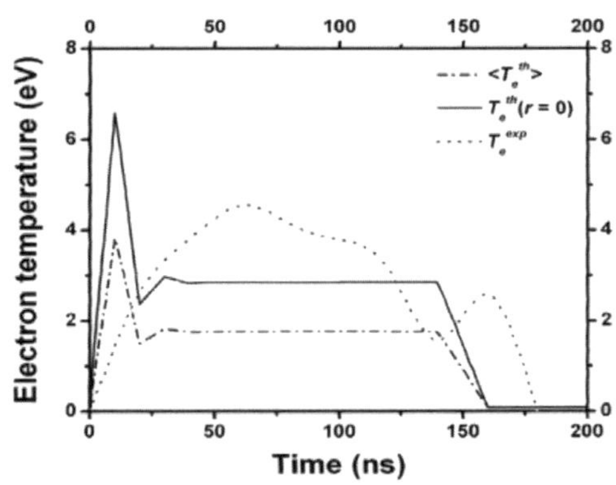

Figure 5. Electron temperature experimentally and theoretically determined in NPLD in Ne (16,7 Torr) used for excitation of a DUV Cu^+ Ne-CuBr laser.

4. Conclusions

Using results obtained by the analytical solution of the steady-state heat conduction equation for electrons and determining new thermal conductivity with one-parameter scanning procedure, 2D (r, t) model is developed through a numerical method and MATWAB software for determination of space-time resolved electron temperature in NPLDs used for excitation of the high-power DUV Cu^+ Ne-CuBr, NIR Cu^+ He-CuBr, the MIR Sr He-$SrBr_2$ lasers and the DUV Cu^+ Ne-CuBr laser.

The discrepancy seen in figure 5 between the experimental results [7] and the new theoretical results are maybe due to the approximation of an uniform excitation pulse over the time. Simulations with time-dependent excitation pulses are in progress.

Acknowledgements

The third author is supported by the Bulgarian Science Fund for Sofia University Foundation under Grant No. 75/2015.

References

[1] Carman R J, Brown J W and Piper J A 1994 *IEEE J. Quantum Electronics* **30**/8 1876-94
[2] Vuchkov N K, Astadjov D N and Sabotinov N V 1994 *IEEE J. Quantum Electronics* **30**/3 750-8
[3] Temelkov K A, Vuchkov N K, Freijo-Martin I. and Ekov R P 2010 *J. Phys.* D: *Appl. Phys.* **43** 075206
[4] Temelkov K A, Slaveeva S I and Vuchkov N K 2011 *IEEE Transactions on Plasma Sci.* **39**/3 831-5
[5] Slaveeva S I, Temelkov K A and Vuchkov N K 2012 *Comptes Rendus de l'Académie Bulgare des Sciences* **65**/8 1043-8
[6] Chernogorova T P, Temelkov K A, Slaveeva S I and Vuchkov N K 2015 *Proc. SPIE* **9447** 944719
[7] Temelkov K A and Vuchkov N K 2014 *IEEE Transactions on Plasma Sci.* **42**/12 3938-41

Theoretical study of thermal conductivities of various gas mixtures through the generalized Lennard-Jones interaction potential for application in gas-discharge lasers

K A Temelkov[1], S I Slaveeva and Yu I Fedchenko

Metal Vapor Lasers Laboratory, Georgi Nadjakov Institute of Solid State Physics, Bulgarian Academy of Sciences, 72 Tzarigradsko Chaussee, 1784 Sofia, Bulgaria

E-mail: temelkov@issp.bas.bg

Abstract. Thermal conductivities of helium, neon, bromine, and hydrogen are calculated on the basis of the *(12-6)* Lennard-Jones interaction approximation. Where necessary for a more precise approximation, a generalized *(n-m)* Lennard-Jones interaction potential is used. Thermal conductivities of binary gas systems are calculated and compared through two different empirical methods for the case of gas discharges in He, Ne, and Ne-He mixtures with small admixtures of bromine and hydrogen. A new simple method is proposed for the thermal conductivity determination for the 3- and 4-component gas mixtures of our interest.

1. Introduction

Scientists often require reliable values of the physical properties of gases essential to the design of many kinds of scientific and industrial equipment. Handbooks provide convenient data sources. In spite of the enormous amount of data collected and correlated over the years [1-4] and the impressive developments in molecular theory and thermodynamics as well [3, 4], the rapid advance of technology into new fields seems always to maintain a significant gap between demand and availability, i. e. the scientist frequently finds a need for physical properties which have not been measured and which cannot be calculated using the existing theory.

One of the problems in the physics of gaseous discharges, gas-discharge lasers, and plasma in general is the determination of the thermophysical properties of gases and gas mixtures, namely, thermal conductivity, diffusion coefficient, etc. In particular, in gas-discharge laser physics thermal conductivity thoroughly determines the gas temperature distribution, and hence, the heavy-particle processes, thermal mode, stability of the laser operation, active particle concentration, etc., as we demonstrated in [5].

The elementary "hard-sphere" gas model assumed all atoms or molecules to be noninteracting rigid spheres of certain diameter and mass moving randomly at a mean velocity. Deviation from the hard-sphere kinetic theory led to studies on the interaction of atoms or molecules based on the realization that molecules attract at intermediate separation and repel when they come very close. The semiempirical *(12-6)* interaction potential function of Lennard-Jones describes attraction and repulsion in an approximate quantitative fashion. Though this potential function is based upon rather tenuous

[1] To whom any correspondence should be addessed.

theoretical grounds and has been widely criticized, it has been extensively used and modified. Where necessary, a generalized *(n-m)* Lennard-Jones (L-J) interaction potential has been used. For some gases, a simplified modification *(n-6)* L-J interaction potential has been employed for the first time in [6].

In view of the enormous prospect for application of several high-power metal atom and ion lasers in precise material processing and characterization of different materials, including soft and hard biological tissues, a detailed theoretical study on the thermal conductivity of various gases, both pure and mixtures, is carried out under gas-discharge conditions that are optimal for laser operation on the corresponding metal atom and ion transition. In order to reduce the operating temperature, metal halides are used instead of the elemental metal.

Thermal conductivities of helium, neon, bromine, and hydrogen are calculated on the basis of *(12-6)* L-J interaction potential. Varying the repulsive parameter n, a new modified *(7.5-6)* L-J interaction potential (M L-J) is used for the first time to calculate the thermal conductivity of helium. Thermal conductivities of binary gas systems are calculated through the empirical method of Brokaw and, where possible, are compared for the first time with ones estimated via the combining (mixing) rules for the case of various gas mixtures. Due to the complex character of the existing formulae for the thermal conductivity of a k-component mixture, for example ones based on the Wassiljewa's equation and its modifications, a new simple method is proposed for the thermal conductivity determination for the 3- and 4-component gas mixtures of our interest.

2. Results and discussion

Following [3] the thermal conductivity $k_{L\text{-}J}$ for *(12-6)* L-J approximation is expressed as follows:

$$k_{L-J} = 0.083264 \frac{T_g^{\frac{1}{2}}}{\mu^{\frac{1}{2}} \cdot \sigma^2 \cdot \Omega_V \left(\frac{k_b T_g}{\varepsilon_0} \right)}, \quad (1)$$

where μ is the particle mass in amu, T_g is the gas temperature in K, σ is the inter-particle distance in Å,

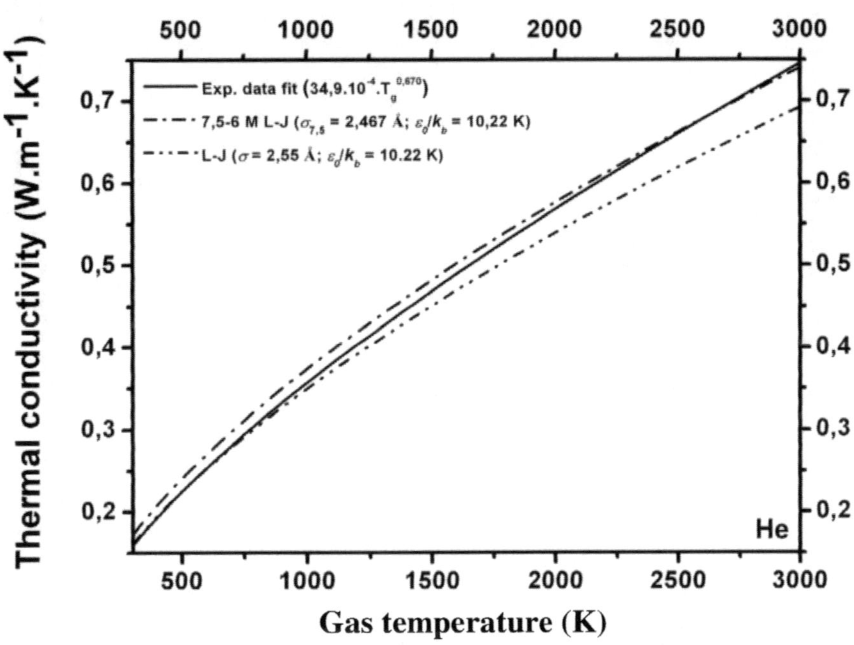

Figure 1. Thermal conductivities as a function of gas temperature for He, considering fit of experimental data, L-J and M L-J approximations.

at which the potential energy is zero, ε_0 is the potential well depth, Ω_V is collision integral, which depends on the interaction potential, k_b is the Boltzmann constant. Ω_V for the L-J potential the following expression is used [3]: $1/\Omega_V = 0.697\left[1+0.323\ln\left(k_b T_g/\varepsilon_0\right)\right]$. It is well known that the *(12-6)* L-J interaction potential function is not suited for quantum gases, namely He, H_2 and D_2. Using the *(n-6)* M L-J interaction potential and varying the repulsive parameter *n* from 6.5 to 30 with a step of 0.5, it is found that a *(7.5-6)* M L-J interaction potential provides the best approximation of the existing experimental data for the thermal conductivity of helium. In figure 1, the thermal conductivities are shown as a function of the gas temperature for helium, considering allometric fit of the experimental data taken from [1], *(12-6)* L-J and *(7.5-6)* M L-J approximations.

L-J interaction potential function is not suited for quantum gases, namely He, H_2 and D_2. Using the *(n-6)* M L-J interaction potential and varying the repulsive parameter *n* from 6.5 to 30 with a step of 0.5, it is found that a *(7.5-6)* M L-J interaction potential provides the best approximation of the existing experimental data for the thermal conductivity of helium. In figure 1, the thermal conductivities are shown as a function of the gas temperature for helium, considering allometric fit of the experimental data taken from [1], *(12-6)* L-J and *(7.5-6)* M L-J approximations.

To use L-J or M L-J interaction potentials, reliable values are required of the two potential parameters (σ and ε_0), which are experimentally determined or theoretically through well-known empirical relations with the critical volume (V_c) and the critical (T_c) or boiling (T_b) temperature. In the all-too-frequent situation, in which no experimental or theoretical values of the needed constants (σ, ε_0, V_c, T_c and T_b) are at hand, rigid (hard) sphere (R Sph) approximation must be used for thermal conductivity estimation. For this approximation, σ is the R Sph diameter (sum of atomic radii of the interacting particles) in Å and Ω_V is unity in (1).

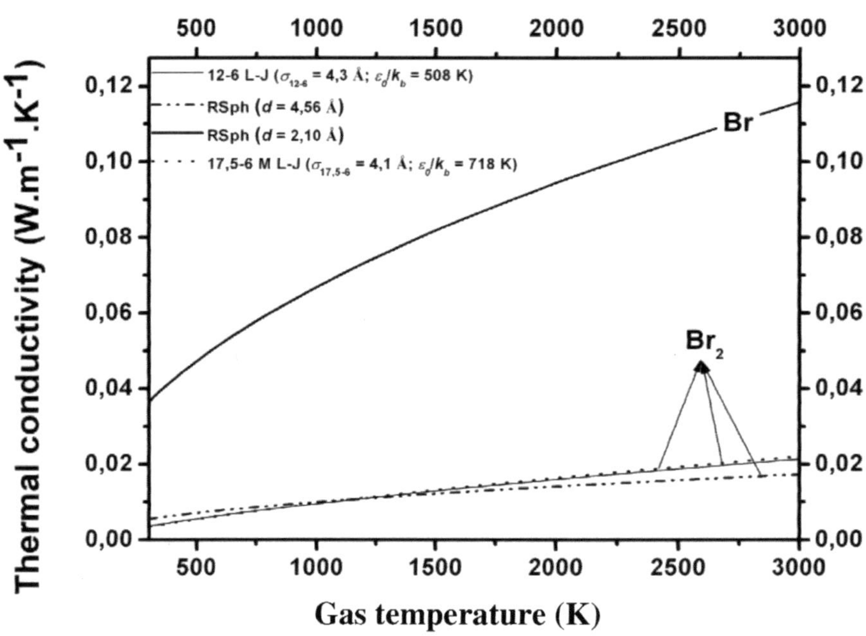

Figure 2. Thermal conductivities as a function of the gas temperature for atomic and molecular bromine.

Fgure 2 presents the thermal conductivities as a function of the gas temperature for atomic and molecular bromine. Reliable experimentally determined values of the two potential parameters and critical constants exist only for molecular (Br_2) bromine. In this figure our results for the thermal conductivity obtained with *(12-6)* L-J interaction potential are also compared with the one obtained with the *(17.5-6)* M L-J interaction potential used in [6]. The discrepancy is negligible. The two potential parameters, which have been experimentally determined [3] and are used in our research, do not require modification of the *(12-6)* L-J interaction potential. From the viewpoint of chemistry, the dissociation of Br_2 molecules to Br atoms is a reversible reaction with rate constants for direct (dissociation of Br_2 molecules) and reverse (recombination of Br atoms) reactions, which mainly depend on the gas temperature. The pressure starts influencing the thermal conductivity at values higher than 1 atm (0.1 MPa). Unfortunately, from the viewpoint of plasmachemistry, the electron-molecule impacts lead to the dissociation prevailing over the recombination under the gas-discharge conditions characterized by an electron temperature of about one order of magnitude higher than the gas temperature. Both temperatures are defined through mean energy or velocity values of the Maxwellian distribution function for the respective particles, i.e. electrons or heavy particles. Using spectroscopic extinction measurements, a detailed study on the Br_2 and Br atom concentration and pressure in the discharge zone has been carried in [7]. In figure 3, the thermal conductivities of the binary gas system Br-Br_2 are plotted as calculated on the basis of the empirical method of Brokaw [3]. Applying the Brokaw's method for binary systems two or three times, the thermal conductivities of 3- and 4-component gas mixtures are obtained. In figure 3, the thermal conductivities are also given for Ne, considering allometric fit of the experimental data taken from [1], and for Ne-Br-Br_2 as well.

Thermal conductivities of gases containing H_2 molecules and H atoms are shown in figure 4 with two different allometric fitting functions, as is made in [5]. This is because of the dissociation of hydrogen molecules, which increases rapidly at gas temperatures above 1600 K, resulting in the intensification of heat extraction from the discharge zone. Thermal conductivities of Ne-H_2-Br-Br_2 and Ne-H-Br-Br_2 are also presented in this figure.

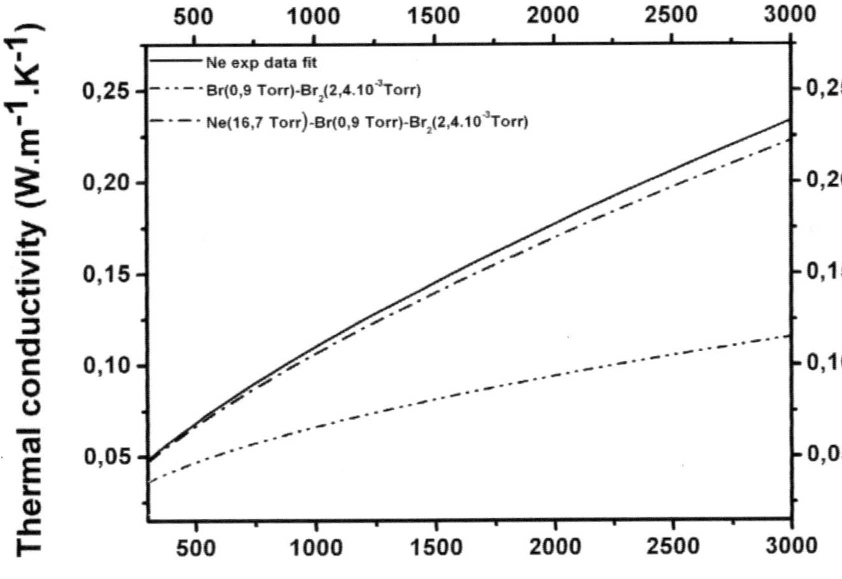

Figure 3. Thermal conductivity as a fuction of the gas temperature for Ne, considering fit of experimental data, Br-Br_2, Ne-Br- Br_2 gas mixtures.

Sometimes certain amount of Ne is added to the buffer-gas He for additional gas-discharge heating, i.e. an additional increase in the gas temperature. Figure 5 presents thermal conductivities of He, Ne, and several Ne-He mixtures, namely, 5 Torr–40 Torr, 10 Torr–35 Torr, and 15 Torr–30 Torr, as a function of the gas temperature using Brokaw's method for binary gas mixtures and the method of combining rules (Ne-He L-J and Ne-He M L-J). Allometric fit of the experimental data, *(12-6)* L-J and *(7.5-6)* M L-J approximations are used for pure He thermal conductivity. A modification of the *(12-6)* L-J interaction potential is not required for pure Ne.

Thermal conductivities of gas mixtures between Br-Br_2 and He(10 Torr), Ne(16.7 Torr), He(45 Torr), and the abovementioned Ne-He mixtures are shown in figure 6.

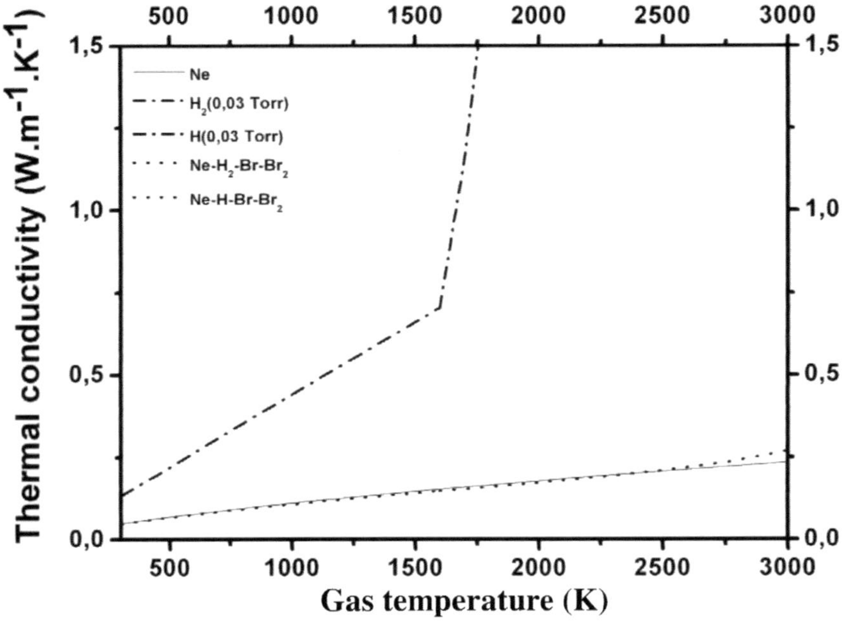

Figure 4. Thermal conductivities of H_2, H, Ne-H_2-Br-Br_2, and Ne-H-Br-Br_2 gases and mixtures.

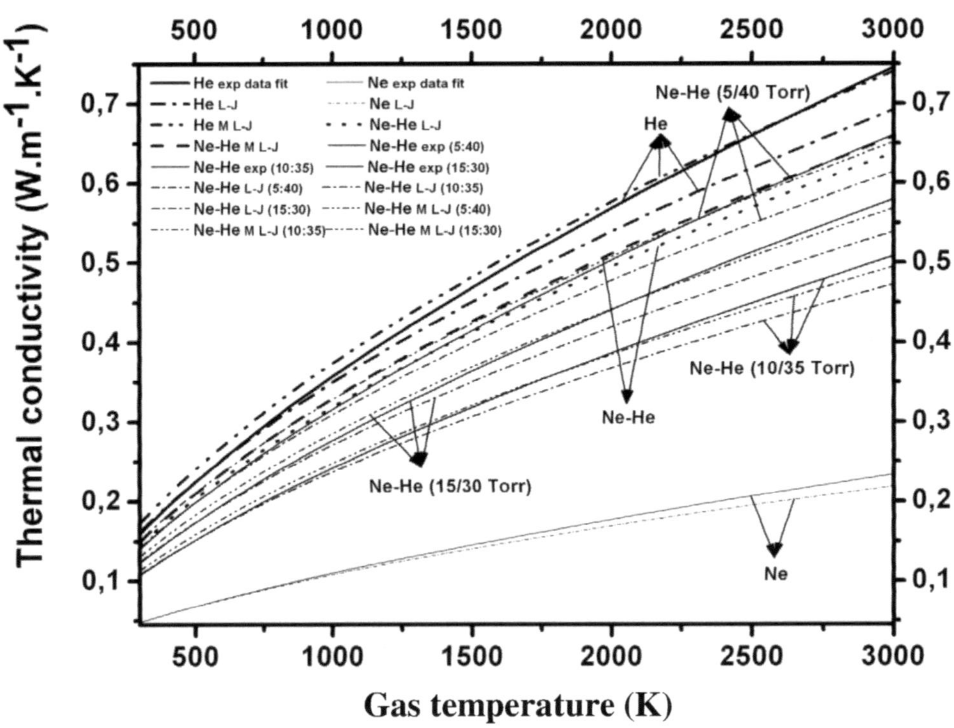

Figure 5. Thermal conductivities of He, considering fit of experimental data, L-J and M L-J approximations, Ne, and Ne-He mixtures.

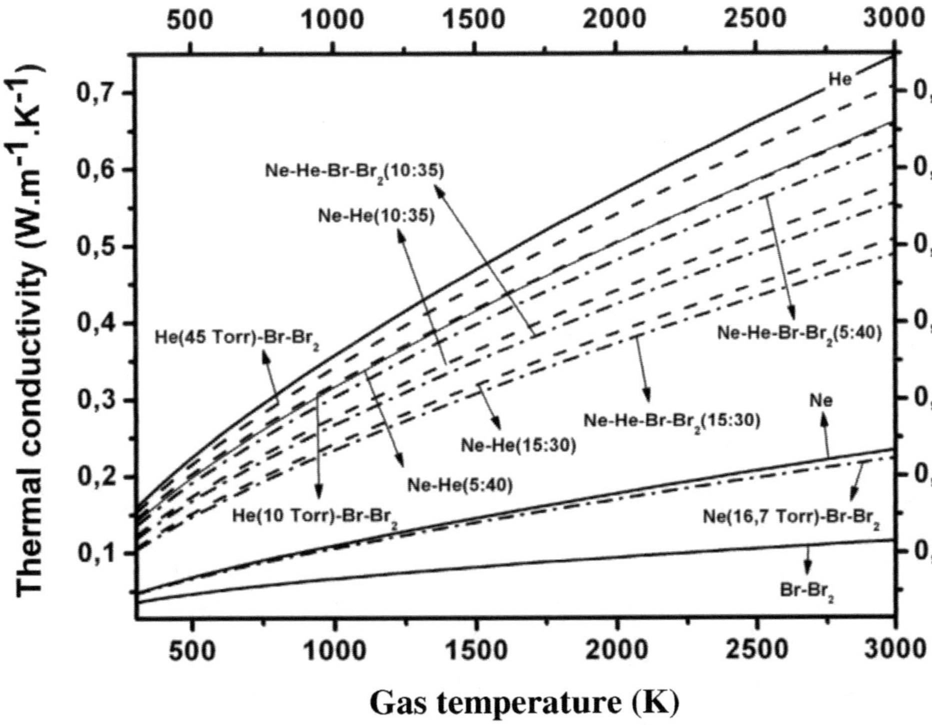

Figure 6. Thermal conductivities of gas mixtures between Br-Br_2 and He, Ne, and Ne-He mixtures.

3. Conclusions

The thermal conductivity of helium, calculated by the new *(7.5-6)* M L-J interaction potential, is a better approximation than the *(12-6)* L-J and R Sph approximations. The empirical method of Brokaw is widely used for binary systems, because it is not dependent on the way of thermal conductivity determination for pure components, namely experimental, (M) L-J, R Sph, etc. Combining rules show promise when one of the pure components is at low pressure in comparison to the other one.

Applying two or three times the empirical method of Brokaw for binary systems, a new simple method is developed for calculation of thermal conductivities of various 3- and 4-component gas mixtures used as active media of several metal atom and ion lasers. Utilization of Wassiljewa's equation for calculation of thermal conductivity of *k*-component mixture is in progress.

References

[1] Handbook 1976 *Tables of physical constants* ed Kikoin I K (Moscow: Atomizdat)
[2] Handbook 1991 *Tables of physical constants* ed Grigoryev I S and Meylihov E Z (Moscow: Energoatomizdat) pp 340-4
[3] Reid R C and Sherwood T K 1966 *The properties of gases and liquids* (New York: McGraw-Hill Book Company)
[4] Hasted J B 1964 *Physics of atomic collisions* (London: Butterworths)
[5] Temelkov K A, Vuchkov N K, Freijo-Martin I and Ekov R P 2010 *J. Phys.* D: *Appl. Phys* **43** 075206
[6] Damyanova M, Zarkova L and Hohm U 2009 *Int. J. Thermophys.* **30** 1165-78
[7] Astadjov D N, Vuchkov N K, Petrash G G and Sabotinov N V 1989 *Metal Vapor and Metal Halide Vapor Lasers* ed G G Petrash (New York: Nova Science Publishers)

Monte Carlo estimates of edge particle sources in TJ-II plasmas

D López-Bruna[1,3], Tsv Popov[2] and E de la Cal[1]

[1]Laboratorio Nacional de Fusión, CIEMAT,
40 Avenida Complutense, 28040 Madrid, Spain
[2]Faculty of Physics, St. Kl. Ohridski University of Sofia,
5 J. Bourchier blvd. 1164 Sofia, Bulgaria

E-mail: daniel.lopezbruna@ciemat.es

Abstract. Three-dimensional calculations of the electron source in plasmas of the TJ-II stellarator (Madrid, Spain) are performed using the Monte Carlo code EIRENE. When possible, the results are compared with diagnostic measurements in equivalent coordinates. Examples are shown for the Hα light evolution during a plasma collapse, CX fluxes, neutrals distributions along diagnostic chords and line radiation emissivities.

1. Introduction

The evaluation of sources in magnetic confinement fusion devices is a complex task. In most cases, the source terms do not conform to flux surfaces and three-dimensional (3D) calculations are required. Particle sources, e.g., involve the motion of neutrals through the vacuum vessel (VV) and the contained plasma, together with atomic physics and interaction with the first wall. In medium size or large tokamak plasmas, one-dimensional calculations can give a fair approximation of the neutrals distribution and the corresponding electron source [1], although detailed calculations near the divertor demand 3D calculations. Matters are commonly more complicated in stellarators, like the TJ-II Heliac [2], where the plasma-wall interaction is strong in a complicated VV geometry. For instance, in addition to the possible insertion of poloidal limiters, typical TJ-II plasmas are limited by the "hard core", a metallic case that protects the central conductors all along the toroidal coordinate.

In this paper we present calculations that require 3D neutrals distributions inside the VV. They are necessary for comparison with experiments that involve specific geometries, like gas puffing valves, diagnostic chords or solid angles through which radiation is collected. Examples are given for (i) the evolution of Hα light during a radiative collapse following the L-H transition, (ii) CX spectra at a specific time of a discharge, (iii) electron source from ionization of neutrals in relation with electric probe measurements and (iv) helium 667 mm light recorded near a gas-puffing valve in He plasmas.

2. Calculation methods and results

The calculations presented here were conducted using the EIRENE code [3] adapted to the TJ-II geometry [4]. Aside from the geometrical inputs for the TJ-II VV, valves, NBI ports and diagnostic

[3] To whom any correspondence should be addressed.

viewing lines, there are three decisive inputs that come from the plasmas under study: gas puffing, plasma distributions of density and temperature, and ion outflows. The plasma distributions are provided as radial profiles, which means that such magnitudes are flux-surface functions. The outflows, or number of particles per second escaping from the plasma, are then obtained from the ion distribution and the net particle confinement time, τ_p. In the TJ-II, this magnitude has been estimated experimentally [5], but in the absence of dedicated experiments it can be also estimated through the use of transport codes coupled with EIRENE if some experimental information, like recycling coefficient [6], Hα [7] or CX spectra [8] can be provided. It is often the case that other sources, like NBI flows, must be taken into account. One difficulty here is that the NBI deposition and shine-through depend on the neutrals distribution itself, which calls for a coupling of different codes to transport calculations [9]. For TJ-II studies of particle and energy transport, we commonly use transport models for the density evolution that allow reproducing the experimental density profiles. The particle sources are updated with enough frequency using the evolving plasma profiles, τ_p and NBI deposition profiles. In this way, the obtained sources are not only compatible with all the provided experimental information, but also with transport. For example, if the plasma is in steady state, the source and confinement times provided are compatible with such a steady state, which may not happen if the evaluation of sources is decoupled from transport.

Figure 1 gives an idea of the geometry of the TJ-II device. It shows the distributions of neutral hydrogen molecules and atoms obtained using parameters from TJ-II discharge #34517, a low-density (~1.1×10^{19} m^{-3}) NBI plasma without additional gas puffing. The structure of the TJ-II ports can be fairly well seen in the distribution of molecules; and the different TJ-II sub-sectors, eight per each of the four periods, can be seen in both distributions. This stresses the importance of the geometry to evaluate diagnostic measurements as in the examples that follow.

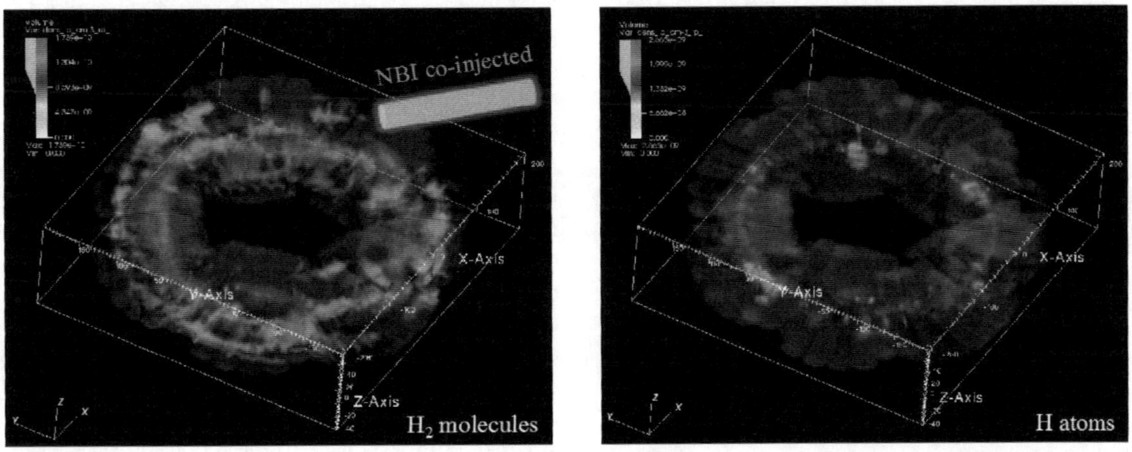

Figure 1. Space density of hydrogen molecules (left, visible in 0.5–1.0×10^{10} m^{-3} range) and atoms (right, 1.0–1.5×10^{9} m^{-3}) based on TJ-II #34517 plasma parameters following 24×10^{6} trajectories. Note TJ-II ports filled with molecules and structure of TJ-II sectors of the VV also visible in the atoms distribution. 3D plots are produced with the *VisIt* software [10].

2.1. Limiter insertion and wall coating

Figure 2.a shows the effect of inserting two poloidal limiters under the conditions of high-density NBI plasma. Two monitors are simulated: one looking at one of the limiters, the other one far from them. When the limiters reach the LCFS position (vertical line) the monitors behave oppositely. The signal increases considerably for the monitor oriented towards one limiter, while it decreases for the other monitor due precisely to the plasma flow intercepted by the limiters. Figure 2.b shows the atom density along the top line of the perpendicular CX diagnostic for identical calculations except that the coating is changed from B to Li. The electron density along the chord is also shown to give an indication of the portion of the chord intersecting the plasma. The higher recycling inherent to B-coating is manifested mainly in the atom density near the LCFS on the diagnostic side of the chord.

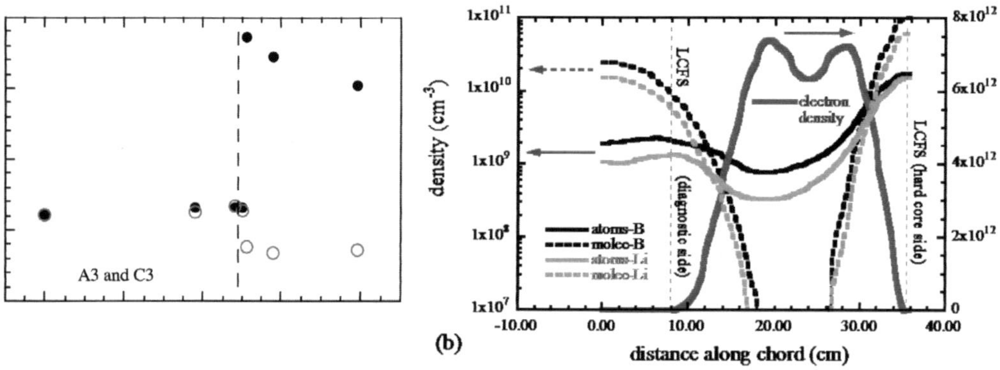

Figure 2. (a) Calculated effect of the insertion of two limiters (sectors A3 and C3) on the Hα light at two locations in NBI plasma: TJ-II sector C4 looking at one limiter; and sector D4 far from both limiters. (b) Effect of coating, B or Li, on the neutrals density along a chord that intersects the plasma near its center for an ECH plasma (τ_p = 5 ms). The electron density along the chord is shown for reference.

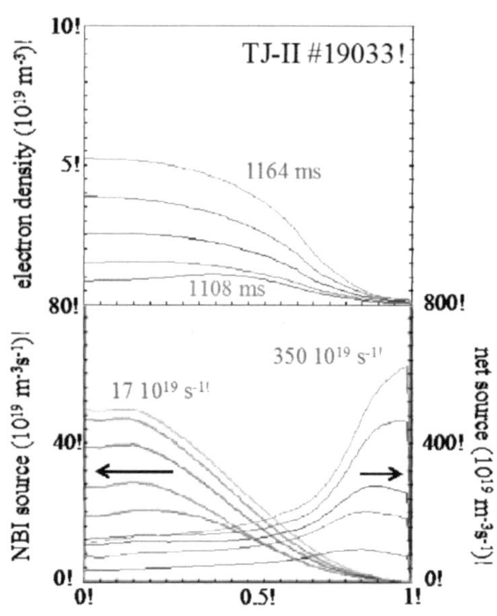

Figure 3. Calculated evolution of the density profile (top) matching experimental measurements, and corresponding electron source profiles from coupled Monte Carlo codes (bottom). Note the different scales for the NBI (left) and total (right) electron sources. Net integral values are indicated for the final profiles (t = 1164 ms).

2.2. Sources and diagnostics in NBI plasmas

Calculations that include particle sources from NBI require a previous estimate of the spatial distribution of neutrals. These distributions are then used to take CX losses into account in Monte Carlo methods [11] for NBI beams, again adapted to the TJ-II (see [12] and references therein), so the corresponding particle source is included in a consistent way despite possible uncertainties, such as gas desorption from direct impact on the NBI protection tiles [13]. Once the neutrals distribution is obtained (e.g. figure 1), the NBI contribution to the net particle source can be calculated. Figure 3 shows that the deposition of particles coming from the neutral beams is small in comparison with the recycling sources. The calculations belong to TJ-II discharge #19033 at even times from approximately the start of NBI (1090 ms) to the time of maximum diamagnetic energy (1165 ms, $W_{dia} \approx 3$ kJ) following the increasing density profiles. The discharge corresponds to balanced NBI up to ≈ 0.6 MW of deposited power according to the calculations. The integral values of the source profiles, indicated for t = 1164 ms, show that the net electron source coming from NBI deposition is about 20 times smaller than the recycling source in this case.

The behavior of Hα emissivity as recorded by different monitors during the radiative collapse can be explained based on the calculated sources as explained above. Figure 4.a shows the time trace of the Hα monitor located in sector A3 of the TJ-II device in a discharge similar to the one in figure 3, where an L-H transition happens close to the radiative collapse due to the increasing density (see [8] for details). The simulated evolution of the intensity for the same Hα monitor matches approximately the experimental one: it decreases at the L-H transition, but the cooling produced by the almost immediate collapse makes the neutrals density grow giving rise to a large increment of Hα light. CX spectra were recorded during the discharge, so that a comparison with calculated values could also be performed (figure 4.b). In this case, two thermal populations were required to explain the curved energy distribution in a logarithmic scale: 0.4% of the ions have a temperature ~ T_e, close to three times the bulk ion temperature.

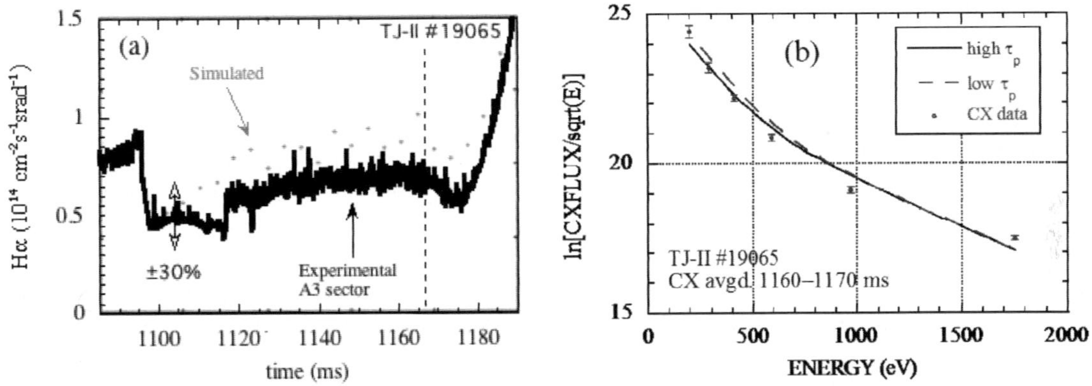

Figure 4. (a) Evolution of Hα light during a radiative collapse following the L-H transition (vertical dashed line) and the calculated intensity for the same Hα monitor. The double arrow indicates the uncertainty in Hα calibration. (b) CX spectrum obtained from a neutral particle analyzer and corresponding EIRENE calculation considering two thermal populations. The lines correspond to two τ_p values differing by 50%.

2.3. Edge sources and probes

Recent work has shown bi-Maxwellian distributions of electrons in the edge of tokamak and stellarator plasmas [14]. One population seems to be the expected thermal component while the other, cooler, is found with energies in the order of ~ 5 eV, but with a higher density. The calculation of neutrals distribution in the sector where the electric probe is inserted (arrow in figure 5.a) shows an approximate flux-surface distribution of the source density term except near the hard-core region, where the plasma is limited by the wall. The flux-surface averaged source density is plotted in figure 5.b, together with the measured electron density that corresponds to each of the two populations mentioned above: high (squares) and low (triangles) temperature. Figure 5.c represents the atom and molecule distributions along a vertical chord passing through the probe. For reference, the electron density is also given for the same chord in the plasma region (between vertical dashed lines). A small shady rectangle indicates the range covered in figure 5.b. Note that the molecule density decreases as the chord approaches the LCFS from outside the plasma, opposite to the atom density.

2.4. Line radiation emissivities

Finally we show a first comparison with measurements taken by a tangential camera set so as to collect the emissivity of He-lines from the plasma region above the helium-puffing valve. Appropriate ratios of He lines are related with the electron density and temperature – this technique can provide their spatio-temporal evolution [15]. The data in figure 6 were collected in ECRH TJ-II plasma where He was puffed at 2 cm from the LCFS ($r = a$). The figure corresponds to the emissivity of the 667 nm

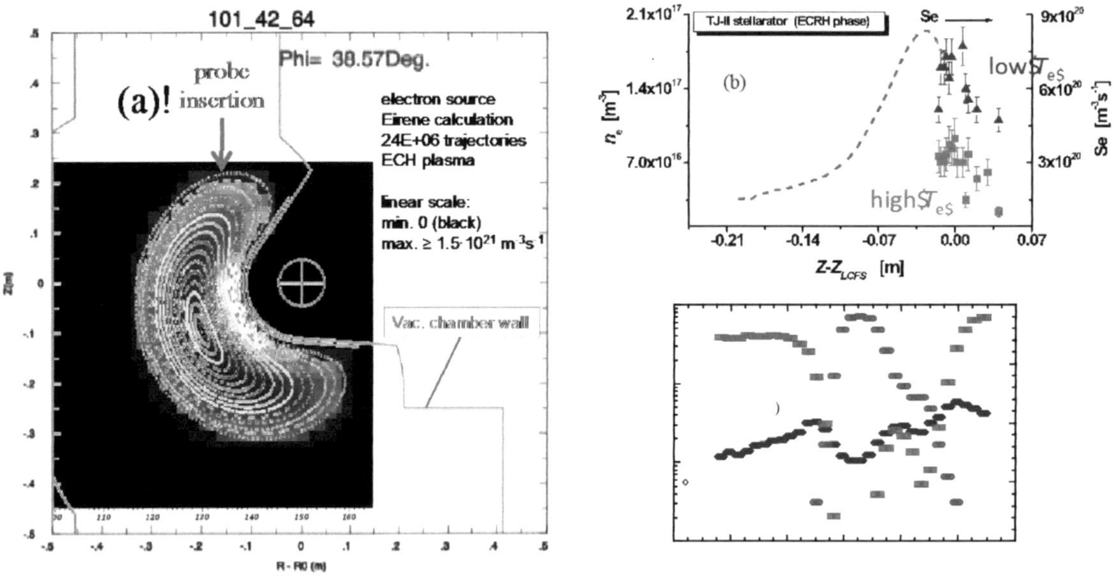

Figure 5. Calculation for a TJ-II discharge in ECH phase. (a) Electron source density distribution at the toroidal angle that corresponds to the location of the electric probe; (b) corresponding radial profile including measurements of the density of thermal and sub-thermal electrons from the probe [14]; and (c) calculated atom and molecule densities along a vertical chord that passes along the probe insertion axis – the n_e profile along the chord is also shown.

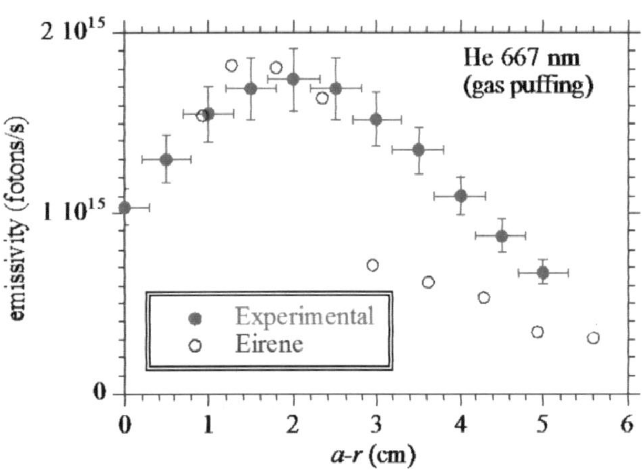

Figure 6. Emissivity of line radiation from He (667 nm) as a function of vertical distance to the LCFS above the He-puffing valve (red dots), and EIRENE calculation for the corresponding simulated diagnostic (blue circles). The experimental error bars represent the uncertainty in the localization of the LCFS and 10 % for the measurements.

He-line received in viewing lines that pass vertically above the gas puffing valve. We overplot the emissivities calculated for the equivalent numerical diagnostic using the experimental density and temperature profiles as input for EIRENE. The experimental emissivities are not calibrated; they were rescaled to match the calculated values. Therefore the confinement time in the calculation is not critical and it was fixed to a reasonable value, 5 ms.

Ongoing research is dedicated to refine the measurements and calculations similar to those shown here, due to their importance to magnetic fusion research. The development of new numerical diagnostics that require neutral distributions, such as Hα spectral analysis [16,17], is encouraged.

3. Conclusions

Calculations that require 3D distributions of neutrals around the plasma-edge region are calculated, based on coupled transport and Monte Carlo codes, for the plasma parameters and geometry of the TJ-II stellarator. They are compared with experimental results (evolution of Hα light during a radiative collapse following the L-H transition, CX spectra, helium 667 mm light recorded near a gas-puffing valve in He plasmas) or used to interpret data (Hα emissivities in different monitors, neutrals densities along diagnostic chords, electron source from ionization of neutrals in relation with electric probe measurements).

References

[1] Rehker S and Wobig H 1973 *Plasma Phys.* **15** 1083
[2] Alejaldre C et al 1990 *Fusion Technol.* **17** 131–9
[3] Reiter D 2005 The EIRENE Code User Manual (including: B2-EIRENE interface) URL http://www.eirene.de/html/manual.html
[4] Guasp J and Salas A 2006 Manual for the Use of Eirene at Ciemat Laboratorio Nacional de Fusión, Ciemat, Madrid URL http://www.eirene.de/html/relevant reports.html
[5] Tabarés F L, Brañas B, García-Cortés I, Tafalla D, Estrada T and Tribaldos V 2001 *Plasma Phys. Control. Fusion* **43** 1023
[6] Vargas V I et al. 2009 Density Dependence of Particle Transport in ECH Plasmas of the TJ-II Stellarator *Informes Técnicos Ciemat 1162 Ciemat, Madrid, Spain*
[7] Velasco J L 2015 Perturbative Particle Transport Experiments with Pellet Injection 20th International Stellarator-Heliotron *Proc. Workshop* (Greifswald, Germany) p.131 (P1S4-44)
[8] López-Bruna D et al. 2013 *Plasma Phys. Control. Fusion* **55** 015001
[9] López-Bruna D, Reynolds J M, Cappa A, Martinell J, García J and Gutiérrez-Tapia C 2010 Programas Periféricos de ASTRA para el TJ-II *Informes Técnicos Ciemat 1201 Ciemat, Madrid, Spain*
[10] Childs H et al. 2013 *VisIt: An End-User Tool For Visualizing and Analyzing Very Large Data High Performance Visualization–Enabling Extreme-Scale Scientific Insight* (Lawrence Berkeley Natl. Lab. USA) pp 357–72
[11] Lister G 1985 *FAFNER: A Fully 3-D Neutral Beam Injection Code Using Monte Carlo Methods* Tech. Report IPP 4/222 Max Pank Institute für Plasmaphysik, Garching, Germany
[12] Guasp J, Fuentes C and Liniers M 2001 Dinámica de impurezas durante la inyección de haces neutros en el TJ-II (simulación) *Informes Técnicos Ciemat 981 Ciemat, Madrid, Spain*
[13] Liniers M et al. 2015 NBI Related Gas Sources in the TJ-II Stellarator *Proc. 42nd EPS Conf. Plasma Phys.* (Lisbon: European Physical Society) P4.168 http://ocs.ciemat.es/EPS2015PAP/pdf/P4.168.pdf
[14] Popov Tsv et al. 2015 *Plasma Phys. Control. Fusion* **57** 115011
[15] de la Cal E, Guasp J and the TJ-II Team 2011 *Plasma Phys. Control. Fusion* **53** 085006
[16] Rapisarda D, Zurro B, Tribaldos V and Baciero A 2007 *Plasma Phys. Control. Fusion* **49** 309
[17] Petrović Z Lj and Phelps A V 2009 *Phys. Rev.* E **80** 066401

Radial profile of the electron energy distribution function in RF capacitive gas-discharge plasma

M Dimitrova[1,5], Tsv Popov[2], N Puac[3], N Skoro[3], K Spasic[3], G Malovic[3], F M Dias[4] and Z Lj Petrovic[3]

[1]Acad. E. Djakov Institute of Electronics, Bulgarian Academy of Sciences, 72 Tzarigradsko Chaussee, 1784 Sofia, Bulgaria
[2]Faculty of Physics, St. Kl. Ohridski University of Sofia, 5 J. Bourchier blvd. 1164 Sofia, Bulgaria
[3]Institute of Physics, University of Belgrade, 118 Pregrevica, 11080 Belgrade, Serbia
[4]Instituto de Plasmas e Fusão Nuclear, Instituto Superior Técnico, Universidade de Lisboa, 1049-001 Lisboa, Portugal

E-mail: miglena.dimitrova@gmail.com

Abstract. This paper reports experimental results on low-pressure argon capacitive RF discharge (parallel-plate capacitively-coupled plasma – CCP) under different conditions, namely, gas pressure in the range $3 \div 30$ Pa and RF power in the range $10 \div 100$ W. The *IV* characteristics measured were processed by two different second-derivative probe techniques for determination of the plasma parameters and the electron energy distribution function. The radial profiles of the main plasma parameters are presented.

1. Introduction

Low-temperature (non-equilibrium) plasmas are the basis of a number of technologies, old, current and future. The success story is, of course, plasma etching, which, together with photo-lithography, is the basis for miniaturization of integrated circuits; other major applications include surface alloying, thin-film deposition, plasma displays, modification of the properties of polymers and organic materials. Recently, the most promising has seemed to be the field of plasma-based medical applications. However, non-equilibrium plasmas are difficult to describe by universal theories, so that joint efforts are needed in diagnostics and modelling in order to understand their properties and use that knowledge to control, design and optimize applications. The importance of the research is that it will facilitate applications while focusing the scientific effort on the fundamental aspects of non-equilibrium plasmas [1]. Joint experimental and modelling efforts are required to make any general, fundamental conclusions about non-equilibrium plasmas.

This paper reports experimental results on studying a capacitive low-pressure argon RF discharge for a range of conditions [2], such as gas pressure in the range $3 \div 30$ Pa and power ranging from 10 W to 100 W. The current-voltage (*IV*) characteristics obtained by a Langmuir probe were processed by two different second-derivative probe techniques [3] to determine the plasma parameters and the

[5] To whom any correspondence should be addressed.

electron energy distribution function (EEDF). The radial profiles of the main plasma parameters were also constructed.

2. Langmuir probe measurements in a RF capacitively coupled system

The RF system consisted of a chamber with parallel plate electrodes powered by a RF source with a driving frequency $f = 13.56$ MHz. The discharge in this experiment was ignited at an electrode distance of 0.07 m between the top powered and the bottom grounded electrode both with a diameter of 0.11 m. The transmitted RF power was measured at the source and always stayed below 1 % of transmitted power. A Hiden's ESPION Langmuir probe system with a motion stage was used to measure the *IV* characteristics. The probe tip with a diameter of 0.15 mm and a length of 10 mm was placed in the plasma with the motion stage allowing motion along the radial direction at $L = 0.02$ m above the bottom electrode. The *IV* characteristics' data were recorded using the ESPION software and then exported for further analysis.

In the "classical regime", which corresponds to our experimental conditions, the probe operates in the absence of a magnetic field and at low gas pressures in the range of 0.1 Pa to 100 Pa [4]. Then the electron probe current of the *IV* characteristic is expressed by the well-known formula [4]:

$$I_e(U) = -\frac{2\pi e S}{m^2}\int_{eU}^{\infty}(W - eU)f(W)dW \;, \tag{1}$$

where e and m are the electron charge and mass, S is the probe area, $W = \frac{1}{2}mc^2 + eU$ is the total electron energy in the probe sheath and c is the electron velocity. Here the probe is negatively biased by a potential U_p and U is the probe potential with respect to the plasma potential U_{pl} ($U = U_p - U_{pl}$). $f(\varepsilon)$ is the isotropic electron energy probability function (EEPF) [1], normalized by:

$$\frac{4\pi\sqrt{2}}{m^{3/2}}\int_0^{\infty} f(W)\sqrt{W}dW = \int_0^{\infty} f(\varepsilon)\sqrt{\varepsilon}d\varepsilon = n\,. \tag{2}$$

The EEPF can be determined by using the Druyvesteyn's formula [5]:

$$f(\varepsilon) = \frac{2\sqrt{2m}}{e^3 S}\frac{d^2 I_e(U)}{dU^2}\,. \tag{3}$$

The second derivatives obtained from the measured *IV* characteristics were evaluated by two techniques. The first one consists in adjacent averaging, smoothing and differentiating twice the measured *IV* characteristics. The instrumental function of the differentiation technique is triangular with a half-width equal to the step of change of the probe bias [6]. The other differentiating technique is based on the convolution of data with an adaptive, differentiating filter, whose instrumental function is automatically adjusted so that noise and distortion, i.e., error, are locally kept at about the same level. Details on how noise and distortion can be readily evaluated can be found in [3], as well as in the references therein. In this work, as before, we continue using (1 + cos (…)) kind of filters for their flexibility on how the full-width-to-half-maximum (FWHM) can be adjusted to run-time, hence how readily the noise-to-error ratio can be kept constant.

Figures 1 and 2 represent examples of the EEDFs evaluated at the center of the discharge chamber at 50 W discharge power and gas pressure of 6 Pa and 27 Pa. The solid black lines present results from direct differentiation. The distortions between 0 and 10 eV cannot be explained by the influence of the differentiation method's instrumental function and the probe size [6]. A possible reason of the appearance of additional distortions is the insufficient ratio Σ between the surface area of the measuring (S_p) and reference (S_r) probes of the HIDEN probe circuit. Usually, it is accepted that it has to be:

$$\Sigma = \frac{S_r}{S_p} > \sqrt{\frac{M}{m}} . \tag{4}$$

Here M is the ion mass. For precise measurements, a more severe criterion for the ratio Σ must be used [7]:

$$\sum = \frac{S_r n_r (T_r)^{1/2}}{S_p n_p (T_p)^{1/2}} G_\alpha \geq 10^4 . \tag{5}$$

In this equation, T and n are the electron temperatures and densities at the positions of the measuring (p) and reference (r) probes. The factor G_α is tabulated in [7] for different probe and reference probe geometries.

On the other hand, the reference probe is flush mounted on the probe holder (close to the probe tip) with surface area not enough to satisfy the condition (5). To compensate for the insufficient ratio Σ in the second technique, an effective resistance R_{eff} was added to the probe circuit. Results with $R_{eff} = 200\Omega$ (red curve) and $R_{eff} = 500\Omega$ (blue curve) are presented in figure 1. As the effective resistance R_{eff} is increased, the EEDF evaluated approaches the Maxwellian with an electron temperature of 5 eV up to the energy of the first excited level of argon (11.56 eV). At higher energies, the EEDF deviates from Maxwellian due to the inelastic electron-atom collisions and the non-equilibrium nature of the discharge. We should point out that the distortions of the blue curve in the range $0 \div 3.5$ eV correspond to the distortions due to the instrumental function and the influence of the probe size [6].

The same considerations are related to the results recorded at a higher pressure presented in figure 2. Here, the effective resistances are $R_{eff} = 2k\Omega$ (red curve) and $R_{eff} = 5k\Omega$ (blue curve). It can be seen that the blue curve is not appropriate. The electron temperature in this regime of the discharge is 6 eV.

We have to point out that to solve the problem mentioned above, additional experiments with different reference probe areas are planned for the near future.

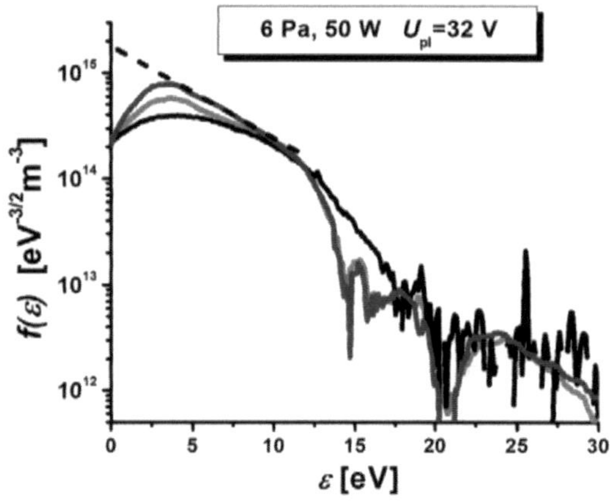

Figure 1. EEPF obtained by direct differentiation of the measured IV (solid black line); red curve – obtained by introducing a 200 Ω effective resistance in the probe circuit; and the blue curve, 500Ω.

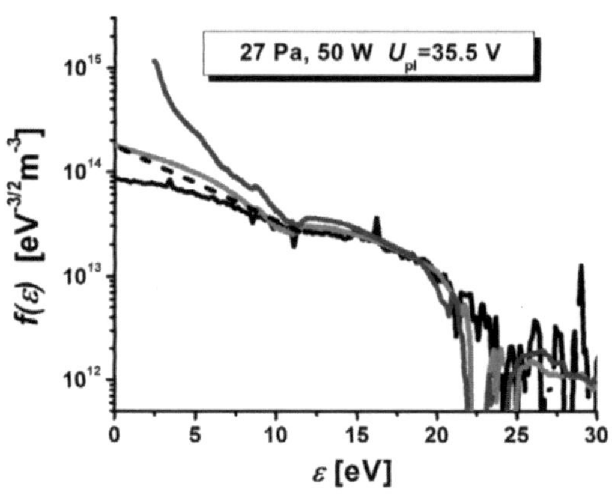

Figure 2. EEPF obtained by direct differentiation of the measured IV (solid black line); red curve – obtained by introducing a $2k\Omega$ effective resistance in the probe circuit; and the blue curve, $5k\Omega$.

3. Experimental results

Below we present the main plasma parameters evaluated from the *IV* characteristics measured at different radial positions. The radial distributions of the plasma potential at different discharge RF power values and different gas pressures are presented in figures 3 a) and 3 b). The accuracy of the evaluation is ~10%. In figure 3, the U_{pl} is higher at the center of the discharge. This effect is more pronounced at higher gas pressures (figure 3 b)). At lower pressures, the U_{pl} is almost constant. The dashed lines indicate the size of the electrodes. As the discharge power is increased, the plasma potential values increase by about 10 V.

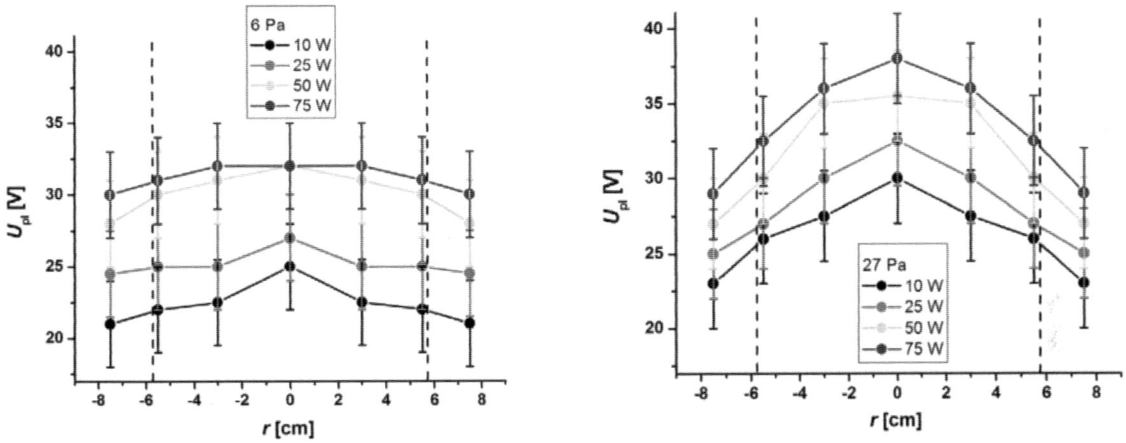

a) b)

Figure 3. Radial distributions of the plasma potential at different discharge RF power values and different gas pressures.

As we mentioned, the EEPF can be approximated by a Maxwellian distribution up to the energy of the first excited level of argon. An example of the resulting electron temperatures are presented in figure 4 a), b).

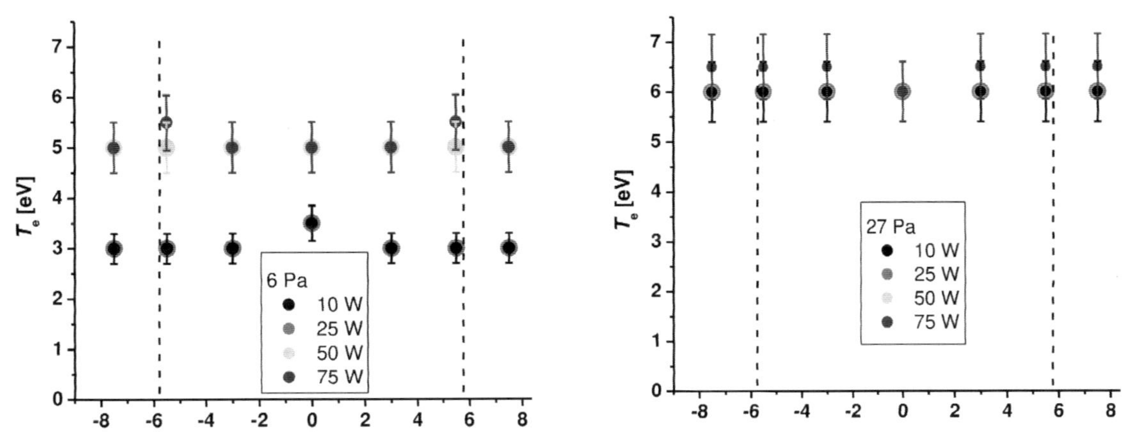

a) b)

Figure 4. Radial distributions of the electron temperatures at different discharge RF power values and different gas pressures.

Figure 5 shows the corresponding radial profiles of the electron densities at different discharge RF power values and different gas pressures.

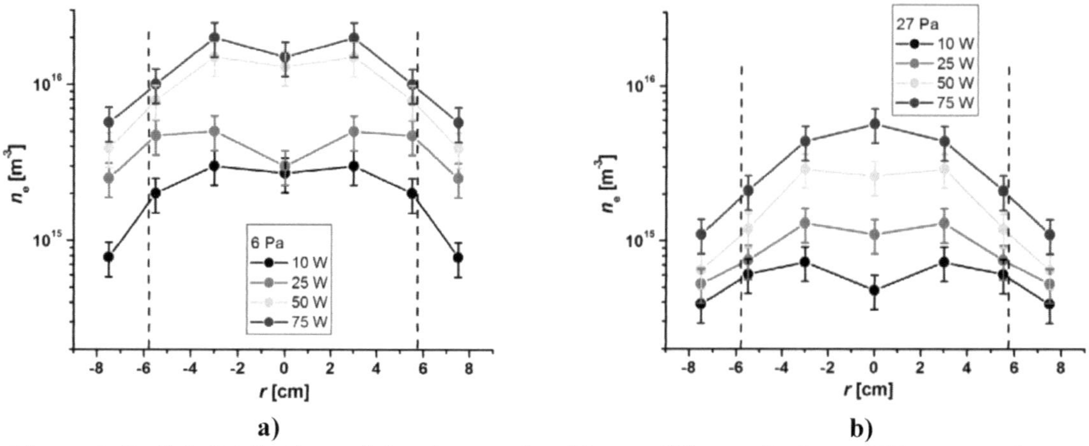

Figure 5. Radial distributions of the electron densities at different discharge RF power values and different gas pressures.

4. Conclusions

This paper reports experimental results on a capacitive low-pressure argon RF discharge at different conditions, as gas pressure in the range $3 \div 30$ Pa and RF power within $10 \div 100$ W.

The current-voltage (*IV*) characteristics measured were processed by two different second derivative probe techniques for determination of the plasma parameters and the electron energy distribution function (EEDF). The radial profiles of the plasma potential and the electron temperatures and densities at different discharge RF powers and different gas pressures are presented.

Acknowledgements

This research has been supported by the JOINT RESEARCH PROJECTs between the Academy of Sciences and Arts of Serbia (SASA) and the Institute of Electronics BAS BG, and the Portuguese FCT – Fundação para a Ciência e Tecnologia, under Project UID/FIS/50010/2013. Additional funding for experiment is provided by SASA project 155 and MPNTR of Serbia ON171037 and III41011.

References

[1] Godyak V A and Demidov V I 2011 *J. Phys. D: Appl. Phys.* **44** 233001
[2] Godyak V A, Piejak R B and Alexandrovich B M 1992 *Plasma Sources Sci. Technol.* **1** 36-58
[3] Dias F M and Popov Tsv K 2007 *J. Phys.: Conf. Series* **63** 012005
[4] Demidov V I, Kolokolov N B and Kudriavtsev A A 1996 *Probe Methods of Diagnostics of Low Temperature Plasma* (Moscow: Energoatomizdat) (in Russian)
[5] Druyvesteyn M J 1930 *Z. Phys.* **64** 781-98
[6] Popov Tsv K, Dimitrova M, Dias F M, Tsaneva V N, Stelmashenko N A, Blamire M G and Barber Z H 2006 *J. Phys.: Conf. Series* **44** 60–9
[7] Chang J-S 1973 *J. Phys. D: Appl. Phys.* **6** 1674

Plasma interaction with tungsten samples in the COMPASS tokamak in ohmic ELMy H-modes

M Dimitrova[1,2,5], V Weinzettl[1], J Matejicek[1], Tsv Popov[3], S Marinov[3], S Costea[4], R Dejarnac[1], J Stöckel[1], J Havlicek[1] and R Panek[1]

[1]Institute of Plasma Physics, Academy of Sciences of the Czech Republic v.v.i., 3 Za Slovankou, 182 00 Prague 8, Czech Republic
[2]Acad. E. Djakov Institute of Electronics, Bulgarian Academy of Sciences, 72 Tzarigradsko Chaussee, 1784 Sofia, Bulgaria
[3]Faculty of Physics, St. Kl. Ohridski University of Sofia, 5 J. Bourchier Blvd., 1164 Sofia, Bulgaria
[4]Institute for Ion Physics and Applied Physics, University of Innsbruck, 25 Technikerstr., A-6020 Innsbruck, Austria

E-mail: dimitrova@ipp.cas.cz

Abstract. This paper reports experimental results on plasma interaction with tungsten samples with or without pre-grown He fuzz. Under the experimental conditions, arcing was observed on the fuzzy tungsten samples, resulting in localized melting of the fuzz structure that did not extend into the bulk. The parallel power flux densities were obtained from the data measured by Langmuir probes embedded in the divertor tiles on the COMPASS tokamak. Measurements of the current-voltage probe characteristics were performed during ohmic ELMy H-modes reached in deuterium plasmas at a toroidal magnetic field B_T = 1.15 T, plasma current I_p = 300 kA and line-averaged electron density $n_e = 5\times10^{19}$ m^{-3}. The data obtained between the ELMs were processed by the recently published first-derivative probe technique for precise determination of the plasma potential and the electron energy distribution function (EEDF). The spatial profile of the EEDF shows that at the high-field side it is Maxwellian with a temperature of 5 – 10 eV. The electron temperatures and the ion-saturation current density obtained were used to evaluate the radial distribution of the parallel power flux density as being in the order of 0.05 – 7 MW/m^2.

1. Introduction

Plasma-material interaction is an important issue for the plasma facing components in current and future fusion devices. The plasma particles incident on the material may result in a number of phenomena, such as implantation, diffusion, retention, sputtering, erosion, ionization of the eroded particles in the plasma and their re-deposition, structural modifications of the material, etc. These will affect both the plasma operation and the performance and lifetime of the plasma facing components. The nature and extent of the interaction depends both on the state of the material surface and on the local plasma parameters. For proper understanding of the interaction, these parameters need to be known.

[5] To whom any correspondence should be addressed.

This paper reports experimental data obtained on the COMPASS tokamak [1] by Langmuir probes (LPs) embedded in the divertor tiles [2] during ohmic ELMy H-modes reached in deuterium plasmas. The measured current-voltage (*IV*) characteristics were processed by the recently published [3,4] first-derivative probe technique (FDPT) for precise determination of the plasma potential and the EEDF. The electron temperatures and densities obtained were used to evaluate the radial distribution of the parallel power flux density and the electron pressure.

Under the experimental conditions described below, we investigated the plasma interaction with tungsten samples with or without a so-called He fuzz, resulting from an exposure of tungsten to He plasma at elevated temperatures.

2. Experiment description

The samples were placed at different vertical positions z (figure 1) in the scrape-off layer at the high-field side (HFS) with two different orientations of the sample set: 4 samples tangential to the separatrix surface (toroidally spaced) and 3 samples sloped toroidally by 50 degrees (radially spaced).

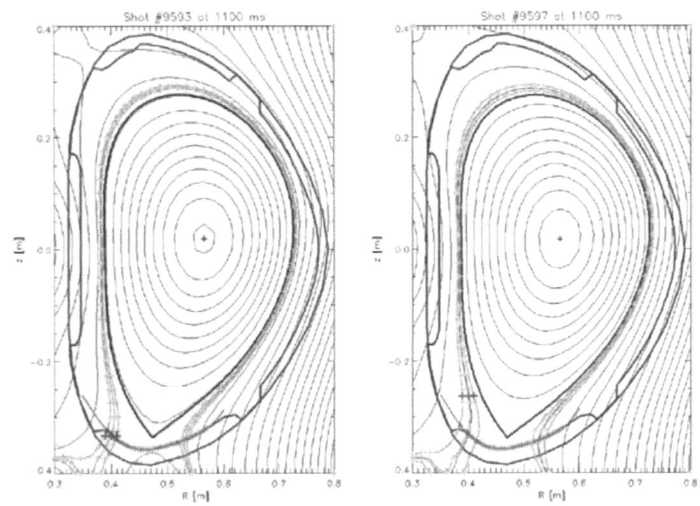

$z = -0.333$ m $Z = -0.263$ m

Figure 1. Magnetic surface mapping using EFIT.

Figure 1 is an example where the equilibrium FITing code (EFIT) reconstructions of the magnetic field surfaces are used to visualize the position of the sample. The red crosses indicate the positions of the centers of the exposed samples. The red triangles are their mappings to the divertor Langmuir probes at HFS. The orange curves show the corresponding magnetic surfaces.

The parallel power flux density at the samples position was evaluated by means of the measured data from the available divertor probes. It is assumed that the parallel power flux density remains the same at the same magnetic surface.

3. Langmuir probe measurements in the divertor region of the COMPASS tokamak

Langmuir probes (LPs) are relatively simple, inexpensive and robust diagnostic tools that are compatible with the tokamak scrape-off layer (SOL) conditions. They are readily deployable in large arrays built into the edge structures, such as limiters and divertor tiles. Because of their simplicity and the ability to perform local measurements of the electron energy distribution function (EEDF) and the plasma potential, LPs are frequently used in magnetized plasmas to characterize the edge plasma parameters [2,4,5].

The divertor probe system in the COMPASS tokamak consists of 39 single graphite Langmuir probes poloidally embedded in the divertor tiles; they are oriented parallel to the magnetic field and provide profiles with a spatial resolution in the poloidal direction down to 5 mm [2]. Each probe has an area of $S = 56\times10^{-6}$ m^2. The probes are biased with respect to the tokamak chamber wall by a triangular voltage $U_p(t)$ and swept at a frequency of 1 kHz [6].

The probe measurements were performed during ohmic ELMy H-mode shots, with toroidal magnetic field $B_T = 1.15$ T and plasma current $I_p = 300$ kA. Figure 2 is an example for shot #9348 (line-averaged electron density $n_e = 7\times10^{19}$ m^{-3}) of the plasma current, the D-alpha signal and the signal from Langmuir probe 1 (LP#1), which is at a radial position 0.396 m in the tokamak vessel.

Journal of Physics: Conference Series **700** (2016) 012008 doi:10.1088/1742-6596/700/1/012008

The probe works in an ion-saturation mode (negative voltage applied of –100 V) for monitoring the H-mode. A good agreement is seen between the spectroscopy signals (the black curve) and those from the probe (the red curve); moreover, the structure of the ELMs is seen more clearly from the probe signal.

The data obtained between the ELMs were processed by the FDPT for precise determination of the plasma potential and the EEDF.The FDPT for evaluating the plasma parameters in tokamak plasma was published and discussed in detail in [3].

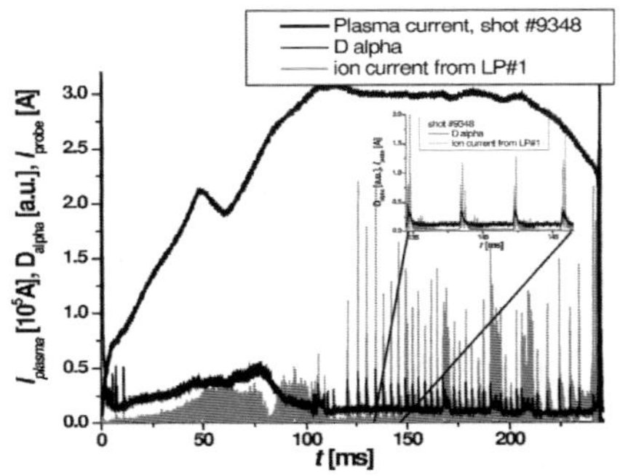

Figure 2. Temporal profiles of the plasma current, D-alpha signal and ion-saturation current from LP#1, shot 9348.

4. Results and discussion

Results of the evaluation of the poloidal distribution of the ion-saturation current density, J_{sat}, and the plasma potential, U_{pl}, during two ohmic ELMy H-mode deuterium deuterium discharges are presented in figure 3 a and 3 b. The discharges differ slightly in the line-averaged electron density $n_e = 7\times10^{19}$ m^{-3} (#9348 – black symbols) and $n_e = 5\times10^{19}$ m^{-3} (#9593 – red symbols). An important fact is that discharge #9348 is without a sample inside the chamber. During shot #9593, the tungsten samples were inserted at a position $R = 0.39 – 0.41$ m, $z = –333$ mm. In figure 2 a one can see the positions of the strike points indicated by dashed lines as provided by EFIT during the steady-state phase of the discharges, at 140 ms (see figure 2). The positions for shot #9348 are for inner strike point $R = 0.45$ m and for outer, $R = 0.485$ m. For shot # 9593, these are respectively $R = 0.453$ m and $R = 0.486$ m. The J_{sat} profiles agree with the results from the EFIT reconstruction within 5 – 7 mm.

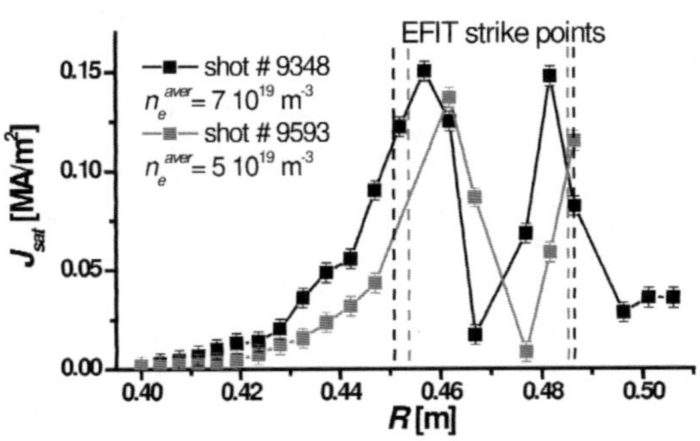

Figure 3 a). Spatial distribution of the ion saturation current density in H-mode.

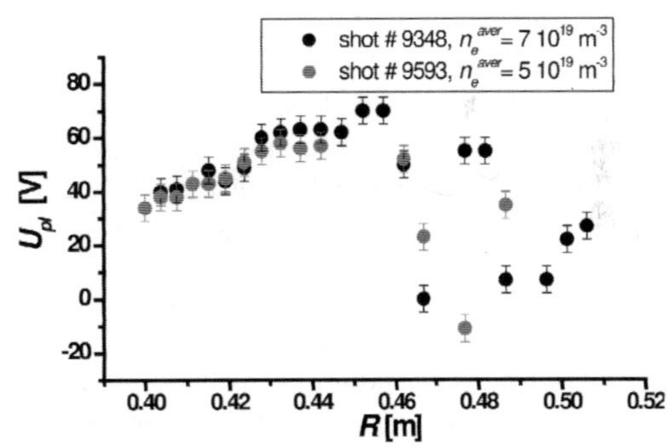

Figure 3 b). Spatial distribution of the plasma potential in H-mode.

In the divertor region during the shots in question, it was found that the EEDF at the HFS is Maxwellian with a temperature of 4 – 10 eV (figure 4 a)). In the private flux region (PFR), it differs from the Maxwellian, but can be approximated by a bi-Maxwellian distribution. In figure 4 a, the triangles represent the temperature of the low-energy electron fraction, while the squares, the temperature of the high-energy electrons in the bi-Maxwellian EEDF. The dots indicate the temperature of the Maxwellian EEDF. It is seen that the bi-Maxwellian distribution appears around the inner strike point, a position evaluated from the probes, namely, a maximum signal on J_{sat}. Figure 4 b) presents the corresponding electron densities (represented by the same symbols as used in figure 4 a).

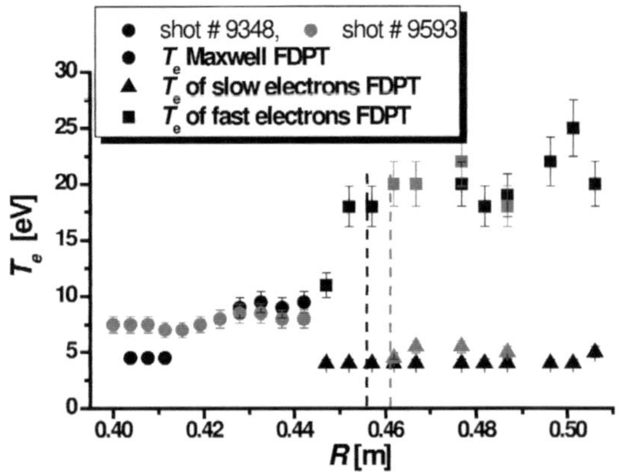

Figure 4 a). Spatial distribution of the electron temperature T_e.

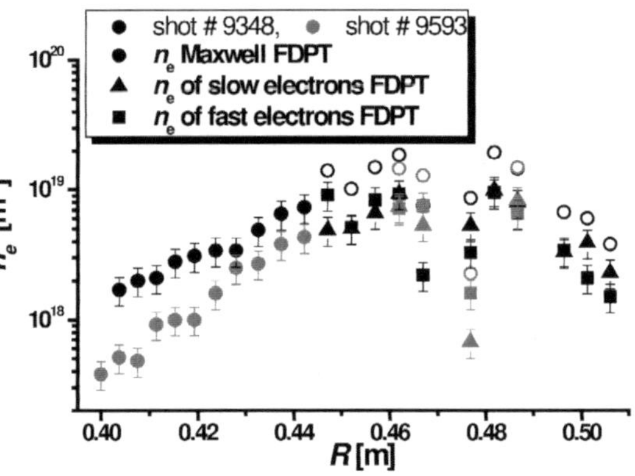

Figure 4 b). Spatial distribution of the electron densities n_e.

Using the results for the electron temperatures and densities obtained by probe measurements, one can calculate the parallel power flux density by using the following equation [7]:

$$Q_{\parallel}^{e} = 7.5 T_e \Gamma_{se} \, . \tag{1}$$

Here we assumed that $T_i = T_e$ and the particle flux density $\Gamma_{se} = J_{sat}/e$, where J_{sat} is the ion-saturation current density.

Figure 5 shows the calculations of the parallel power flux density for different shots when the tungsten sample is at different z positions in the plasma. It is seen that the position of the sample does not affect the plasma parameters in the divertor. The heat flux on the samples corresponds to the open magnetic surface which crosses the divertor at the radial position 0.385 – 0.415 m. Then, the parallel power flux density on the sample surfaces is about 0.02 – 0.5 MW/m^2. On HFS during H-mode, the power flux density profile increases exponentially as $Q_{II}^{e} = Q_0 \exp(-R_{sep}/\lambda_q)$, where the decay length in the inner divertor is λ_q = 0.012 m and Q_0 =2 MW/m^2 (the dashed black line in figure 5) for the discharges when the sample interacts with the plasma. The higher values in the spatial profiles for shot #9348 of the electron densities and of the power flux density are due to the higher line-averaged density.

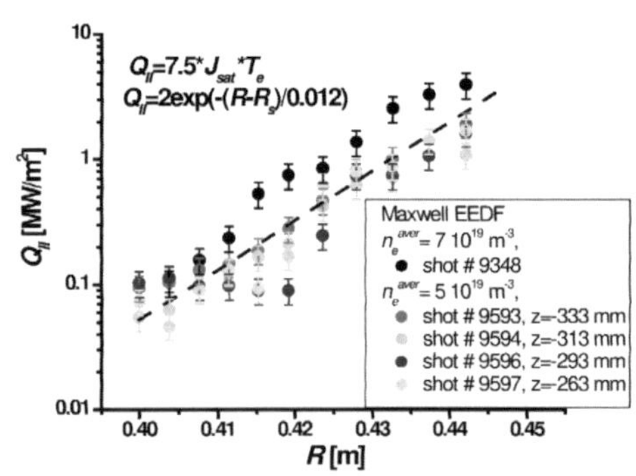

Figure 5. Radial distribution of the parallel power flux density.

5. Results of the plasma interaction with tungsten samples

Under the experimental conditions described, we studied the plasma interaction with tungsten samples, with or without a pre-grown He fuzz. He fuzz is a nano-scale fibrous structure that forms on a tungsten surface exposed to helium plasma. Its low density, together with the tendency to increase the occurrence of unipolar arcing, may make it susceptible to (locally) enhanced erosion in tokamak conditions, especially during transient events such as ELMs [8]. Four tungsten samples after an exposure in COMPASS (in an orientation roughly parallel with the tokamak wall) are shown in figure 6. Arc traces on the fuzzed samples are clearly visible. These consisted of localized melting of

Figure 6. Exposed tungsten samples in a graphite holder. Sample 3 was only ground tungsten, samples 1, 2 and 4 had He fuzz grown under different conditions. Arc traces on the fuzzy samples are seen, mainly in the vertical direction.

the fuzz structure and a reduction of its thickness in the affected areas. However, the melting did not penetrate through the fuzz thickness into the bulk. A significantly higher number of arc traces was observed after an exposure with sample surfaces roughly perpendicular to the tokamak wall. Due to the frequent occurrence of hot dust particles originating from the holder, it was not possible to isolate individual arcing events and correlate them with particular ELMs. This is a subject of an ongoing investigation.

6. Conclusions

This paper reports experimental results of plasma interaction with tungsten samples with or without pre-grown He fuzz. Under the conditions investigated, arcing was observed on the fuzzy tungsten samples, resulting in localized melting of the fuzz structure that did not extend into the bulk.

The parallel power flux densities were obtained from data measured by Langmuir probes embedded in the divertor tiles on the COMPASS tokamak. Measurements of the current-voltage (*IV*) probe characteristics were performed during ohmic ELMy H-modes reached in deuterium plasmas. The data obtained between the ELMs were processed by the first-derivative probe technique to evaluate the electron temperatures and densities.

Acknowledgements

This research has been supported by the Czech Science Foundation grant No. 14-12837S, by the International Atomic Energy Agency (IAEA) Research Contract No 17125/R0, R1 and R2 as a part of the IAEA CRP F13014 on "Utilisation of a Network of Small Magnetic Confinement Fusion Devices for Mainstream Fusion Research", by the JOINT RESEARCH PROJECT between the Institute of Plasma Physics v.v.i., AS CR and the Institute of Electronics BAS BG; and by MSMT Project # LM2011021.

References

[1] Panek R, et al 2015 *Plasma Phys. Control. Fusion* **58** 014015

[2] Dimitrova M, Dejarnac R, Popov Tsv K, Ivanova P, Vasileva E, Kovačič J, Stöckel J, Havlicek J, Janky F and Panek R 2014 *Contrib. Plasma Phys.* **54**/3 255–60

[3] Popov Tsv K, Ivanova P I, Stockel J and Dejarnac R 2009 *Plasma Phys. Control. Fusion* **51** 065014

[4] Popov Tsv K, Dimitrova M, Ivanova P, Hasan E, Horacek J, Dejarnac R, Stöckel J, Weinzettl V and Kovačič J 2014 *Contrib. Plasma Phys.* **54**/3 267–72

[5] Popov Tsv K, et al 2015 *Plasma Phys. Control. Fusion* **57** 115011

[6] Mitov M, Bankova A, Dimitrova M, Ivanova P, Tutulkov K, Djermanova N, Popov Tsv K, Stöckel J and Dejarnac R 2012 *J. Phys.: Conf. Series* **356** 012008

[7] Stangeby P C 2000 *The Plasma Boundary of Magnetic Fusion Devices (Series in Plasma Physics)* (IoP)

[8] Aussems D U B, Nishijima D, Brandt C, van der Meiden H J, Vilémová M, Matějíček J, de Temmerman G, Doerner R P and Lopes Cardozo N J 2015 *J. Nucl. Mater.* **463** 303-7

Electron energy distribution function in the divertor region of the COMPASS tokamak

M Dimitrova[1,2,4], E Hasan[2,3], P Ivanova[2], E Vasileva[2], Tsv Popov[3], R Dejarnac[1], J Stöckel[1] and R Panek[1]

[1]Institute of Plasma Physics, Academy of Sciences of the Czech Republic, 3 Za Slovankou, 182 00 Prague 8, The Czech Republic
[2]Acad. E. Djakov Institute of Electronics, Bulgarian Academy of Sciences, 72 Tsarigradsko Chaussee, 1784 Sofia, Bulgaria
[3]Faculty of Physics, St. Kl. Ohridski University of Sofia, 5 J. Bourchier Blvd., 1164 Sofia, Bulgaria

E-mail: dimitrova@ipp.cas.cz

Abstract. The plasma parameters during an L-mode hydrogen discharge in the COMPASS tokamak with a toroidal magnetic field $B_T = 1.15$ T, line-averaged electron density $n_e = 6\times10^{19}$ m^{-3} and a plasma current variation from 209 kA to 100 kA were studied in the divertor region. The electron energy distribution function for 209 kA at the high-field side and the private region is Maxwellian with a temperature in the range of 5 – 9 eV, while around the outer strike point and the low-field side it is bi-Maxwellian with a low-energy electron group (4 – 5 eV) and higher energy electrons (10 – 20 eV). As the plasma current decreases, the appearance of the bi-Maxwellian EEDF is shifted towards the low-field side; at plasma current of 100 kA, the EEDF is Maxwellian in the whole divertor region.

1. Introduction

The experimental knowledge of the electron energy distribution function (EEDF) is of great importance in understanding the underlying physical processes that occur at the edge of tokamak plasmas, such as the interaction of the plasma with the tokamak wall, turbulence and the formation of the edge transport barrier.

The plasma parameters during an L-mode hydrogen discharge in the COMPASS tokamak [1] with a toroidal magnetic field $B_T = 1.15$ T and a line-averaged electron density $n_e = 6\times10^{19}$ m^{-3} with a variation of the plasma current from 209 kA to 100 kA were studied in the divertor region by means of 39 Langmuir probes embedded in the divertor tiles. For precise determination of the plasma potential and evaluation of the real EEDF, the current-voltage *(IV)* characteristics measured were processed using the recently published [2-5] first-derivative probe technique (FDPT).

2. Langmuir probe measurements in the divertor region of the COMPASS tokamak

The probe measurements were performed during L-mode hydrogen plasmas (discharge # 9009 and #9008) with a toroidal magnetic field $B_T = 1.15$ T and a line-averaged electron density $n_e = 6\times10^{19}$ m^{-3}.

[4] To whom any correspondence should be addressed.

Figure 1 presents the temporal profile of the plasma current during the discharge. The time points, (respectively plasma currents) when the plasma parameters were evaluated are marked by different colors.

The divertor probe system in the COMPASS tokamak consists of 39 single graphite Langmuir probes poloidally embedded in the divertor tiles providing data with a 5 mm spatial resolution [3]. The probes are biased with respect to the tokamak chamber wall by a triangular voltage $U_p(t)$ swept at a frequency of 1 kHz supplied by a KEPCO 100-4M power supply. The ramp-up/ramp-down phases last 0.5 ms, so that the plasma current can be assumed constant during each sweep cycle. The probe potential and the probe current versus time are recorded by the tokamak DAQ [6]. The *IV* probe characteristics are constructed using the data recorded at different moments during the discharge.

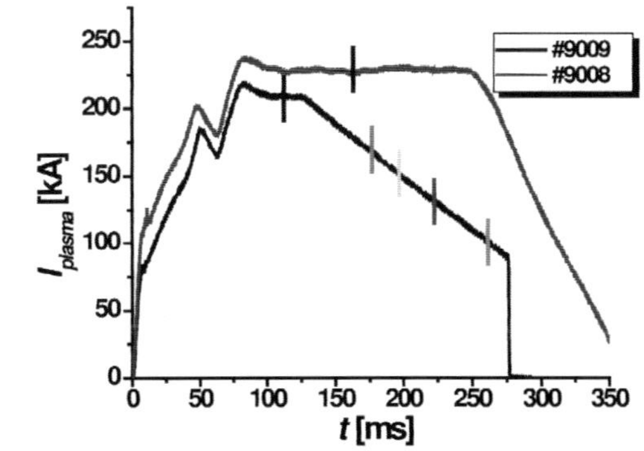

Figure 1. Temporal profiles of the plasma current for discharge #9009 and #9008.

Figure 2 a) presents the radial profile of the ion saturation current density, J_{sat}. The different colors (the same as in figure 1) correspond to the different plasma current values. The radial positions of the maximums at different discharge currents J_{sat} correspond to the positions of the outer strike points provided by the Equilibrium FITing code (EFIT) reconstructions. The EFIT reconstructions of the magnetic field surfaces with accuracy of about 5 mm (figure 2 b)) show that as the plasma current decreases, the position of the outer strike point changes within a few millimetres. That is why the outer strike point positions are presented as a region here and below. In contrast, the position of the inner strike point is shifted towards the high-field side by about 5 cm.

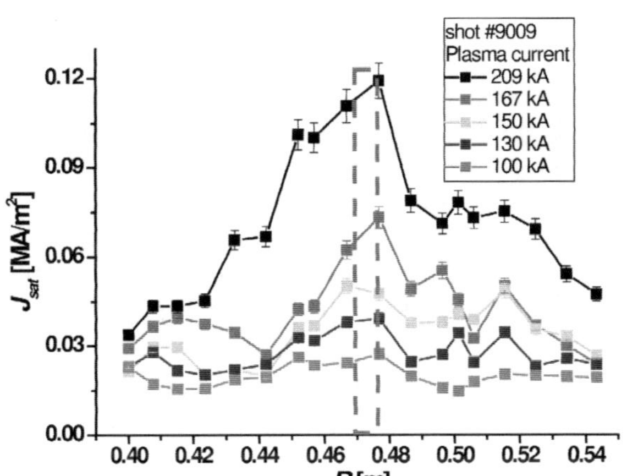

Figure 2 a). Spatial distribution of the ion saturation current density.

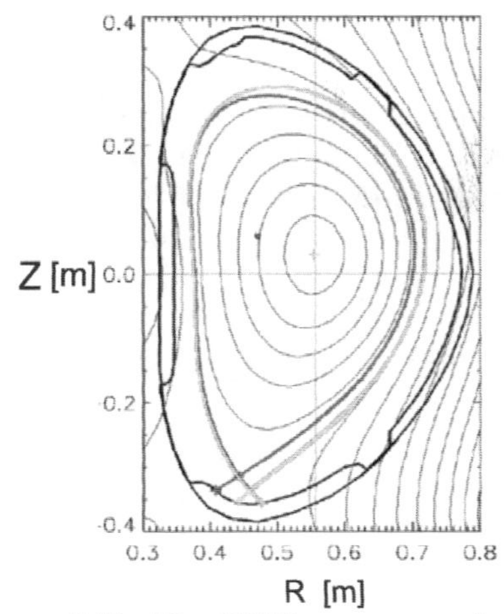

Figure 2 b). The EFIT reconstructions of the magnetic field surfaces - separatrix at 209 kA (blue) and 150 kA (red curve).

Further, the FDPT was applied to process the *IV* characteristics in order to obtain the plasma potential and the EEDF (respectively, the electron temperatures and densities) [2-5]. For example, figure 3 a) shows the electron energy probability function (EEPF) on probe #12 (at radial position $R = 0.4421$ m) when the plasma currents is $I_{pl} = 210$ kA for shot #9009.

The EEPF has the same information about the electron gas as EEDF and is frequently used to represent measured probe data. The EEPF, presented in a semi-log scale, allows a quick visualization of a departure of the measured EEPF from the Maxwellian distribution, which is a straight line in this representation [7]. The EEPF on the high-field side (HFS) is Maxwellian. Using the slope of the EEPF in a logarithmic scale, one can evaluate the electron temperature at (8.5±0.8 eV).

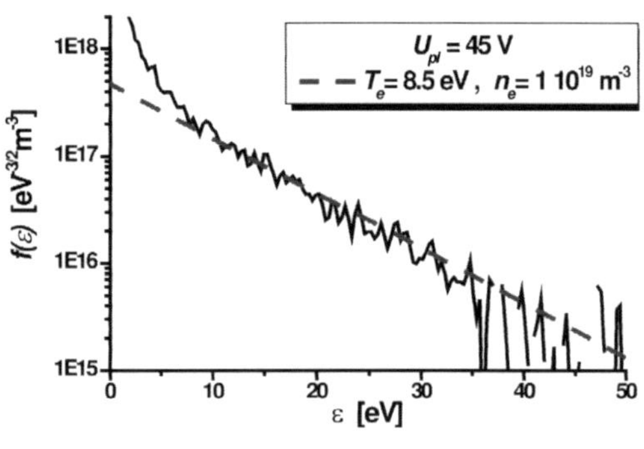

Figure 3 a). Experimental EEPF (black line) and the Maxwell model line (dashed) for LP#12, R = 0.4421 m.

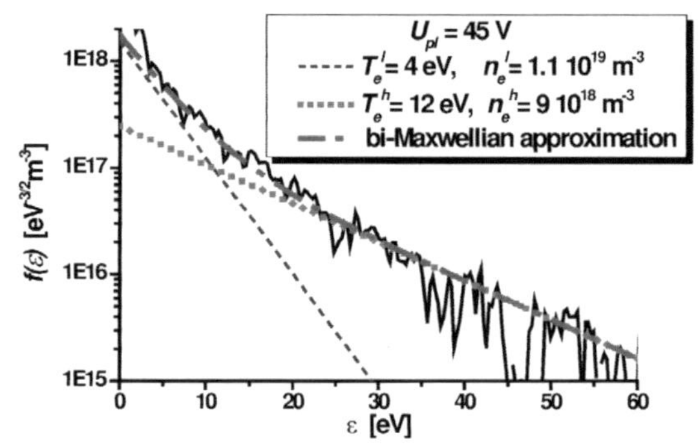

Figure 3 b). Experimental EEPF (black line) and the sum of the model ones for two electron temperatures (dashed dotted line) for LP#15, R = 0.4569 m.

The electron density is calculated using the expression for normalization of the EEDF; in this case $n_e = 1\times10^{19}\mathrm{m}^{-3}$. Around the outer strike point, the EEPF is found to be non-Maxwellian but can be approximated by a bi-Maxwellian one – i.e., by a sum of two Maxwellian EEPFs with two different temperatures. An example of a bi-Maxwellian EEPF measured by probe #15 (R = 0.4569 m) is presented in figure 3 b). The low-temperature electron group (blue line, T_e^l = 4±0.4 eV) has a density higher than that of the high-temperature one (red line, T_e^h = 12±0.1 eV). The influence of the two-temperature EEDF on the parallel heat flux density, as well as a discussion on the origins of the two temperatures, is presented in detail in our recent publication [4].

3. Results and discussion

We present below the main plasma parameters for discharge #9009 using the IV characteristics. The results of the evaluation of the poloidal distribution of the floating potential and the plasma potential during L-mode hydrogen discharges are presented in figures 4 a) and 4 b). The colors correspond to the different plasma current values. The position of the outer strike points (presented as a region as was mentioned above) from the reconstruction of the magnetic surfaces by the Equilibrium FITing code (EFIT) is also shown in the figures. It is seen that at higher plasma currents the values of the

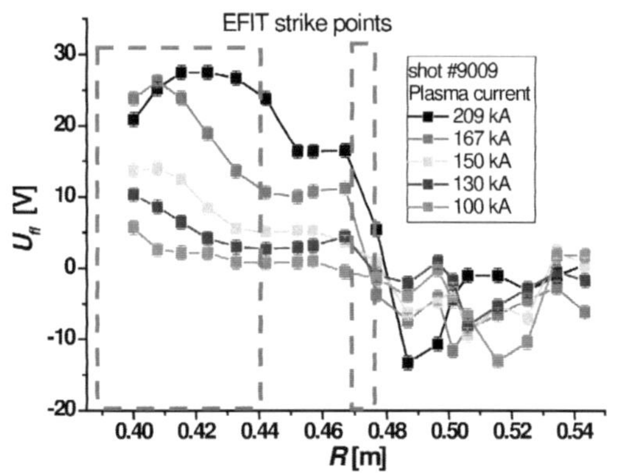

Figure 4 a). Poloidal distribution of the floating potential at different plasma currents.

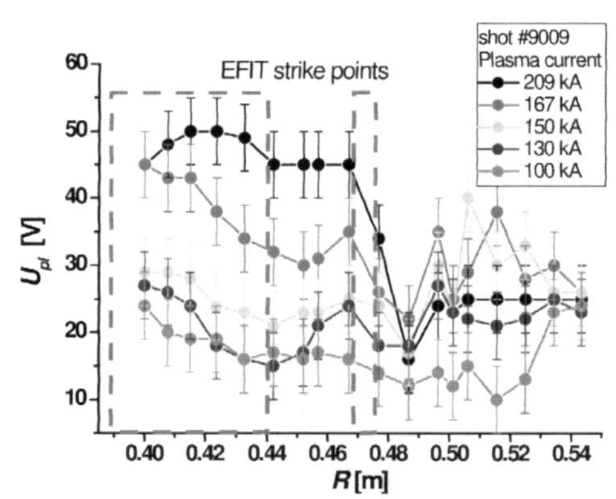

Figure 4 b). Poloidal distribution of the plasma potential at different plasma currents.

potentials are increasing, which is noticeable on the HFS and in the private flux region. Thus, for a difference of 100 kA, the plasma potential value rises by a factor of two, from 20 V to 50 V. However, on the low-field side (LFS), the effect of the plasma current is less obvious, because the plasma is much more turbulent, so that similar values of about 30 V are evaluated.

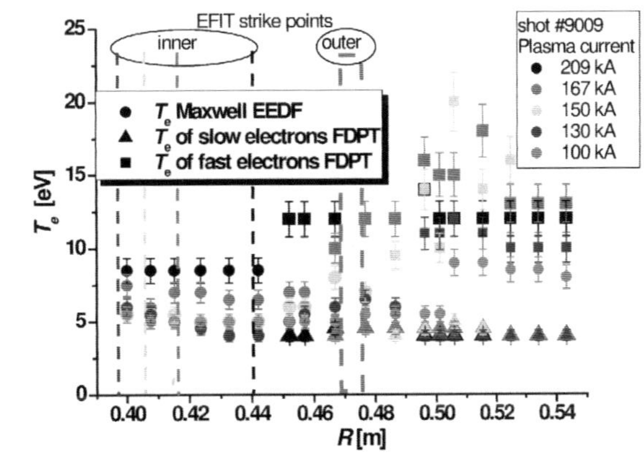

Figure 5. Poloidal distribution of the electron temperature T_e for different discharge currents for shot #9009.

Figure 5 shows the poloidal distribution of the electron temperature. The dots correspond to the Maxwellian EEDFs. The triangles and squares indicate the low and high temperatures of the bi-Maxwellian EEDFs, respectively. The position of the two strike points from the EFIT are indicated by colored dashed lines. It is seen that, at plasma current of 210 kA, the EEDF changes to bi-Maxwellian at 2 cm from the outer strike point in the private flux region. When the plasma current decreases, the appearance of the bi-Maxwellian EEDF is shifted to the LFS direction. At 100 kA, the EEDF is Maxwellian in the whole divertor region with temperatures of 5–9 eV.

Using the EEDF, the electron densities are determined for both electron groups. In figure 6 a), an example is given for the poloidal profile of the electron densities at 210 kA represented (here and below) by the same symbols as used in figure 5. Both electron groups exhibit similar values of the densities. In figure 6 b), the dependence is presented of the total electron densities on the plasma current.

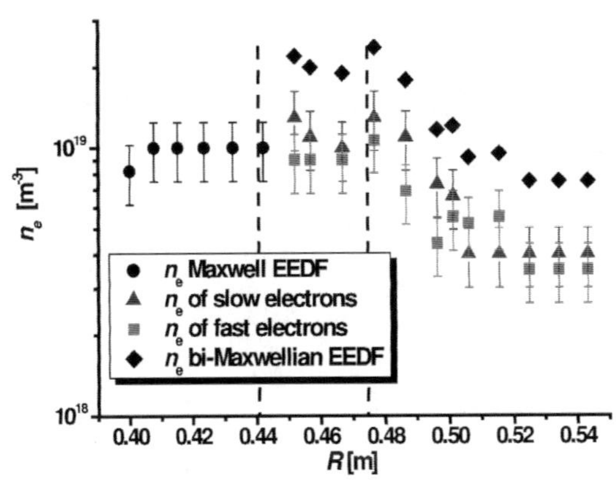

Figure 6 a). Poloidal distribution of the electron densities n_e at discharge current of 210 kA for shot #9009.

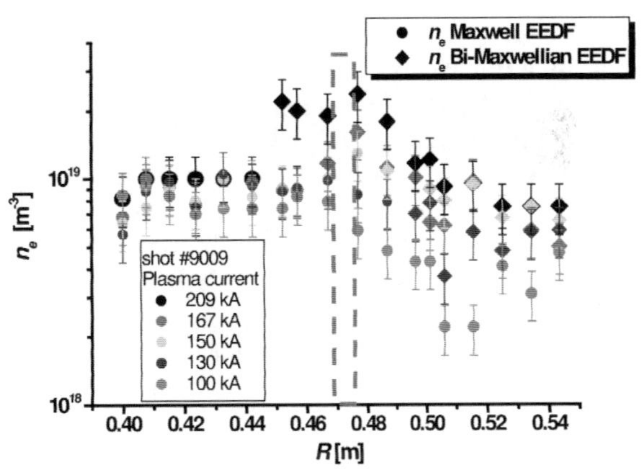

Figure 6 b). Poloidal distribution of the total electron densities n_e for different discharge currents for shot #9009.

For a comparison, we present below (figure 7) results from shot #9008 with a constant discharge current of 230 kA and a line average density of 5×10^{19} m^{-3} during the steady-state phase of the discharge. The data refer to time 162 ms.

The EEDF in the private flux region is Maxwellian, while in the LFS and HFS outside of the strike points it is bi-Maxwellian. The distance between the strike points is ~5 cm. This is in agreement with our previous studies [3,8].

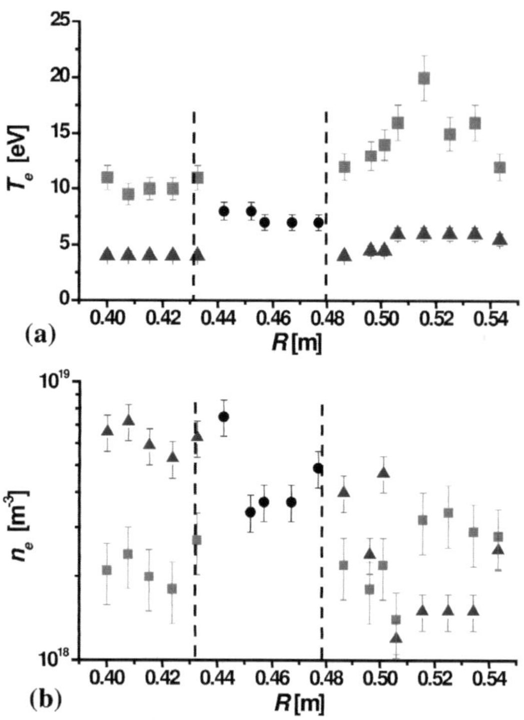

Figure 7. Poloidal distribution of (a) the electron temperature T_e and (b) the electron densities n_e during discharge #9008.

4. Conclusions

This paper reports the electron energy distribution function (EEDF) in the divertor of the COMPASS tokamak as determined by the first-derivative probe technique. The current-voltage probe characteristics measured were processed by this technique, which gives information about the plasma parameters. The effect of the plasma current on the plasma parameters was studied.

The poloidal profile of the electron temperatures show that at a plasma current of 100 kA the EEDF is Maxwellian with temperatures in the range of 5 – 9 eV. As the plasma current increases, the behavior of the EEDF changes. It was found that in hydrogen plasmas in the vicinity of the strike points the EEDF can be approximated by a bi-Maxwellian distribution, with a low-energy electron group (4 – 5 eV) and higher energy electrons (10 – 20 eV). The dependence of the strike point position when the bi-Maxwellian distribution appears was also studied. When the plasma current decreases, the appearance of the bi-Maxwellian EEDF is shifted to the LFS direction.

Acknowledgements

This research has been supported by the International Atomic Energy Agency (IAEA) Research Contract No 17125/R0, R1 and R2 as a part of the IAEA CRP F13014 on "Utilisation of a Network of Small Magnetic Confinement Fusion Devices for Mainstream Fusion Research", by the JOINT RESEARCH PROJECT between the Institute of Plasma Physics v.v.i., AS CR and the Institute of Electronics BAS BG; by a project BG051 PO001-3.3.06-0057 and by MSMT Project # LM2011021.

References

[1] Panek R, et al 2015 *Plasma Phys. Control. Fusion* **58** 014015

[2] Popov Tsv K, Ivanova P I, Stockel J and Dejarnac R 2009 *Plasma Phys. Control. Fusion* **51** 065014

[3] Dimitrova M, Dejarnac R, Popov Tsv K, Ivanova P, Vasileva E, Kovačič J, Stöckel J, Havlicek J, Janky F and Panek R 2014 *Contrib. Plasma Phys.* **54**/3 255–60

[4] Popov Tsv K, et al 2015 *Plasma Phys. Control. Fusion* **57** 115011

[5] Popov Tsv K, Dimitrova M, Ivanova P, Hasan E, Horacek J, Dejarnac R, Stöckel J, Weinzettl V and Kovačič J 2014 *Contrib. Plasma Phys.* **54**/3 267–72

[6] Mitov M, Bankova A, Dimitrova M, Ivanova P, Tutulkov K, Djermanova N, Popov Tsv K, Stöckel J and Dejarnac R 2012 *J. Phys.: Conf. Series* **356** 012008

[7] Godyak V A and Demidov V I 2011 *J. Phys. D: Appl. Phys.* **44** 233001

[8] Ivanova P, Dimitrova M, Vasileva E, Popov Tsv K, Dejarnac R, Stöckel J, Imríšek M, Hacek P and Panek R 2016 *J. Phys.: Conf. Series* (6th Int. Workshop & Summer School on Plasma Phys. 2014, Kiten, Bulgaria) at press

Development of problem-oriented software packages for numerical studies and computer-aided design (CAD) of gyrotrons

M Damyanova[1,3], S Sabchevski[1], I Zhelyazkov[2], E Vasileva[1], E Balabanova[1], P Dankov[2] and P Malinov[2]

[1]Institute of Electronics, Bulgarian Academy of Sciences,
72 Tsarigradsko Chaussee, 1784 Sofia, Bulgaria
[2]Faculty of Physics, St. Kliment Ohridski University of Sofia,
5 James Bourchier blvd., 1164 Sofia, Bulgaria

E-mail: milena@ie.bas.bg

Abstract. Gyrotrons are the most powerful sources of coherent CW (continuous wave) radiation in the frequency range situated between the long-wavelength edge of the infrared light (far-infrared region) and the microwaves, i.e., in the region of the electromagnetic spectrum which is usually called the THz-gap (or T-gap), since the output power of other devices (e.g., solid-state oscillators) operating in this interval is by several orders of magnitude lower. In the recent years, the unique capabilities of the sub-THz and THz gyrotrons have opened the road to many novel and future prospective applications in various physical studies and advanced high-power terahertz technologies. In this paper, we present the current status and functionality of the problem-oriented software packages (most notably GYROSIM and GYREOSS) used for numerical studies, computer-aided design (CAD) and optimization of gyrotrons for diverse applications. They consist of a hierarchy of codes specialized to modelling and simulation of different subsystems of the gyrotrons (EOS, resonant cavity, etc.) and are based on adequate physical models, efficient numerical methods and algorithms.

1. Introduction

Among the various sources of coherent CW (continuous-wave) radiation operating in a region of the electromagnetic spectrum situated between microwaves and light (and, thus, between electronics and photonics), the gyrotrons are characterized by the highest value of a figure of merit defined as Pf^2, where P and f are the output power and frequency, respectively. In recent years, the gyrotrons developed worldwide have spanned a broad range from sub-THz to THz frequencies and have contributed enormously to bridging the so-called THz gap (a.k.a. T-gap) providing powerful radiation for many applications. Amid them are some well-known as, for example, heating and current drive of fusion plasmas, thermal treatment (e.g., ceramics sintering) but also many novel and emerging usages in advanced spectroscopic techniques (notably ESR and DNP-NMR spectroscopy), sensing and inspection, radars, medical technologies, just to name a few [1–3]. The development of gyrotrons

[3] To whom any correspondence should be addressed.

relies heavily on modelling and simulation by means of numerical experiments [4, 5]. Since the gyrotron tube is a complex vacuum device that includes several subsystems (electron-optical system, EOS; electrodynamical system; quasi-optical system, etc.) there is no a single physical model that emulates its operation. Therefore, following the principle of decomposition, each subsystem is being described by its own hierarchy of physical models (ranging from simplified linearized and analytical to the most sophisticated self-consistent) that are realized in various numerical codes for computer-aided design (CAD) and optimization of gyrotrons. Our work on modelling and simulation of gyrotrons has been motivated by our participation in several joint research projects carried out in collaboration with FIR UF (Research Center for Development of Far-Infrared Region at the University of Fukui, Japan), IHM-KIT (Institute for Pulsed Power and Microwave Technology at Karlsruhe Institute of Technology, Germany) and CRPP-EPFL (Centre de Recherches en Physique des Plasmas École Polytechnique Fédérale de Lausanne, Switzerland) and includes: (i) formulation of adequate and informative physical models; (ii) selection of efficient numerical methods and algorithms and implementation (programming) of the models in computer codes (stand-alone programs, computational modules, subroutine libraries) using advanced programing techniques and integrated development environments (IDE); (iii) maintenance of diverse computational infrastructure on which both the developed applied software and the underlying programing blocks (numerical libraries, compilers, IDE, and so on) are installed; (iv) numerical experiments for analysis, CAD and optimization of gyrotrons for different applications. In this paper, we present the current status and the functionality of the problem-oriented software packages for numerical studies, CAD and optimization of high-performance gyrotrons for different applications.

2. Current status of the simulation tools for analysis, CAD and optimization of gyrotrons

The hierarchy of the developed simulation tools is shown in Figure 1. It includes several levels that correspond to the topics listed above. Here, we are focused on the software level, which is represented by a variety of numerical codes structured in the problem-oriented software packages GYROSIM (which stands for GYROtron SIMulation), GO&ART (Geometric Optics and Analytical Ray Tracing), and GYREOG (GYRotron Electron-Optical System Simulation) that are being developed and executed on our computational platforms (see the hardware level).

The package GYROSIM includes numerical codes for simulation of both the electron-optical system (EOS) and the electrodynamical system (resonant cavity) of gyrotrons. Its components GUN-MIG and GUN-CUSP are static PIC (Particle-in-Cell) codes assigned to ray tracing (trajectory analysis) of axially-symmetric EOS with magnetron injection guns (MIG). They are based on a self-consistent physical model formulated in 2.5D (i.e., in a phase space with three momentum coordinates and two spatial dimensions), which takes into account both the initial velocities and the space-charge effects. The dynamical part of the model includes fully-relativistic equations of motion, while the electrostatic field distribution (the field part of the model) is described by the Poisson equation in polar coordinates with the corresponding Dirichlet and Neumann boundary conditions. The field solver uses the finite difference method with a successive over-relaxation (FDM-SOR) and a special procedure for optimization of the parameter of relaxation. Additionally, the convergence of the iterative process, which leads to the self-consistent solution, is speeded up by using a sequence of meshes with decreasing steps (i.e., beginning with a coarser grid) as well as by a gradual increase of the extracted current. The emission model can handle both the space-charge limited and the temperature-limited regimes of the cathode operation. The magnetic field produced by a system of solenoids and additional coils for its adjustment is calculated using the Biot–Savart law for the current loops of the magnets. Another option, realized in both GUN-MIG and GUN-CUSP, is to calculate the off-axis components of the magnetic field using a paraxial series expansion and its values (calculated or measured experimentally) on the axis. The MIG-CUSP code is specialized to the simulation of EOS that forms uniaxial (a.k.a. axis-encircling) electron beams for large orbit gyrotrons (LOG) using a magnetic cusp (or, more generally, magnetic field reversal) and has additional components for precise calculation and interpolation of the magnetic field in the non-adiabatic region.

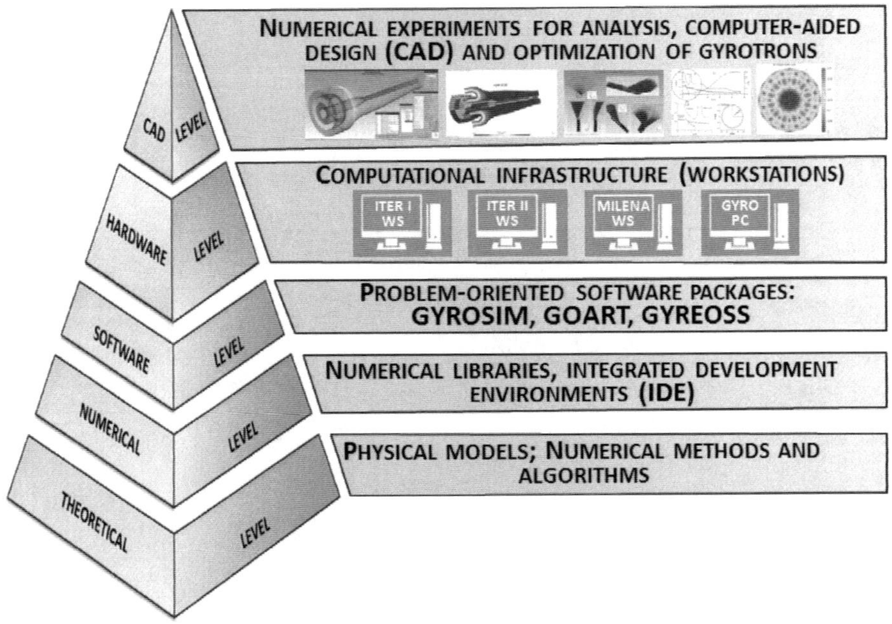

Figure 1. Hierarchy of the tools for modelling and simulation of gyrotrons.

GYROSIM includes also a hierarchy of codes for analysis of the cavity resonator. The first group of such codes consists of computer programs that are based on simplified (linear and cold-cavity) models and are being used at the initial stage of the CAD for (i) identification of the modes spectrum (eigenfrequencies); (ii) evaluation of the quality factors; (iii) calculation of the starting currents; (iv) plotting the mode patterns and dispersion diagrams; and most notably for (v) preliminary selection of the operating mode and identification of the possible competing modes. For a more detailed analysis of the beam–wave interaction (energy transfer) and thus for calculation of the efficiency and output power a set of codes based on more adequate physical models are used. The CCC-CAV code uses a cold-cavity field profile in order to trace the dynamics of electrons and to calculate the energy balance of the interaction. In the single mode time independent CS-SMTI code, a self-consistent physical model (a modified version of the one formulated by Fliflet *et al.*, [6]) is implemented. For studying the modes interactions (e.g., modes competition, modes cooperation, mode switching, and excitation of multiple modes) the multimode time-dependent CS-MMTD code is used. Its model is applicable to an arbitrary number of modes operating at the fundamental and/or the second harmonic of the cyclotron frequency. The most recent extensions to the GYROSIM package are the computational modules for studying the finite-bandwidth resonances (spectral analysis) of high-order axial modes (HOAM) in open gyrotron cavities [7]. They include a novel solver of the inhomogeneous Helmholtz equation with radiation boundary conditions (which equation governs the axial profile of the field amplitude of the excited cavity mode) as well as subroutines for plotting and analysis of several resonance curves calculated from a sequence of field profiles obtained by scanning the frequency. The results of the numerical experiments give a better physical insight on the spectrum of the HOAM and support the interpretation of the continuous frequency tunability (manifested in many experiments) as a result of successive excitation of a sequence of axial modes. As an illustration, in Figure 2 some results are shown from numerical experiments for CAD of a 395 GHz gyrotron for the DNP-NMR spectroscopy carried out using the codes of GYROSIM package.

GO&ART package consists of several codes (RAYS, COMODES, and TRACE) for analysis of quasi-optical components (Vlasov and Denisov type launchers, reflectors and phase-correcting mirrors, and so on) as well as systems based on them (e.g., internal mode converters and transmission lines).

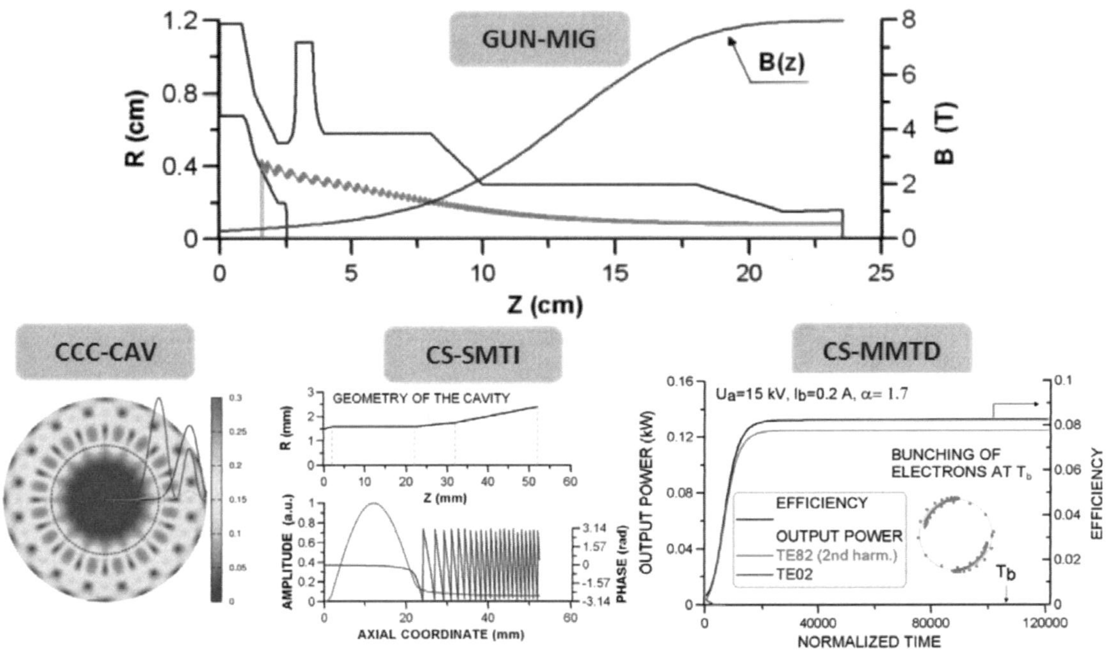

Figure 2. Simulation results for a 395 GHz gyrotrons operating on a $TE_{8,2}$ mode at the second harmonic of the cyclotron frequency. Upper panel: trajectory analysis of the EOS. Lower panel: field pattern and coupling factors for co- and counter-rotating modes; amplitude and phase of the field inside the cavity; multimode simulation using CS-MMTD.

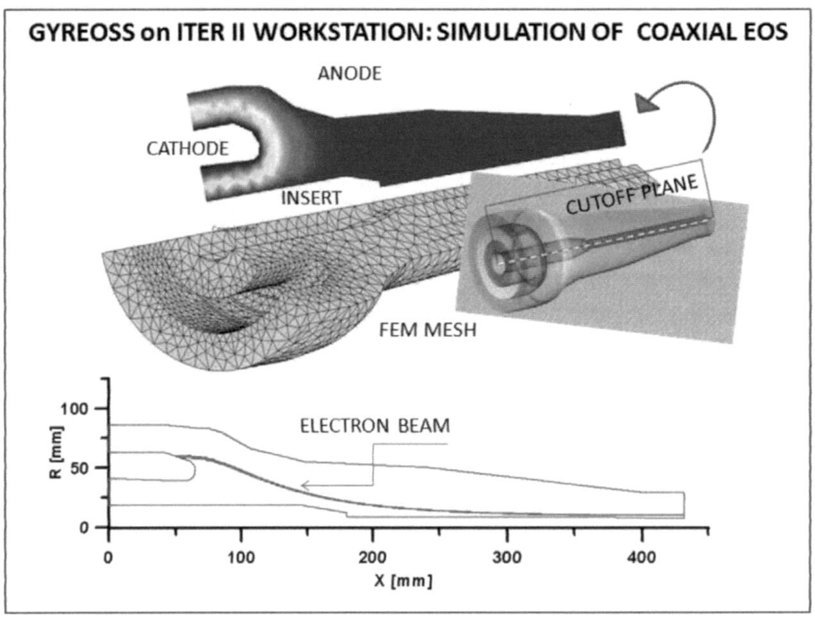

Figure 3. Simulation of a coaxial EOS of 2 MW/170 GHz gyrotron. Shown are the tetrahedral mesh and electrostatic potential distribution (upper panel) and the trajectories of the electron beam (lower panel).

GYREOSS is a problem-oriented software package for trajectory analysis (ray tracing) and PIC simulations of the EOS of gyrotrons in both 2D and 3D [8]. The field solver of GYREOSS utilizes the finite element method for solving the boundary value problem for the Poisson equation and is realized as a FreeFEM++ script for the integrated FreeFEMM++-cs environment [9]. In the relativistic particle pusher of GYREOSS, a second-order leap-frog Boris–Bunemann scheme for tracing the motion of the particles is implemented. For allocation (scattering) of the space charge to the nodes of the computational mesh, an efficient charge-conserving algorithm is used. The overall structure and the program implementation of the computational modules allow the current serial (sequential) version to be parallelized and a future work in this direction is planned. An example, illustrating a typical output of the simulations carried out by GYREOSS is shown in figure 3.

3. Conclusions

The problem-oriented packages for numerical analysis, CAD and optimization of gyrotrons presented in this paper are being upgraded regularly in order to adapt them to the ever changing hardware and software environments (computational infrastructure, operating systems of the workstations, novel versions of the underlying numerical libraries, compilers). Alongside with such maintenance, they are undergoing constant improvements aiming to a better program implementation, debugging and optimization. More important, however, is that their structure allows an easy extensibility through embedding of new computational modules that add more physical models and functionality. Another characteristic feature of the simulation packages is that they are highly portable to different platforms despite the fact that they are heterogeneous, i.e., implemented in different algorithmic languages (FORTRAN, C, C++, scripts for various IDE). The problem-oriented packages developed have already been used successfully for numerical studies and CAD of various gyrotrons for novel applications [2, 3, 5].

References

[1] Thumm M 2015 State-of-the-art of high power gyro-devices and free electron masers (Update 2014) *KIT Scientific Reports* 7693 (KIT Scientific Publishing Karlsruhe) pp 1–172

[2] Idehara T and Sabchevski S 2012 Development and aapplications of high-frequency gyrotrons in FIR FU covering the sub-THz to THz range *J. Infrared, Millimeter and Terahertz Waves* **33** 667–694

[3] Glyavin M Y, Idehara T and Sabchevski S P 2015 Development of THz gyrotrons at IAP RAS and FIR UF and their applications in physical research and high-power THz technologies *IEEE Trans. on Terahertz Sci. Technol.* **5** 788–797

[4] Sabchevski S, Zhelyazkov I, Thumm M, Piosczyk B, Tran T M, Hogge J-Ph and Pagonakis J G 2007 Recent evolution of the simulation tools for computer aided design of electron-optical systems for powerful gyrotrons *Computer Modelling in Engin. & Sci.* **20** 203–220

[5] Damyanova M, Kern S, Illy S, Thumm M, Sabchevski S, Zhelyazkov I and Vasileva E 2014 Modelling and simulation of gyrotrons for ITER *J. Phys.: Conf. Series* **516** 012028

[6] Fliflet A W, Read M E, Chu K R and Seeley R 1982 A self-consistent field theory of gyrotron oscillators: application to a low Q gyromonotron *Int. J. Electronics* **53** 505-521

[7] Sabchevski S P and Idehara T 2015 A numerical study on finite-bandwidth resonances of high-order axial modes (HOAM) in a gyrotron cavity *J. Infrared, Millimeter and Terahertz Waves* **36** 628–653

[8] Damyanova M, Sabchevski S, Zhelyazkov I, Vasileva E, Balabanova E, Dankov P and Malinov P 2014 Current status of the development of the problem-oriented software package GYREOSS *J. Phys.: Conf. Series* **514** 012056

[9] Hecht F 2013 *FreeFem++* 3rd edition Version 3.23 (Laboratorie Jacques-Louis Lions, Université Pierre et Marie Curie Paris) pp 1–392 http://www.freefem.org/ff++/ftp/freefem++doc.pdf

Journal of Physics: Conference Series **700** (2016) 012011 doi:10.1088/1742-6596/700/1/012011

Thermophysical properties of CF_4/O_2 and SF_6/O_2 gas mixtures

M Damyanova[1,4], U Hohm [2], E Balabanova[1] and D Barton[3]

[1]Institute of Electronics, Bulgarian Academy of Sciences, 72 Tsarigradsko Chaussee 1784 Sofia, Bulgaria.
[2]Institut für Physikalische und Theoretische Chemie der TU Braunschweig, Hans-Sommer-Str. 10, D-38106 Braunschweig, Germany.
[3]Theoretische Organische Chemie, Organisch-Chemisches Institut and Center for Multiscale Theory and Computation, Westfälische Wilhelms-Universität Münster, Corrensstraße 40, 48149 Münster, Germany.

E-mail: milenadam@gmail.com

Abstract. Fitting formulae are presented for the calculation of the second interaction virial coefficients, mixture viscosities and binary diffusion coefficients for CF_4/O_2 and SF_6/O_2 gas mixtures in the temperature range between 200 K and 1000 K. The data recommended are obtained from the isotropic (n-6) Lennard-Jones intermolecular interaction potentials of the pure substances by using the Hohm-Zarkova-Damyanova mixing rules. In general, a good agreement is observed between our results and the experimental and theoretical data found in the literature.

1. Introduction

The determination of reliable thermophysical data for pure gasses and gas mixtures is an important but in many cases a cumbersome experimental and computational task. Such data related to the transport (viscosity, diffusion) and equilibrium (second virial coefficient) properties are needed in many practical applications. The fluorinated gases and gas mixtures, e.g. CF_4/O_2 and SF_6/O_2, are widely used for dry etching of silicon, some silicides, polymers etc. [1-3]. Also, there is evidence that the mixture SF_6/O_2 plays a significant role in studying the convective and diffusive intrapulmonary gas transport [4].

Despite the intensive use of the above mentioned mixtures in practical applications, there is still little knowledge about their thermophysical properties – second virial coefficients (interaction B_{12} and mixture B_{mix}), viscosities (interaction η_{12} and mixture η_{mix}), and diffusion coefficient D_{12}, especially in wide ranges of temperature and composition.

The published thermophysical data for both mixtures are very scarce and, in addition, they are given only in small temperature ranges. For example, we have found only three experimentally measured values for the interaction second virial coefficients, B_{12}, for both mixtures [5, 6] and the data are in the very narrow temperature range between 290 K and 320 K. There are no calculated values for B_{12}. Experimental data of the viscosity η_{mix} of both binary mixtures are measured with a high accuracy (uncertainty < 1%) only by Kestin et al. [7]. Using these measurements of the viscosity, the authors calculated the binary diffusion coefficients D_{12} for both mixtures. The uncertainty of their D_{12} values is

[4] To whom any correspondence should be addressed.

about 2 %. Haghighi et al. [8, 9] published calculated η_{mix} and D_{12} values for equimolar CF_4/O_2 and SF_6/O_2 mixtures. They claim an uncertainty of their data of 1 % for viscosities and 5 % for diffusion coefficients. Experimental data for D_{12} are available only for the SF_6/O_2 mixture [10-12]. However, for both mixtures we have found calculated values of D_{12} [4, 7-10].

In this paper we show calculations of the thermophysical properties of CF_4/O_2 and SF_6/O_2 gas mixtures in the temperature range 200 -1000 K and present recommended fitting formulae for the calculation of the intermolecular potential parameters, and equilibrium and transport properties.

2. Theoretical background

The thermophysical properties of the gas mixtures depend on the intermolecular interaction potentials between the different kinds of mixture molecules. Each binary interaction between the molecules is described by our previously suggested [13] isotropic (*n*-6) Lennard-Jones temperature-dependent potential model (LJTDP), (1). In the case of pure gases, the indexes in (1) are the same, i.e.1 = 2.

$$U_{12}(R,T) = \frac{\varepsilon_{12}(T)}{n_{12}-6}\left[6\left(\frac{R_{m12}(T)}{R}\right)^{n_{12}} - n_{12}\left(\frac{R_{m12}(T)}{R}\right)^{6}\right], \tag{1}$$

where: R_{m12} is the equilibrium distance, ε_{12} is the potential well depth and n_{12} is the repulsive parameter.

The intermolecular potential parameters for pure CF_4 and SF_6 were obtained earlier by Zarkova and Hohm [13], whereas those for O_2 were given by Damyanova *et al.* [14].

The temperature-dependent potential parameters for interactions between unlike molecules in the mixtures (R_{m12} and ε_{12}) are calculated by means of the mixing rule elaborated by Hohm-Zarkova-Damyanova (HZD) [15].

$$\left(R_{m12}(T)\right)^{n_{12}-6} = \frac{\left\{0.5\left[\left(\varepsilon_1(T)\right)^{\varphi}\left(R_{m1}(T)\right)^{\psi} + \left(\varepsilon_2(T)\right)^{\varphi}\left(R_{m2}(T)\right)^{\psi}\right]\right\}^{1/\varphi}}{\left(\varepsilon_1(T)\varepsilon_2(T)\right)^{1/2}\left(R_{m1}(T)R_{m2}(T)\right)^{3}}, \tag{2}$$

$$\varepsilon_{12}(T) = \left(\varepsilon_1(T)\varepsilon_2(T)\right)^{1/2}\frac{\left(R_{m1}(T)R_{m2}(T)\right)^{3}}{\left(R_{m12}(T)\right)^{6}}\frac{2\alpha_1\alpha_2\left(C_6^{11}C_6^{22}\right)^{1/2}}{C_6^{11}\alpha_2^2 + C_6^{22}\alpha_1^2}, \tag{3}$$

where $\varphi = \dfrac{1}{1+n_{12}}$ and $\psi = \dfrac{n_{12}}{1+n_{12}}$.

The repulsive parameter n_{12} is calculated as the arithmetic mean of the neat substances parameters.

Applying the mixing rule HZD for calculation of ε_{12} (eq.3), the dipole polarizability α and the dispersion-interaction energy constants C_6 of the pure substances have to be known.

All molecular parameters used for calculations in this work are given in table 1.

Table 1. Pure gas values for mass, dipole polarizabilities, dispersion-interaction energy constants and zero temperature potential parameters used in this work; k_B is the Boltzmann constant.

Substance	Molar mass/ (g/mol)	$(\varepsilon_{11}(0)/k_B)$/ K	10^{10} R_{m11} (0)/ m	n	α/ a.u.[a]	C_6/ a.u.[a]
CF_4	88.005	328.40 [13]	4.329 [13]	52.71 [13]	19.10 [16]	254.0 [18]
SF_6	146.1	417.80 [13]	5.041 [13]	34.76 [13]	30.04 [16]	585.8 [19]
O_2	32.0	185.53 [14]	3.460 [14]	23.20 [14]	10.585 [16]	61.57 [17]

[a] Conversion factors are for α: 1 a.u.=$1.64878\cdot10^{-41}$ $C^2m^2J^{-1}$ and C_6: 1 a.u. = $9.57345\cdot10^{-80}$ Jm^6.

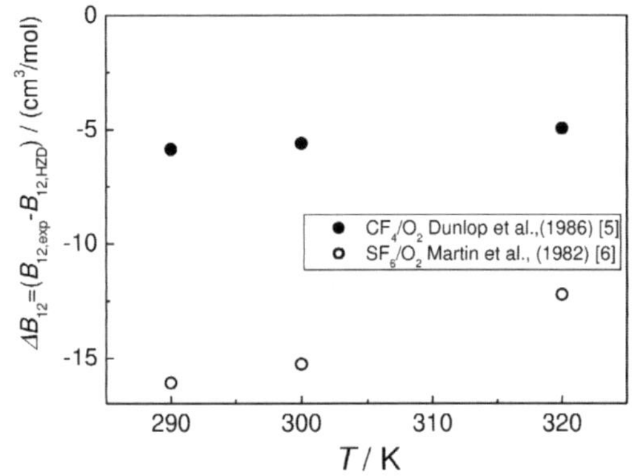

Figure 1. Absolute deviations between measured $B_{12,exp}$ [5, 6] and calculated $B_{12,HZD}$ interaction second virial coefficients of this work for CF_4/O_2 and SF_6/O_2 gas mixtures.

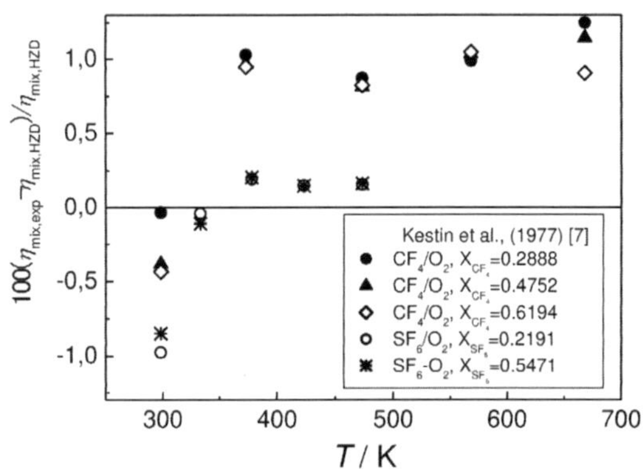

Figure 2. Relative deviations between our data $\eta_{mix,HZD}$ and measured viscosities $\eta_{mix,exp}$ [7] for different mole fractions of the binary mixtures CF_4/O_2 and SF_6/O_2.

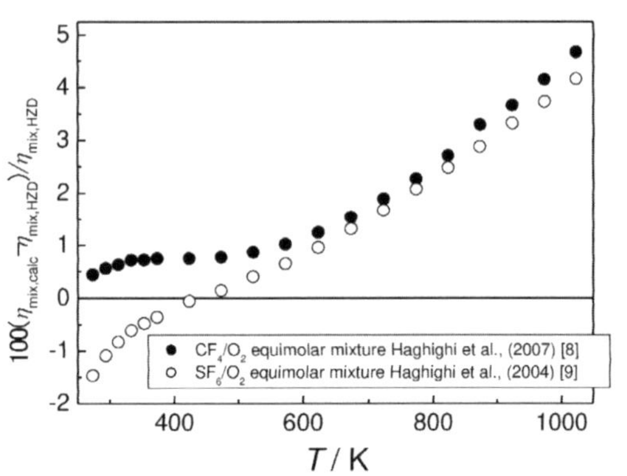

Figure 3. Relative deviations between our data $\eta_{mix,HZD}$ and calculated viscosities $\eta_{mix,calc}$ [8,9] for equimolar binary mixtures of CF_4/O_2 and SF_6/O_2.

3. Results and discussion

Based on the LJTDP model – equation 1, and the HZD mixing rule – equations 2 and 3, all thermophysical properties for CF_4/O_2 and SF_6/O_2 gas mixtures are calculated. In figures 1-5, the calculated data of this work are compared to those existing in the literature.

In figure 1, the absolute deviations between our results for B_{12} and the data from the literature are plotted. A slightly better agreement is observed for the CF_4/O_2 mixture. The relative deviations for both mixtures are in the range of 20 %.

In figures 2 and 3, the relative deviations are presented between the data of this work and the experimentally determined and calculated data from the literature.

The comparison between the binary diffusion coefficients obtained in this work and the data from the literature is shown in figures 4 and 5.

For a fast evaluation and compact representation, our recommended values of $R_{m12}(T)$, $\varepsilon_{12}(T)$, $B_{12}(T)$, $\eta_{12}(T)$ and $D_{12}(T)$ in the temperature range between 200 K and 1000 K are presented in the form of fitting formulae. The equilibrium distance $R_{m,12}(T)$ is fitted to the dimensionless expression

$$R_{m12}(T)/(10^{-10}\mathrm{m}) = R_{m12}(0) + A_1\exp\left(-\frac{A_2}{(T/\mathrm{K})}\right) + A_3\exp\left(-\frac{A_4}{(T/\mathrm{K})}\right).$$

The well depth $\varepsilon_{12}(T)$ can be calculated by $\varepsilon_{12}(T) = \varepsilon_{12}(0)\left(\frac{R_{m12}(0)}{R_{m12}(T)}\right)^6$. The zero temperature constants for the CF_4/O_2 gas mixture are given as

$R_{m12}(0)/(10^{-10}\mathrm{m}) = 3.90226$ and

$(\varepsilon_{12}(0)/k_B)/\mathrm{K} = 233.272$.

For the SF_6/O_2 gas mixture they are

$R_{m12}(0)/(10^{-10}\mathrm{m}) = 4.28103$ and

$(\varepsilon_{12}(0)/k_B)/\mathrm{K} = 239.149$.

In the case of the thermophysical properties of the mixtures the following formulae are used:

$$B_{12}(T)/(\mathrm{cm^3/mol}) = \sum_{k=1}^{5} A_k\left(\frac{T}{\mathrm{K}}\right)^{1-k},$$

$$\eta_{12}(T)/(\mu\mathrm{Pa\cdot s}) = \sum_{k=1}^{5} A_k\left(\frac{T}{\mathrm{K}}\right)^{k-1} \quad \text{and}$$

$$D_{12}(T)/(10^{-6}\mathrm{m^2/s}) = \sum_{k=1}^{5} A_k\left(\frac{T}{\mathrm{K}}\right)^{k-1}.$$

The fitting constants are given in table 2. Here ΔP is the standard error of the fit.

Table 2. Fitting constants

	Property	A_1	A_2	A_3	A_4	A_5	ΔP
CF_4/O_2	$R_{m12}(T)$	0.065063	745.759	0.206229	2003.72	-	$1.3\cdot10^{-5}$
	$B_{12}(T)$	62.528	$-2.4517\cdot10^{4}$	$-2.06341\cdot10^{5}$	$-2.24496\cdot10^{8}$	$7.33932\cdot10^{9}$	0.0024
	$\eta_{12}(T)$	-0.35780	$7.41461\cdot10^{-2}$	$-5.40712\cdot10^{-5}$	$2.72262\cdot10^{-8}$	$-5.71621\cdot10^{-12}$	0.0048
	$D_{12}(T)$	-1.04543	$1.14184\cdot10^{-2}$	$1.22530\cdot10^{-4}$	$-5.45932\cdot10^{-8}$	$1.31318\cdot10^{-11}$	0.00599
SF_6/O_2	$R_{m12}(T)$	$2.197297\cdot10^{-4}$	-428.364	0.245725	850.655	-	$4.5\cdot10^{-4}$
	$B_{12}(T)$	76.811	$-2.6582\cdot10^{4}$	$-5.13901\cdot10^{6}$	$7.00884\cdot10^{8}$	$-6.94799\cdot10^{10}$	0.047
	$\eta_{12}(T)$	-0.69580	$6.62836\cdot10^{-2}$	$-4.56954\cdot10^{-5}$	$2.02217\cdot10^{-8}$	$-3.21890\cdot10^{-12}$	0.0064
	$D_{12}(T)$	-0.72638	$7.00914\cdot10^{-3}$	$1.04792\cdot10^{-4}$	$-5.08581\cdot10^{-8}$	$1.37097\cdot10^{-11}$	0.0050

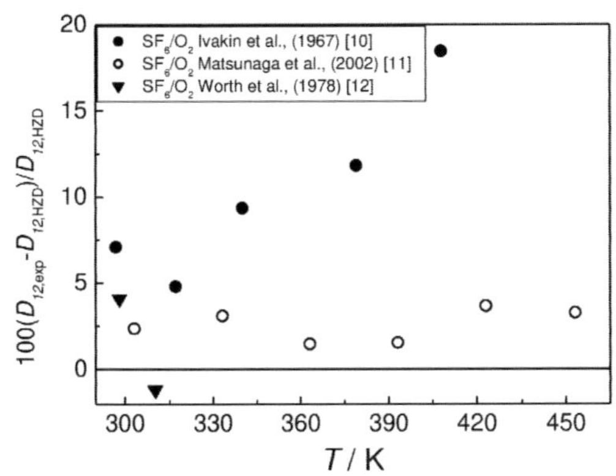

Figure 4. Relative deviations between experimentally determined $D_{12,exp}$ [10-12] and calculated $D_{12,HZD}$ binary diffusion coefficients of this work for SF_6/O_2 gas mixture.

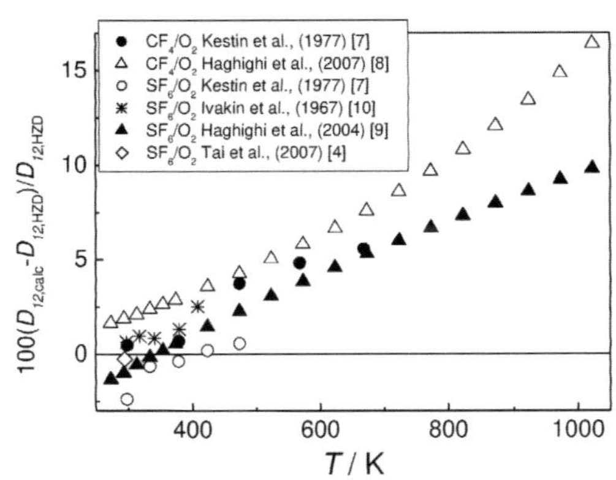

Figure 5. Relative deviations between calculated $D_{12,calc}$ [4,7-9] and binary diffusion coefficients of this work, $D_{12,HZD}$, for CF_4/O_2 and SF_6/O_2 gas mixtures.

4. Conclusions

Our proposed fitting formulae can be used for a fast calculation of the potential parameters, interaction second virial coefficient B_{12}, viscosity η_{12} and diffusion coefficient D_{12} for the binary mixtures CF_4/O_2 and SF_6/O_2 in the temperature range between 200 K and 1000 K. A comparison with the available experimental data suggests an accuracy of 20 % for B_{12} values better than 1 % for η_{mix} and 13 % for D_{12}, (see figures 1, 2, and 4). In figures 3 and 5, a nearly linearly increasing mismatch between our results presented in this paper and other theoretical data of the literature is observed. In the temperature range between 300 K and 1000 K, the deviations do not exceed 5 % in the case of the viscosities. We conclude that in this case the different underlying theories lead to comparable results. However, in the case of the binary diffusion coefficients, the relative deviations grow up to 15 % and the calculated results are significantly different. Only experiments carried out at higher temperatures would be capable of deciding between the different theories.

References

[1] Grigoryev Yu N and Gorobchuk A G 2007 *Russ. Microelectr.* **36** 321-332
[2] Chow T P and Fanelli G M 1985 *J. Electrochem. Soc.* **132** 1969-1973
[3] Meichsner J, Schmidt M, Schneider R and Wagner H 2012 *Nonthermal Plasma Chem. and Phys.* (CRC Press, Boca Raton)
[4] Tai R C, Chang H and Farhi L 1980 *Resp. Physiology* **40** 253-267
[5] Dunlop P J, Bignell C M and Robjohnes H L 1986 *Ber. Bunsenges. Phys. Chem.* **90** 351-353
[6] Martin M L, Trengove R D, Harris K R and Dunlop P J 1982 *Ber. Bunsenges. Phys. Chem.* **86** 626-627
[7] Kestin J, Khalifa H E, Ro S T and Wakeham W A 1977 *Physica* **88A** 242-260
[8] Haghighi B, Djavanmardi A H and Papari M M 2007 *Int. J. Comp. Methods* **4** 59-80
[9] Haghighi B, Djavanmardi A H, Papari M M and Najafi M 2004 *J. Theor. Comp. Chemistry* **3** 69-90
[10] Ivakin B A, Suetin P E and Plesovski W P 1967 *J. Techn. Phys.* **37** 1913-1914
[11] Matsunaga N, Hori M and Nagashima A 2002 *Nippon Kikai Gakkai Ronbunshu, B-hen* **68** 550-555
[12] Worth H, Nusse W and Piiper J 1978 *Resp. Physiology* **32** 15-26
[13] Zarkova L and Hohm U 2002 *J. Phys. Chem. Ref. Data* **31** 183-216
[14] Damyanova M, Hohm U, Balabanova E and Zarkova L 2010 *J. Phys.: Conf. Ser.* **223** 012012
[15] Hohm U, Zarkova L and Damyanova M 2006 *Int. J. Thermophys.* **27** 1725-1745
[16] Hohm U 2013 *J. Mol. Struct.* **1054-1055** 282-292
[17] Kumar A, Meath J W, Bündgen P and Thakkar A J 1996 *J. Chem. Phys.* **105** 4927-4937
[18] Bulanin M O and Kislyakov I M 2002 *Opt. Spectr.* **93** 27-32
[19] Kumar A, Fairley G R G and Meath W J 1985 *J. Chem. Phys.* **83** 70-77

Automation of the electron-beam welding process

E Koleva[1, 2,1], V Dzharov[1], M Kardjiev[1] and G Mladenov[1]

[1]Emil Djakov Institute of Electronics, Bulgarian Academy of Sciences,
72 Tzarigradsko Chaussee, 1784 Sofia, Bulgaria
[2]University of Chemical Technology and Metallurgy,
8 St. Kliment Ohridski blvd., 1756 Sofia, Bulgaria

E-mail: eligeorg@abv. Bg

Abstract. In this work, the automatic control is considered of the vacuum and cooling systems of the located in the IE-BAS equipment for electron-beam welding, evaporation and surface modification. A project was elaborated for the control and management based on the development of an engineering support system using existing and additional technical means of automation. Optimization of the indicators, which are critical for the duration of reaching the working regime and stopping the operation of the installation, can be made using experimentally obtained transient characteristics. The automation of the available equipment aimed at improving its efficiency and the repeatability of the obtained results, as well as at stabilizing the process parameters, should be integrated in an Engineering Support System which, besides the operator supervision, consists of several subsystems for equipment control, data acquisition, information analysis, system management and decision-making support.

1. Introduction

Over the years, the electron beam became a flexible and cost-effective manufacturing tool. Due to the deep penetration in the work-piece, the electron beam is able to generate deep and narrow welds with a minimal thermally affected zone without the use of welding consumables. The high vacuum required by the method prevents the heated and melted material from oxidizing and being affected by atmospheric contamination. With the development of advanced computer control, the number of electron-beam applications has increased significantly. In what concerns new applications of the electron-beam welding (EBW) technologies [1], the EBW plants turned into complex equipment containing highly stabilized power sources and electronic units, reliable and effective vacuum systems, technological chambers with precision 3D manipulators, thus becoming truly software controlled and programmed manufacturing tools with high efficiency and excellent reproducibility. The technological data gathered during the process enable one to monitor the quality and support the improvement of the testing process of the manufactured components; these data can be recorded for future analysis of the relations between the adjusted process parameters and the quality and stability of the welds.

In this paper, some aspects of the development of an Engineering Support System (ESS) are considered. The first step of this process is the choice of the ESS architecture and specification of the functionality of the implemented interactive sub-systems. The automation of the equipment control

[1] To whom any correspondence should be addressed.

sub-system is considered in more detail in view of a practical realization that will help the implementation of a fully integrated ESS.

2. Engineering support system (ESS)

A fully integrated ESS can be defined as an information system that supports the analysis, design, implementation and operation of the technological process and its elements, based on knowledge, defined principles and recommended practices. The main aim of developing an ESS for EBW is to integrate and organize the knowledge on the process and to use this knowledge to improve the modeling and control capabilities, their efficiency, adaptability, flexibility and re-configurability, as well as to achieve integration with other technological processes that can be performed on the same equipment, as surface modification, evaporation and thin-layer deposition, selective electron-beam melting, etc.

Usually, the electron-beam installations operate in a semi-automatic control mode. Even in the cases of highly-sophisticated computer controlled processes, the automation (including vacuum pump system, vacuum pressure control, cooling system control, manipulator control, processor-based high-voltage and emission-current control, electron-beam motion control and its characterization, PC-based automatic beam-power distribution, data acquisition and archiving etc.), operator supervision of the process and manual tuning are still required. The best practice shows [2-4] that modeling and simulation, when combined with a Decision Support System (DSS), can provide an organization of the tools necessary to make good business-case decisions by integrating the models, the data that drive the simulations and the optimization tools. A way to integrate the DSSs (combined with Data Collection and Acquisition, Information Analysis and System Management agents) with the automation (and/or manual) control and observation during the performance of EBW is the development of an ESS.

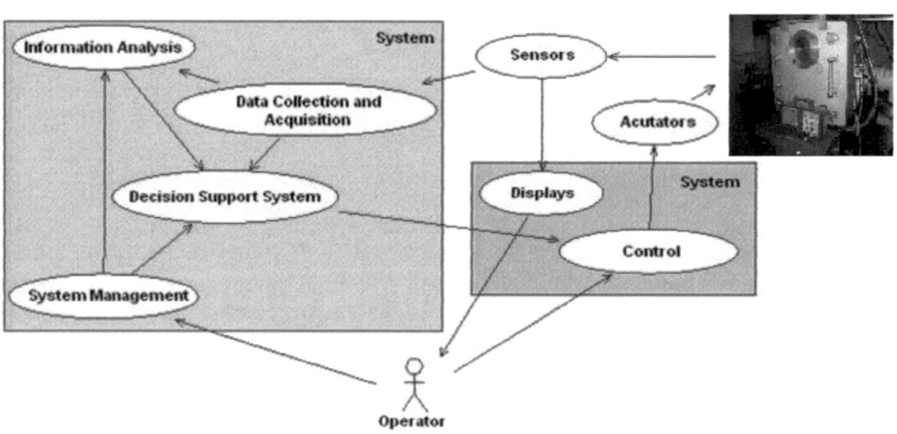

Figure 1. Engineering Support System architecture as a 'use-case' diagram.

The particular ESS architecture chosen [4] is presented as a UML use-case diagram in figure 1. The basic use-cases of the EBW ESS are: Data Collection and Acquisition (DCA), Information Analysis (IA), Decision Support System (DSS) and System Management (SM). They should be integrated with the System for ***Equipment Control* (EC)** in order to realize a working ESS.

The IA sub-system integrates:

- analytical modeling of the different processes – providing the DSS or the user with the control algorithm with calculated data (weld geometry parameters, weld defects, temperature distribution, etc.),
- statistical modeling and quality management and data analysis,
- information modeling,

- data-base management – different materials, models, experimental data, regime conditions, etc.,
- process parameters optimization.

The Decision Support System implements the results from the DCA, IA and SM sub-systems in order to provide the decision makers, the operators and managers with key information that enables them to make more efficient and consistent decisions. The System Management sub-system involves the definition of requirements and parametric constraints.

The System for ***Equipment Control*** **(EC)** (automated and/or manual) is usually designed by the EB equipment manufacturers. It includes sources of actual measurement of data for the state of the EBW equipment by sensors or other measuring devices – ***Monitoring agents*** **(MoA)**, such as: vacuum pump system, vacuum pressure control, processor-based high-voltage and emission-current control, PC-based automatic beam-power distribution control, data acquisition and archiving etc. The ***Control Agents*** **(CA)** – actuators – execute the control algorithms. During the real-time control, the CA interfere with the EC block by some industrial controller device or following operator manipulation with CA. It is possible to apply different control algorithms (PID, fuzzy, adaptive etc.) after receiving the signals from the measuring devices thus implementing state feedback regulators. The MoAs follow the system behavior after applying the recommended method for designing and control. In the case of standard EBW control systems, the described two agents, together with the Data Collection and Acquisition sub-system are available. If all requirements are satisfied, then MoAs update the database by newly achieved results (solutions) for next re-use and application. Otherwise MoA and CA have to repeat their operations. The loop is performed at regular time intervals.

In order to automate the EC system of the EBW equipment considered, development of EBW control algorithms are needed, involving existing and additional technical means for automation, identification of transitional processes, control of static processes, development of control models and Programmable Logical Controller (PLC) based algorithms.

3. EBW control algorithm

From a system engineering point of view, the EBW is a complex system of processes that can be engineered to accomplish specific business and technological objectivities. The main features of an EBW plant, which must be taken into consideration by the ESS development, are the following: vacuum level (5×10^{-4}hPa), power density (2 kW/cm^2) available for welding of different materials, local superheating, precise control of the beam and welded samples motion, as well as automation of the process.

The EBW vacuum and cooling systems control algorithm, as a sequence of operations of starting and shutting down the EBW equipment, timing and performance criteria, as well as the actions in two major emergency situations – a water supply interruption and electricity cuts are presented in figure 2. The timing and performance criteria for each separate operation, if they are available, are presented above the corresponding block.

When starting the EBW installation, it is necessary to take into account the ambient temperature, which is a condition for the initial heating of the rotary pump (RP1) oil. If the temperature is above 20 °C, we move to the next operation and start the RP1 on a gas ballast mode. The check for availability of electricity and water must be carried out continuously by appropriate technical means.

In order to improve the operation of the EBW installation in emergency situations, we recommend building of sound and light alarms, air-cooling for the diffusion pump and the use of alternative supply (UPS).

4. Experimental conditions

Following the control algorithm, experimental starting and stopping of the installation for electron-beam welding were performed. There are two processes, which limit the shortening of the starting and shutting-down time – reaching the necessary level of the vacuum in the chamber at the start and cooling the diffusion pump when stopping the EBW equipment. This is why they are investigated here in order to make further improvements.

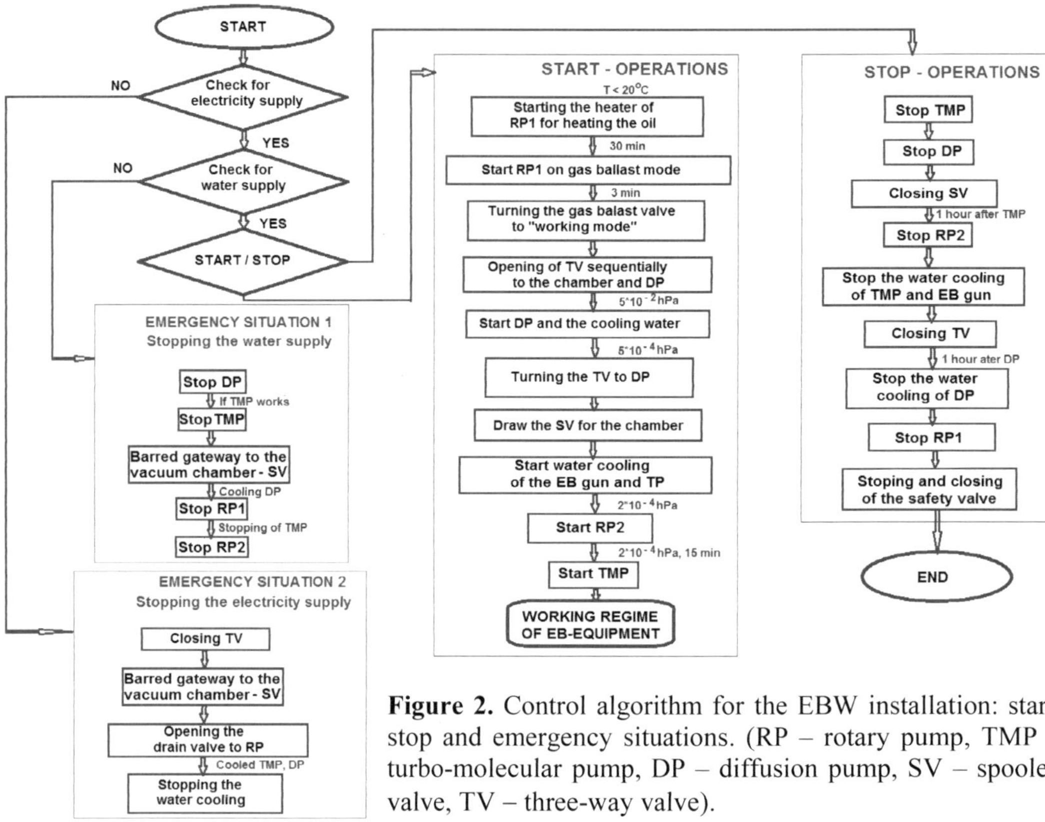

Figure 2. Control algorithm for the EBW installation: start, stop and emergency situations. (RP – rotary pump, TMP – turbo-molecular pump, DP – diffusion pump, SV – spooled valve, TV – three-way valve).

The temperature of the diffusion pump and the readings from the sensors for high and low vacuum level were considered. The total duration of the experiment was 5 hours and 15 minutes. This time did not include the realization of the technological process of electron-beam welding of materials, evaporation or surface modification, but merely the time needed for the equipment to reach the corresponding operating mode followed by the shuting down procedure. The optimization and management of this time interval is associated with both optimal management of the technical means available and the introduction of new ones with more flexible management and with better parameters or support functions. Figure 3 is a graph of the input pulses which represent the starting or stopping of the pumps - diffusion pump (DP), turbo-molecular pump (TMP) and two rotary pumps (RP1 and RP2).

The results of the measurements of the temperature of the diffusion pump (DP, in °C), the pressure in the zone of the sensor for low pressure (LP, in hPa) and the pressure in the zone of the sensor for high pressure (HP, in hPa), depending on the working time, are presented at figures 4, 5 and 6, correspondingly.

5. Identification of the transitional processes

The transitional functions of the temperature of the diffusion pump (DP), the pressure in the zone of the sensor for low pressure (LP) and the pressure in the zone of the sensor for high pressure (HP) are estimated on the basis of the experimental data obtained. The gains k ($k = \Delta y = y(0) - y(\infty)$) and step input signal $u(t) = 1(t)$, the time constants T and time delays τ (defined from the beginning of the experiment) are defined.

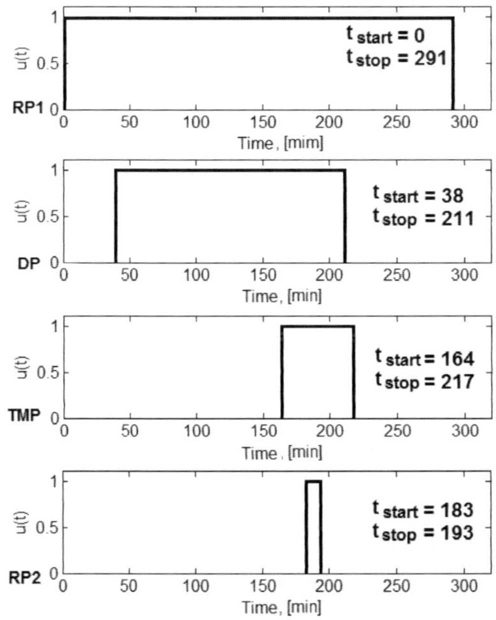

Figure 3. Input signals representing the starting or stopping of the pumps – diffusion pump (DP), turbo-molecular pump (TMP) and two rotary pumps (RP1 and RP2).

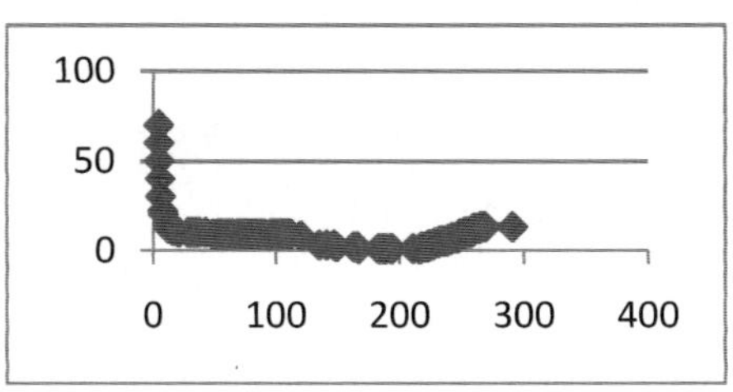

Figure 4. Temperature of the diffusion pump (DP), °C, depending on the working time.

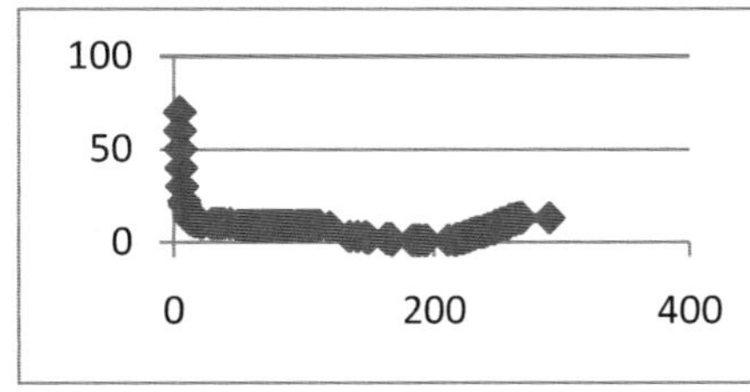

Figure 5. The pressure in the zone of the sensor for low pressure (LP) in hPa, depending on the working time.

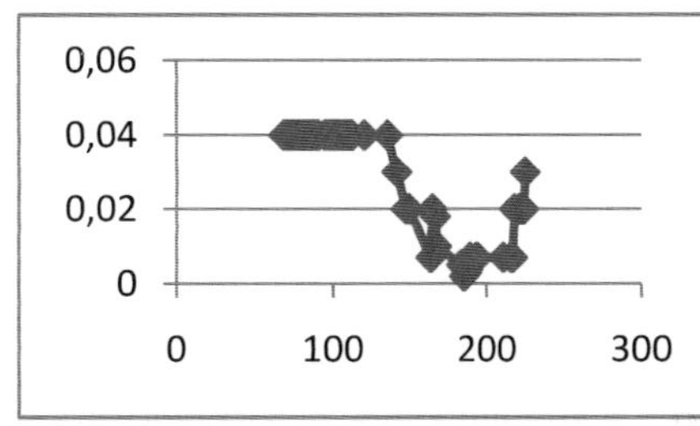

Figure 6. The pressure in the zone of the sensor for high pressure (HP) in hPa, depending on the working time.

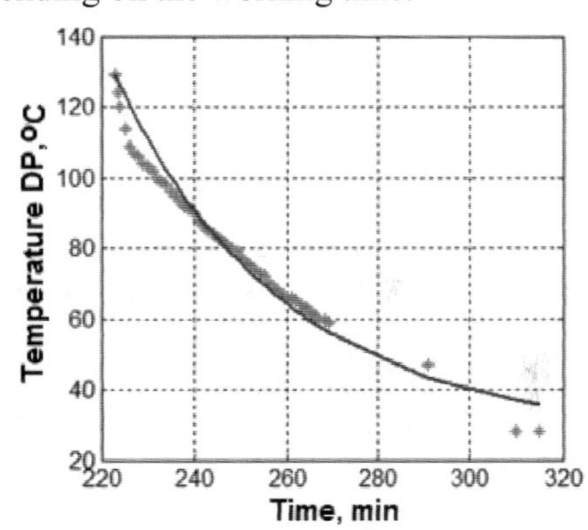

Figure 7. The transitional function of the temperature of the DP during stopping the EBW installation (stars – measured values, continuous – estimated values).

After defining the time constants *T*, their values are optimized through a minimization utilizing the scanning method of the criterion root-mean-squared error RSME:

$$RMSE = \sqrt{\frac{\sum_{i=1}^{n}(y^*_i - y_i)^2}{n}},$$

where y^*_i and y_i are respectively the experimental and the estimated values using the equations for the corresponding transitional function. The results obtained are presented in table 1.

Figure 7 presents as an example the transitional function of the temperature of the DP during stopping the EBW installation (stars – measured values, continuous – estimated values).

Table 1. Estimated parameters for transitional processes during EBW.

	$y(0)$	k	T	τ	$y(t)$	RMSE
Temperature DP -start Temperature DP – stop	28 28	108 101	18. 8 36	49 223	$k\left[e^{-\frac{t-\tau}{T}}\right]$	3. 5241 4.7248
High pressure – start	0. 007	0. 033	13. 5	135	$k\left[e^{-\frac{t-\tau}{T}}\right]$	0. 0021
Low pressure - start (before starting the DP)	10	60	1. 6	4. 5	$k\left[e^{-\frac{t-\tau}{T}}\right]$	2. 6845
Low pressure - start (after starting the DP)	1	8	22	112	$k\left[e^{-\frac{t-\tau}{T}}\right]$	0. 6132

6. Conclusions

The main aim of developing an ESS for EBW plants is to integrate and organize the knowledge for the EBW processes and plants and to use this knowledge to improve the modeling and control capabilities, their efficiency, adaptability, flexibility and re-configurability.

The ESS is intended for experts and technical staff in manufacturing departments, as well as for learning and training purposes or increasing the qualification in this area of university students or management staff. During the creation and verification of the ESS algorithm, a clarification of strategies and control mechanisms will be conducted so that the ESS could be used for dealing with critical situations.

This paper presents the chosen architecture of the ESS and some aspects of the specification of the functionality of the implemented interactive sub-systems. The automation of the control sub-system of the equipment is considered in more detail. The practical realization of the considered EBW algorithm using PLC will help the implementation of a fully integrated ESS.

Acknowledgements

The work was supported financially by the Bulgarian National Science Fund under contract BIn-5/2009.

References

[1] Koleva G E and Mladenov M G 2011 Experience on electron beam welding *Practical Aspects and Applications of Electron Beam Irradiation* eds M R Nemtanu and M B India (Transworld Research Network) pp 94-134

[2] Sheng-Tun L, Huang-Chin H and I-Wei S 2003 An ontology based knowledge management system for the metal industry *Proc. Int. Ebeam. Conf.* (20-24 May 2003 Budapest Hungary)

[3] Bowen J P and Hinchey M G 2005 Ten commandments revisited: a ten-year perspective on the industrial application of formal methods *Proc. FMICS'05* (5-6 Sept. 2005 Lisbon Portugal)

[4] Koleva E, Batchkova I and Velev K 2008 Engineering support system for modeling and control of the process EBMR of mMetals *Proc. Int. Ebeam. Conf.* (26-28 Oct. 2008 Reno USA)

Journal of Physics: Conference Series **700** (2016) 012013 doi:10.1088/1742-6596/700/1/012013

Electron beam characterization by a tomographic approach

E Koleva[1,2,3], G Mladenov[1], D Todorov[1], L Koleva[2] and M Kardjiev[1]

[1]Emil Djakov Institute of Electronics, Bulgarian Academy of Sciences, 72 Tzarigradsko Chaussee, 1784 Sofia, Bulgaria
[2]University of Chemical Technology and Metallurgy, 8 Kliment Ohridski blvd., 1756 Sofia, Bulgaria

E-mail: eligeorg@abv.bg

Abstract. In this paper, experimental data are analyzed for the integral current density distribution when the electron beam parameters are varied, namely, focusing current, beam current, venelt voltage and the distance to the measuring device. The 3D beam radial current density distribution is reconstructed by implementing a tomographic approach. The characterization of the electron beam is considered in connection with the estimation of the following parameters: the radial and the angular beam distribution standard deviations, the position of the beam focus and the beam emittance.

1. Introduction

The characterization of the radial and the angular space distributions of particle trajectories (or the respective current distributions) in powerful electron beams is an important scientific and practical task, connected with improving the quality of the electron beam technologies, developing standards for electron beam welding machines and permitting transfer of a concrete electron beam welding technology from one equipment to another. To apply the advantages of electron beam welding, it is necessary to know the properties of the electron beam. There only exist standards for measurements of the electron beam current and accelerating voltage as beam characteristics, applicable to the acceptance inspection of an electron beam welding machine or in research. These parameters cannot characterize the quality of the electron beam in terms of the ability to be transported over long distances and to be focused into a small spot with a minimum of divergence.

Tomography is a technique for reconstruction of a two-dimensional object image from a set of its one-dimensional projections, measured as an array of line integrals (or slices) of the object studied. This technique is widely used in different scientific areas, starting from medical applications, material sciences etc. A Fourier transform from real to frequency space and a consequent inverse Fourier transform permits to reconstruct the beam cross-section current density image (beam radial intensity profile) with its asymmetry features. In the last decade, profile measurement of intense electron beams by an enhanced modified Faraday cup was proposed. We developed further this tomography evaluation to obtain the emittance of the beam, containing the current distribution of the beam in the studied cross-section, together with the angular distribution of the beam electrons. The emittance (or its reciprocal value – the brightness normalized to one volt) is an invariant value along the beam

[3] To whom any correspondence should be addressed.

and could be used as a standard characteristic, for prognosis and optimization, as well as for transfer of technologies from one electron beam equipment to another.

Measurements of the accelerating voltage, beam current, focus coil current, and vacuum level provide little insight into the characteristics of the electron beam used to for welding. The profile of the beam has been monitored lately by the enhanced modified Faraday cup (EMFC) diagnostic tool, proposed by J. Elmer and co-workers [1,2]. The benefit of integrating this diagnostic tool into future process control regimes is that the beam radial current distribution characteristics, as the peak power density, full width at half maximum and full width at $1/e^2$ values, can be quantified. The results show that machine performance, in terms of these measured beam characteristics, varies over time. Testing has shown that the variability of the beam characteristics can be considerably decreased by using the EMFC diagnostic tool. With the implementation of this diagnostic tool in the process control procedures, every electron beam weld in approximately 90 welds over an 18-month time frame, met all of the requirements defined in the weld process specification and passed all of the post-weld quality control checks [3].

For the weld quality assurance, the right choice is important of the beam focus position with respect to the work-piece surface. The interaction of the beam with the welded material, different for various metals and for changed beam focus positions, is a limiting factor in guaranteeing the right choice of the sharp focus by a subjective operator's decision. The visualization and focusing of the beam spot could be done by directing the beam on a block of refractory metal. When the metal becomes hot, the operator adjusts the beam focusing current until the spot's optical brightness appear to be at maximum. The sharp focus beam position, evaluated more precisely by EMFC in the absence of beam-sample metal interaction, is also inapplicable to controlling the electron beam welding due to changes arising from beam scattering and from gas focusing of beam electrons in the metal vapors and the generated plasma in the interaction beam/metal-work-piece zone [4, 5].

There are several approaches for electron beam characterization [6-17], but generally they are divided into two big groups: i) assuming a Gaussian distribution of the beam current density and ii) without this assumption.

The measurement technique, using a Faraday cup and few radial slits in a disc, on which the monitored electron beam is rotated, in fact measures the integrated electron beam radial current density distributions by integrating the current passing along these thin slits in projections of the beam intensity, taken at equally spaced angles around the beam (figure 1).

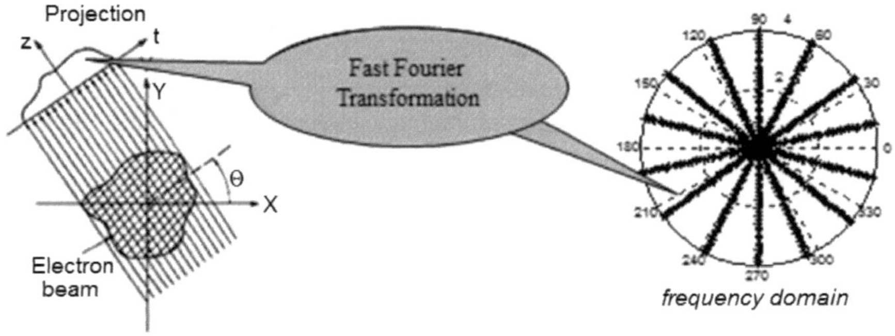

Figure 1. Tomographic approach – measurement of integral current densities at different angles and obtaining a set of radial distribution images in the frequency domain for each angle θ.

To apply the computer tomographic approach for estimation of the radial current density distribution, a refractory metal disk (W) with several radial slits is needed. The voltage signal is measured and the integral radial current density distributions are determined at the different angles of the slits (see figure 2 and figure 3).

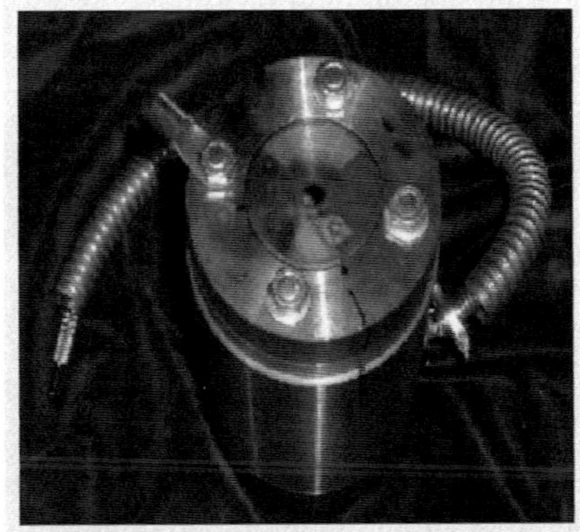

Figure 2. Experimental measuring device.

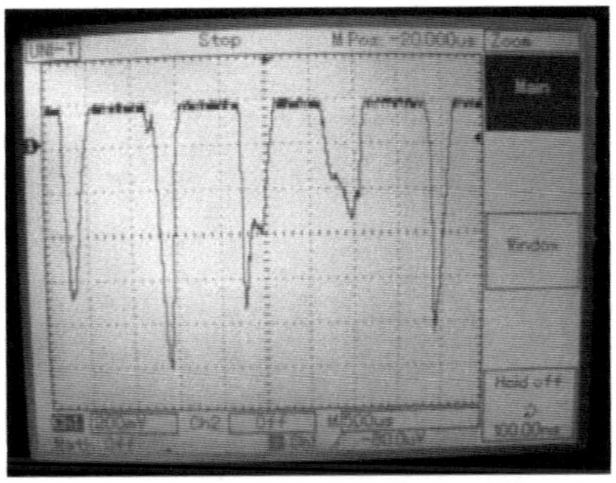

Figure 3. The measured voltage from five slits.

2. Experimental conditions

Our measuring devise had eight slits, two of them situated perpendicularly to each other:

a) $\theta = 0^o$; b) $\theta = 51^o$; c) $\theta = 90^o$; d) $\theta = 102^o$; e) $\theta = 153^o$; f) $\theta = 204^o$; g) $\theta = 255^o$; h) $\theta = 306^o$.

The integral current density distributions at different angles were obtained when the following parameters were varied:

I_f – focusing current, I_e – the beam current, U_v – venelt voltage (see table 1). This set of experiments was repeated for three values of the distance to the measuring device H – 330 mm, 340 mm and 350 mm, measured from the top of the vacuum chamber (72 experiments). The electron beam installation had a power of 2 kW, the pressure in the vacuum chamber being 7×10^{-5} hPa. The tungsten disc used was 4 mm thick and the slit width was 0.1 mm.

Figure 4 presents the experimental signal for experiment No. 3 (table 1) at $I_f = 221$ mA, $I_e = 6$ mA, $U_V = -50$ V and $H = 350$ mm.

Table 1. Experimental conditions.

No.	I_f [mA]	I_e [mA]	U_v [V]
1	221	2	-50
2	221	4	-50
3	221	6	-50
4	225	2	-50
5	225	4	-50
6	225	6	-50
7	229	2	-50
8	229	4	-50
9	229	6	-50
10	233	2	-50
11	233	4	-50
12	233	6	-50

No.	I_f [mA]	I_e [mA]	U_v [V]
13	221	2	-40
14	221	4	-40
15	221	6	-40
16	225	2	-40
17	225	4	-40
18	225	6	-40
19	229	2	-40
20	229	4	-40
21	229	6	-40
22	233	2	-40
23	233	4	-40
24	233	6	-40

3. Radial current density distribution image reconstruction

Based of the tomographic approach, the 3D beam radial current density distribution was reconstructed from its one-dimensional (1-D) projections in three cross-sections of the electron beam:

$$j_\theta(\theta, t) = \int_{-\infty}^{+\infty} j(x, y) dz .$$

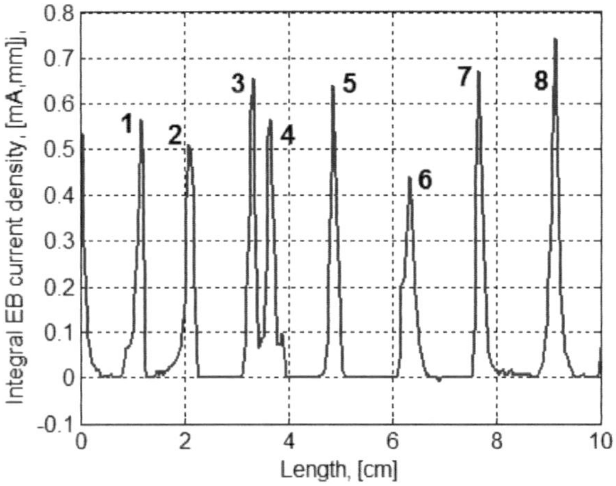

Figure 4. The experimental signal for experiment No. 3, $H = 350$ mm and degrees: 1 - 0° ; 2 - 51°; 3 - 90°; 4 - 102°; 5 - 153°; 6 - 204°; 7 - 255°; 8 - 306°.

The t axes coincide with the slices of the measuring device situated at different measurement angles θ (see figure 1). The collection of these projection functions at all angles θ is called Radon transform. In order to reconstruct the images, the Fourier slice theorem was used. The slice theorem proves that the 1D Fourier transform of the projection function $j_\theta(\theta,t)$ is equal to the 2D Fourier transform of the image evaluated on the line that the projection was taken on.

The algorithm implemented was the following [18-20]:

Step 1. Obtaining 1-D projections of a 2-D radial current density distribution at all possible angles in three cross-sections of the electron beam.

Step 2. Processing 1-D projections of a 2-D radial current density distribution at all possible angles in three cross-sections of the electron beam (Radon transform, digital signal processing).

Step 3. 1-D Fourier transform of the obtained projections to the frequency domain for each cross-section of the beam.

Step 4. Filtering the signals – by direct convolution or using the fast Fourier transform.

Step 5. Interpolation in the frequency domain.

Step 6. Initial image reconstruction by inverse 2D Fourier transform.

Step 7. 3D current density distribution modeling using the reconstructed images in three cross-sections of the beam – implementing spline-functions.

Figure 5 shows the reconstructed radial current density distributions in 2D and 3D view for $I_f = 221$ mA, $I_e = 2$ mA, $U_v = -50$ V (experiment 1, table 1) and $H = 330$ mm as an illustration of the implementation of the algorithm described.

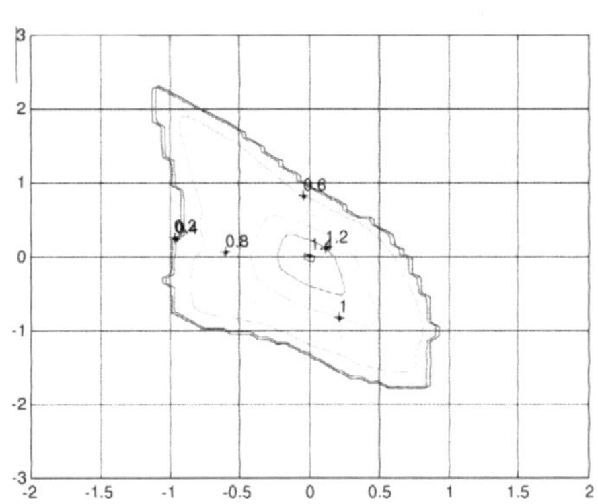

a)

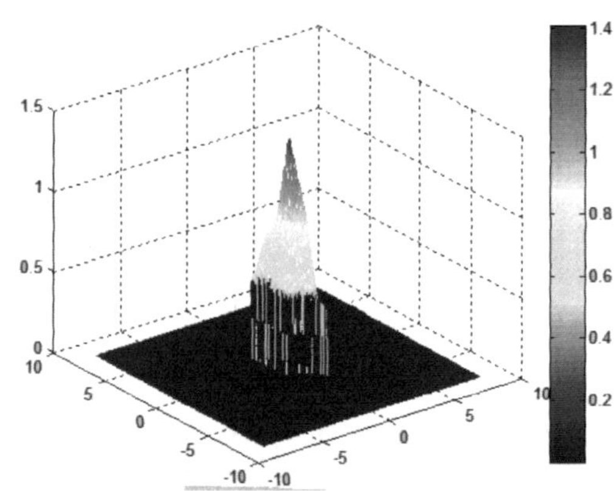

b)

Figure 5. the reconstructed radial current density distributions in 2D (a) and 3D (b) view for $I_f = 221$ mA, $I_e = 2$ mA, $U_v = -50$ V (experiment 1, table 1) and $H = 330$ mm.

4. Electron beam characterization

Initially, for each of the beam cross-sections and each of the electron beam parameters set (table 1), some beam current density characteristics were calculated (table 2) using the data obtained from the measuring device, consisting of a tungsten entrance disc with radial slits and a Faraday cup, (shown on figure 2) and utilizing our computer program.

Table 2. Electron beam current density characteristics.

	$H = 330$ mm				$H = 340$ mm				$H = 350$ mm			
No.	$d_{b,1}$ [mm]	σ_1 [mm]	$j_{xy,1}$	$j_{\theta,max,1}$	$d_{b,2}$ [mm]	σ_2 [mm]	$j_{xy,2}$	$j_{\theta,max,2}$	$d_{b,3}$ [mm]	σ_3 [mm]	$j_{xy,3}$	$j_{\theta,max,3}$
1	5.2	0.77	0.2976	0.1720	6	0.80	0.3184	0.1560	6.8	1.26	0.2400	0.0920
2	5.6	0.63	0.3184	0.2120	6.8	0.68	0.9696	0.5600	5.6	0.85	0.9280	0.4640
3	5.2	0.82	1.1264	0.5600	5.2	0.52	1.1600	0.9000	5.2	0.68	0.9920	0.6400
4	4.8	0.67	0.1472	0.0820	4.8	0.64	0.3024	0.1920	4.4	0.44	0.0762	0.0600
5	6.8	0.68	0.7360	0.5140	4.0	0.63	0.8992	0.6160	5.2	0.52	0.3968	0.3200
6	6.0	1.05	0.8140	0.8920	4.4	0.79	1.2320	0.6200	8.4	0.84	1.2768	0.7280
7	3.6	0.55	0.1848	0.1320	4.4	0.52	0.2368	0.1720	5.6	0.56	0.0845	0.0576
8	5.2	0.52	0.5120	0.4920	4.4	0.75	0.6112	0.3440	7.6	0.76	0.3600	0.2000
9	3.2	0.48	0.9600	0.7520	4.0	0.67	1.7680	1.0200	9.2	0.92	0.6016	0.3440
10	3.6	0.55	0.1208	0.0980	2.8	0.43	0.2368	0.2200	8.4	1.56	0.0960	0.0288
11	7.2	0.97	0.6528	0.2720	6.4	0.64	0.7208	0.6720	9.2	1.35	0.4715	0.3560
12	3.6	0.38	0.7968	0.7760	2.8	0.42	0.9364	0.9600	8.0	0.80	0.3772	0.2400
13	6.4	0.84	0.1699	0.0800	5.2	0.94	0.2640	0.1240	4.8	0.80	0.7960	0.6400
14	6.8	1.27	0.6272	0.2880	6.0	0.86	0.7392	0.3920	4.0	0.69	0.3936	0.2240
15	4.8	0.70	1.2288	0.7200	5.6	1.00	1.3440	0.5400	6.0	0.86	0.8928	0.4000
16	4.4	0.77	0.1872	0.0940	5.2	0.57	0.3336	0.2440	5.2	0.52	0.0930	0.0824
17	4.8	0.59	0.6208	0.4560	6.0	0.71	0.6752	0.4000	6.0	0.84	0.3288	0.1600
18	5.6	0.63	1.2544	0.8160	6.4	0.75	1.1120	0.5400	3.6	0.55	0.4512	0.3480
19	7.6	0.76	0.2512	0.2100	4.0	0.56	0.2336	0.1560	7.2	1.28	0.1531	0.0712
20	6.8	0.68	0.5760	0.4160	5.6	0.56	0.9376	0.8320	4.8	0.53	0.1094	0.0912
21	6.0	0.63	1.2960	0.8160	2.8	0.52	1.1520	0.8400	5.6	1.05	0.6768	0.3560
22	3.6	0.37	0.1304	0.1420	3.6	0.46	0.2304	0.2120	5.2	0.61	0.1066	0.0824
23	6.4	0.88	0.4128	0.2280	7.2	0.72	1.0368	0.6480	6.8	1.13	0.5936	0.2680
24	6.4	0.94	0.8928	0.3840	4.4	0.72	1.0484	0.8600	4.4	0.51	0.3968	0.3440

The signal range, corresponding to the beam diameter, was calculated as:

$$R = x_{max} - x_{min=d_b}\,.$$

The double indefinite integral of the radial beam current density distribution, integrated over x and y, and representing the beam current, is:

$$j_{xy} = \int_{-\infty}^{\infty}\int_{-\infty}^{\infty} j(x, y)\, dt\, dz\,,$$

σ is the standard deviation (beam radius on half maximum of signal); $j_{\theta,max}$ is the maximum value of the integral current density distribution, as measured for each of the beam cross-sections and each of the electron beam parameters set. The values for the standard deviation σ_i are estimated by minimization of the following loss function:

$$Q_\alpha = \sum_{i=1}^{n} (j_{\theta i}(t) - \hat{j}_{\theta i}(t))^2\,,$$

where $j_{\theta i}(t)$ and $\hat{j}_{\theta i}(t)$ are correspondingly the observed and estimated discrete (from 1 to n) values for every individual integral signal of the radial electron beam current distribution.

Assuming symmetrical current density distribution and averaging the results presented in table 2 for all measurement angles, estimation of the beam focus position, the radial and angular standard deviations σ_{oi} and σ'_{oi}, using the equation:

$$(\sigma_i)^2 = (\sigma_{oi})^2 + (z_{oi} - z_o)^2 (\sigma'_{oi})^2,$$

written under the condition of zero value of the co-variance between x and x' in the canonic position of the emittance diagram, one can find $\sigma_{x'oi}$ at measured σ_{xi} . Since $\sigma_i^2 = \sigma_{oi}^2 + \Delta_{o-i}^2 \sigma_{oi}'^2$, then:

$$\sigma_{oi}'^2 = \frac{\sigma_i^2 - \sigma_{oi}^2}{\Delta_{o-i}^2}.$$

The emittance and the standard deviations are related:

$$\varepsilon_p = C \sigma_o \sigma'_o,$$

where the coefficient C can be calculated as:

$$C = [-2\ln(1-p)]^{1/2}.$$

For value of the coefficient equal to 9, the whole beam current is considered (without reduction of the part of the beam current $p = I_m / I_b$, for example, to 80%).

Table 3. Electron beam characteristics.

No.	H_{focus} [mm]	σ_0 [mm]	σ'_0 [mrad]	ε [mm.mrad]
1	334.48	0.71	0.0671	0.4282
2	331.63	0.63	0.0312	0.1763
3	341.77	0.51	0.0545	0.2506
4	332.78	0.68	0.0297	0.1806
5	324.26	0.69	0.0175	0.1080
6	343.54	0.77	0.0529	0.3654
7	339.26	0.52	0.0194	0.0908
8	345.55	0.77	0.0372	0.2605
9	322.79	0.43	0.0299	0.1155
10	335.50	0.23	0.1088	0.2296
11	337.73	0.60	0.0986	0.5321
12	334.26	0.32	0.0465	0.1356
13	339.22	0.94	0.0459	0.3887
14	349.32	0.69	0.0552	0.3424
15	341.62	1.01	0.0621	0.5614
16	347.55	0.51	0.0327	0.1511
17	300.71	0.39	0.0151	0.0531
18	338.89	0.75	0.0461	0.3121
19	336.66	0.47	0.0891	0.3806
20	347.82	0.53	0.0241	0.1144
21	336.32	0.45	0.0692	0.2824
22	326.29	0.36	0.0207	0.0675
23	337.52	0.70	0.0712	0.4475
24	369.16	0.25	0.0231	0.0526

The transformations of coordinate x and x' in the draft space (that is free from forces external to the beam) are given in a matrix expression as:

$$\begin{pmatrix} x \\ x' \end{pmatrix}_2 = \begin{pmatrix} 1 & L \\ 0 & 1 \end{pmatrix} \begin{pmatrix} x \\ x' \end{pmatrix}_1 .$$

There index 1 stands for the cross-section at $z = z_1$ before the draft region with length L, and index 2, at $z = z_2$. L is the distance between these two cross-sections, measured along the beam axis.

The results calculated for the focus position (H_{focus}) measured with respect to the top wall of the vacuum chamber, the radial (σ_o) and the angular (σ'_o) standard deviations and the emittance value are presented in table 3. The average of the emittance values is 0.2511.

5. Conclusions

The management of the electron beam welding quality directed to optimizing the process parameters is an important way to improve the use of the expensive equipment and to make the EBW process more efficient in terms of consuming materials, time and energy.

The beam emittance, as well as the beam profile, are significant and appropriate characteristics of the beam quality. The measurement of these characteristics will: (i) assist in standardizing the electron optical systems, (ii) provide adequate conditions for welding production quality control by keeping a high reproducibility of the welds (iii) support the attempts to transfer a concrete technology from one welding machine to another and (iv) assist in implementing expert systems for the operator's choice of suitable regimes for producing desirable welds.

Acknowledgements

The work was supported financially by the Bulgarian National Science Fund under contract BIn-5/2009.

References

[1] Elmer J W, Teruya A T and O'Brien D W 1993 *Welding J.* **72**/11 493
[2] Elmer J W and Teruya A T 1998 *Sci. Technol. Welding and Joining* **3**/2 51
[3] Elmer J W and Palmer T A 2008 *J. Manuf. Sci. Eng.* **130** 041008
[4] Ledovskoy V P, Mladenov G M and Potsar A A 1970 *IEEE Conf. Publication* **70** 510
[5] Ledovskoy V P and Mladenov G M 1970 *J. Technical Phys.* (in Russian) **40**/10 2260
[6] Koleva E, Vutova K and Mladenov G 2005 *Proc DAE-BRNS Symp. Electron Beam Technol. Appl. SEBTA* (Sept. 2005 Mumbai India) 307-22
[7] Koleva E and Mladenov G 2005 *IEEE CPMT Annual School Lectures* **25**/1 3-6
[8] Koleva E, Menhard C, Loewer T and Mladenov G 2006 *Electronics and Electrical Eng.* (in Bulgarian) **5-6** 51-60
[9] Koleva E G, Mladenov G M 2006 *Russian Phys. J.* **11** 49-53
[10] Mladenov G and Koleva E 2007 *Proc. 7th Int. Conf. Beam Technol.* (Halle Germany) 85-92
[11] Koleva E and Mladenov G 2009 *Proc. 1st Int.. Conf. Technol. and Equipment Electron Beam Welding* (May 2008 St. Petersburg Russia) 99-109
[12] Koleva E, Mladenov G, Garkova K and Kardjiev M *Electronics and Electrical Eng.* (in Bulgarian) **5- 6** 64-9
[13] Mladenov G and Koleva E 2010 In *Welding: Processes, Quality, and Applications* ed R J Klein Mechanical Engineering -Theory and Applications (New York Nova Sci. Publishers) 1-99
[14] Koleva E, Vutova K, Mladenov G and Todorov D 2010 *Proc. 16th Int. SHCE and 10th Int. CMM* (19-24 Sept. 2010 Tomsk Russia) 31-4
[15] Koleva E and Mladenov G 2011 In *Experience on electron beam welding* eds M Nemtanu and B M Trivandrum India Transword Research Network Practical Aspects and Applications of Electron Beam Irradiation 95-133
[16] Mladenov G M, Koleva E G and Vutova K Z 2011 *Int. Rev. Mechanical Eng.* **5**/2 235-43

[17] Mladenov G and Koleva E 2009 *Proc. 6th Int. Conf. Beam Technol. Laser Appl.* (Politechnical University Publ.) (23-25 Sept. 2009 St. Petersburg Russia) 115-21
[18] Kuchment P and Uhlmann G 2012 *The Radon and X-Ray Transforms* (Department of Mathematics, University of Washington NSF supported Summer School)
[19] Hall G, Trevor J Terrell and Murphy L 1992 *Image Processing and Transputers* ed H C Webber (IOS Press) 47-72
[20] Hielscher R, Potts D, Prestin J, Schaeben H and Schmalz M 2008 *The Radon transform on SO(3): a Fourier slice theorem and numerical inversion* (IOP Publishing Ltd UK Inverse Problems 24 025011 21)

Model of a surface-wave discharge at atmospheric pressure with a fixed profile of the gas temperature

M Nikovski[1,3], Zh Kiss'ovski[1] and E Tatarova[2]

[1] Faculty of Physics, St. Kl. Ohridski University of Sofia, 5 J Bourchier blvd., 1164 Sofia, Bulgaria
[2] Institute of Plasmas and Nuclear Fusion, Instituto Superior Técnico, University of Lisbon, 1049-001 Lisbon, Portugal

E-mail: mon_1992@abv.bg

Abstract. We present a 3D model of a surface-wave-sustained discharge at 2.45 GHz at atmospheric pressure. A small plasma source creates a plasma column in a dielectric tube and a plasma torch is observed above the top. The plasma parameters and the axial profile of the gas temperature are significantly changed in the presence of the substrate above the plasma torch. The Boltzmann equation for electrons under the local approximation is solved, together with the heavy particle balance equations at a fixed axial profile of the gas temperature. The model of this finite length plasma column includes also the dispersion relation of azimuthally-symmetric surface waves. A detailed collisional-radiative model is also implemented for argon discharge at atmospheric pressure, which includes 21 rate balance equations for excited Ar atoms [(Ar($1s_5$-$1s_2$), Ar($2p_{10}$-$2p_1$), Ar(2s3d), Ar(3p)], for positive Ar^+ and Ar_2^+ ions and for excited molecules. The changes in the EEDF shape and the mean electron energy along the plasma column are investigated and the axial structures of the discharge and plasma parameters are obtained.

1. Introduction

The surface-wave discharges at atmospheric pressure are an object of intensive investigations due to their applications for environmental and industrial purposes. Detailed modelling of the discharges can provide insight into the importance of various processes, their rate constants, collision frequencies of the electrons, the effective temperature and concentration of electrons [1], and some basic plasma characteristics (gas temperature, population density, etc.). It also provides a better understanding of the regime of operation, distribution of the components of the EM field, and gives guidelines for optimizing the operating parameters. Modelling these discharges is difficult because of the large number of reactions between the charged particles and neutrals and the complexity of the processes at atmospheric pressure, even in argon gas [2-4].

This study presents a self-consistent model of a plasma column sustained by surface waves in a small-radius ceramic capillary of a portable microwave plasma source [5] at a fixed profile of the gas

[3] To whom any correspondence should be addressed.

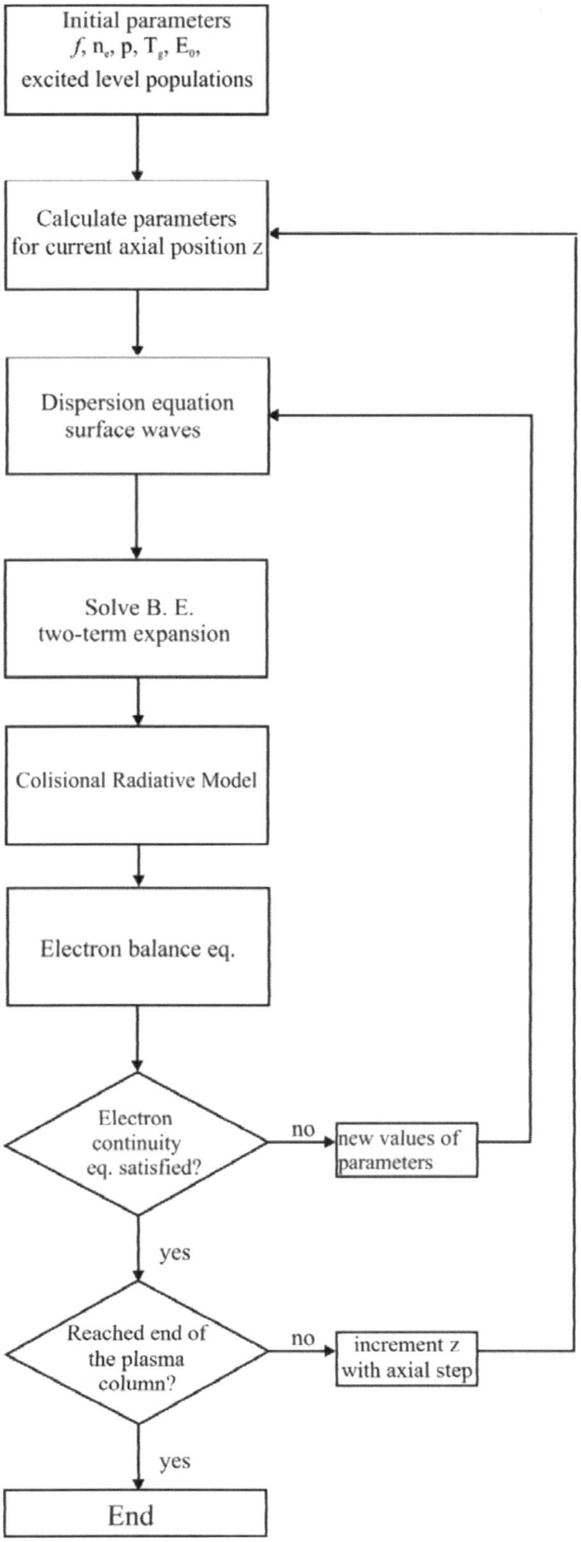

Figure 1. Flowchart of the model.

temperature. The plasma parameters and the axial profile of the gas temperature are significantly changed in the presence of the metal substrate above the plasma torch. In this case, the electron density increases, while the profile of the gas temperature depends on the thermal conductivity of the metal and the cooling of the gas by the substrate [6]. This experiment is modelled under the assumption for a fixed profile of the gas temperature along the plasma column with a different trend. The Boltzmann equation for electrons under the local approximation is solved, together with the heavy particle balance equations at this fixed profile of the gas temperature.

2. Description of the model

Figure 1 shows a flowchart of the theorerical model presented in this study. The self-consistent model includes as a first step a solver for the dispersion relation of azimuthally-symmetric surface waves sustaining the discharge in a high-permittivity ceramic tube (ε_d = 9.3). The dispersion equation of surface waves is solved for the case of a thin-cylinder approximation in cylindrical coordinates, with waves propagating in axial direction along the plasma column. The components of the azimuthally symmetric mode (TM_{00}) are E_z, E_r and H_φ with a maximum of the field amplitude at the plasma-dielectric interface. The plasma column is assumed homogeneous in radial direction and the plasma dielectric constant is calculated using an averaged value of the electron concentration n_e and an effective value of the electron-neutral collision frequency ν_{en}. By applying the boundary conditions across the three structure domains (plasma-dielectric-air), the dispersion relation is solved numerically via an iterative procedure. The values for the attenuation coefficient α and the phase constant β are obtained for given initial values of the electron concentration and the electron-neutral collision frequency. The distribution of the electromagnetic field in radial direction is obtained. The value of the effective electric field at the local position is calculated as $E_0 = \sqrt{E_z^2 + E_r^2}$ .The next part of the program is a numerical solver for the Boltzmann equation for electrons under the local approximation and a two-term approach for the EEDF. This equation is solved for the stationary regime of the discharge taking into account elastic and inelastic collisions, Penning ionization, etc. in the collisional integral.

After obtaining the effective electron temperature and concentration, all populations of the argon atom excited states, and the frequencies of excitation, ionization, de-excitation, etc. are calculated. These values are taken into account in the heavy particle balance equations. The populations of the argon atoms excited levels are calculated by the collisional-radiative model for argon plasma [7]. The following heavy particles are included in the model: Ar s-levels ($1s_2$-$1s_5$), the Ar 2p-levels ($2p_1$-$2p_{10}$), blocks for Ar(2s3d) and higher levels, atomic Ar^+ and molecular Ar_2^+ ions and excited molecules Ar_2*. The kinetic processes considered in the Boltzmann equation are: elastic collisions and inelastic collisions (excitation, de-excitation, ionization, step-wise ionization, dissociative recombination for Ar_2^+ and three-body recombination, etc., and Penning ionization for the excited molecules) (table 1). In this manner, the local parameters close to the exciter of SW waves are obtained.

Table 1. Rate balance equations considered in the model.

No.	Rate balance equation
1.-4.	4 1s-levels – Ar ($1s_5$-$1s_2$)
5.-14.	10 2p-levels – Ar ($2p_{10}$-$2p_1$)
15.	Ar (2s3d)
16.	Ar (3p)
17.	Ar (higher levels)
18.	Ar^+ ions
19.	Ar_2^+ ions
20.	Ar_2^* molecules
21.	Equation for the quasi-neutrality of the plasma

3. Results

In this study, results are presented for the EEDF at three possible distributions of the gas temperature in the plasma column: constant value (line 1) and two linear profiles with increasing slopes (lines 2 and 3) (figure 2). The Boltzmann equation is solved via the iterative finite difference method, so that the electron energy distribution function (EEDF) at these three distributions is obtained. The shape of EEDF is close to the Maxwellian, but its slope in a semi-logarithmic scale increases with the increase of the gas temperature gradient (figure 3).

The mean electron energy (effective electron temperature T_e), calculated from the EEDF, as well as the electron concentration n_e, decrease along the plasma column (figure 4 and figure 5).

By solving the dispersion relation of azimuthally-symmetric surface waves, values of the attenuation constant α and the phase constant β are obtained and compared at the different profiles of the gas temperature (figure 6). It is observed that the attenuation constant approaches the phase constant much faster in the presence of a significant temperature gradient along the axis. Figure 7 shows the effective electric field which decreases monotonically along the plasma column.

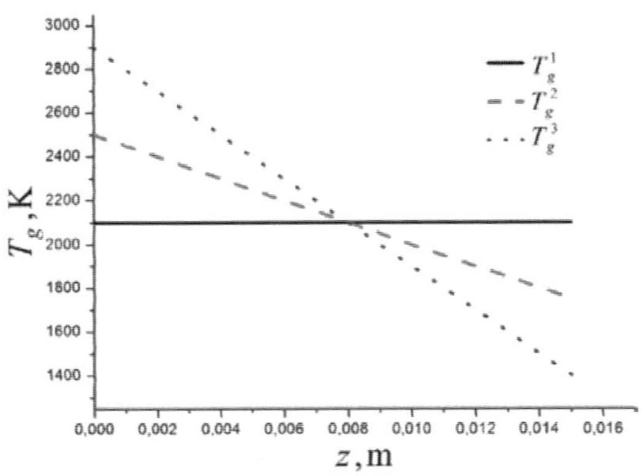

Figure 2. Axial profiles of the gas temperature T_g.

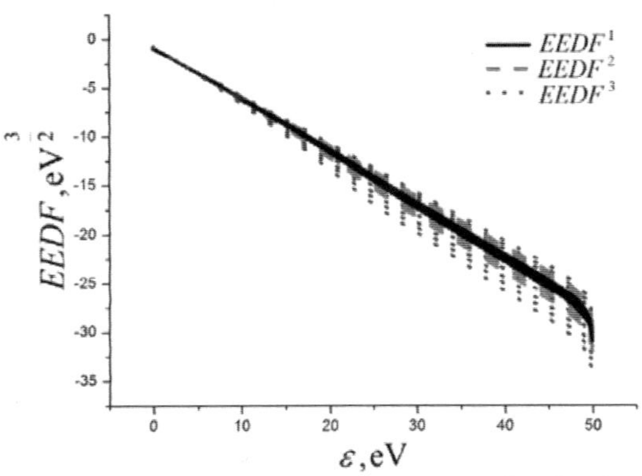

Figure 3. Electron energy distribution function for different profiles of T_g.

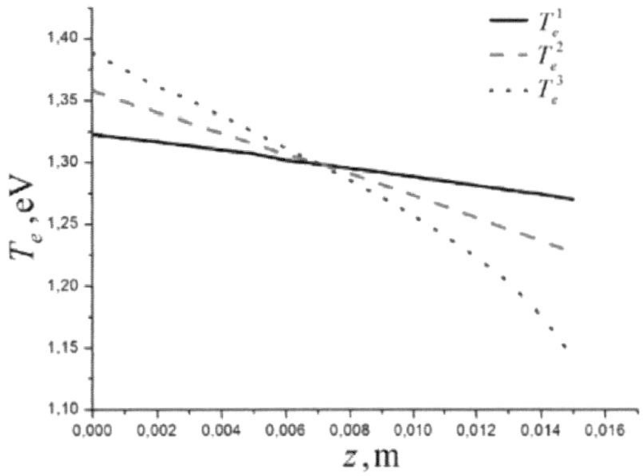

Figure 4. Electron temperature T_e along the plasma column.

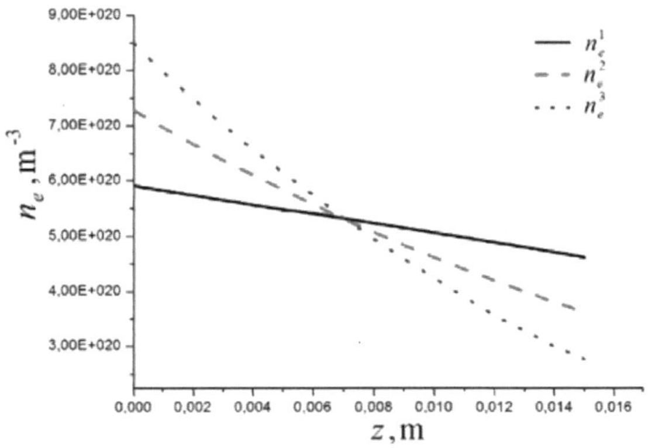

Figure 5. Electron concentration n_e along the plasma column.

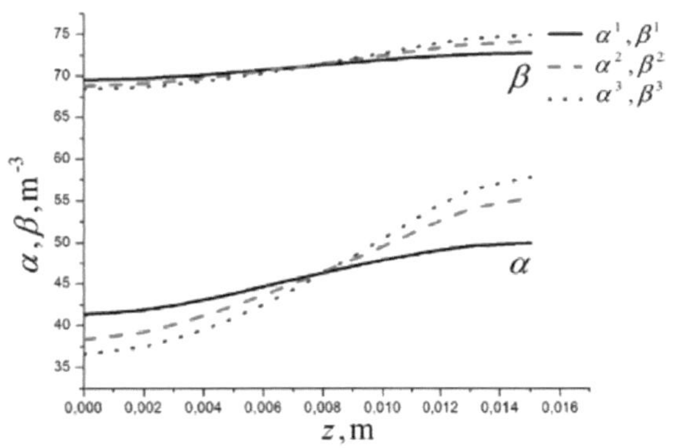

Figure 6. Attenuation constant α and phase constant β along the plasma column.

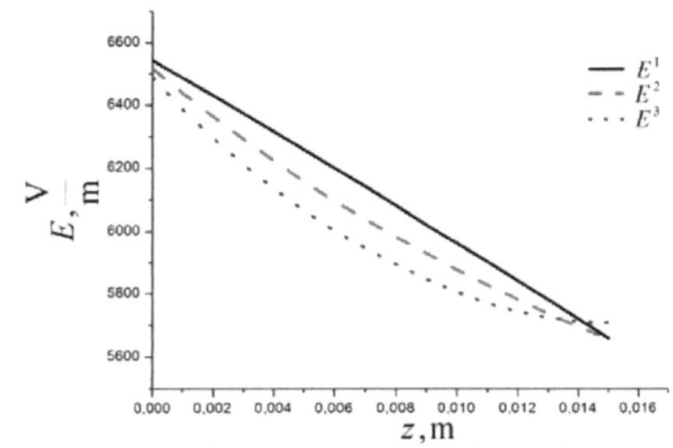

Figure 7. Effective electric field E along the plasma column.

4. Conclusions

A model of a surface-wave discharge at atmospheric pressure in argon gas in a high-permittivity ceramic tube ($\varepsilon_d = 9.3$) is used for three fixed profiles of the gas temperature. The Boltzmann equation for electrons is solved, together with the detailed collisional-radiative model for argon plasma at atmospheric pressure. The values are obtained of the plasma parameters along the axis of the plasma column at increasing inclination profile of T_g. They show a steeper decrease with the distance, while the attenuation constant α and the phase constant β have a steeper increase.

References

[1] Moisan M and Ferreira C M1993 *Microwave Discharges – Fundamentals and Applications* (Nato Science Series)

[2] Tatarova E, Dias F M, Felizardo E, Henriques J, Pinheiro M J, Ferreira C M and Gordiets B 2010 *J. Appl. Phys.* **108** 123305

[3] Henriques J, Ivanov A, Tatarova E, Kiss'ovski Zh and Ferreira C M 2011 *Bulletin APS* **56** 35

[4] Pencheva M, Petrova T, Benova E and Zhelyazkov I 2008 *J. Phys: Conf. Series* **44** 11

[5] Kiss'ovski Zh, Kolev M, Ivanov A, Lishev St and Koleva I 2009 *J. Phys.* D*: Appl. Phys.* **42** 182004

[6] Kiss'ovski Zh, Djermanova N, Mitev D and Vachkov V 2014 *J. Phys: Conf. Series* **514** 012007

[7] Zhu X and Pu Y 2010 *J. Phys.* D*: Appl. Phys.* **43** 015204

Journal of Physics: Conference Series **700** (2016) 012015 doi:10.1088/1742-6596/700/1/012015

Stray magnetic field influence on the CPT resonance in a coated Rb vacuum cell

E Taskova[1], E Alipieva and G Todorov

Emil Djakov Institute of Electronics, Bulgarian Academy of Sciences, 72 Tzarigradsko Chaussee, 1784 Sofia, Bulgaria

E-mail: taskova@ie.bas.bg

Abstract. Interaction of a resonant laser beam with an atomic absorption medium creates population redistribution and interference between atomic levels. This anisotropy of the medium is experimentally observed as coherent population trapping (CPT) or electromagnetically induced transparency (EIT). Due to the small sub-natural width of the CPT and EIT resonances, they find wide applications in metrology, quantum optics, atom cooling. A non-compensated stray magnetic field (SMF) can change the shape and sign of the resonance or destroy it completely. In this work, we present an experimental and theoretical investigation of the influence of a stray magnetic field on the CPT resonances obtained on Zeeman sublevels of the D_1 line of ^{87}Rb in a paraffin-coated vacuum cell. The role is clarified of the polarization moments with different rank in creating the integral registered fluorescent signal in the presence of a stray magnetic field. It is shown that a transverse magnetic field plays an important role in changing the shape of the signal.

1. Introduction

Coherent population trapping (CPT) is an interference effect arising in a three-level atomic system after interaction with two coherent fields which connect the ground levels to a common upper level [1]. In a Hanle-configuration, the CPT resonance is due to the interference between the Zeeman sub-levels of a single hyperfine level. The effect manifests itself as a dark resonance (lack of fluorescence), or as a peak of transmission. The resonance width is determined by the lifetime of the coherent state. As the ground states are long-lived, the CPT resonances are ultra narrow. Widths of 1-10 Hz have been measured [2].

The narrow resonance and good signal/noise ratio attracts the attention of the scientists and the CPT phenomena find many applications – in magnetometry, quantum optics, atom cooling. The role of the factors which change the shape and width of the resonance is increasing respectively. As our signal is obtained through a controlled scanning of a magnetic field around its zero value, a stray magnetic field can change the shape and sign of the resonance or destroy it completely. Earlier, the influence of the stray magnetic field on Hanle CPT was investigated in an uncoated vacuum cell [3].

In a coated cell, the long lifetime of the coherent state makes possible the transfer of coherence by multi-photon processes because of mixing of the contribution of polarization moments with different ranks. They have different relaxation constants and are influenced in a different way by the laser

[1]To whom any correspondence should be addressed.

Journal of Physics: Conference Series **700** (2016) 012015 doi:10.1088/1742-6596/700/1/012015

power and the stray magnetic field [4]. In a previous work [5], we investigated the CPT resonance dependence on the laser power and the experimental geometry in a coated cell when a polarized fluorescence is registered. A satisfying agreement between the experiment and the theoretical modeling was achieved when an additional dc magnetic field was added to the computation model.

In the present work, we studied the connections of the polarization moments having different ranks with the tensor components, which form the observable integral fluorescence, in the presence of a stray magnetic field.

2. Experiment

2.1. The set-up

The experimental set-up is shown in figure 1(a). It consists of a single-mode diode laser (DL) oscillating at λ = 794.76 nm, a Rb cell and a system of registration and control of the experiment. The laser beam with diameter 1.5 mm passes through an optical isolator (I) to avoid the optical feedback and a polarization rotator to change the laser polarization direction when necessary. An additional linear polarizer is inserted in order to obtain pure linear light polarization. The laser power is attenuated by means of grey filters and measured by a Thorlabs power meter. Then the laser beam passes through the vapor cell enclosed in a 3-layer magnetic shield which isolates the stray magnetic field. Additional Helmholtz coils (not shown in figure 1) are mounted in the magnetic shield to compensate the residuum magnetic field. The vacuum cell is paraffin-coated, containing ^{87}Rb, 25 mm long with diameter 20 mm. The laser frequency is controlled by monitoring the fluorescence from a second Rb vapor cell. All measurements are performed at room temperature (25 °C). A dc magnetic field B_{scan}, created by a solenoid, is applied collinearly to the laser beam. The current through the solenoid is controlled by a digital function generator Rigol.

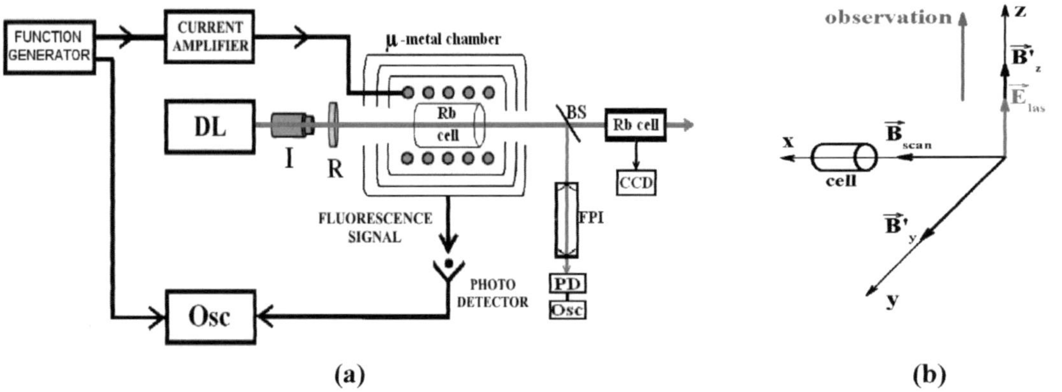

Figure 1. (a) the experimental set-up; (b) geometry of the experiment.

Our experiment was performed on the $5^2S_{1/2} \rightarrow 5^2P_{1/2}$ transition (F = 2→F = 1) of the ^{87}Rb D_1 line. The ground state coherence was created by a linearly polarized laser beam. The fluorescence from the cell was detected by a photodiode mounted so as to gather the light emitted in the **0z** direction (figure 1(b)). The fluorescence dependence on the dc magnetic field $\mathbf{B}_{scan}$ = ±60 mG was stored in a Tektronix oscilloscope. The resonance is centered at zero magnetic field where the levels of the $5^2S_{1/2} \rightarrow 5^2P_{1/2}$ transition are degenerated. The fluorescence signal shape dependence on the laser power and the transverse magnetic field was studied.

In figure 2, the CPT resonances are shown obtained at different laser power excitation, from P_{las}=5 μW to P_{las}=19 mW. The shape and sign of the resonance change as the laser power is increased. Two resonances were chosen for theoretical modeling – at low and high power, having opposite signs.

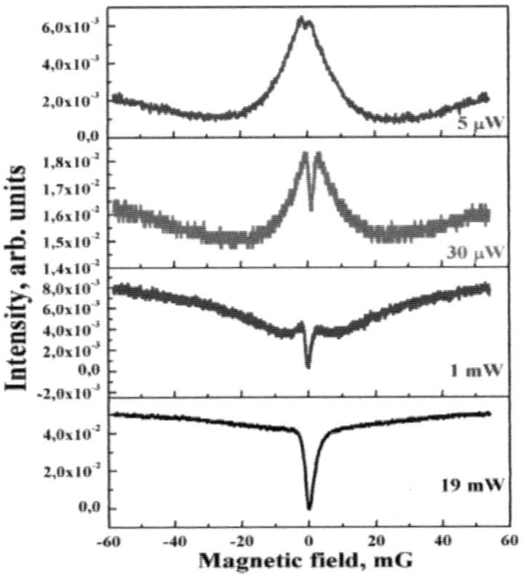

Figure 2. CPT resonance detected in parallel to $\mathbf{E}_{las}$ direction at different laser powers P_{las}.

3. Theoretical modeling

The theoretical model in the present work used a basic system of equations for the component of the statistical operator, $\hat{\rho}$, adapted to describe the interaction of an atomic system with a static (scanned) magnetic field $\mathbf{B}_{scan}$, an additional arbitrarily oriented magnetic field $\mathbf{B}'$ and a linearly polarized strong laser field $\mathbf{E}(\omega_0, t)$ resonant to the atomic transition with total quantum number F_φ, F_f. Using the irreducible tensor operator (ITO) formalism and the adopted calculation scheme [6], solutions for arbitrary polarization moments (PM) ρ_q^κ can be obtained, depending on the scanned magnetic field, for different parameters: – Rabi frequency, relaxation constants, including spontaneous emission transfer, and stray magnetic field $\mathbf{B}'$.

As the quantization axis **0z** is chosen along the electrical vector of the laser light $\mathbf{E}_{las}$ and the magnetic field $\mathbf{B}_{scan}$ is scanned along the laser beam propagation **0x**, only two components of the stray magnetic field $\mathbf{B}'$ are of importance.– $\mathbf{B}'_y$ and $\mathbf{B}'_z$. A stray magnetic field $\mathbf{B}'_y$ orthogonal to the scanned magnetic field $\mathbf{B}_{scan}$ and $\mathbf{E}_{las}$ is equivalent to introducing a variable with the $\mathbf{B}_{scan}$ ellipticity in the laser field. As a result, orientation moments ρ_q^1, $\rho = f, \varphi$ are created. The dynamics of the tensor components describing the modification of the orientation of the upper f_0^1 and lower $\varphi_{0,\pm1}^1$ level is shown in figure 3. It is seen that only the component of the transverse magnetic field directed along the **0y** is of importance.

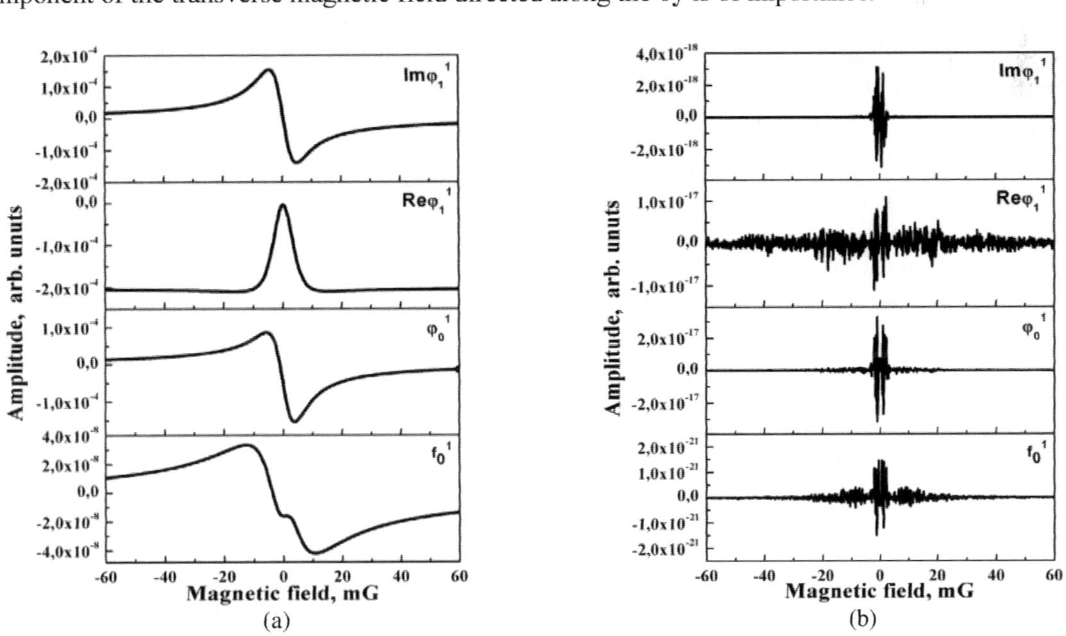

Figure 3. Dependence of the orientation moment on the scanned magnetic field in the presence of a transverse magnetic field: a) directed along the **0y** axis; b) directed along the **0z** axis.

Detailed analyses of the numerical solutions for all tensor components with rank 0 and 2 of the upper and lower state allow the visualization of the dynamics of their behavior depending on the laboratory magnetic field (figure 4).

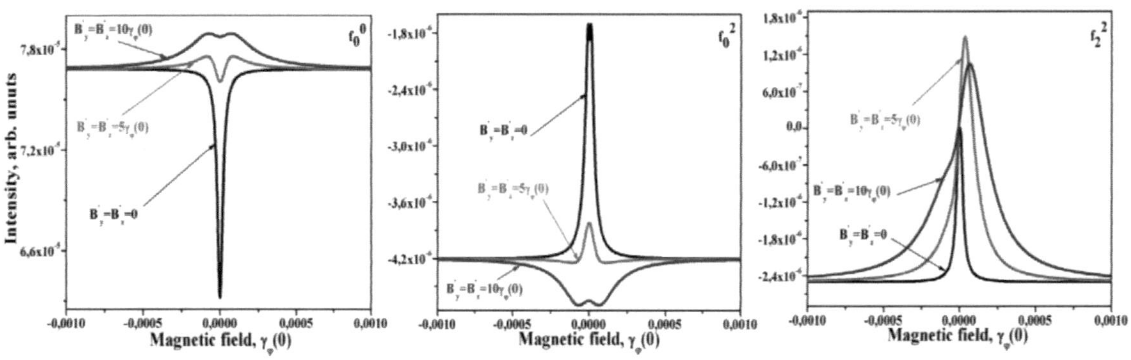

Figure 4. Behavior of population f_0^0, longitudinal f_0^2 and transverse f_2^2 alignment in the presence of a transverse magnetic field.

It is seen that as the transverse magnetic field is increased, the amplitude of the population f^0_0 and the narrow resonance in the longitudinal alignment f^0_2 of the upper state decrease. As a result, the sign of the resonance in the non-polarized fluorescence changes. The behavior of the population and the longitudinal alignment of the lower level is similar. The transverse alignment f^2_2 dependence on the stray magnetic field is non-monotonic. The physical reason for this behavior is clear. The transverse magnetic field creates orientation and at the same time destroys the lower level alignment. This leads to a decrease in the amplitudes of the longitudinal and transverse alignment resonances of the upper level. This process competes with the spontaneous transfers of coherency and population; with the increase of the laser power, its efficacy decreases.

The non-polarized spontaneous emission intensity $I_{f\varphi}(\vec{e}_n)$ in a chosen direction $\vec{e}_n$ from the upper (f, F_f) to the lower (φ, F_φ) level is described by

$$I_{f\varphi}(\vec{e}_n)=C_0(-1)^{F_f+F\varphi}(2F_f+1)^{-1/2}\left\|d_{f\varphi}\right\|^2\sum(2\kappa+1)\begin{Bmatrix}1&1&\kappa\\F_f&F_f&F_\varphi\end{Bmatrix}\sum_q(-1)^q f_q^\kappa \mathrm{T}^\kappa_{-q}(\vec{e}_n)\quad,\qquad(1)$$

where $\mathrm{T}^\kappa_{-q}(\vec{e}_n)$ is the observation tensor [7].

The integral fluorescent signal was obtained using the formula for the non-polarized fluorescence. Two atomic sub-ensembles were included in the theoretical model. The first is responsible for the wide pedestal. The lifetime of these atoms is determined by the time of flight. The second ensemble of atoms forms the narrow resonance and their lifetime is determined by the relaxation of the laser-created coherent state. The Rabi frequency and the value of the transverse magnetic field were varied. To examine the transverse magnetic field influence on the resonance shape, two cases were chosen – at a laser power P_{las} –30 μW and at P_{las} – 700 μW. The experimental records of the signal are shown in figure 5: a) – in the presence of a transverse magnetic field; b) with a compensated transverse magnetic field.

It is seen that the transverse magnetic field destroys the narrow structure centered at the zero point. Interpretations of the experimental results are made with numerical modeling of the experiment.

Computed signals are in a good agreement with the experiment.

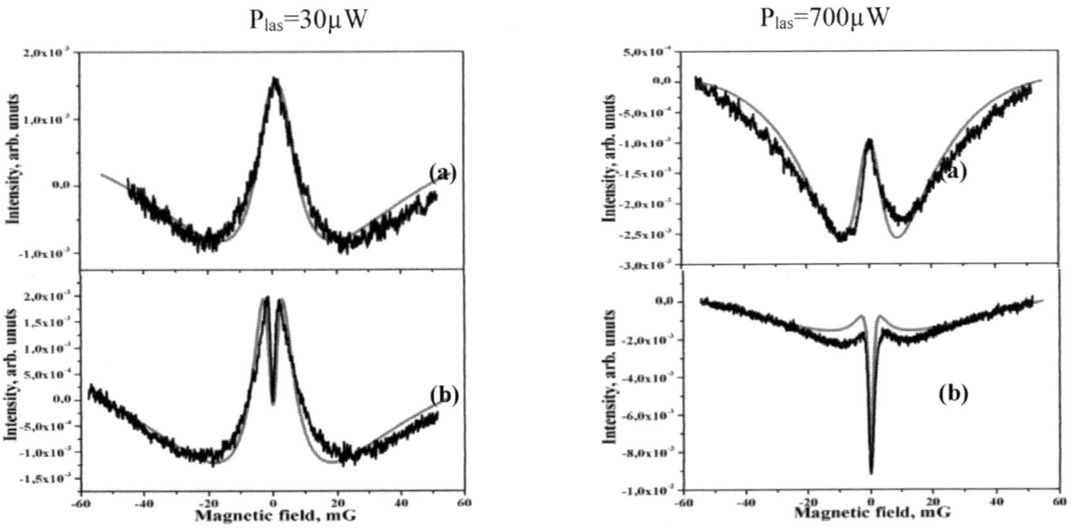

Figure 5. a) in presence of a transverse magnetic field; b) with a compensated transverse magnetic field.

4. Conclusions

A transformation from bright to black CPT resonance, depending on the laser power and a transverse magnetic field, was registered on the D_1 ^{87}Rb line in a paraffin-coated cell. A model for numerical calculation of the CPT in Hanle-configuration was applied, which uses the ITO formalism and takes into account the influence of a stray magnetic field. The role was clarified of the orientation in creating the integral experimentally observed signal in the presence of a stray magnetic field. The comparison of the theoretical shape of the resonances with the experiment shows a good agreement.

Acknowledgements

This work has received partial funding from the 7th EU Framework Programme: "Coherent optics sensors for medical applications-COSMA" (Grant Agreement No: PIRSES-GA-2012-295264).

References

[1] Arimondo E 1996 *Prog. Opt.* **35** 257-354
[2] Graf M T, Kimball D F, Rochester S M, Kerner K, Wong C, Budker D, Alexandrov E B, Balabas M V and Yashchuk V V 2005 *Phys. Rev.* A **72** 023401
[3] Huss A, Lammegger R, Windholz L, Alipieva E, Gateva S, Petrov L, Taskova E and Todorov G 2006 *JOSA* B **23**/9 1729-36
[4] Klein M, Novikova I, Phillips D F and Walsworth R L 2006 *J. Mod. Opt.* **53**/16-17 2583-91
[5] Taskova E, Alipieva E and Todorov G 2012 *J. Phys.: Conf. Series* **356** 012030
[6] Polischuk V, Domelunksen V, Alipieva E and Todorov G 2012 *Bulg. J. Phys.* **39** 150-64
[7] Alexandrov E, Chaika M and Hvostenko G 1991 *Interference of atomic states* (Springer-Verlag)

Journal of Physics: Conference Series **700** (2016) 012016 doi:10.1088/1742-6596/700/1/012016

A self-consistent one-dimensional multi-fluid model of the plasma-wall transition in the presence of two species of negatively biased particles

T Gyergyek[1,2,3] and J Kovačič[2]

[1]University of Ljubljana, Faculty of Electrical Engineering, 25 Tržaška, 1000 Ljubljana, Slovenia
[2]Jožef Stefan Institute, 39 Jamova, 1000 Ljubljana, Slovenia

E-mail: Tomaz.gyergyek@fe.uni-lj.si

Abstract. A one-dimensional three-fluid model of the plasma-wall transition in front of a large, negative, planar electrode is developed. The plasma contains one species of singly charged positive ions, Maxwellian electrons and a second group of singly charged negative particles. These particles can be either electrons with temperature different than that of the basic electron population or singly charged negative ions. The case with a second electron population with a high temperature is analyzed and transitions between the low and the high solutions are studied.

1. Introduction

Plasmas that contain more than one population of negatively charged particles are of considerable importance in many areas of plasma physics. Numerous studies on sheath formation in front of a negative electrode immersed in this type of plasmas can be found in the literature due to the great practical interest – see, e.g., the long list of references in [1] and [2]. A common approach in these studies is that the positive ions are treated by fluid equations, while for the negative particles the Boltzmann relation is assumed [2]. In this work, however, the electrons and the second negative particle species are also treated by their own set of continuity and momentum exchange equations.

2. Model

It is assumed that the continuity equations and equations of motion are valid for all three species of charged particles:

$$\frac{\partial n_i}{\partial t}+\nabla\cdot\left(n_i\mathbf{u}_i\right)=S_i,\quad \frac{\partial n_e}{\partial t}+\nabla\cdot\left(n_e\mathbf{u}_e\right)=S_e,\quad \frac{\partial n_2}{\partial t}+\nabla\cdot\left(n_2\mathbf{u}_2\right)=S_2, \tag{1}$$

$$m_i n_i\left(\frac{\partial \mathbf{u}_i}{\partial t}+\left(\mathbf{u}_i\cdot\nabla\right)\mathbf{u}_i\right)=n_i e_0\mathbf{E}-\nabla p_i+\mathbf{A}_i-m_i\mathbf{u}_i S_i, \tag{2}$$

$$m_e n_e\left(\frac{\partial \mathbf{u}_e}{\partial t}+\left(\mathbf{u}_e\cdot\nabla\right)\mathbf{u}_e\right)=-n_e e_0\mathbf{E}-\nabla p_e+\mathbf{A}_e-m_e\mathbf{u}_e S_e, \tag{3}$$

[3] To whom any correspondence should be addressed.

$$m_2 n_2 \left(\frac{\partial \mathbf{u}_2}{\partial t} + \left(\mathbf{u}_2 \cdot \nabla \right) \mathbf{u}_2 \right) = -n_2 e_0 \mathbf{E} - \nabla p_2 + \mathbf{A}_2 - m_2 \mathbf{u}_2 S_2. \tag{4}$$

The meaning of the symbols is the following: m_i is the ion mass, m_e is the electron mass, m_2 is the mass of the second negatively charged particle species, n_i, n_e and n_2 are the respective particle densities, $\mathbf{u}_i$, $\mathbf{u}_e$ and $\mathbf{u}_2$ are the flow velocities, e_0 is the elementary charge, S_i, S_e and S_2 are the source terms, discussed below, $\mathbf{A}_i$, $\mathbf{A}_e$ and $\mathbf{A}_2$ are collision terms. In this work the elastic collisions are neglected. The electric field $\mathbf{E}$ is related to the potential Φ by $\mathbf{E} = -\nabla\Phi$ and the potential profile $\Phi(\mathbf{r})$ is determined by the Poisson equation:

$$\nabla^2 \Phi = -\frac{e_0}{\varepsilon_0} \left(n_i(\mathbf{r}) - n_e(\mathbf{r}) - n_2(\mathbf{r}) \right). \tag{5}$$

Here ε_0 is the permittivity of the free space. Isothermal flow of ions, electrons and second species of negative particles is assumed and the gradient pressure terms are expressed using the ideal gas law:

$$\nabla p_i = kT_i \nabla n_i, \; \nabla p_e = kT_e \nabla n_e, \; \nabla p_2 = kT_2 \nabla n_2. \tag{6}$$

Here k is the Boltzmann constant.

The source terms S_i, S_e and S_2 give the difference between the number of created and annihilated charged particles of the respective species per unit volume and per unit time. They can be functions of space (and time) and in models, like the one presented in this work, it is usually assumed that they are given functions of space. Ionization and annihilation can have various physical mechanisms. In this work, it is assumed that the source terms are constants independent of the space coordinates. Of course, the consistency of the model requires that $S_i = S_e + S_2$ since in the steady state the same number of positive and negative particles must be created and annihilated in any selected unit volume. The source terms are therefore written in the following form:

$$S_i = \frac{\alpha}{\tau}, \; S_e = \frac{\beta}{\tau}, \; S_i = \frac{\gamma}{\tau}. \tag{7}$$

Here τ is effective time between two consecutive creations of a charged particle, while α, β and γ give the number of created particles of respective species in unit volume. Note that α, β, γ, and τ are all positive constants and their values are such that the subtraction of annihilated particles from the created particles has already been taken into account. Steady state is assumed, so all partial derivatives over time are neglected. It is assumed that on one side the plasma is bounded by a large planar electrode (collector), which is perpendicular to the x axis. The model is one-dimensional, so gradient and Laplace operators are replaced by

$$\nabla \to \frac{d}{dx}, \; \nabla^2 \to \frac{d^2}{dx^2}, \tag{8}$$

and the electric field has one component only given by:

$$E = -\frac{d\Phi}{dx}. \tag{9}$$

When (6) – (9) are taken into account, and the collision terms $\mathbf{A}_i$, $\mathbf{A}_e$ and $\mathbf{A}_2$ are neglected, equations (1) – (5) are written in the following form:

$$\frac{d}{dX}\left(N_i V_i \right) = \varepsilon s_i, \; \frac{d}{dX}\left(N_e V_e \right) = \varepsilon s_e, \; \frac{d}{dX}\left(N_2 V_2 \right) = \varepsilon s_2, \tag{10}$$

$$N_i V_i \frac{dV_i}{dX} = -N_i \frac{d\Psi}{dX} - \Theta \frac{dN_i}{dX} - \varepsilon V_i s_i, \tag{11}$$

$$N_e V_e \frac{dV_e}{dX} = \frac{N_e}{\mu} \frac{d\Psi}{dX} - \frac{1}{\mu} \frac{dN_e}{dX} - \varepsilon V_e s_e, \tag{12}$$

$$N_2 V_2 \frac{dV_2}{dX} = \frac{N_2}{\xi}\frac{d\Psi}{dX} - \frac{\Omega}{\xi}\frac{dN_2}{dX} - \varepsilon V_2 s_2, \tag{13}$$

$$\frac{d^2\Psi}{dX^2} = N_e(X) + N_2(X) - N_i(X). \tag{14}$$

If one is interested only in the quasi-neutral pre-sheath region, the Poisson equation (14) is replaced by the following neutrality condition:

$$N_e(X) + N_2(X) - N_i(X) = 0. \tag{15}$$

The following variables have been introduced:

$$\lambda_D = \sqrt{\frac{\varepsilon_0 k T_e}{n_0 e_0^2}},\ c_0 = \sqrt{\frac{kT_e}{m_i}},\ L = c_0\tau,\ \varepsilon = \frac{\lambda_D}{L},\ \mu = \frac{m_e}{m_i},\ \xi = \frac{m_2}{m_i}, \tag{16}$$

$$N_i = \frac{n_i}{n_0},\ N_e = \frac{n_e}{n_0},\ N_2 = \frac{n_2}{n_0},\ s_i = \frac{\alpha}{n_0},\ s_e = \frac{\beta}{n_0},\ s_2 = \frac{\gamma}{n_0},\ \Psi = \frac{e_0\Phi}{kT_e}, \tag{17}$$

$$V_i = \frac{u_i}{c_0},\ V_e = \frac{u_e}{c_0},\ V_2 = \frac{u_2}{c_0},\ \Theta = \frac{T_i}{T_e},\ \Omega = \frac{T_2}{T_e},\ X = \frac{x}{\lambda_D}. \tag{18}$$

Here n_0 is the plasma density in the unperturbed region far from the collector and c_0 is called the normalizing velocity; it is not the same as the ion sound velocity c_S, which is in normalized units (16) – (18) given by [1]:

$$V_S = \frac{c_S}{c_0} = \sqrt{\vartheta + \Theta}, \tag{19}$$

where ϑ is the "screening temperature" normalized to T_e and is given by [3,1]:

$$\vartheta = \frac{N_e + N_2}{\frac{dN_e}{d\Psi} + \frac{dN_2}{d\Psi}}. \tag{20}$$

3. Results

In the top graphs of figure 1, solutions of the system (10) – (14) are shown for the following parameters: $\mu = \xi = 1/1836$ (protons), $\Omega = 25$, $\Theta = 0$, $s_i = s_2 = 1$, $s_e = 0$, $\varepsilon = 10^{-5}$; and boundary conditions, given at $X = 0$, $N_i(0) = 1$, $N_1(0) = 0.736$, $N_2(0) = 0.264$, $V_i(0) = 10^{-7}$, $V_e(0) = V_2(0) = 0$, $\Psi(0) = 0$ and $d\Psi/dX(0) = 0$. The selection of the parameters is not motivated so much by some specific physical situation, but rather by our intention to illustrate some properties of the model. For example, in a recent paper [4], Langmuir probe measurements in COMPASS tokamak and TJ-II stellarator have been presented, where it has been shown that in both machines in the vicinity of the last closed flux surface the electron distribution function is bi-Maxwellian with the higher electron temperature 3 – 5 times larger than the lower one and the density of the lower electron temperature population several times larger than the density of the high temperature electron population. The system (10) – (14) is integrated in the positive X direction towards the collector. The system (10) – (14) has three singularities [5]. The integration of the system (10) – (14) breaks down when either $V_i(X)$ drops below $\sqrt{\Theta}$, or $V_e(X)$ exceeds $\mu^{-1/2}$, or $V_2(X)$ exceeds $\sqrt{\Omega/\xi}$. In the case shown in Fig. 1, the last condition is fulfilled. In plot (*a*) the potential profile $\Psi(X)$ is shown, in graph (*b*) density profiles $N_i(X)$, $N_1(X)$ and $N_2(X)$ are displayed, in graph (*c*) the profile of the screening temperature $\vartheta(X)$ is presented and in graph (*d*) the profiles of the ion velocity $V_i(X)$ and ion sound velocity $V_S(X)$ are shown. All the profiles exhibit oscillations. Because of this, the ion velocity $V_i(X)$ and ion sound velocity $V_S(X)$ profiles have many intersections and it is not possible to say which of these points represents the sheath edge. Thus, the system (10) – (13) and (15) is solved with the same parameters

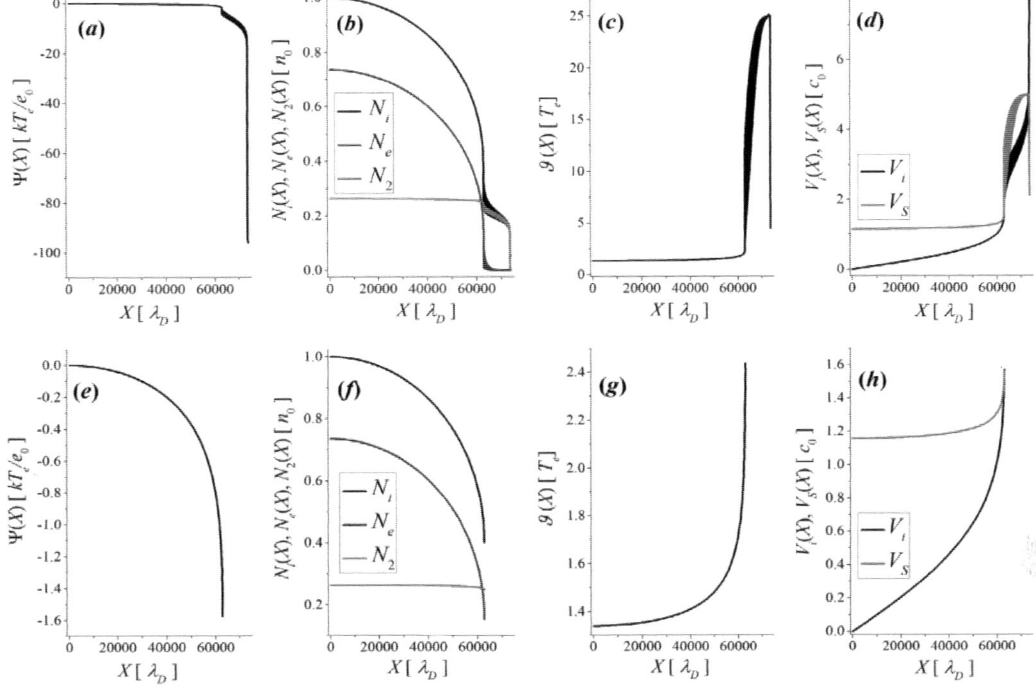

Figure 1. Solutions of the system (10) – (14) in the top graphs and of the system (10) – (13), (15) in the bottom graphs. Parameters and boundary conditions are the same in both cases: $\mu = \xi = 1/1836$, $\Omega = 25$, $\Theta = 0$, $s_i = s_2 = 1$, $s_e = 0$, $\varepsilon = 10^{-5}$, $N_i(0) = 1$, $N_1(0) = 0.736$, $N_2(0) = 0.264$, $V_i(0) = 10^{-7}$, $V_e(0) = V_2(0) = 0$, $\Psi(0) = 0$ and $d\Psi/dX(0) = 0$.

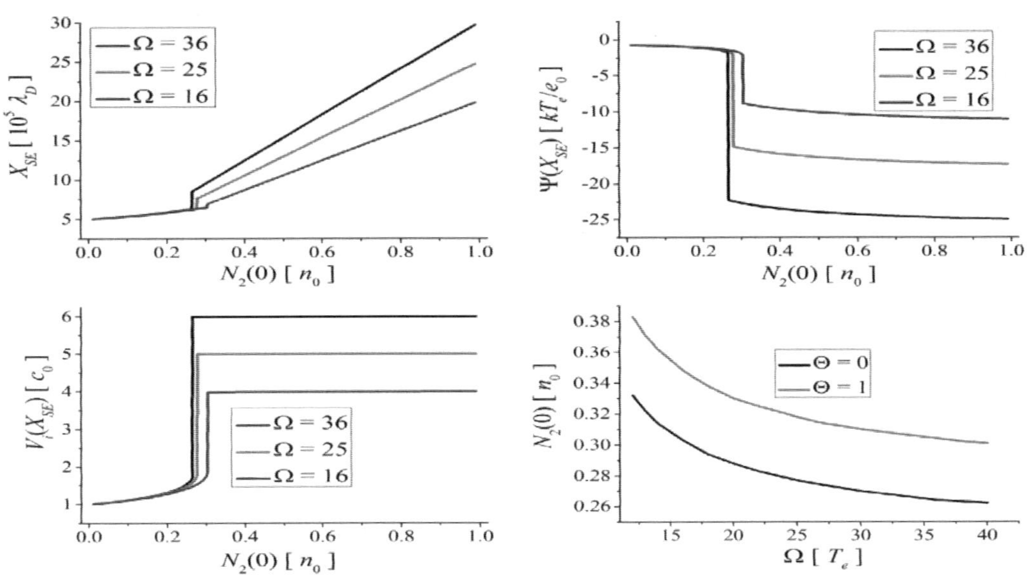

Figure 2. The sheath edge position X_{SE}, the respective potential $\Psi(X_{SE})$ and the ion velocity at the sheath edge $V_i(X_{SE})$, obtained from the system (10) – (13) and (15), plotted versus $N_2(0)$ at three different values of Ω. In the bottom right plot $N_2(0)$, where the jump between the low and the high solution occurs, is plotted versus Ω for two different ion temperatures Θ.

and boundary conditions – bottom plots (*e*) – (*h*). The system (10) – (13) and (15) also has several singular points. Their analysis is beyond the scope of this work. Here we are only interested in the singularity of the system (10) – (13) and (15), which occurs when the Bohm criterion [6], $V_i = V_S$, is fulfilled – plot (*h*). The position, where this occurs, is the sheath edge and is labeled by X_{SE}.

Figure 2 presents the sheath edge position X_{SE}, the respective potential $\Psi(X_{SE})$ and the ion velocity at the sheath edge $V_i(X_{SE})$, obtained from the system (10) – (13) and (15), plotted versus $N_2(0)$ at three different values of Ω. The other parameters and boundary conditions are: $\mu = \xi = 1/1836$, $\Theta = 0$, $s_i = s_2 = 0.1$, $s_e = 0$, $\varepsilon = 10^{-5}$, $N_i(0) = 1$, $N_1(0) = 1 - N_2(0)$, $V_i(0) = 10^{-7}$, $V_e(0) = V_2(0) = 0$, $\Psi(0) = 0$ and $d\Psi/dX(0) = 0$. All the curves exhibit sharp jumps at the transition between the "low" and the "high" solution. In the bottom right plot $N_2(0)$, where the jump between the low and the high solution occurs, is plotted versus Ω for two different ion temperatures Θ.

4. Conclusions

A three-fluid one-dimensional model of the plasma-wall transition in front of a planar negative electrode immersed in plasma that contains two species of negatively biased particles is presented. The case with two-temperature electrons is analyzed. Depending on the ratio between the densities and temperatures of the two groups of electrons, the system of equations (10) – (14) can predict multivalued Bohm criterion, i.e. several positions and potentials of the sheath edge. But if the Poisson equation (14) is replaced by the neutrality condition (15), the integration of the system (10) – (13) with (15) in most cases breaks down at the sheath edge, where the plasma neutrality breaks down. The model can also be used for the analysis of plasma that contains electrons and singly charged negative ions.

Acknowledgements

This work has been partially supported by Grant No. P2- 0073 of the Slovenian Research Agency and CEEPUS CIII-AT-0063-10-1415 mobility scheme.

References

[1] Gyergyek T, Jurčič-Zlobec B, Čerček M and Kovačič J 2009 *Plasma Sources Sci. Technol.* **18** 035001

[2] Fernandez Palop J I, Ballesteros J, Hernandez M A and Morales Crespo R A 2007 *Plasma Sources Sci. Technol.* **16** S76

[3] Riemann K U 1995 *IEEE Trans. Plasma Sci.* **23** 709

[4] Popov Tsv, Dimitrova M, Pedrosa M A, López-Btuna D, Horacek J, Kovačič J, Dejarnac R, Stöckel J, Aftanas M, Böhm P, Bilkova, Hidalgo C and Panek R, 2015 *Plasma Phys. Control. Fusion* **57** 115011

[5] Gyergyek T and Kovačič J 2015 *Physics of Plasmas* **22** 093511

[6] Bohm D 1949 Minimum ionic kinetic energy for a stable sheath in *The characteristics of electrical discharges in magnetic fields,* eds. A Guthrie and R K Wakerling (McGraw-Hill, New York) Chap. 3 77-86

Simultaneous poloidal measurements using new magnetically driven reciprocating probes in COMPASS

R Dejarnac[1,4], J P Gunn[2], M Dimitrova[1], M Hron[1], R Panek[1], J-Y Pascal[2], C Saragosti-Chausy[3], P Tamain[2] and the COMPASS team

[1]Institute of Plasma Physics, Czech Academy of Sciences, 182 00 Prague, Czech Republic
[2]CEA, IRFM, F-13108 Saint Paul-Lez-Durance, France
[3]Ecole Nationale Superieure des Arts et Metiers ParisTech, F-75013 Paris, France

E-mail: dejarnac@ipp.cas.cz

Abstract. Particles and heat transport in the scrape-off layer (SOL) of tokamaks is not yet fully understood. COMPASS is a small-size tokamakp where the edge plasma is well diagnosed in view of studying the competition between the parallel and the cross-field transport in the SOL. In order to better characterize SOL dynamics, in particular the poloidal asymmetry of the main parameters' radial profiles, two new in-situ magnetically driven reciprocating manipulators have been recently installed in COMPASS. These manipulators, the so-called pecker probes, are two additional poloidal measurement points to the existing two (vertical and horizontal) reciprocating manipulators. The pecker probes are located at the low field side of COMPASS at ±47.5° with respect to the outer mid-plane and are equipped with identical tunnel probe heads, providing simultaneous measurements of the ion saturation current density J_{sat}, the electron temperature T_e and the parallel Mach number $M_{//}$ with high temporal resolution. In this paper, a detailed description of the pecker probe system in COMPASS is described and first measurements are presented.

1. Introduction

The COMPASS tokamak is well suited [1] for studying the particle and heat transport in the scrape-off layer (SOL). It is a small and flexible machine where it is easy to use several types of probes to characterize the SOL physics [2-7]. An advantage of COMPASS is its being equipped with two reciprocating manipulators, one vertical from the top of the machine and one horizontal at the outer mid-plane, allowing two independent and simultaneous measurement points in the poloidal plane. In order to increase this spatial resolution from two to four measurement points, two new in-situ linear, magnetically driven reciprocating manipulators, based on the novel technique developed in [8], have been recently installed and commissioned in COMPASS ports in the same poloidal plane at ±47.5° with respect to the outer mid-plane. These manipulators, the so-called pecker probes (PPs) because of the resemblance with a woodpecker, are based on the rotation of an energized coil in the tokamak total magnetic field. The coil is fixed on a rotating axis and provides a leverage which is converted to linear

[4] To whom any correspondence should be addressed.

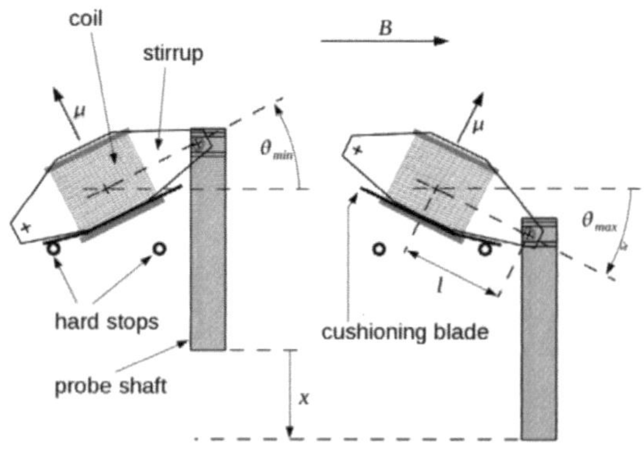

Figure 1. Schematic of PP geometry. The vertical direction corresponds to the tokamak minor radius, along which the probe moves at a distance x. On the left, the probe is drawn fully retracted (coil angle $\theta = \theta_{min}$, probe position $x = 0$ mm). On the right, it is fully inserted (coil angle $\theta = \theta_{max}$, probe position $x = x_{max}$). The magnetic moment μ produced by the coil current and the lever arm l are indicated. $\theta = 0$ corresponds to the coil axis being perpendicular to the magnetic field B, when the probe is half way between its minimum and maximum positions [8].

motion. As described in [8], *"upon application of a voltage V, the torque* $\tau_{coil} = \mu B cos\theta$ *causes the coil to rotate to align the magnetic moment* $\mu = NIA$ *with the total magnetic field B"*, with N and A being the total number of windings and the cross-section of the coil, respectively, with I being the total current in the drive circuit. The schematic of the PP system principle taken from [8] is reproduced in figure 1. The two PPs are equipped with identical tunnel probe (TP) heads allowing simultaneous fast DC measurements of the absolutely calibrated ion-current density J_{sat} and the electron temperature T_e [9,10]. Since the two TPs are mounted back-to-back, it is also possible to retrieve the parallel Mach number $M_{//}$ using the two independent J_{sat} measurements on both the ion and electron sides. As a matter of fact, thanks to the current advancement of numerical techniques and computational capabilities, first-principle simulations of SOL turbulence are now reaching predictive capabilities. A recent joint effort, described as a synergetic numerical-experimental approach to fundamental aspects of turbulent transport in the tokamak edge, has been started within the EU fusion community using state-of-the-art numerical codes [11-14] to unveil the physics behind the SOL dynamics. The validation of these simulations results with, in the first step, experimental measurements in the simplest configurations is a crucial point for the understanding of SOL plasma dynamics and predictive simulations of future devices. The COMPASS tokamak is part of this numerical-experimental validation, especially with its access to highly-resolved measurements of poloidal asymmetry only accessible thanks to the new PP system.

In this paper, the PP system developed for the COMPASS tokamak (design and commissioning) is described in section 2. First measurements of the J_{sat} profiles in the COMPASS SOL measured by one TP mounted on the upper PP and by a single Langmuir tip mounted on the horizontal reciprocating manipulator are presented in section 3. The vertical manipulator was not available at that time since it was out of the vacuum vessel for a massive upgrade. The two PPs were commissioned separately by the only one specific power supply available at the IPP Prague, a KEPCO 36V-12A, therefore only one PP could be used at a time. The upper PP was chosen to be compared with the horizontal manipulator for these J_{sat} radial profile measurements. The general conclusion is presented in section 4.

2. The COMPASS pecker probes: design and commissioning

The equation describing the coil rotation (equation (9) in [8]) is solved numerically in order to identify the main feature of the PP dynamics and to optimize the coil geometry. The dimensions of the COMPASS PP are limited by the space available in the tokamak port where it is installed. As an example, a photograph of one of the two PPs is shown in figure 2 before its installation inside the chamber. The PP is fixed on the bottom flange which is screwed to the tokamak port. The carbon shield faces the plasma and the coil itself and the mechanical support structure are hidden inside the port. The optimized design of the COMPASS PP system gives the following: coil length $L^{coil} = 64$ mm, coil internal radius $R^{coil,intern} = 24$ mm, length of the lever arm $L^{lever\ arm} = 76$ mm, total

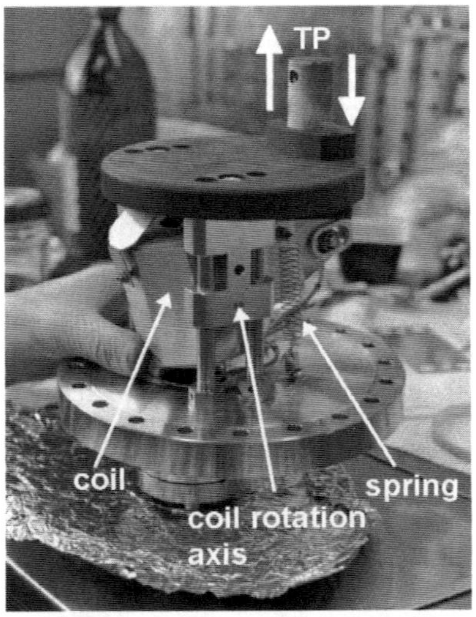

Figure 2. Photograph of one of COMPASS PPs before being installed in the vessel.

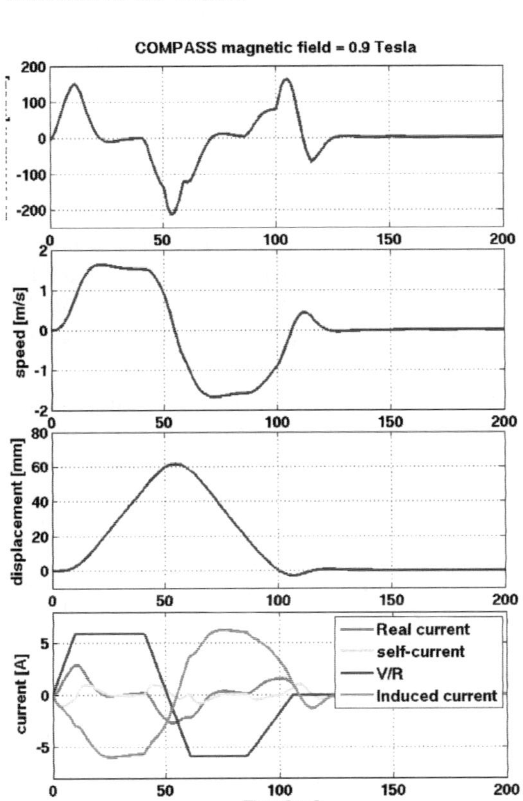

Figure 3. Simulation results for COMPASS PP with $N = 420$ & $B_T = 0.9$T. From top panel: acceleration, speed and displacement of the probe. The bottom panel shows the different currents in the system.

resistance of the system $R^{tot} = 2.6\ \Omega$. Therefore, only the number of windings N of the coil needs to be estimated for the COMPASS toroidal field value $B_T = 0.9$ T at the PP location, i.e., at the low field side, in order to have a moving speed of ~ 1 m/s and to keep an acceleration around 100 m/s^2. As explained in [8], the linear probe speed is proportional to the current in the circuit induced by the probe rotation (equations (20)-(22) in [8]). The induced current is simply the difference between the real current, as it would be measured by an Ampere-meter, and the theoretical current V/R^{tot}. The probe position is obtained by simply integrating the speed from the beginning of the reciprocation. A current due to the self-inductance of the coil is also present in the system. This current is generally of low intensity and can be neglected [8]. The response of an optimized system for COMPASS PP with $N = 420$ windings is shown in figure 3. Predictive calculations show that the acceleration is slightly higher than originally wanted, but still acceptable. The optimum voltage to be applied to the PP is $V^{opt} = 15.6$ V. A positive voltage moves the probe inside the plasma and by reversing the polarity of the applied voltage, keeping the same magnitude, the probe is retracted back to its parking position. For a smooth landing, the voltage is slowly decreased towards zero in the middle of its way back. It can be seen that the current due to the coil self-inductance is indeed negligible.

The two PPs were installed and commissioned in the COMPASS tokamak in April and July 2015, respectively. Figure 4 shows the measured currents *(top panel)* in the coil of PP#2, the bottom PP (the one located at -47.5° with respect to the outer mid-plane) and the displacement of the probe *(bottom panel)* calculated as the integral of the speed, which is directly proportional to the measured induced current, during the vacuum discharge #10481. The currents are compared to the predictive currents from the solution of equation (9) in [8] for the same applied voltage of 15.6 V. Note that the current is different from the expected value $V/R^{tot} = 15.6/2.6 = 6$ A. This is because the total resistance is changing from shot to shot due to heating of the wires. To avoid this uncertainty, the total resistance is measured before each movement as V/I when the probe is at parking position and the system is in steady state; in this case during the time interval $1045 < t < 1060$ ms in figure 4. In this particular discharge, the resistance was $R^{tot} = 2.9\ \Omega$. Calculations are made without *(thin dashed curves)* and with *(thin curves)*

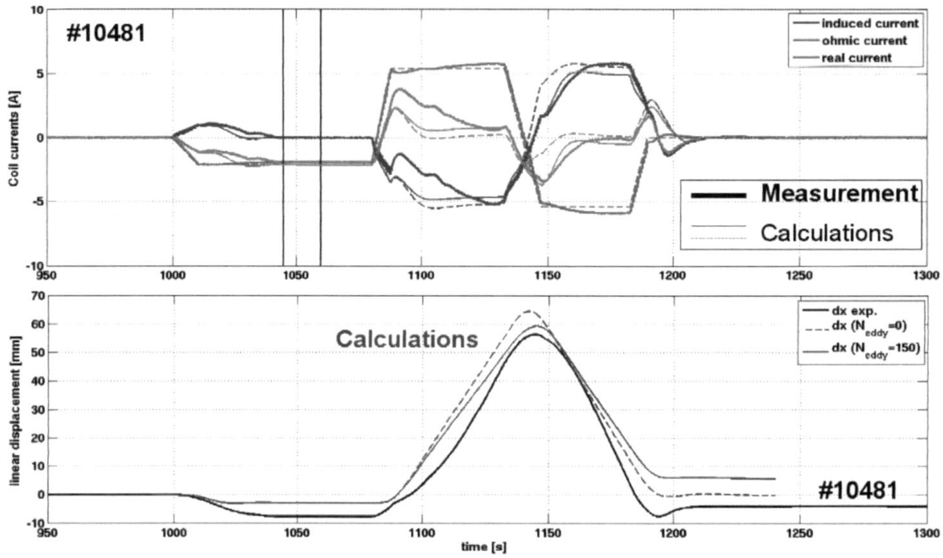

Figure 4. Comparison of measured currents in the PP#2 coil (top panel) and of the resulting probe position (bottom panel) with simulation results without and with eddy currents, characterized by the equivalent windings number N_{eddy}. Here, $N_{eddy} = 0$ (thin dashed curves) and $N_{eddy} = 150$ (thin full curves).

an estimate of eddy currents flowing into the system. As it is well explained in [8]: *"For an eddy current to flow in a closed loop inside a moving conductor, there must be a time varying magnetic flux through its cross section. This can occur if a component of the object's velocity is parallel to the gradient of a spatially non-uniform magnetic field, or if it rotates about an axis that is not parallel to a magnetic field"*. Therefore, eddy currents must flow in the PP conducting structure. They can be expressed *"as being due to the back electro-motive force in a second, unbiased copper coil of the same geometry as the drive coil, with a number of windings* N_{eddy}*"* [8]. It can be seen in figure 4 that the eddy currents with an equivalent winding number $N_{eddy} = 150$ are affecting the response of the system and that the consequent simulated currents are closer to measured currents than without eddy currents *(top panel)*. Even if the probe position seems to be too damped on its way back to parking compared to reality *(bottom panel)*, the curve with $N_{eddy} = 150$ is closer to the real movement of the probe. This number is an optimum number found for best matching with experimental curves. Simulations with larger numbers of equivalent windings start to strongly diverge from the experimental curves. In general, one observes a very good agreement between the experimental measurements and the simulation results, which makes us being confident in our simulations and in understanding which type of currents dominate the probe motion.

3. First simultaneous poloidal measurements in COMPASS using the pecker probe system

The first simultaneous poloidal measurements in COMPASS were made using only one PP and the horizontal reciprocating manipulator (HRCP), which is 135° toroidally separated. At present, only one power supply (KEPCO 36V-12A), specific for the PPs, is available at IPP Prague. Therefore, only one PP can be powered at a time. For the comparison with the HRCP, we chose the top PP#1, which is located at +47.5° with respect to the outer mid-plane. The PP#1 is equipped with two TPs mounted back to back, each of them looking at either the ion- or the electron-side of the plasma. Each TP is biased negatively by a biasing voltage $V^{bias} = -200$ V in order to collect the ion saturation current, I_{sat}. Due to a failure in the power line to the TP at the electron-side, only the data from the ion-side will be presented here. The COMPASS TP consists of a hollow conducting cylinder, the "tunnel", of 6 mm in diameter and in length, and a conducting "backplate" at one end, which is electrically insulated from the tunnel. The tunnel axis is parallel to the magnetic field and the plasma flows into the open orifice.

The ion saturation current is distributed and measured on both collectors; their sum gives the total I_{sat}. The advantage of the TP is to be immune to sheath expansion effects due to its concave geometry. Indeed, due to its shape, it provides an absolutely calibrated measurement of the parallel ion current density [9,10]. The resulting J_{sat} is simply: $J_{sat,TP} = I_{sat,TP}/r_{TUN}^2$, r_{TUN} being the tunnel radius.

The HRCP is equipped with a classical Langmuir probe of cylindrical shape with dimensions $L^{LP} = 1.5$ mm and $r^{LP} = 0.45$ mm. In first approximation, the collecting area of the LP is taken as the geometric area of the probe, i.e., its projection along **B**. The measurements presented in this paper were performed in the diverted ohmic discharge #9788. Both reciprocations were made during the steady-state of the discharge with a constant plasma current $I_P = 200$ kA and a line averaged density $n_e = 4\times10^{19}$ m^{-3}. Figure 5 shows both the LP *(top panel)* and the TP *(bottom panel)* positions with respect to the separatrix in time. It is seen that the two manipulators do not reach the same absolute radial position. The HRCP goes 10 mm deeper in the plasma and reaches the sepratrix. The two probes are also not synchronised. The deepest position occurs at different times, for the HRCP being ~30 ms earlier than for the PP. Figure 6 *(left)* shows the magnetic reconstruction of the equilibrium in shot #9788 at the time

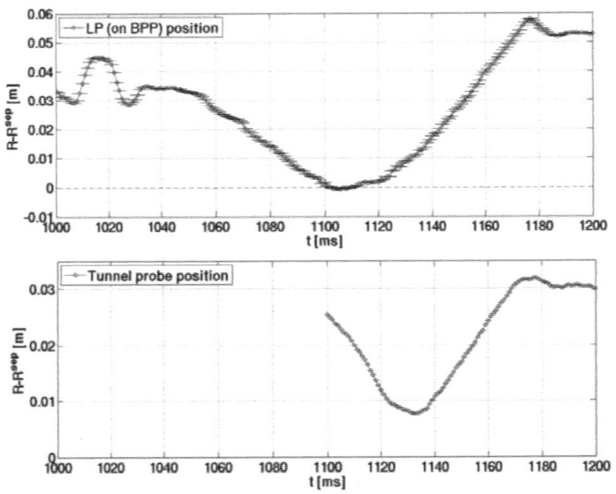

Figure 5. HRCP (top panel) and TP (bottom panel) positions as a function of time during shot #9788.

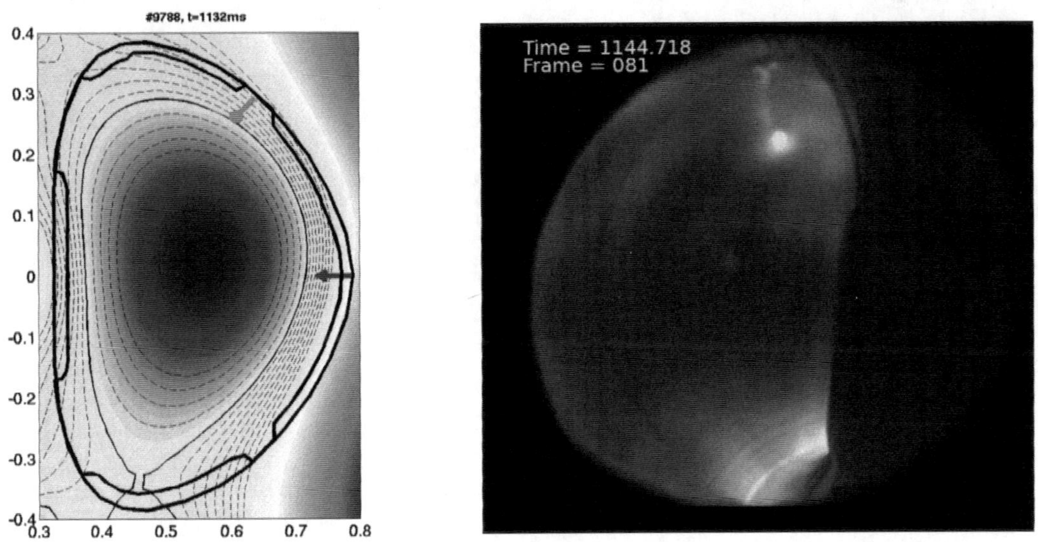

Figure 6. (Left) Magnetic reconstruction of the COMPASS equilibrium during shot #9788 at the time of the TP deepest insertion in the plasma. The PP and HRCP positions are marked by a pink dot and a blue arrow, respectively. (Right) a snapshot from the tangential, fast visible camera shows the PP interaction with the plasma later on during its way back to parking position.

$t = 1132$ ms corresponding to the maximum insertion of the PP. The positions of both the PP and the HRCP are marked with a pink dot and a blue arrow, respectively. On the right side of figure 6, a snapshot from the tangential, fast visible camera shows the PP interaction with the plasma later on during its way back to parking position.

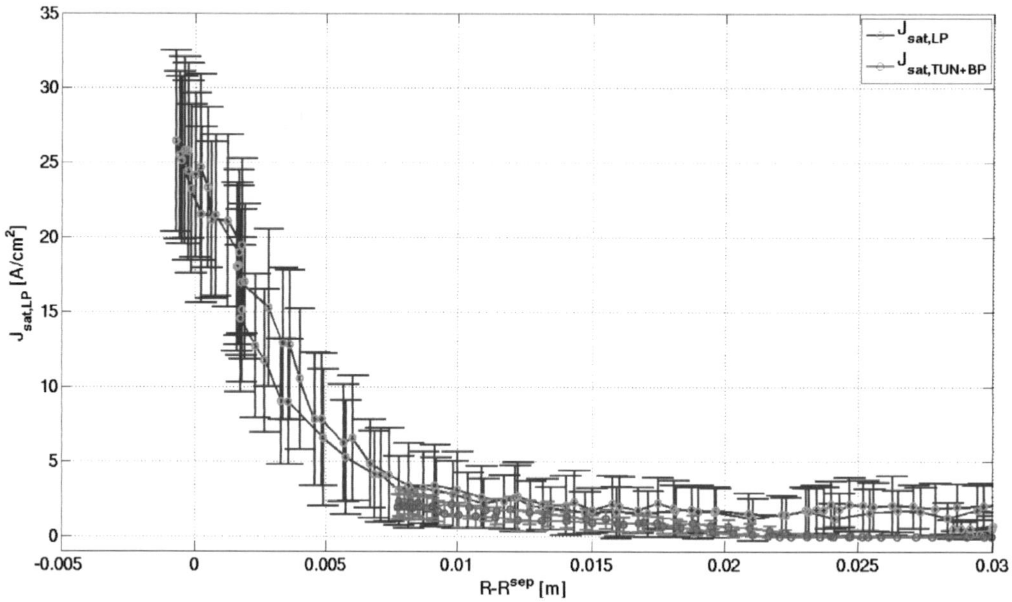

Figure 7. Radial profiles of the ion saturation current density measured by the TP mounted on the PP and the LP mounted on the HRCP. The distance to separatrix for the TP data is mapped to outer mid-plane.

Radial profiles of the ion saturation current density measured by both probes are presented in figure 7. In the case of the PP, the distance to separatrix is mapped to the outer mid-plane. It can be seen that $J_{sat,TP}$ is lower by 35-40% compared to $J_{sat,LP}$. This might be due to a poloidal asymmetry; however, it is well known that small (and even bigger) LPs are prone to sheath expansion effects [15]. It has been demonstrated experimentally that the collecting area of such probes are from 2 to 4 times larger than their geometrical projection along **B** [16]. In our case, if we assume $J_{sat,TP}$ to be the real value of the ion current density, the underestimation of the LP collecting area is by a factor 1.6. However, the TP backplate is prone to secondary electron emission (s.e.e.) from ion bombardment. In the case of the tunnel, this effect is cancelled by the fact that every electron emitted is immediately re-collected by the tunnel. For the backplate, the situation is the opposite. Every electron leaving the tunnel follows field lines which are parallel to the tunnel surface. The backplate s.e.e. rate is unfortunately unknown, but can be as high as 80% [17]. Consequently, the current measured at the backplate is overestimated, which tends to increase the difference between $J_{sat,TP}$ and $J_{sat,LP}$ and to increase the error on the LP collecting area. Interestingly, it can be seen that the level of fluctuations is lower for the TP measurements than for the classical LP. It can be explained by the TP measuring the true J_{sat} fluctuations, whilst the LP is contaminated by the sheath expansion current, which depends on other plasma parameters such as the electron temperature, making it impossible to distinguish between density and temperature fluctuations [10]. The results reported in this paper are the very first results obtained by the PP system on COMPASS proving its functionality. This paper is a basis for future publications with all the COMPASS manipulators fully functional in the very near future.

4. Conclusions

The Institute of Plasma Physics in Prague has a long tradition of probes measurements in tokamaks since 1980's; this expertise continues now with the COMPASS tokamak. In order to increase the poloidal coverage by probes, two new in-situ, linear reciprocating manipulators, the pecker probes, have been designed, installed and commissioned in COMPASS, bringing to four the number of poloidal measurement points together with the two existing reciprocating manipulators. The pecker probes are located at ±47.5° with respect to the outer mid-plane in the same poloidal cross-section, 135° away for the two reciprocating manipulators in the toroidal direction. The pecker probe system is based on the rotation of an energized coil in the tokamak total magnetic field. Predictive simulations of the COMPASS pecker probes are in good agreement with the measurements. Each pecker probe is equipped with two identical tunnel probes, mounted back-to-back, allowing fast measurements of the absolutely calibrated ion current density, the electron temperature and the parallel Mach number. First measurements of the ion saturation current density at two poloidal locations are presented here. The comparison is only between one pecker/tunnel probe and a classical Langmuir probe mounted on the horizontal reciprocating manipulator because only one power supply for the pecker probe coil was available at that time and the vertical reciprocating manipulator was out of the machine for upgrade. The measurements confirmed the problem of sheath expansion around small convex Langmuir probes. A second power supply to power the two PPs at the same time has been purchased, so that in a near future a full poloidal coverage with four measurement points will be performed in COMPASS. This paper presents the very new pecker probe system installed in COMPASS and proves its functionality before future more complete studies of simultaneous multi-poloidal characterization of the SOL unveiling the physics behind SOL dynamics.

References

[1] Weinzettl V, et al. 2011 *Fusion Eng. Design* **86** 1227-31
[2] Adámek J, Horáček J, Seidl J, Müller H W, Schrittwieser R, Mehlmann F, Vondráček P and Pták S 2014 *Contr. Plasma Phys.* **54**/3 279-84
[3] Dejarnac R, Dimitrova M, Komm M, Schweer B, Terra A, Martin A, Boizante G, Gunn J P and Pánek R 2014 *Fusion Eng. Design* **89** 2220-4
[4] Komm M, Kocan M, Carralero D, Muller H, Stockel J and Rea C 2014 *Proc. 41th EPS Conf.* P5.026
[5] Kovarik K, Spolaore M, Ďuran I, Stöckel J and Vianello N 2014 *Proc. 41th EPS Conf.* P5.025
[6] Dimitrova M, Dejarnac R, Popov Tsv K, Ivanova P, Kovačič J, Stöckel J, Havlíček J, Janky F and Pánek R 2014 *Contr. Plasma Phys.* **54**/3 255-60
[7] Dejarnac R, et al. 2015 *J. Nucl. Mater.* **463** 381-4
[8] Gunn J P and Pascal J-Y 2011 *Rev. Sci. Instr.* **82** 123505
[9] Gunn J P, Pánek R, Stöckel J, van Oost G and van Rompuy T 2005 *Czechoslovak J. Phys.* **55** 255-63
[10] Gunn J P, Dejarnac R and Stöckel J 2016 Simultaneous DC measurements of ion current density and electron temperature using a tunnel probe *J. Phys.:Conf. Series* (this volume)
[11] Dudson B D, Umansky M V, Xu X Q, Snyder P B and Wilson H R 2009 *Comp. Phys. Comm.* **180** 1467-80
[12] Naulin V, Windisch T and Grulke O 2008 *Phys. Plasmas* **15** 012307
[13] Ricci P, Halpern F D, Jolliet S, Loizu J, Mosetto A, Fasoli A, Furno I and Theiler C 2012 *Plasma Phys. Control. Fusion* **54** 124047
[14] Tamain P, Ghendrih Ph, Tsitrone E, Grandgirard V, Garbet X, Sarazin Y, Serre E, Ciraolo G and Chiavassa G 2010 *J. Comp. Phys.* **229** 361-78
[15] Dejarnac R, Gunn J P, Stockel J, Adamek J, Brotankov J and Ionita C 2007 *Plasma Phys. Control. Fusion* **47** 1791-1808
[16] Gunn J P, et al. 2013 *J. Nucl. Mater.* **438** 184-8
[17] Gunn J P 2012 *Plasma Phys. Control. Fusion* **54** 085007

Journal of Physics: Conference Series **700** (2016) 012018 doi:10.1088/1742-6596/700/1/012018

Simultaneous DC measurements of ion current density and electron temperature using a tunnel probe

J P Gunn[1,3], R Dejarnac[2] and J Stöckel[2]

[1]CEA, IRFM, F-13108 Saint Paul Lez Durance, France
[2]Institute of Plasma Physics, Czech Academy of Sciences, 182 00 Prague, Czech Republic

E-mail: Jamie.Gunn@cea.fr

Abstract. The tunnel probe is a concave Langmuir probe designed to operate in strongly magnetized plasma. Due to its shape, the tunnel probe is immune to sheath expansion effects and thus provides absolutely calibrated measurements of the parallel ion current density. A two-dimensional, self-consistent kinetic model is employed to model the flow of charges within the cavity of the tunnel probe. The calculation predicts that the distribution of the ion flux onto the inner conductors depends on the electric field inside the tunnel, which in turn depends on the electron temperature. Therefore, if the tunnel is divided into two negatively biased collectors, it is possible to use the simulation results to determine the electron temperature from the measured ion current ratio. This means that a DC-biased tunnel probe can be used to provide fast, simultaneous measurements of the parallel ion current density and the electron temperature without collecting a single electron. Measurements in the CASTOR and Tore Supra tokamaks agree well with the numerical simulations.

1. Introduction

Almost invariably in tokamaks, Langmuir probes of convex geometry, such as short bits of cylindrical wire, are employed. Frequently-cited probe diameters from the literature range from 0.5 to 3 mm, and it is often incorrectly assumed that they operate in the strongly magnetized regime. In the context of probe measurements in fusion devices, "strongly magnetized" implies that the ion Larmor radius r_L and Debye length λ_D are assumed to be negligible compared to the probe size, so that the ion current to a negatively biased probe is proportional to the flux of guiding centers intersecting the geometrical projection of the probe along magnetic field lines, independent of the magnitude of the applied voltage V_{bias}. However, an empirical comparison of various probe geometries showed that the ion current, even that collected by 5 mm diameter domed probes, does not saturate [1] with negative V_{bias}. It was experimentally demonstrated [2] that the collecting area of small probes can be more than 2 to 4 times their geometrical projection along $\boldsymbol{B}$. The effective collecting area of most convex probes used in tokamaks is thus highly uncertain, perhaps depending on the very quantities one wants to measure. Many theories have been elaborated for ion collection by convex probes in plasmas whose degree of magnetization varies from zero to infinite. Early on, such theories were applicable only to very specific, sometimes asymptotic parameter regimes. Now with modern computing power, three

[3] To whom any correspondence should be addressed.

dimensional simulations of spherical probes in magnetized plasmas are being performed (see [3] and references therein). Nonetheless, interpretation of convex probe data is rather cumbersome, as one must constantly take account of the particular parameter regime in which the probe operates (e.g. thin sheath vs. thick sheath, flowing vs. stagnant, magnetized vs. unmagnetized), find an appropriate model if it exists, and either extract calibration curves from figures that were printed half a century ago [e.g. 4], or be able to make the calculations oneself.

To avoid this problem, a new kind of concave Langmuir probe called the "tunnel probe" (TP) has been invented [5,6]. It consists of a hollow conducting cylinder, the "tunnel" that is closed at one end by an electrically isolated "backplate" (figure 1). The tunnel axis is parallel to the magnetic field. Plasma flows into the open orifice and the ion flux is distributed between the tunnel and the backplate. Kinetic modelling shows that despite its strong influence on the distribution of the ion flux inside the tunnel, the electric field does not penetrate into the plasma. Therefore, ion streamlines leading up to the mouth of the tunnel are totally unperturbed. That is, the effective collecting area of a concave Langmuir probe embedded in a nearly planar object is almost exactly equal to the geometrical cross-section of its orifice projected along the field lines, which means that such a probe is absolutely calibrated for the measurement of the parallel ion current density $J_{//,i}$, independent of the other plasma parameters. An experimental indication that this probe has a well-calibrated effective collecting area is that the total ion current saturates perfectly (see upper panel of figure 5). In contrast to convex probes, which are difficult to calibrate because of the expansion of the sheath electric field into the plasma, concave probes are immune to such problems because the sheath electric field is contained inside the cavity. Inside the probe, however, the distribution of the ion current between the tunnel and backplate depends on the electric field, which is determined by the electron temperature T_e via the Boltzmann relation. This means that a TP biased to a negative DC voltage can be used to provide fast, simultaneous measurements of $J_{//,i}$ (the sum of currents to tunnel and backplate) and T_e (related to the ratio of the currents).

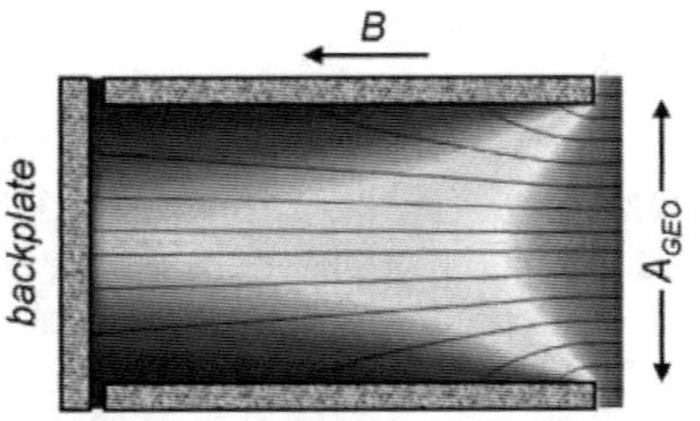

Figure 1. Section of TP along the cylindrical axis. Color indicates ion density (red is denser than blue). Streamlines of the ion flow are indicated by the black curves in the axial-radial plane. The effective collecting area of the probe is equal to the geometrical projection A_{GEO} of its opening along $\boldsymbol{B}$.

2. Kinetic model of the tunnel probe

As described in [7], experimental and numerical investigations of shielded probes in the Tore Supra tokamak [8] revealed that the magnetized sheath strongly affects the flow of charges through a small orifice whose axis is parallel to the magnetic field. It was found that the transmission factor of the orifice depends on the local plasma parameters. Since the sheath thickness scales with λ_D and V_{bias}, the fraction of the incident current collected by the walls of the orifice can vary. The TP was thus proposed in order to exploit this dependence. By making kinetic simulations for a wide range of expected conditions, it was hoped that calibration factors could be derived which would allow the TP to make simultaneous, DC measurements of $J_{//,i}$ and T_e.

Several prototype TPs of different geometries were built and tested over a few years in the CASTOR tokamak [9]. In CASTOR, the magnetic field was $B = 1$ T and the plasmas were hydrogen. To calibrate the TP, the XOOPIC code [10] was used to make a large number of simulations for expected plasma conditions. The qualitative behavior of the tunnel-to-backplate ion current distribution conformed to the prediction of the calculations, although the quantitative agreement was not so good. Typically, for a given $J_{//,i}$ and T_e, the measured fraction of the ion current collected by the tunnel was significantly lower than expected. This behavior is now understood to be related to the boundary conditions for particle injection onto the simulation domain, which have been improved. For the CASTOR runs we injected unshifted, half-Maxwellian distributions, but this led to the formation of a source sheath and a significant drop in the density at the tunnel entrance. Since the calibration depends on that density, as well as T_e and V_{bias}, the calibration factors would require a complicated transformation to relate the current density, which is directly measured, to the sheath edge density n_{se}, a quantity that must be calculated.

A PIC code that works in cylindrical geometry (on the R-Z plane) has been developed to produce calibration curves for the TP. The code is named "PICCYL", PIC code in cylindrical geometry. Standard PIC techniques such as second order leap-frog particle advance, first order grid weighting, non-uniform particle weighting, etc. have been implemented to optimize the code, and will not be discussed here. The novel aspects of the code involve, first of all, its normalization. PICCYL is designed according to the physics-based scales that determine the dynamics of the magnetized sheath, as described in [11]. The two principal parameters that contain all the physics are the Debye length $\lambda_D = \sqrt{\varepsilon_0 T_e / e n_{se}}$, and the magnetization strength $\xi = \omega_{pi} / \omega_{ci} = \sqrt{m_i n_{se} / \varepsilon_0 B^2}$. This makes PICCYL extremely useful for interpreting measurements in different tokamaks, with different magnetic fields, different plasma species, and different TP geometries, using a single normalized database.

The theory of Mach probes [12] is crucial to understanding how the TP works. Since the local densities at the entrances of the two TPs on either side of the Mach probe can be different in the presence of a parallel flow, the local Debye lengths can also be different. One has to account for these differences when attempting to use the TP to make DC T_e measurements. Often in the field of plasma-wall interactions, solutions are found on one of two asymptotic scales. To resolve the details of plasma flow to a probe, we are obliged to model non-neutral plasmas on the Debye scale. However, it is necessary to prescribe physically consistent boundary conditions for the plasma entering the sheath which rigorously satisfy the Bohm criterion. These boundary conditions are provided by a collisionless kinetic model [13] that solves the presheath equations in the opposite limit of vanishing λ_D, and predicts the ion distribution function approaching the probe up until the sheath edge as a function of the background ion flow speed and the temperature. In PICCYL, the ions are launched from an injection plane towards the TP with a parallel speed distribution given by the kinetic model [13]. The distribution corresponding to $T_i = 2T_e$ has been used for most of our simulations, since that is a typical value for ohmic discharges in Tore Supra as measured by a retarding field analyzer [14,15]. Since the kinetic distribution satisfies the Bohm criterion, there is no artificial source sheath in front of the injection plane, and the effective local λ_D in the quasineutral plasma at the entrance of the tunnel is what we intend it to be. For a given $J_{//,i}$ the sheath edge density in PICCYL is higher than that obtained when injecting an unshifted, half-Maxwellian in XOOPIC. The local sheath at the tunnel surface is thus thinner, meaning that fewer ions are deflected from their guiding center trajectories, resulting in a low ion current to the tunnel. We shall see below that the improved boundary conditions remove the discrepancy observed in the first CASTOR experiments.

A series of simulations was completed in order to construct a database that spans a sufficiently wide parameter space to cover the range of $J_{//,i}$, T_e, B, V_{bias}, and ion mass number A typically encountered in the tokamak SOL (figure 2). The magnetization strength ξ was varied from 1 to 18.5 and the tunnel radius, from 10 λ_D to 170 λ_D. For each (ξ,r_{TP}) pair, 5 to 10 values of the probe bias voltage φ (normalized to T_e) were simulated to be able to construct I-V characteristics over the range $-200 \le V_{bias} - V_f \le 0$ V. The advantage of the normalization scheme is immediately obvious, because it allows us to reduce a 5-dimensional parameter space to a 3-dimensional one. This database covers experiments in the CASTOR tokamak (hydrogen plasma with B = 1 T), and the Tore Supra tokamak (hydrogen, deuterium, or helium plasma with $2 < B < 4$ T). Not only is it used to interpret past experimental results, but to design TPs for future experiments.

A random selection of ion orbits is shown in figure 3 for the case of $r_{TP} = 40\ \lambda_D$ and $\xi = 5.0$, corresponding to $J_{//,I}$ = 0.66 A/cm^2 and T_e = 20.7 eV for deuterium plasma in Tore Supra with B = 3.5 T. It can be seen that when an ion enters a region of strong radial electric field, it experiences

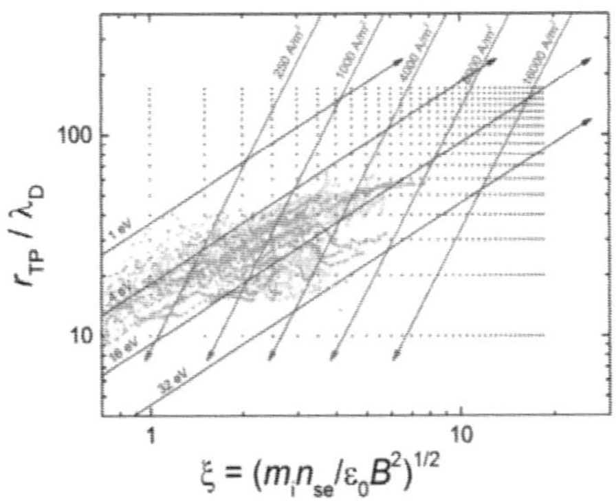

Figure 2. TP calibration database. Values of normalized tunnel radius and magnetization strength are indicated by gray points. For each point, from 5 to 10 simulations were performed with different V_{bias}. In total, 5246 simulations were made. For illustration, contours of constant $J_{//,i}$ (red lines) and T_e (blue lines) are superimposed for the particular case of Tore Supra (deuterium ions with local magnetic field B = 3.5 T). The cloud of cyan (magenta) points are results of standard fitting to I-V characteristics, i.e. the sum of tunnel plus backplate currents, measured on the downstream (upstream) sides of the probe during an experiment dedicated to SOL flow studies.

an azimuthal EXB drift while accelerating exponentially towards the tunnel surface, with the Larmor gyration destroyed. It is because the scale-length of the electric field gradient is shorter than r_L that the magnetic moment is no longer a conserved quantity. Ions are demagnetized and can deviate from guiding center trajectories. Incidentally, it is for this reason that the ion current to convex probes is much higher than that predicted by the guiding center approximation.

3. Comparison between PICCYL simulations and measurements

The first test of the TP concept was made with a series of prototypes in the CASTOR tokamak. The TPs were mounted on a manipulator that could be moved radially between discharges. We define the experimental current ratio as the fraction of the total current which flows to the tunnel $R_{c,exp} = I_{TUN}/(I_{TUN}+I_{BP})$. The resulting PICCYL calibration curves are shown in figure 4. The typical evolution of the measurements is superimposed for CASTOR shot 13172. The analysis procedure is straightforward. One calculates $J_{//,i}$ from the sum of the two ion currents divided by the cross-section of the tunnel, and the current ratio is used to calculate T_e by interpolation within the numerical results. During the plasma current ramp-up phase at the beginning of the discharge, the ratio is very high and the density low. For extremely small $J_{//,i}$, Debye shielding is weak and the potential distribution in the tunnel is essentially the same as the vacuum one. The current measurements give no useful information about T_e. This is evident from the theoretical results which all converge as $J_{//,i}$ tends to zero. Fortunately, such low densities only occur during the plasma current ramp-up phase at the beginning of the discharge. The points smoothly follow the low-temperature contours until the hot flat top phase when fully developed turbulence sets in. This phase appears as a cloud of points bracketed by the ranges $20 \leq T_e \leq 40$ eV and $0.6 \leq J_{//,i} \leq 1.3$ A/cm^2.

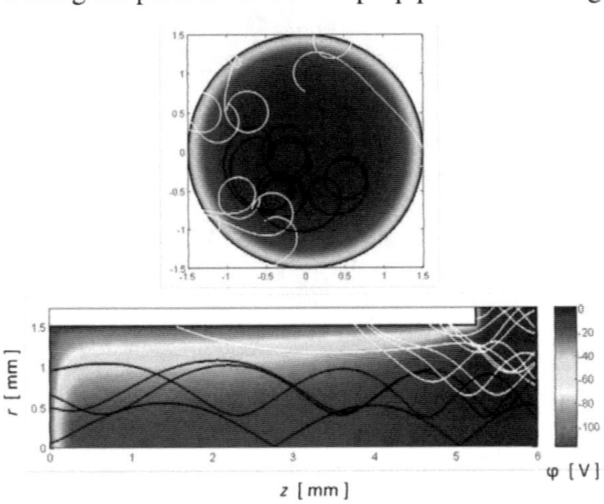

Figure 3. Ion orbits for $r_{TP} = 40\ \lambda_D$ and $\xi = 5.0$. The black curves and white curves correspond to ions that strike the backplate and tunnel, respectively. The top panel is the projection along the tunnel axis, and the bottom panel is the r-z plane with the axis at $r = 0$. Color indicates electric potential in V.

The physics governing the TP is fundamentally different than that of a classical LP. The voltage applied on the LP is swept in order to measure a restricted part of the electron distribution function. The

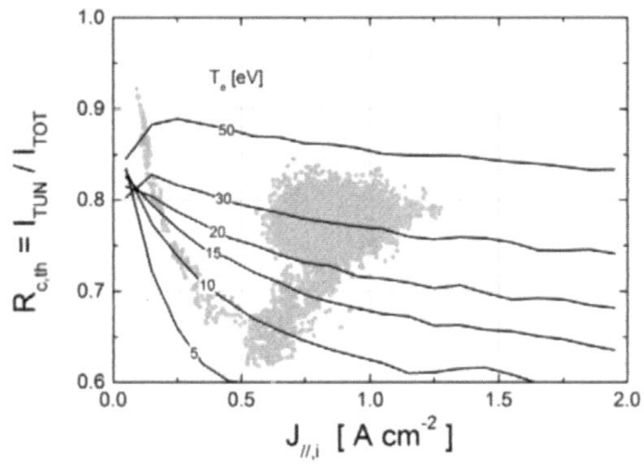

Figure 4. Theoretical tunnel-to-total current ratio predicted by the PICCYL code for the 5 mm diameter TP in hydrogen plasma in CASTOR assuming V_{bias} = −100 V and $B = 1$ T. Experimental results from a DC-biased probe (CASTOR shot 13172) are superimposed (grey points).

TP, on the other hand, is biased to a fixed potential that is sufficiently negative to repel all electrons. *The temperature of the electrons is measured even though none are collected.* It is necessary to compare the two methods. As a validation exercise, we examine measurements in CASTOR. Two copper tunnels were implemented, one facing the ion direction and the other the electron direction. The tunnel diameter was 5.0 mm and its depth was 5.0 mm. The voltage on all conductors was swept in order to measure the I-V characteristics. The averaged data that were acquired by the ion side TP on shots 16204 during the pure ohmic phase, and 16214 during the electrode bias phase, are shown in figure 5. On shot 16204 the probe was situated at r = 83 mm and on shot 16214 it was placed at r = 63 mm. We focus on these specific data because $J_{//,i}$ was found to be roughly 0.5 A/cm^2 in both cases, while the

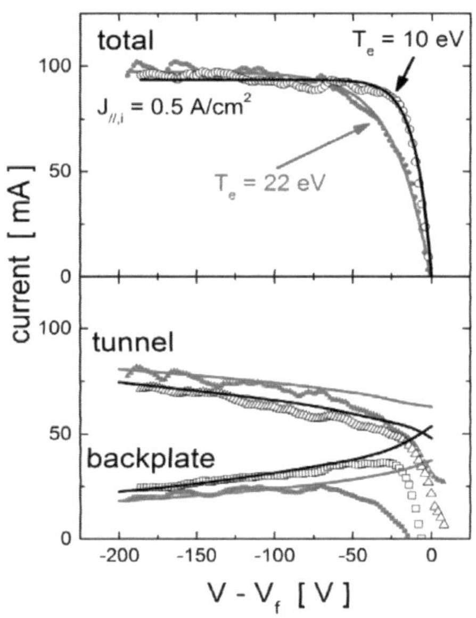

Figure 5. The upper panel shows I-V characteristics obtained by summing the tunnel and backplate currents together for CASTOR shots 16204 (open black circles) and 16214 (full red dots). The raw tunnel and backplate currents are shown in the lower panel. Theoretical ion currents predicted by PICCYL are superimposed on the raw data.

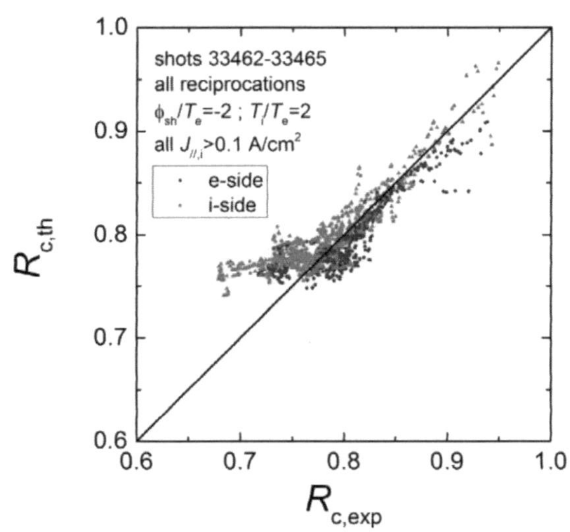

Figure 6. Comparison of theoretical tunnel-to-total current ratios calculated by the PICCYL code with the measured ratios from each I-V characteristic having $J_{//,I}$ > 0.1 A/cm^2. The ratios were taken 100 V below floating potential. For all cases, sheath potential $\phi_{sh} = -2T_e$ and ion temperature $T_i = 2T_e$ were assumed. Blue dots and red triangles correspond to electron-side and ion-side tunnels, respectively.

electron temperatures differed by about a factor of two. An exponential function is fit to the data to obtain estimates of T_e = 10 eV and T_e = 22 eV using the classical swept LP technique. Then the ion currents to the backplate and the tunnel are calculated by extrapolation within the PICCYL database. The theoretical ion currents agree quite well with the measurements both in absolute value and in their voltage dependence.

4. Measurements in the Tore Supra tokamak

The Mach probe used in an experiment dedicated to SOL flow studies in Tore Supra [16] was in fact composed of two back-to-back TPs. The SOL profiles exhibited a wide range of $J_{//,i}$, T_e, and $M_{//}$ and thus provide a challenging test of the TP theory. The profiles of $J_{//,i}$ and T_e were measured on both sides of the probe, with the core density increased shot by shot until the radiative density limit. The radiated power fraction evolved from approximately 40-50 % on the first discharge, to 100 % on the fourth discharge, resulting in an order of magnitude variation of T_e. The variation of the plasma contact point with respect to the region of ballooning-type radial transport lead to strong variations of the parallel flow. From each I-V characteristic the resulting $J_{//,i}$ and T_e are used to calculate the local values of ξ and r_{TP}/λ_D, and the theoretical I-V characteristic is constructed by interpolation within the normalized database. The theoretical and experimental current ratios are compared for $V_{bias} - V_f = -100$ V for all $J_{//,I} > 0.1$ A/cm^2 in figure 6. Independently of which side of the Mach probe produced each measurement, and independently of the reciprocation, there is a fairly decent correlation between theory and experiment. One can remark that the theoretical ratio is a bit higher than the measurement for the smallest values, which correspond to the lowest measurements of T_e. The normalized database was built using the unique value $\tau = T_i/T_e = 2$. However, it has already been observed in Tore Supra (and many other devices) that the ion-to-electron temperature ratio in the SOL can vary strongly with the local plasma density. At low densities, the two tend to be decoupled due to low ion-electron collision rate. At higher densities, the collision rate can be sufficient to bring about temperature equipartition. Therefore, even though no T_i measurements were made, we hypothesize that τ decreases significantly as the core density increases in our experiment. As a result of the smaller ion Larmor radius, less current flows to the tunnel than is predicted by the PICCYL simulations.

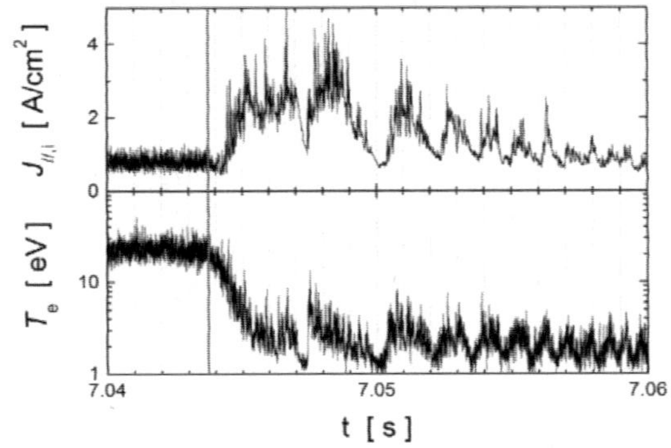

Figure 7. Ion current density and electron temperature measured by the ion-side tunnel collector on the vertical reciprocating probe during discharge TS47771 with I_p = 1.2 MA. A supersonic gas pulse impacted the plasma at the time indicated by the red line. During the 20 ms shown here, the probe was at the deepest position of its plunge, nearly immobile, about 7 mm from the LCFS. The gas-pulse-triggered MHD activity, manifested by the series of long time-scale bursts of ion saturation current which lasted a few ms, accompanied by significant cooling of the SOL.

5. Concluding remarks

It was convincingly demonstrated that for simple $J_{//,i}$ measurements the TP is superior to classical convex Langmuir probes because it is not subject to sheath expansion effects. For turbulence measurements, this means that the TP measures the true $J_{//,i}$ fluctuations, whereas the classical probe signals are contaminated by sheath expansion current which depends on the other plasma parameters. Standard Langmuir probes cannot separate density fluctuations from temperature fluctuations. New schemes to measure temperature fluctuations without fast voltage sweeping have been proposed, usually involving probe arrays (for example triple probes) or more sophisticated tipes of probs, for example combined Langmuir and ball-pen probes. Given the positive results of the present study, the TP method can be proposed as a new technique that can be added to the arsenal of tools available for measuring plasma turbulence. What makes the TP potentially interesting is that the current density and temperature fluctuations can be made simultaneously at the same location, using cheap DC electronics.

As an illustration of the kind of results one can obtain with the TP, we show measurements that were made with the ion-side tunnel on the vertical reciprocating probe during a supersonic molecular beam injection [17,18] on Tore Supra deuterium discharge 47771 with plasma current 1.2 MA (figure 7). A supersonic gas pulse was released from an injector on the high field side of the tokamak. The tunnel and backplate were biased to –150 V and the DC currents were measured at a rate of 1 MHz. The magnetic field at the probe location was 3.5 T. Following the arrival of the gas pulse, the local density increased several times, and the SOL cooled to a few eV. MHD activity was also triggered, visible as the long-time scale pulses which last a few ms.

Acknowlegements
This work, supported by the European Communities under the contract of Association between EURATOM and CEA, was carried out within the framework of the European Fusion Development Agreement. The views and opinions expressed herein do not necessarily reflect those of the European Commission.

References

[1] Dejarnac R, Gunn J P, Stöckel J, Adámek J, Brotánková J and Ionita C 2007 *Plasma Phys. Control. Fusion* **47** 1791
[2] Gunn J P, et al. 2013 *J. Nucl. Mater.* **438** 184
[3] Hutchinson I H and Patacchini L 2010 *Plasma Phys. Contrl. Fusion* **52** 124005
[4] Laframboise J G 1966 Theory of spherical and cylindrical langmuir probes in a collisionless, maxwellian plasma at rest *UTIAS Report No. 100* (University of Toronto, Institute for Aerospace Studies)
[5] Gunn J P, et al. 2002 *Czech J. Phys.* **52** 1107
[6] Gunn J P, Panek R, Stöckel J, van Oost G and van Rompuy T 2005 *Czech. J. Phys.* **55** 255
[7] Gunn J P 2001 *Phys. Plasmas* **8** 1040
[8] Saoutic B, Chatelier M and De Michelis C 2009 Special issue on Tore Supra *Fusion Sci. Technol.* **56** 1079
[9] Stöckel J, et al. 2007 *J. Phys.: Conf. Series* **63** 012001
[10] Verboncoeur J P, Langdon A B and Gladd N T 1995 *Comp. Phys. Comm.* **87** 199
[11] Gunn J P 1997 *Phys. Plasmas* **4** 4435
[12] Hutchinson I H 1987 *Phys. Fluids* **30** 3777
[13] Chung K-S and Hutchinson I H 1988 *Phys. Rev.* A **38** 4721
[14] Kočan M, Gunn J P, Pascal J-Y, Bonhomme G, Devynck P and Ďuran I 2009 *J. Nucl. Mater.* **390–391** 1074
[15] Kočan M, et al. 2011 *J. Nucl. Mater.* **415** 1133
[16] Gunn J P, et al. 2007 *J. Nucl. Mater.* **363-365** 484
[17] Pánek R, Gunn J P, Bucalossi J, Ďuran I, Geraud A, Hron M, Loarer T, Pégourié B, Stöckel J and Tsitrone E 2005 *J. Nucl. Mater.* **337-339** 530
[18] Tamain T, et al. 2013 *J. Nucl. Mater.* **438** 148

Journal of Physics: Conference Series **700** (2016) 012019 doi:10.1088/1742-6596/700/1/012019

WO_3 nanopaticles and PEDOT:PSS/WO_3 composite thin films studied for photocatalytic and electrochromic applications

Stefan Ivanov Boyadjiev[1,3], **Bruno Manduca**[2], **Júlia Szűcs**[2] **and Imre Miklós Szilágyi**[1,2]

[1] MTA-BME Technical Analytical Chemistry Research Group, Szent Gellért tér 4, H-1111, Budapest, Hungary
[2] Budapest University of Technology and Economics, Department of Inorganic and Analytical Chemistry, 4 Műegyetem rakpart, H-1111 Budapest, Hungary

E-mail: boiajiev@gmail.com

Abstract. WO_3 is a widely studied material for electrochromic and photocatalytic applications. In the present study, WO_3 nanoparticles with a controlled structure (monoclinic or hexagonal) were obtained by controlled thermal decomposition of hexagonal ammonium tungsten bronze in air at 500 °C and 600 °C, respectively. The formation, morphology, structure and composition of the as-prepared nanoparticles were studied by powder X-ray diffraction (XRD), transmission electron microscopy (TEM), and scanning electron microscopy combined with energy-dispersive X-ray spectroscopy (SEM-EDX). The photocatalytic activity of the monoclinic and hexagonal WO_3 nanoparticles was studied by decomposing methyl orange in aqueous solution under UV light irradiation. In order to study the electrochromic properties of the WO_3 nanoparticles, as well to introduce them for self-cleaning photocatalytic surface applications, thin films were prepared from the WO_3 particles together with a conductive polymer. For this, PEDOT:PSS was used, which gives excellent opportunities for obtaining transparent and conductive thin films, suitable for both electrochromic and photocatalytic applications. By spin-coating, transparent PEDOT:PSS/WO_3 composite thin films were prepared, on which cyclic voltammetry measurements were performed, and the coloring and bleaching states were studied. Our initial results for the PEDOT:PSS/WO_3 composite thin films are promising, suggesting that such composites, after further development, might be successfully used in electrochromic devices and photocatalysis.

1. Introduction

Tungsten trioxide (WO_3) is one of the most studied inorganic materials with a wide potential of technological applications in fields like large scale electrochromic devices as displays, smart windows and optical switching coatings [1, 2]. In addition, WO_3 can be also used for many other applications, such as photocatalysis [3-5], water splitting [6], gas sensors for various toxic and flammable gases [7-14], and catalysis [15].

[3] To whom any correspondence should be addressed.

WO_3 is one of the most widely researched photocatalysts because of its very good reductive and oxidative proprieties combined with the excellent abilities to absorb both visible and UV light, having a bandgap of 2.4-2.8 eV. Its photo-generated hole (h^+) is oxidizing enough to generate $^{\bullet}OH$ radical from water, but the photo-excited electron (e^-) cannot reduce O_2, which decreases its photocatalytic activity. However, this is compensated by the fact that WO_3 absorbs visible light in the most intensive part of the solar spectrum.

WO_3 has several crystalline modifications, but those with better photocatalytic proprieties are the monoclinic (m-) WO_3 (m-WO_3) and the hexagonal (h-) WO_3 (h-WO_3). The h-WO_3 has a three- and six-dimensional channel-system along its structure, which is stabilized by cationic impurities, e.g. NH_4^+, in the hexagonal channels [16, 17]. The m-WO_3, on the other hand, is completely oxidized, and therefore has a better photocatalytic activity [3]. Both h-WO_3 and m-WO_3 can be prepared by thermal annealing under different conditions. We optimized the annealing conditions of hexagonal ammonium tungsten bronze, $(NH_4)_xWO_{3-y}$ (HATB), through previous thermal studies (TG/DTA) in order to obtain the WO_3 samples with appropriate crystal structures and compositions [18]. Furthermore, both the monoclinic and hexagonal WO_3 have excellent electrochromic properties.

In order to enhance the photocatalytic activity, the catalyst particles should be small to have a high specific surface area and to offer a high number of surface active centers by unit mass. Dispersion of WO_3 particles during photocatalytic treatment of wastewater or any polluted liquids shows excellent photoactivity, but requires other troublesome and high-cost post-filtering processes to separate the nano-sized catalyst from the treated liquid. Thus, especially for long-term commercial applications, the powder photocatalyst needs to be immobilized onto a support, which is transparent and preferably conductive or semi-conductive, and which does not disturb the photo-electron transitions. One of the most widely studied and used materials for immobilizing the catalytic particles, i.e. resins, are usually insulators, which is a drawback, though their positive side is their low price. In our study, we proposed that a thin film of conductive polymer, as PEDOT:PSS, with dispersed photocatalytic nanoparticles could be also applied in photocatalysis.

Such a transparent conductive composite thin film containing WO_3 nanoparticles is promising also for electrochromic applications. PEDOT:PSS itself possesses electrochromic properties; these could be improved by adding nanoparticles of WO_3, which show an even stronger electrochromic effect. Electrochromic studies of such composite PEDOT:PSS/WO_3 films have not been previously performed, and knowledge is needed on how these materials can be combined in order to be used in electrochromic devices.

2. Experimental

Hexagonal and monoclinic WO_3 nanoparticles were prepared by annealing hexagonal ammonium tungsten bronze (HATB), $(NH_4)_xWO_{3-y}$, in air at 500 and 600 °C, respectively. $(NH_4)_xWO_{3-y}$ was prepared by the partial reduction of ammonium paratungstate tetrahydrate, $(NH_4)_{10}[H_2W_{12}O_{42}].4H_2O$ (APT), in H_2 for 6 h at 400 °C [19]. At temperatures over 450-500 °C, WO_3 crystallizes in a hexagonal, while over this temperature, predominantly in a monoclinic structure [20].

The crystalline structure of the samples was confirmed by XRD using a P'Analytical X'pert Pro MPD X-ray diffractometer using Cu Kα irradiation. Transmission electron microscopy (TEM) images were taken by an FEI Morgagni 268D device. The SEM-EDX study was performed by a JEOL JSM-5500LV scanning electron microscope (SEM).

The photocatalytic activity of the samples was tested by studying the degradation of methyl orange (MO) in aqueous solution under UV light. The absorbance of the MO solution, proportional to the MO concentration, was measured by UV-Vis spectroscopy. 1 and 3 mg of the WO_3 particles were placed inside a quartz cuvette, after that 3 mL of the MO solution (1 mg/mL) was pipetted into the cuvette. It was then sealed (with Parafilm), and left in the dark for 30 min, so that an adsorption equilibrium of the dye on the surface of the particles can be reached. This was required since the absorbance decays during the first 30 minutes in dark due to the adsorption of MO on the surface of the catalyst, which might lead to a false photocatalytic degradation result, if not taken into account. In order to avoid this error, the measurement started after this waiting time of 30 min without any illumination. Then the cuvette was placed between two parallel 18 W black lights (Osram) providing UV light irradiation for 240 minutes, and the absorbance was measured every 30 minutes.

Spin-coating was used to prepare PEDOT:PSS/WO_3 composite thin films on glass and ITO transparent conductive (TCO) glass substrates. Dispersion of the 200 mg WO_3 nanoparticles into 1 mL commercial PEDOT:PSS and 3 mL water for spin-coating (Sigma-Aldrich) was prepared and mixed in an ultrasonic bath. A homemade spin-coating apparatus was used for the deposition, and the process parameters were tuned and controlled for obtaining homogenous films with thickness around a few hundred nm.

Cyclic voltammetry experiments were performed in a standard three-electrode set-up. The cell used Pt as a counter electrode and a saturated calomel electrode as a reference electrode, together with the studied PEDOT:PSS/WO_3 films on ITO glass sheets. The electrodes were immersed in a 1 mol/L $LiClO_4$ electrolyte dissolved in propylene carbonate (PC).

3. Results and discussion

The morphology of the prepared WO_3 samples was studied by TEM and SEM. Both the as-prepared hexagonal and monoclinic WO_3 powders consisted of smaller particles and aggregates of these particles with different size (figure 1a)). The performed TEM imaging revealed that the WO_3 particles were built of grains with a size of about 50-80 nanometers (figure 1b)). Such structures are both beneficial for photocatalysis and electrochromism, as the open structure of the particle aggregates means a higher specific surface area, and for electrochromic applications it is generally better to apply particles with a smaller size, in order to increase the transparency in bleached mode.

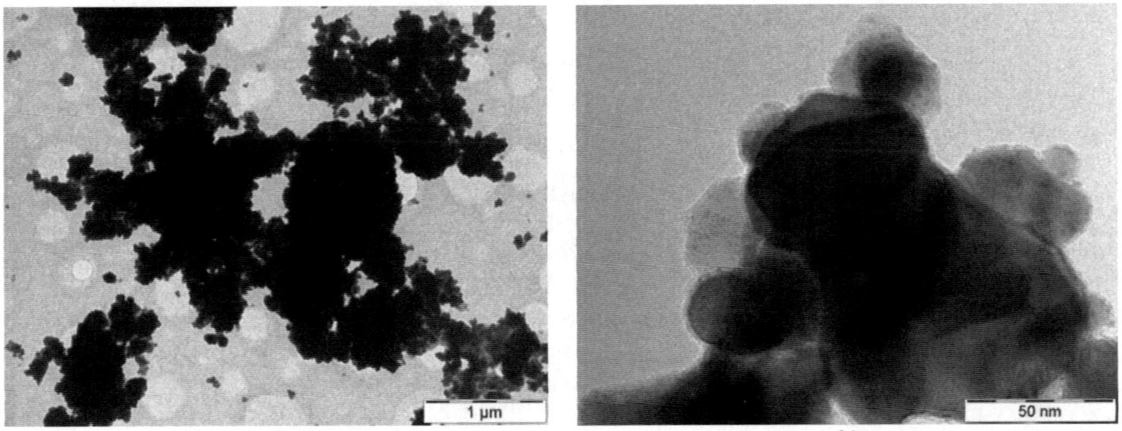

a) b)

Figure 1. TEM images of h- and m-WO_3 nanoparticles: a) aggregate of h-WO_3 particles; b) m-WO_3 nanoparticles.

The hexagonal and monoclinic crystalline systems were confirmed by the XRD study (figure 2). The HATB samples (ICDD 42-0452) annealed to 600 °C showed the typical patterns of monoclinic WO_3 (ICDD 43-1035), while those annealed to 500 °C exhibited the hexagonal WO_3 pattern (ICDD 85-2460). These results indicated that the samples were made up by pure m-WO_3 and h-WO_3 phases. Also, from the XRD study it can be concluded that both the h- and m-WO_3 particles were highly crystalline, which was important, since from our previous studies it was determined that WO_3 with a higher degree of crystallinity showed better both photocatalytic [3] and electrochromic [21] properties.

The performed EDX study confirmed the formation of pure m-WO_3. The signal of nitrogen originating from residual NH_4^+ was not observed, thus proving the full decomposition of HATB. The hexagonal WO_3 powders contained traces of NH_4^+, the N signal in this case was difficult to be clearly distinguished from the noise. Still, their presence is expected, as they are important to stabilize the channel structure.

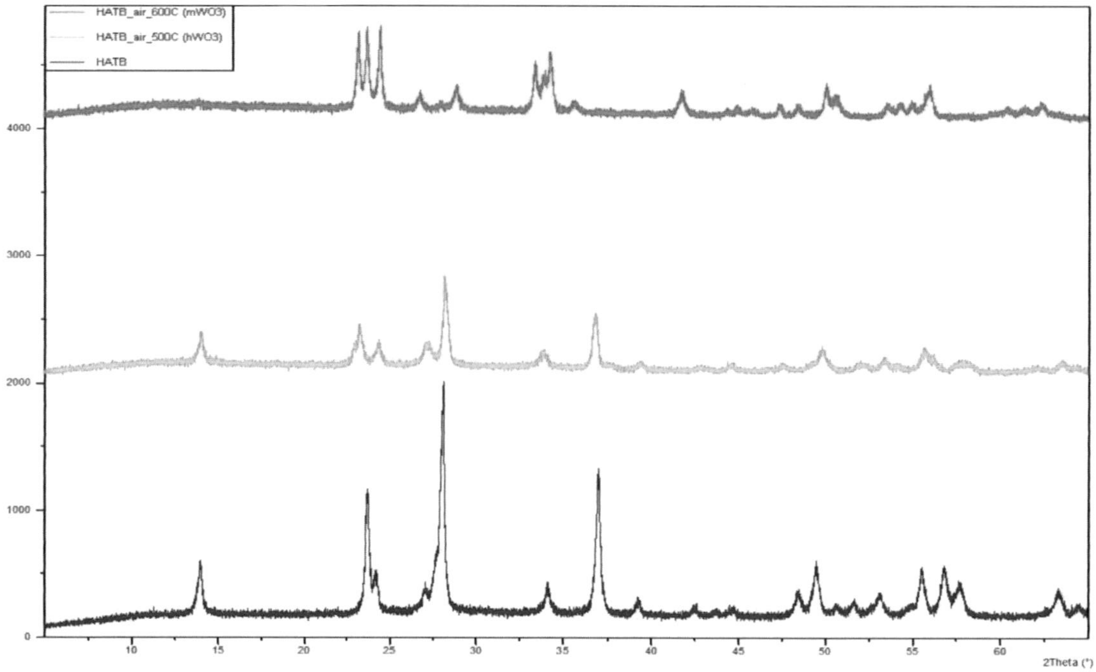

Figure 2. Powder XRD patterns of h- and m-WO_3 nanoparticles, compared with the initial HATB before annealing.

The photocatalytic self-oxidation of methyl orange (photolysis) was tested by filling a cuvette with MO solution without any catalyst and irradiating it by UV light for 240 minutes, measuring the absorbance every half an hour. After 4 hours, the MO concentration decreased by almost 4 %. These values were taken as a reference and compared with the values obtained when the h-WO_3 and m-WO_3 samples were used as photocatalysts. Methyl orange was selected as a probe molecule, as its absorption maximum (465 nm) is exactly where WO_3 has a minimum in its UV-Vis spectrum. MO is one of the most widely spread dyes for testing the photocatalytic activity of WO_3 [22]. Figure 3 shows the results from the photocatalytic measurements for 1 mg and 3 mg of h- and m-WO_3 samples in MO. The better photocatalytic activity was found for the m-WO_3 sample showing that MO degradation is accelerated by ~15 % in the presence of 3 mg m-WO_3 photocatalyst. The results were expected, as it was previously revealed that the m-WO_3 is a better photocatalyst [3]. The h-WO_3 is more reduced than the m-WO_3 because it has cationic impurities responsible for the channel-system structure, thus it possesses a lower photocatalytic activity. It was also shown that the amount of photocatalyst clearly influences the degradation of MO, since when 3 mg catalysts were used, the photocatalytic activity was higher.

For immobilizing these photocatalysts onto glass substrates, several issues had to be taken into account. The most important is that the specific surface area decrease (which is unavoidable) should be reduced as much as possible. This means that the nanoparticles should not be stuck into the polymer film, which at the same time has to possess high enough adhesion to both the substrate and the particles and should be stable in the environment of the application. Also, it is preferable that the film be a semiconductor with bandgap energies similar to the photocatalyst, and also to be highly transparent in order not to disturb the photoprocesses. All these requirements are hard to be accomplished simultaneously, so many compromises have to be done. In the present study, we decided to test the applicability of one of the most used conductive polymers, i.e. PEDOT:PSS, for immobilizing photocatalytic nanoparticles. The results showed that when appropriate deposition conditions were applied and the nanoparticles were well dispersed, it was possible to deposit high-quality homogenous composite thin films.

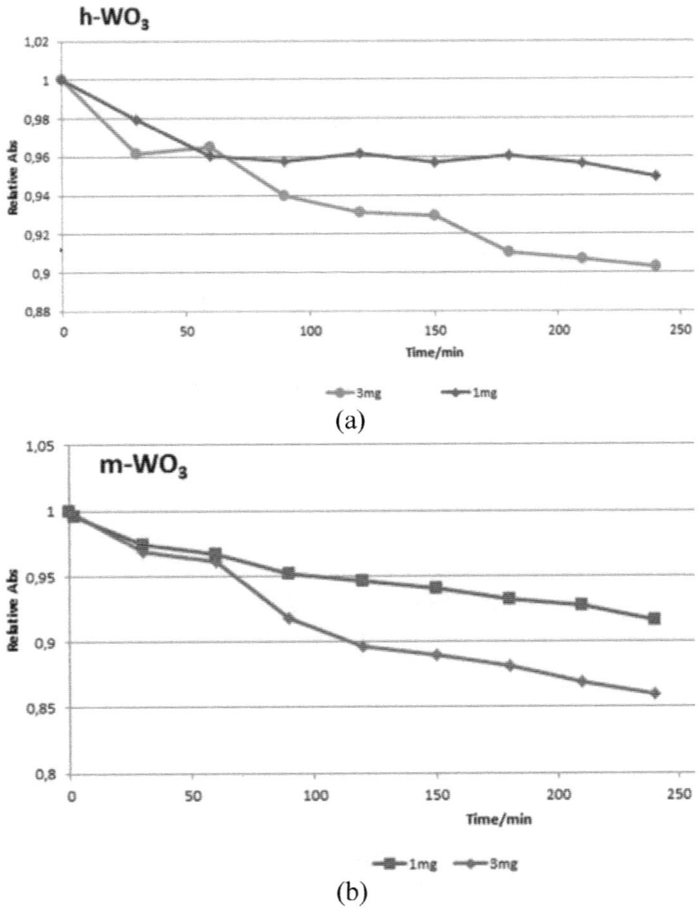

Figure 3. Photocatalytic study of h- and m-WO_3 nanoparticles: a) photocatalytic activity of h-WO_3 nanoparticles; b) photocatalytic activity of m-WO_3 nanoparticles.

In figure 4a), a backscattered electron SEM image of a composite film is presented, where the m-WO_3 nanoparticles are very well distinguishable from the polymer. In such composite thin films, the size of the particles has to be larger, preferably at least twice as the film thickness, so that a big part of the surface stays out of the film. Note that the m-WO_3 nanoparticles were aggregated into micrometer scale blocks. The thicker films were more homogenous, but in them a big part of the nanoparticles remained in the polymer (figure 4b). Thus, we tried to achieve the smaller possible film thicknesses with the applied spin-coating method. A too thin constant and homogenous film is hard to be deposited, since if the spin-coating speed is too high or the viscosity of the dispersion is too low, the result is islands of the composite or a highly inhomogeneous film. Otherwise, with decreasing the spinning speed or increasing the density, the as-deposited films are thicker and the active surface of the particles is decreased.

On the as-prepared composite PEDOT:PSS/WO_3 thin films, cyclic voltammetry tests were performed and the electrochromic effect was observed. Both the pure and composite films showed good coloration (while visually it is stronger for the composite ones) and fast and full bleaching. It seems that the effect is reversible for many cycles, if low voltages, i.e. bellow –/+ 0.8 V, are applied. The measured cyclic voltammetry curve in the range –/+ 3 V is presented in figure 5.

The intercalation of ions into the film happens in the first measured period when the voltage goes from 0 V to negative values. Then the positive Li^+ ions are attracted to the electrode and go into the

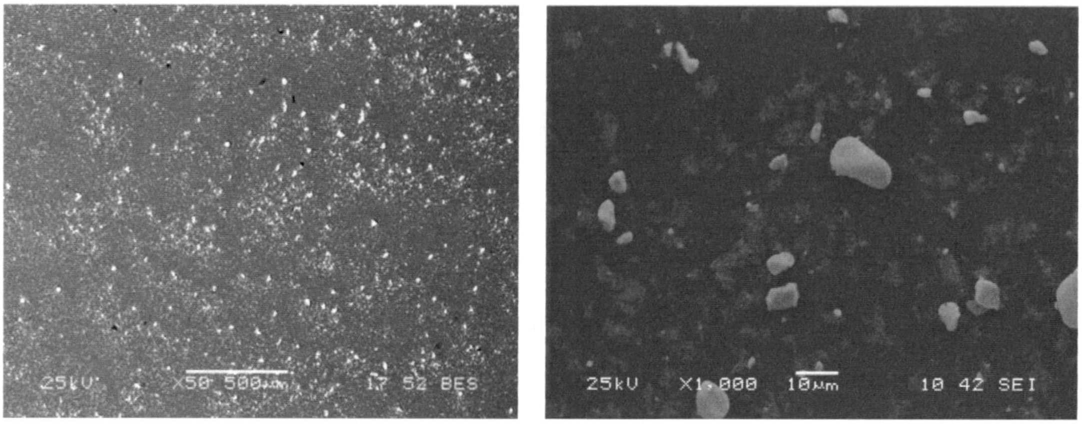

a) b)

Figure 4. SEM images of a composite PEDOT:PSS/m-WO_3 thin film on a TCO glass substrate: a) backscattered electron image; b) secondary electron image.

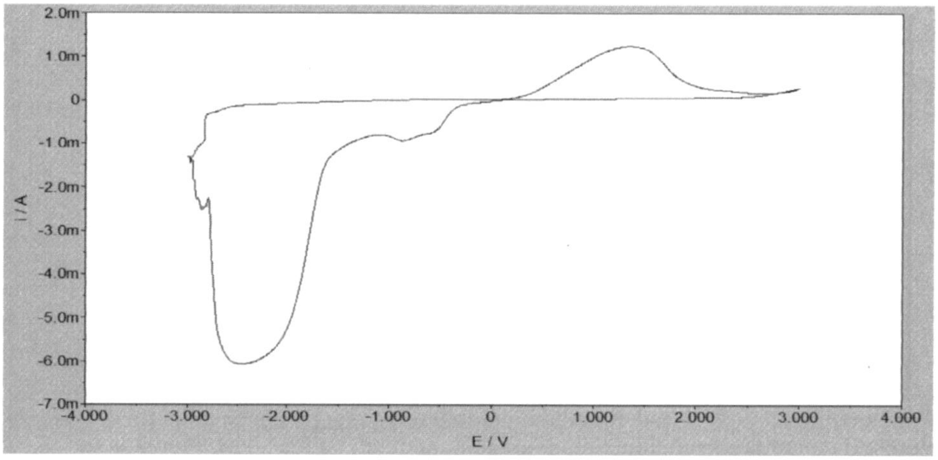

Figure 5. Cyclic voltammetry curve of a composite PEDOT:PSS/m-WO_3 thin film.

free places of the lattice, which consequently leads to the coloration of the film. When the voltage is reversed, the process starts to go backwards and when the applied voltage turns to positive, a large part of the ions de-intercalate, and the film bleaches. The coloration process starts between – 0.25 and 0.3 V, while the bleaching starts shortly after 0 V in the positive voltage regime. For PEDOT:PSS the voltage required for coloration and bleaching was lower than for WO_3. Too high currents can cause destruction of the electrochromic films, so they must be avoided. The left part of the graph is typical but with too a high current peak, which shows the "burning up" of the sample. For this reason, smaller voltages should be applied. We determined that the best interval for measuring is –/+ 0.8 V, where we could perform successful tests with both the pure PEDOT:PSS films and those with WO_3 particles.

There is an existing problem of usage of PEDOT:PSS films for long terms in aqueous or polar liquids environment, as the films are not stable in such. This limits the application of the PEDOT:PSS/WO_3 films to only photocatalysis in gaseous, aerosol or non-polar liquids environments. Regarding the electrochromic applications, the films showed to be stable during the performed tests in the $LiClO_4$/PC electrolyte, but for preparing electrochromic devices the usage of dry electrolyte is recommended.

4. Conclusions

WO_3 nanoparticles with a controlled structure (monoclinic or hexagonal) were obtained by controlled thermal decomposition of $(NH_4)_xWO_{3-y}$ in air at 500 or 600 °C, respectively. By spin-coating of suspensions of the WO_3 nanoparticles in aqueous PEDOT:PSS, composite thin films were deposited onto glass and ITO conductive glass substrates, and their applicability for photocatalytic and electrochromic applications was checked.

The formation, morphology, composition and structure of the as-prepared nanoparticles and nanocomposite films were studied by TEM, SEM-EDX, and XRD. It was also investigated how the structure and composition of the WO_3 particles influenced the photocatalytic properties of the films and how the technology of preparing organic/non-organic composite layers could be optimized for using them in electrochromic devices and photocatalysis.

Our initial experiments on preparing composite PEDOT:PSS/WO_3 and testing its possible application for electrochromism and photocatalysis were successful and although there is much space for further optimization and development, the results are promising for future application of PEDOT:PSS/WO_3 thin films in electrochromic devices and photocatalysis.

Acknowledgements

S. Boyadjiev acknowledges the Postdoctoral Fellowship program of the Hungarian Academy of Sciences (2013-2015). I. M. S. thanks for a János Bolyai Research Fellowship of the Hungarian Academy of Sciences and an OTKA-PD-109129 grant.

References

[1] Granqvist C G 1995 *Handbook of Inorganic Electrochromic Materials* (Amsterdam: Elsevier)
[2] Gesheva K A, Ivanova T, Marsen B, Zollo G and Kalitzova M 2008 *J. Cryst. Growth* **310** 2103
[3] Szilágyi I M, Fórizs B, Rosseler O, Szegedi A, Németh P, Király P, Tárkányi G, Vajna B, Varga-Josepovits K, László K, Tóth A L, Baranyai P and Leskelä M 2012 *J. Catal.* **294** 119
[4] Wicaksana Y, Liu S, Scott J and Amal R 2014 *Molecules* **19** 17747
[5] Szilágyi I M, Santala E, Heikkilä M, Pore V, Kemel M, Nikitin T, Khriachtchev L, Räsänen M, Ritala M and Leskelä M 2013 *Chem. Vapor Dep.* **19** 149
[6] Marsen B, Miller E L, Paluselli D and Rocheleau R E 2007 *Int. J. Hydrogen Energ.* **32** 3110
[7] Guatwei L, Xianju T, Choongyew C, Gaiktin A and Zailani A B M 2011 *J. Mater. Sci. Eng.* **5** 271
[8] Blo M, Carotta M C, Galliera S, Gherardi S, Giberti A, Guidi V, Malagù C, Martinelli G, Sacerdoti M, Vendemiati B and Zanni A 2004 *Sens. Actuat.* B **103** 213
[9] Szilágyi I M, Saukko S, Mizsei J, Tóth A L, Madarász J and Pokol G 2010 *Solid State Sci.* **12** 1857
[10] Szilágyi I M, Wang L, Gouma P I, Balázsi C, Madarász J and Pokol G 2009 *Mater. Res. Bull.* **44** 505
[11] Szilágyi I M, Saukko S, Mizsei J, Király P, Tárkányi G, Tóth A L, Szabó A, Varga-Josepovits K, Madarász J and Pokol G 2008 *Mater. Sc. Forum* **589** 161
[12] Balázsi C, Wang L, Zayim E O, Szilágyi I M, Sedlackova K, Pfeifer J, Tóth A L and Gouma P I 2008 *J. Eur. Ceram. Soc.* **28** 913
[13] Wang L, Pfeifer J, Balázsi C, Szilágyi I M and Gouma P I 2007 *Proc. SPIE* **6769** 67690E
[14] Balázsi C, Wang K, Kalyanasundaram K, Sedlackova K, Zayim E O, Szilágyi I M, Pfeifer J, Tóth A L and Gouma P I 2009 *NATO Science for Peace and Security Series C: Environmental Security; Sensors for Environment, Health and Security* **77**
[15] Firkala T, Fórizs B, Drotár E, Tompos A, Tóth A L, Varga-Josepovits K, László K, Leskelä M and Szilágyi I M 2014 *Catal. Lett.* **144** 831
[16] Szilágyi I M, Madarász J, Pokol G, Hange F, Szalontai G, Varga-Josepovits K and Tóth A L

2009 *J. Therm. Anal. Calorim.* **97** 11

[17] Szilágyi I M, Madarász J, Pokol G, Sajó I, Király P, Tárkányi G, Tóth A L, Szabó A and Varga- Josepovits K 2009 *J. Therm. Anal. Calorim.* **98** 707

[18] Szilágyi I M, Madarász J, Pokol G, Király P, Tárkányi G, Saukko S, Mizsei J, Tóth A L, Szabó A and Varga-Josepovits K 2008 *Chem. Mater.* **20** 4116

[19] Szilágyi I M, Hange F, Madarász J and Pokol G 2006 *Eur. J. Inorg. Chem.* **17** 3413

[20] Szilágyi I M, Pfeifer J, Balázsi C, Tóth A L, Varga-Josepovits K, Madarász J and Pokol G 2008 *J. Therm. Anal. Calorim.* **94** 499

[21] Boiadjiev S, Ivanova T, Rassovska M and Gesheva K A 2007 *J. Optoelectr. Adv. Mater.* **9** 264

[22] Boiadjiev S, Ivanova T, Rassovska M and Gesheva K A 2007 *J. Optoelectr. Adv. Mater.* **9** 264

Investigation of the resistive properties of HTS/manganite bilayers

T Nurgaliev[1], B Blagoev[2,4], V Štrbik[3], Š Chromik[3] and M Sojková[3]

[1]Emil Djakov Institute of Electronics, Bulgarian Academy of Sciences, 72 Tzarigradsko Chaussee, 1784 Sofia, Bulgaria
[2]Georgi Nadjakov Institute of Solid State Physics, Bulgarian Academy of Sciences, 72 Tzarigradsko Chaussee, 1784 Sofia, Bulgaria
[3]Institute of Electrical Engineering, Slovak Academy of Sciences, 9 Dubravska cesta, 84104 Bratislava, Slovak Republic

E-mail: blago_sb@yahoo.com

Abstract. Temperature dependences of the resistivity of manganite $La_{0.7}Ca_{0.3}MnO_3$ (LCMO) films deposited on $LaAlO_3$ and $SrTiO_3$ substrates by RF magnetron sputtering were shown to be successfully simulated in the whole temperature range (covering metal, insulator and metal-insulator transition regions) using a phenomenological phase-coexistence transport model. Quantitative data on the internal parameters of these films were obtained. The possibility was also considered for investigation of individual resistive characteristics and excess conductivity of a high temperature superconducting (HTS) $YBa_2Cu_3O_{7-x}$ (YBCO) thin film in the vicinity of T_C included into a YBCO/LCMO bilayer structure. It was shown that the considered YBCO film in the temperature range from 85.5 K to 114.9 K behaves as a two-dimensional system with respect to the fluctuations in the superconducting order parameter, while a three-dimensional regime is observed in a narrow range of temperatures at $T < 85.5$ K. Such behavior was assumed to be partly due to the FM LCMO component of the bilayer, the spin-polarized charge carriers of which enter into the YBCO film and cause a "breaking" of superconducting pairs in the superconducting gap and pseudogap regimes.

1. Introduction

Manganites $La_{0.7}Ca_{0.3}MnO_3$ (LCMO), $La_{0.7}Sr_{0.3}MnO_3$ (LSMO) show a broad spectrum of physical properties due to the coexistence of spatially separated different electronic and magnetic states [1, 2]. The conduction mechanism of manganites is very complex and dominated by double-exchange interaction accompanied by a strong electron-phonon coupling due to the dynamic Jahn-Teller effect [3-6]. At low temperatures, when the sample is in ferromagnetic (FM) metal state ($T < T_{MI}$, where T_{MI} is the metal-insulator transition temperature), the origin of the electrical resistivity ρ is related to scattering on the crystal defects, impurities (temperature-independent residual resistance) and to electron-electron, electron-phonon and electron-magnon scatterings [1, 2, 6]. At higher temperatures $T > T_{MI}$ the paramagnetic insulating phase becomes dominant, and $\rho(T)$ can be described as the effect of an adiabatic small polaron hopping process. In our recent works [7, 8], we showed that the ρ -T dependence for high-quality LSMO films, except for the temperature region $T \sim T_{MI}$, can be described

[4] To whom any correspondence should be addressed

in the framework of the above mechanisms, while not-optimally prepared polycrystalline LSMO thin films can be described by a model of two intergrain conduction channels [9-10]. A phenomenological phase-coexistence transport model was developed successfully in the last decade [6, 11-13], which allows the description of the resistivity in broad temperature ranges, including the temperature region $T \sim T_{MI}$. In this work, we investigated the LCMO film resistive characteristics in the framework of the phase-coexistence transport model. LCMO films were obtained by RF magnetron sputtering.

It is known that the high temperature superconducting (HTS) cuprates with optimal (corresponding to the maximum critical temperature T_C) and with lower carrier concentration show some specific peculiarities in the electrical, magnetic and optical properties [14-18] at temperatures $T < T^*$ well above the critical temperature T_C. These phenomena are associated with a pseudogap regime opening at $T = T^*$, where onset of fluctuations in the superconductor order parameter set in. The existence of fluctuations leads to an increase of the samples' normal conductivity (i.e. to the appearance of excess conductivity) [15, 17]. Analysis of the individual resistive characteristics of the HTS $YBa_2Cu_3O_{7-x}$ (YBCO) thin film in the YBCO/LCMO structure at $T > T_C$ in terms of excess conductivity was another aim of the work.

2. Experimental

The deposition of FM LCMO film on $LaAlO_3$ (LAO) and $SrTiO_3$ (STO) substrates was performed by off-axis RF magnetron sputtering from a stoichiometric ceramic target. To prepare HTS/FM bilayers, HTS YBCO films were deposited by laser ablation on top of the RF-magnetron sputtered manganite films. Some of the deposition details were described in our previous paper [19]. The thickness of the LCMO films deposited on LAO and STO substrates were 75 nm and 20 nm, respectively. The thickness of the LCMO and the YBCO films in the HTS/FM bilayer, deposited on the LAO substrates, were 75 nm and 90 nm. The standard four-probe method was used for measuring the temperature dependence of the resistance of the thin films and bilayers prepared.

3. Results and discussion

3.1. Investigation of $La_{0.7}Ca_3MnO_3$ thin films resistivity using the phase-coexistence transport model

The temperature dependences of the resistivity ρ(T) of $La_{0.7}Ca_{0.3}MnO_3$ thin films are of metallic behavior ($d\rho/dT>0$) at low temperatures and of semiconducting (or "insulative") ($d\rho/dT<0$) character at higher temperatures (figures 1 and 2). The temperature of maximum resistivity T_{MI} is significantly lower for the LCMO film with a small thickness deposited on the STO substrate than for the thicker LCMO film deposited on the LAO substrate. As a rule, the thin films demonstrate a decrease in T_C and T_{MI} as the thickness decreases [2,5]. Moreover, a tensile strain [1], which LCMO film on STO substrate is exposed to, can contribute to a decrease in T_C as well.

A phenomenological phase-coexistence transport model was applied to describe the temperature dependences obtained of the resistivity in the whole temperature range. The model is based on the assumption that the FM metal and paramagnetic insulator (PI) phases coexist at the transition region in doped manganites and nanoregions, with these phases being distributed randomly in the sample. The total resistivity ρ is assumed to be the product of contributions of the resistivities ρ_{PI} and ρ_{FM} from the PI and FM regions $\rho = \rho_{PI}^{1-\nu}\rho_{FM}^{\upsilon}$, where ν is the volume fraction of the FM phase. Assuming that the Curie temperature distribution for a disordered system is Gaussian, the volume fraction ν is supposed to follow a complementary error function [6,11-13] $\upsilon(T) = (1/2)erfc[(T - T_{CurieM})/\Gamma_0]$, where $2^{1/2}\Gamma_0$ is the parameter of standard deviation of the Gaussian distribution, T_{CurieM} is some mean value of the Curie temperature. The resistivity of the PI phase of LCMO can be described in the framework of the small polaron hopping model $\rho_{PI}(T) = \rho_a T \exp(E_a / k_B T)$, where E_a is the activation energy for hopping conduction, ρ_a is the resistivity parameter, k_B is the Boltzmann constant. The resistivity of the FM phase can be approximated by a polynomial function $\rho_{FM}(T) = \rho_0 + \rho_2 T^2 + \rho_N T^N$, where ρ_0 is the residual resistivity caused by grain boundaries and defects in the crystal structure, $\rho_2 T^2$ is the resistivity related to electron-electron scattering and the third term (with $N \sim 4.5$) represents mainly the contribution to the resistivity from the electron-magnon interaction according to the FM double-exchange theory [1, 2].

Figures 1 and 2 display the simulated results (red solid line) for the resistivity-temperature dependences for the above mentioned LCMO thin films. The parameters used for the simulation are shown in table 1.

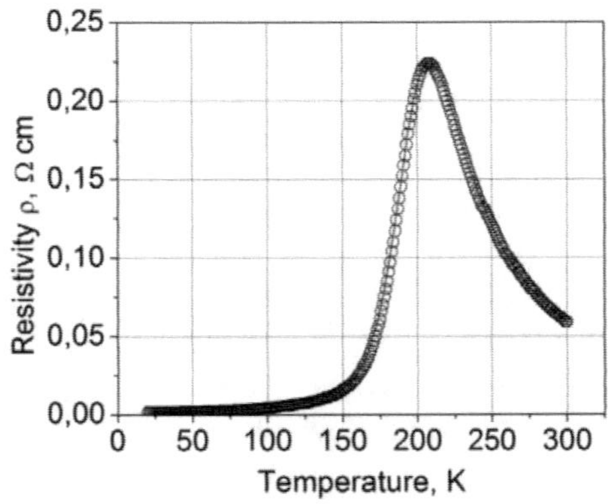

Figure 1. $\rho - T$ dependence in a LCMO film with 75 nm thickness prepared on a LAO substrate. Open circles – experiment, solid line – data modelled using the phase-coexistence transport model (fitting parameters – see table 1).

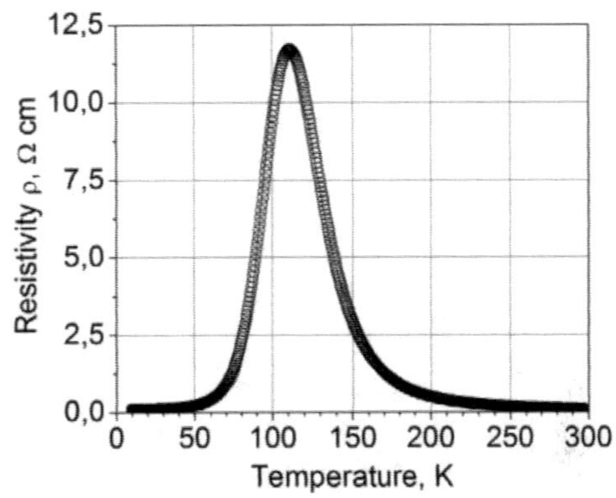

Figure 2. $\rho - T$ dependence in a LCMO film with 20 nm thickness prepared on a STO substrate. Open circles – experiment, solid line – data modelled using the phase-coexistence transport model (fitting parameters – see table 1).

Table 1. Parameters used for simulation of ρ-T dependences in LCMO/LAO and LCMO/STO films in the framework of the phase-coexistence transport model.

Sample	ρ_0, Ωcm	ρ_2, Ωcm/K^2	ρ_4, Ωcm/K^N N	ρ_a, Ωcm/K	T_{CurieM}, K	Γ_0, K	E_a, meV
LCMO/LAO	0.00165	0.1452 10^{-6}	2.211 10^{-12} $N = 4.4$	2.366 10^{-6}	185.5	29	114.43
LCMO/STO	0.1	0.2 10^{-6}	172 10^{-12} $N = 4.635$	5 10^{-6}	97	36.8	107

It is seen that the results of simulation describe well the experimental ρ-T dependences. The LCMO film with a small thickness deposited on the STO substrate is characterized by a high value of the residual resistance ρ_0 and the scattering parameters ρ_2, ρ_4, ρ_a, while the mean transition temperature T_{CurieM} of this sample is significantly lower than that of the LCMO/LAO sample. This is due to the fact that thin films, as a rule, contain more local strains, more defects and impurities, which contribute to activation of the scattering processes and to an increase of the scattering parameters.

3.2. Investigation of a YBCO thin film conductivity in a YBCO/LCMO bilayer

The conductive properties of HTS/FM bilayers are determined by the contribution of the individual conductivities of both HTS and FM layers and by the interaction peculiarities of the layers in the interface. To obtain more detailed information on the resistivity of the YBCO film in the YBCO/LCMO bilayer, we first measured the resistance temperature dependence $R_B(T)$ of the bottom LCMO layer before deposition of the top YBCO layer. The temperature dependence of the resistance

$R_T(T)$ of the top YBCO layer can be determined using the formula $R_T^{-1}(T) \approx R_\Sigma^{-1}(T) - kR_B^{-1}(T)$, where $R_\Sigma(T)$ is the measured *R-T* dependence of the bilayer and the parameter *k* takes into account the resistance change of the bottom LCMO layer during deposition of the top YBCO layer. Figure 3 shows the *R-T* dependence measured in our YBCO/LCMO bilayer (dash dot) and the individual *R-T* dependence of the YBCO film (solid line, red) in this bilayer for temperatures $T > T_C = 83.5$ K, determined in the above way with $k = 1.5$. The *R-T* dependence obtained for the YBCO film is nearly linear and shows that a sharper decrease of *R* with the decrease of *T* occurs when T_C is approached. The extrapolation of the linear part of the above *R(T)* dependence to the low temperature range is depicted by dots (blue) in figure 3.

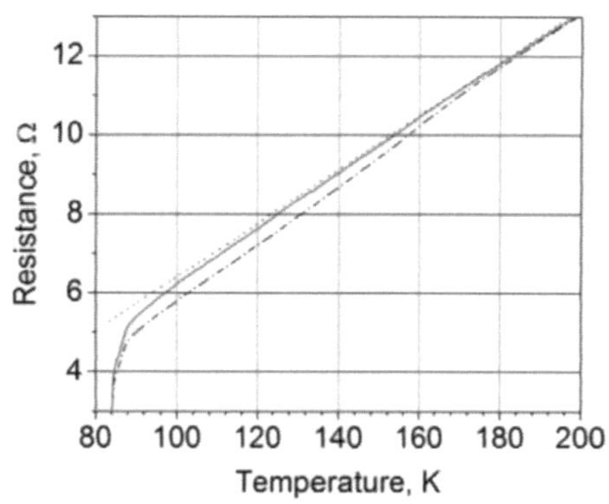

Figure 3. *R-T* dependence measured in the YBCO/LCMO bilayer (dash dot) and the individual *R-T* dependence of YBCO film (solid line, red) in this bilayer. Dots (blue) – the extrapolation results in the linear part of *R(T)* dependence of YBCO film to the low temperature range.

The small nonlinearity in the *R-T* dependence can be associated with the appearance of fluctuation conductivity [15, 17], which causes a more rapid increase in the excess conductivity $\Delta\sigma$ in the sample. This excess conductivity was calculated from the measured *R(T)* and extrapolated $R_{ext}(T)$ dependences as $\Delta\sigma(T) = [R^{-1}(T) - R_{ext}^{-1}(T)]/(4.53t)$, where *t* is the film thickness. Information about the behavior of the excess conductivity $\Delta\sigma(T)$ was obtained using the Aslamazov-Larkin model [20]. For this purpose, the $\ln(\Delta\sigma)$ versus $\ln(\varepsilon)$ dependence was calculated (with $\varepsilon = (T - T_M)/T_M$, $T_M = 84$ K) and plotted (figure 4). The slope *S* of this plot can serve as an indicator of the dimensionality of the excess conductivity: $\Delta\sigma$ is three-dimensional if $S \sim -0.5$ and two-dimensional if $S \sim -1$ [15,17].

It is seen in figure 4 that $S \approx -0.5$ at $-6 < \ln(\varepsilon) < -4$ (temperature range from 84.2 K to 85.5 K) and $S \approx -1$ at $-4 < \ln(\varepsilon) < -1$ (temperature range from 85.5 K to 114.9 K). This means that the YBCO film considered in the temperature range from 85.5 K to 114.9 K behaves as a two-dimensional system with respect to the fluctuations in the superconducting order parameter, while a three-dimensional regime is established at $T < 85.5$ K. We assume that a low value of T_C and a narrow range of temperatures, where $\Delta\sigma(T)$ of the YBCO film demonstrates a three-dimensional character, could partly be a consequence of spin-polarized charge carriers which enter the YBCO film from the FM LCMO component of bilayer and cause a "breaking" of superconducting pairs in both the superconducting gap and pseudogap regimes.

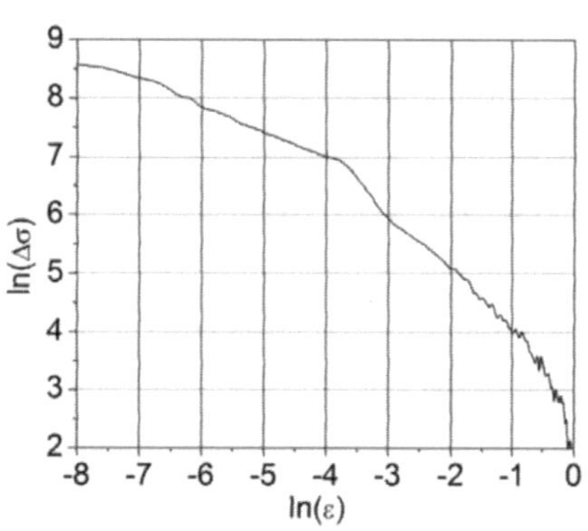

Figure 4. Plot of the ln($\Delta\sigma$) versus ln(ε) dependence.

4. Conclusions

Temperature dependences of the resistivity of manganite LCMO films deposited on LAO and STO substrates by RF magnetron sputtering were shown to be successfully simulated in the whole temperature range (covering the metal, the insulator and the metal-insulator transition regions) using a phenomenological phase-coexistence transport model. Quantitative data on the internal parameters of these films were obtained. The LCMO film with a small thickness deposited on a STO substrate is characterized by a high value of the residual resistance ρ_0 and scattering parameters ρ_2, ρ_4, ρ_a (which is due to the presence of more local strains defects and impurities in thin films). The mean Curie temperature T_{CurieM} of this sample is significantly lower than that of the LCMO/LAO sample. The possibility was also considered for investigating the individual resistive characteristics and excess conductivity of an HTS thin film included into a HTS/FM bilayer structure in the vicinity of T_C. It was shown that the YBCO film considered in the temperature range from 85.5 K to 114.9 K behaves as a two-dimensional system with respect to the fluctuations in the superconducting order parameter, while the three-dimensional regime is observed at $T < 85.5$ K. It was assumed that a low value of T_C and a narrow range of temperatures, where $\Delta\sigma(T)$ of the YBCO film demonstrates a three-dimensional character, could partly be a consequence of spin-polarized charge carriers which enter the YBCO film from the FM LCMO component of the bilayer and cause a "breaking" of superconducting pairs in both the superconducting gap and pseudogap regimes.

Acknowledgment
This work was partially supported by the EU FP7 project INERA, Grant No 316309, and by the Slovak Grant Agency VEGA under projects No. 2/0173/13 and No. 2/0120/14.

References

[1] Haghiri-Gosnet A-M and Renard J-P 2003 *J. Phys.* D: *Appl. Phys.* **36** R127
[2] Dorr K 2006 *J. Phys.* D: *Appl. Phys.* **39** R125
[3] Zener C1951 *Phys. Rev.* **82** 403
[4] Mills A J, Littlewood P B and Shraiman B I 1995 *Phys. Rev. Lett.* **74** 5144
[5] Bibes M, Balcells L I, Valencia S, Fontcuberta J, Wojcik M, Jedryka E and Nadolski S 2001 *Phys. Rev. Lett.* **87** 067210
[6] Phong P T, Manh D H, Bau L V and Lee I-J 2013 *J. Electroceramics* **31** 364
[7] Strbik V, Blagoev B, Mateev E and Nurgaliev T 2014 *J. Phys.: Conf. Series* **514** 012042
[8] Blagoev B S, Nurgaliev T K, Strbik V, Mateev E S and Zaleski A J 2013 *Bulg. J. Phys.* **40** 337
[9] Helman J S and Abeles B 1976 *Phys. Rev. Lett.* **37** 1429
[10] Ming R G, Liu Y S and Ming T Z 2009 *Thin Solid Films* **517** 3748
[11] Alexandrov A S, Bratkovsky A M and Kabanov V V 1999 *Phys. Rev. Lett.* **82** 141
[12] Alexandrov A S, Bratkovsky A M and Kabanov V V 2006 *Phys. Rev. Lett.* **96** 117003
[13] Wang L M, Wang Ch-Y and Tseng Ch-Ch 2012 *Appl. Phys. Lett.* **100** 232403
[14] Nakano T, Oda M, Manabe C, Momono N, Miura Y and Ido M 1994 *Phys. Rev.* B **49** 16000
[15] Prokof'ev D D, Volkov M P and Boikov Yu A 2003 *Phys. Solid State* **45** 1223
[16] Soltan S, Albrecht J and Habermeier U-H 2007 *Physica* C **460-462** 1088
[17] Hassan N, Irfan M, Khan N A, Khan S, Shakoor A, Majid A, Jadoon A M and Hayyat S S 2012 *J. Appl. Phys.* **111** 053914
[18] Loktev V M, Quik R M and Sharapov S G 1999 *Physica* C **314** 233
[19] Blagoev B S, Gostev I G, Nurgaliev T K, Strbik V, Bineva I E, Uspenskaya L, Mateev E S, Neshkov L, Dobrocka E and Chromik S 2014 *J. Phys.: Conf. Series* **514** 012041
[20] Aslamazov L G and Larkin A I 1968 *Sov. Phys. Solid State* **10** 875

Transport properties of $YBa_2Cu_3O_x$ /$La_{0.67}Sr_{0.33}MnO_3$ nanostrips and $YBa_2Cu_3O_x$ /$La_{0.67}Sr_{0.33}MnO_3$/$YBa_2Cu_3O_x$ nanojunctions

V Štrbík[1,3], Š Beňačka[1], Š Gaži[1], M Španková[1], V Šmatko[1], Š Chromik[1], N Gál[1], J Knoška[1,2], M Sojková[1] and M Pisarčík[1]

[1]Institute of Electrical Engineering, Slovak Academy of Sciences, 9 Dúbravská cesta , 841 04 Bratislava, Slovak Republic
[2]Center for Free-Electron Laser Science, DESY, 85 Notkestraβe, 22607 Hamburg, Germany; Department of Physics, University of Hamburg, 149 Luruper Chaussee, 22607 Hamburg, Germany

E-mail: elekstrb@savba.sk

Abstract. A metallic ferromagnet (F) in proximity with a superconductor (S) can transport supercurrent on a long distance through conversion of opposite-spin singlet Cooper pairs (CP) into equal-spin triplet CP (long range triplet component, LRTC), which are not broken by the exchange energy of F. The optimal conditions for the conversion are yet to be clarified; however, it is accepted that the key point to this process include high interface transparency and magnetic inhomogeneity at the SF interface. The aim of our paper is to study SF nanostrips (length of about 1500 nm and width down to 300 nm) and lateral SFS nanojunctions based on high critical temperature $YBa_2Cu_3O_x$ (YBCO) and half-metallic $La_{0.67}Sr_{0.33}MnO_3$ (LSMO) thin films. We applied a focused Ga^+ ion beam (FIB) for patterning the SF nanostrips, as well as lateral SFS nanojunctions, by creating a slot in the nanostrip after removing the YBCO film in the slot along a length of about 200 nm. The temperature dependences of the samples resistance $R(T)$ show critical temperature $T_{Cn} \approx 89$ K of the SF nanostrips; however, the SFS nanojunctions at $T < T_{Cn}$ show a residual resistance $R <$ 100 Ω corresponding to a dirty LSMO ($\rho \approx 10$ mΩ cm) in the slot. The LRTC was not observed in our lateral SFS nanojunctions until now.

1. Introduction

Recent theoretical [1] and experimental [2] studies of equal-spin long-range triplet component (LRTC) in superconductor ferromagnet (SF) proximity systems demonstrate, besides new physical aspects, application possibilities for spintronics, superconducting quantum interference devices (SQUID) and Josephson junctions operating in strong magnetic fields [3,4]. The penetration depth of spin opposite singlet Cooper pairs (CP) into a diffusive normal metal (N) is given at low temperatures by the length scale $\xi_N = (\hbar D_N/2\pi k_B T)^{1/2} \approx 1$ µm, whereas a ferromagnet exchange field $E_{ex} \gg k_B T$ makes the

[3] To whom any correspondence should be addressed.

penetration depth in a diffusive ferromagnet much shorter $\xi_F = (\hbar D_F/2\pi E_{ex})^{1/2} \approx 1$ nm; here T is the temperature, $D_{N,F} = v_F l/3$ are diffusion constants, v_F, the Fermi velocity of carriers, l, the mean-free-path and $\hbar$ and k_B are the Planck and Boltzmann constants. However, under convenient conditions (magnetic inhomogeneity, spin active SF interface, different magnetic orientations in multilayered SF structures), the conversion of singlet CP into LRTC can enhance the ξ_F close to ξ_N. This conversion makes it possible to implement new functional superconducting circuits, and with the present nanoscale technological opportunities, realize submicron SF structures, allowing effective work in magnetic fields up to the critical magnetic field of the superconductor [4]. However, the relative fraction of LRTC in comparison to the singlet CP concentration depends sensitively on the junction parameters, such as the transparency of the SF interface, the amplitude of the exchange energy E_{ex}, the ferromagnet resistivity, the junction geometry and the superconductor's coherence length [5]. The theoretical analyses show that the triplet Andreev reflection amplitude at the interface between half-metal and s-wave superconductor can be significantly enhanced if the F is coupled laterally to the superconductor [6]. This was recently experimentally confirmed using the low temperature superconductor MoGe and half-metal CrO_2 [2]. The realization of high-quality high-T_c SF structures and SFS Josephson weak links manifesting LRTC is a complicated task. The magnetic inhomogeneity in the so-called series geometry must be localized immediately at the SF interface; otherwise the triplet current amplitude exponentially decreases with the distance from the interface. In addition, it has recently been found from theoretical analyses [6] that, in comparison with the serial geometry, the amplitude of the LRTC may be considerably enhanced in the so-called lateral geometry (figure 1a), and the magnetic inhomogeneity (e.g. domain wall below a superconductor) may not necessarily be localized directly at the SF interface. In this paper, we report on the preparation and study of LSMO/YBCO bilayer properties and lateral nanometer YBCO/LSMO/YBCO heterostructures using a focused gallium ion beam (FIB) patterning in view of clarifying whether this technology is a suitable tool for implementing such a technology for such superconductor nanostructures.

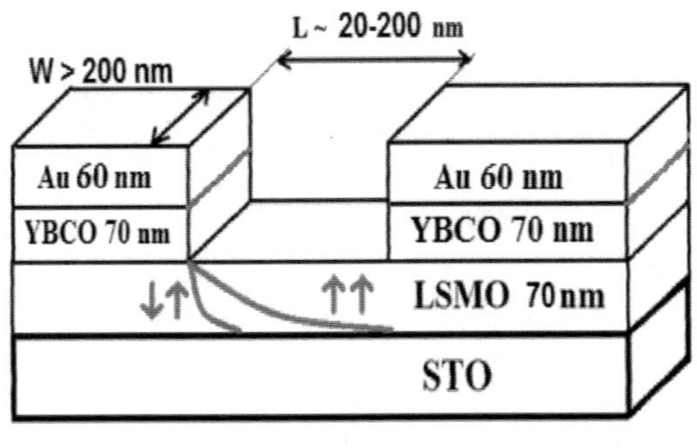

Figure 1a. Drawing of a YBCO/LSMO/YBCO nanostrip with lateral geometry. The length L of LSMO between the YBCO thin films was set up by perpendicular of parallel incident focused gallium ions (35 keV, 1.5 pA).

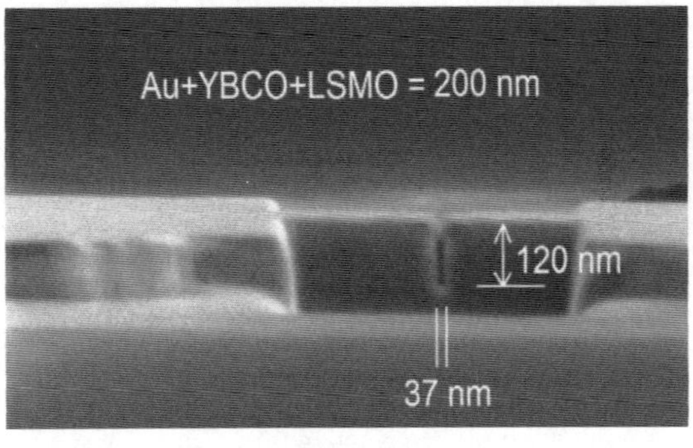

Figure 1b. SEM image of a YBCO/LSMO/YBCO junction of lateral geometry with LSMO length $L \approx 37$ nm and width $W \approx 0.5$ µm. The total thickness of the films is 200 nm, the slot depth, about 120 nm.

2. Experimental

The LSMO/YBCO bilayers were prepared by *in situ* pulsed laser deposition (PLD) on $SrTiO_3$ single crystal substrates. A KrF excimer laser operating at 248 nm, with a pulse width of 20 ns, repetition rate of 10 Hz and energy density of 6 J/cm^2 (spot size on the target ≈ 2 mm^2) was used to grow the LSMO and YBCO films. The substrate holder temperature of 850 °C and oxygen pressure of 53 Pa were set during the deposition. After the deposition, the films were cooled down at a rate of 20 °C/min in O_2 (4×10^4 Pa). The growth rate of the LSMO, as well as of the YBCO, was about 6.5 nm/min. The thickness of LSMO and YBCO layers was in the range from 30 to 70 nm. The bilayers were *ex-situ* protected by a 60-nm thick gold layer and the temperature dependence of the bilayer properties was inspected by a contactless inductive method. The patterning of the LSMO/YBCO bilayer was carried out by photolithography and argon ion beam etching (350 eV, $j \approx 0.1$ mA/cm^2) with cooling of the substrate to about –25°C. The cross section of the lateral geometry is shown in figure 1a. Figure 1b is the SEM image of the realized LSMO/YBCO nanostrips with a subsequent partial or total removal of the YBCO film in a slot of length $L < 200$ nm, using perpendicular or parallel (to the films plane) application of Ga^+ FIB patterning [7]. The properties of the SFS lateral structures were investigated by dc measurements of the temperature dependence of resistivity, the critical current and the current-voltage characteristics.

3. Results and discussions

To ensure a homogeneous critical current density distribution in the cross-section of the superconducting SF thin film nanostrip, the width W of the nanostrips was realized (by FIB patterning) to be smaller than the Pearl's penetration depth $\lambda_\perp = 2\lambda^2/d \approx 700$ nm, where $\lambda \approx 160$ nm is the London penetration depth of YBCO in a-b plane and $d \approx 70$ nm is the YBCO film thickness. The 5 μm distance between the potential ends of the sample consisted of a 3.5 μm long and 5 μm wide microstrip and a 1.5 μm long nanostrip of width W from 0.5 to 0.3 μm, localized in the center of the microstrip. The SF bilayer nanostrip critical temperature was $T_{Cn} \approx 89$ K and the difference in the critical temperatures of the two strips was below 50 mK. The critical current density of the prepared LSMO/YBCO nanostrips, extrapolated to zero temperature $T = 0$, was $J_C(0) \approx 10^7$ A/cm^2 and was only one order of magnitude lower than the theoretical Ginzburg-Landau depairing critical current density $J_C^{GL}(0) \approx 3\times10^8$ A/cm^2 [8], indicating the high quality of the bilayer structure prepared. The most delicate task of the lateral SFS structure preparation is the creation of a very narrow slot by removing

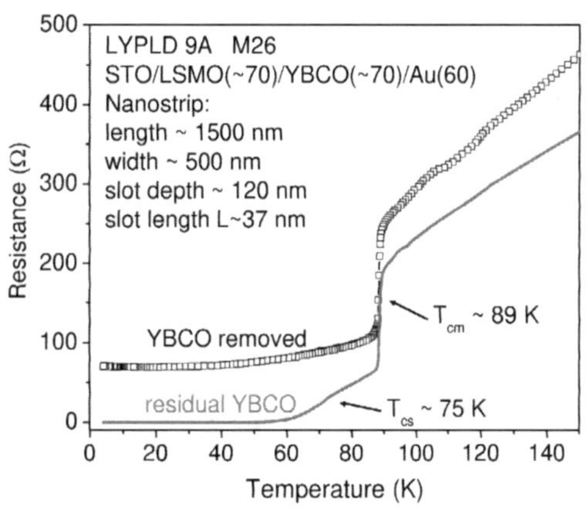

Figure 2a. Temperature dependence of the sample resistance with residual YBCO in the slot (a) and after removing the YBCO from the slot. LSMO resistivity $\rho \approx 10$ mΩ cm (b).

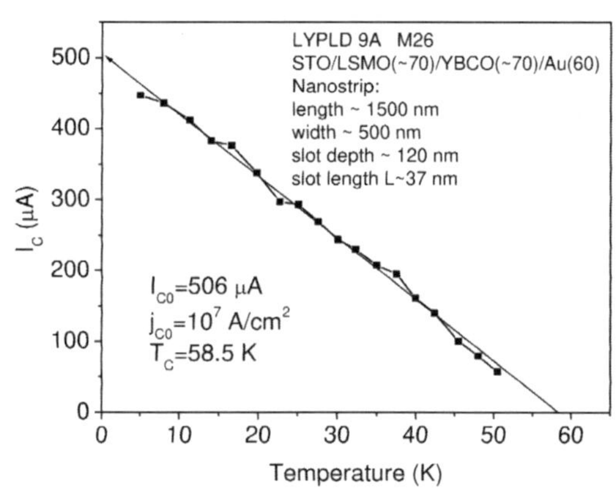

Figure 2b. Temperature dependence of the sample critical current characterizing the properties of a residual YBCO nanostrip in the slot. Critical current density $j_C(0) \approx 10^7$ A/cm^2.

the films (Au,YBCO) above the LSMO, without damaging the LSMO (Ga^+ implantation). Separate experiments, with 28-nm thick LSMO films, showed an enhancement of the initial LSMO resistivity (150 μΩ cm) by more than one order of magnitude with a Ga^+ fluence of 3×10^{13} ions/cm^2 [9]. Figure 2a illustrates the characteristic temperature dependence of the SFS lateral nanojunctions resistance with a residual YBCO in the slot of length ≈ 40 nm. The LSMO/YBCO bilayer microstrip and nanostrip phase transition occurred at $T \approx 89$ K, with the phase transition width being smaller than 1 K. The following cooling down showed a gradual decrease of the sample resistance; the second phase transition of the very thin residual YBCO film in the slot appeared at $T \approx 75$ K with a phase transition width of more than 10 K (due to Ga^+ implantation). The linear temperature dependence of the critical current (figure 2b) is characteristic for high-Tc superconductor thin films [9]. After removing the residual YBCO film, the $R(T)$ was constant with a resistance of about 70 Ω, corresponding to a strongly degraded LSMO with resistivity $\rho \approx 10$ mΩ cm, corresponding to a fluence of about 5×10^{13} ions/cm^2 during the slot patterning (30 keV, 1.5 pA). We assume that this is the main reason why the LRTC was not observed in the lateral nanojunctions.

4. Conclusions

In the work we applied a focused Ga^+ ion beam for the preparation of SF bilayer nanostrips and lateral SFS nanojunctions to investigate the possible presence of a long-range triplet component in the half-metal LSMO. The LSMO/YBCO bilayers were prepared by in-situ PLD deposition on single crystal $SrTiO_3$, typically with thicknesses from 30 to 70 nm. To supress the Ga^+ implantation of the nanostrips, the LSMO/YBCO bilayers were protected by a 60 nm thick Au film. The critical current densities of the bilayer nanostrips ($W \geq 0.4$ μm) were only one order of magnitude lower than the theoretical Ginzburg-Landau value of CP depairing. The critical temperature T_C of the bilayers was only slightly smaller than the T_C of the YBCO single layer. Nevertheless, to form a lateral SFS junction it is necessary to remove the thin films covering the LSMO without degrading the LSMO film. The results showed that 30 keV Ga^+ ions (in perpendicular or parallel application) strongly enhanced the LSMO resistivity due to Ga^+ implantation and, in consequence, reduced the carriers mean-free-path and, thus, the coherence length ξ_F of the half-metal. Another question to be solved was how the implantation influences the magnetic properties of the LSMO thin films. To solve these problems, it is necessary to apply a sufficiently lower ion energy, or other more convenient patterning procedures. Although the calculations in [6] were performed for ballistic lateral half-metal/superconductor junctions, the authors [6] expect that the enhanced triplet proximity effect in lateral geometry also exists in the presence of a disordered half-metal film. A detailed study of the disorder effects in lateral SFS junctions will be the aim of future studies.

Acknowledgement

This work was supported by the Slovak Research and Development Agency under the contract No. APVV-0494-11. We acknowledge the grant agency VEGA for financial support to projects Nos. 2/0173/13, 2/0120/14. This publication is the result of the project implementation: CENTE II, ITMS code 26240120019, supported by the Research & Development Operational Programme funded by the ERDF.

References

[1] Bergeret F, Volkov A and Efetov K 2005 *Rev. Mod. Phys.* **77** 1321
[2] Singh A, Voltan S, Lahabi K and Aarts J 2015 *Phys. Rev.* X **5** 021019
[3] Eschrig M 2011 *Physics Today* **64** 43
[4] Kompaniiets M, Dobrovolskiy O V, Neetzel C, Porrati F, Brötz J, Ensinger W and Huth M 2014 *Appl. Phys. Lett.* **104** 052603
[5] Kawabata S, Asano Y, Tanaka Y and Golubov A A 2013 *J. Phys. Soc. Japan* **82** 124702
[6] Kopferschnidt N and Brouwer P W 2011 *Phys. Rev.* B **83** 014512, 2009 *ibid* **80** 214537
[7] Strbik V, Benacka S, Smatko V, Gazi S, Chromik S, Mateev E, Blagoev B and Nurgaliev T 2012 *J. Phys.: Conf. Series* **356** 012021
[8] Benacka S, Strbik V, Chromik S, Adam R, Darula M and Gazi S 1998 *Fizika Nizkikh Temperatur* **24** 621 (in Russian)
[9] Delin K A and Kleinsasser A W 1996 *Supercond. Sci. and Technol.* **9** 227

Fabrication of hybrid thin film structures from HTS and CMR materials

M Sojková[1,4], V Štrbík[1], T Nurgaliev[2], Š Chromik[1], E Dobročka[1], M Španková[1], B Blagoev[3] and N Gál[1]

[1]Institute of Electrical Engineering, Slovak Academy of Sciences, 9 Dúbravská cesta , 841 04 Bratislava, Slovak Republic
[2]Emil Djakov Institute of Electronics, Bulgarian Academy of Sciences, 72 Tzarigradsko Chaussee, 1784 Sofia, Bulgaria
[3]Georgi Nadjakov Institute of Solid State Physics, Bulgarian Academy of Sciences, 72 Tzarigradsko Chaussee, 1784 Sofia, Bulgaria

E-mail: michaela.sojkova@savba.sk

Abstract. We present the preparation of bilayers from high-temperature superconductors (HTS) and half-metallic ferromagnetic (FM) manganite with a colossal magnetoresistance (CMR). We used $YBa_2Cu_3O_{7-x}$ (YBCO) and $Tl_2Ba_2CaCu_2O_8$ (TBCCO) thin films as a HTS material and $La_{0.67}Sr_{0.33}MnO_3$ (LSMO) film as a CMR material. In the case of YBCO/LSMO, we prepared FM/HTS heterostructure for studying the spin-polarized current injection effect on the electrical properties of the YBCO strip in dc or low-frequency regimes and on the microwave characteristics of the strip. For the first time, we report the preparation of a TBCCO/LSMO bilayer. In some applications, the TBCCO offers better parameters (higher working temperature, lower surface resistance, lower $1/f$ noise) than YBCO.

1. Introduction

Superconducting spintronic devices use the spin-dependent properties of superconductors in superconductor (SC)/ferromagnet (FM) heterostructures. The SC/FM heterostructures are of considerable practical importance. Layered heterostructures could be used for tunable magnetic microwave devices, for dynamic modification of SC critical parameters and for spin-polarized injection [1-3]. The perovskite high T_C superconductors (HTS) and the doped rare-earth manganese perovskite oxides $L_{1-x}A_xMnO_3$ (L = La, Pr..., A = Ca, Sr, Pb...) are promising candidates for preparing such heterostructures. These two perovskites are also interesting because of their ability to form high-quality epitaxial multilayers with each other and they can grow on conventional monocrystalline substrates, such as $SrTiO_3$ (STO), $LaAlO_3$ (LAO), $La_{0.26}Sr_{0.76}Al_{0.61}Ta_{0.37}O_3$ (LSAT), MgO and sapphire [4-7].

In this work, we describe the preparation of two types of HTS/FM bilayers, namely $YBa_2Cu_3O_{7-x}$ (YBCO)/$La_{0.67}Sr_{0.33}MnO_3$ (LSMO) and $Tl_2Ba_2CaCu_2O_{8-y}$ (TBCCO)/LSMO. A standard photolithography process, wet etching and lift-off techniques were used to prepare the YBCO/LSMO heterostructures. The basic electrical characterization of the layers and structures are presented. The

[4] To whom any correspondence should be addressed.

motivation to use TBCCO layer is a higher critical temperature T_C (a higher operating temperature), a lower surface resistance R_s at microwave frequencies and, at a given temperature, lower $1/f$ noise caused by flux motion compared to the YBCO film [8].

2. Experimental

The YBCO and LSMO layers were prepared by laser pulsed deposition (PLD) by the same procedure. A KrF excimer laser operating at 248 nm with a pulse width of 20 ns, repetition rate of 10 Hz and energy density of 6 J/cm^2 (spot size on the target ≈ 2 mm^2) was used to grow the YBCO (LSMO) films. A temperature of the substrate holder of 850 °C and an oxygen pressure of 53 Pa were set during the deposition. After the deposition, the films were cooled down at a rate of 20 °C/min in O_2 (4×10^4 Pa). The thickness of each LSMO and YBCO film on the LAO or MgO substrates was 50 nm. The growth rate of the prepared by laser ablation YBCO and LSMO was about 6.5 nm/min.

The LSMO films were deposited also using an on-axis dc commercial magnetron sputtering system (Lesker, TORUS 2C) from a stoichiometric ceramic target onto one-side polished MgO (0 0 1). The deposition was performed in an Ar (80%) + O_2 (20%) atmosphere at a total pressure of 30 Pa. During the deposition, the substrate was heated to 800 °C. The thickness and the growth rate of the LSMO films were about 500 nm and 8 nm/min, respectively. In order to increase the oxygen content, some LSMO films were subsequently annealed *in-situ* in O_2 (10^4 Pa) at 800 °C for an hour. After annealing, the temperature was lowered down to room temperature at a rate of 4 K/min [9].

The $Tl_2Ba_2CaCu_2O_{8-y}$ thin films were prepared by a two-step process. In the first step, 300-nm thick precursor films with a nominal composition of Ba:Ca:Cu = 2:2:3 were deposited by RF magnetron sputtering from a single target of the same stoichiometry. The films were sputtered at a total pressure of 10 Pa in an Ar atmosphere at room temperature with a deposition rate of 0.3 nm/s corresponding to an RF power of about 100 W. The thallination of the precursor films (the second step) was at annealing temperature of 850 – 860 ºC for 45 minutes in oxygen flow atmosphere [7].

The crystallographic structure and the growth orientation of the obtained films were examined by X-ray diffraction (XRD) using a BRUKER AXS D8 DISCOVER diffractometer with rotating Cu anode (CuK_α). Scanning electron microscopy (SEM) was performed using FEI Inspect F50. The basic electrical characterization of the bilayers and structures were performed by the dc 4-point and a low-frequency inductive (contactless) methods.

The sketch of the HTS/FM heterostructure for studying the spin-polarized current injection effect from FM into HTS is shown in figure 1. The parallel FM strips are connected by a sinusoidal (meander) strip of HTS. The as-prepared ferromagnetic LSMO thin films were patterned using standard photolithography processes and etched by a solution of potassium iodide, hydrochloric and ascorbic acid [7].

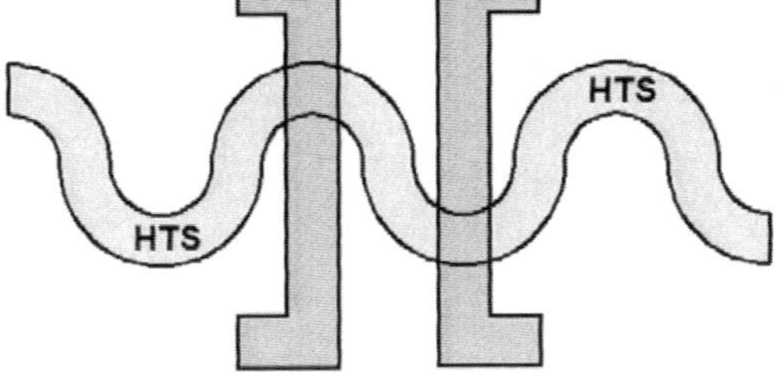

Figure 1. Sketch of a HTS/FM heterostructure.

Subsequently, a superconducting YBCO thin film was deposited on top of the patterned LSMO using PLD. Then the standard photolithography processes and wet etching (0.1 % solution of nitric acid) were used to pattern the YBCO meander. Several bilayer samples consisting of a HTS YBCO strip and FM LSMO electrodes were prepared to study the spin-polarized current injection effect (from the FM LSMO) on the electrical and microwave properties of the YBCO strip. The strip was prepared in the form of a meander to increase its geometrical length and to decrease its resonance frequency in measurements in the regime of microstrip resonator. In this configuration, the sample can be used to study the influence of the current injection effect on the microwave surface impedance in the YBCO structure. The sample was designed to be also suitable for investigation of the effect in low frequency or dc regimes. In this case, injection of the spin-oriented current from the LSMO into the YBCO could affect the critical temperature of the YBCO strip and the critical current density. The normal resistance of the YBCO strip in vicinity of T_C (where a pseudogap exists) could be expected to be somewhat affected by the current injection process as well.

3. Results and discussions

The high-quality LSMO thin films of thickness about 50 nm were deposited on LAO or MgO single crystal substrates by PLD. The crystal structure of the LSMO films grown on the MgO (0 0 1) substrate was investigated by X-ray diffraction techniques θ–2θ scans, ω-scans and φ-scans. The θ–2θ scan of the LSMO films indicated a preferred (0 0 *l*) oriented film growth. The in-plane growth properties were followed using φ-scans of the LSMO (4 0 2) and MgO (4 0 2) planes. We observed four sharp maxima indicating the four-fold symmetry of the LSMO lattice. Our XRD analyses confirmed that the LSMO grown on the MgO substrate fulfills the epitaxial relation LSMO $(0\ 0\ 1)_c[1\ 0\ 0]_c$//MgO (0 0 1) [1 0 0] [10,11].

The resistivity vs. temperature (ρ–T) dependencies of the LSMO film and the LSMO strip are shown in figure 2. The typical feature of the dependence is a broad resistivity peak with a resistivity maximum at a temperature T_{MI}, which corresponds to the metal-insulator transition temperature and is above 400 K. This value is about 40 K higher than the bulk value of the LSMO target ($T_{MIbulk} = 374$ K) [12]. The resistivity of the film is in this case a little higher ($\rho_{max} \approx 25$ mΩ cm) than previously presented [11], but it remains the same after the wet etching as well (figure 2).

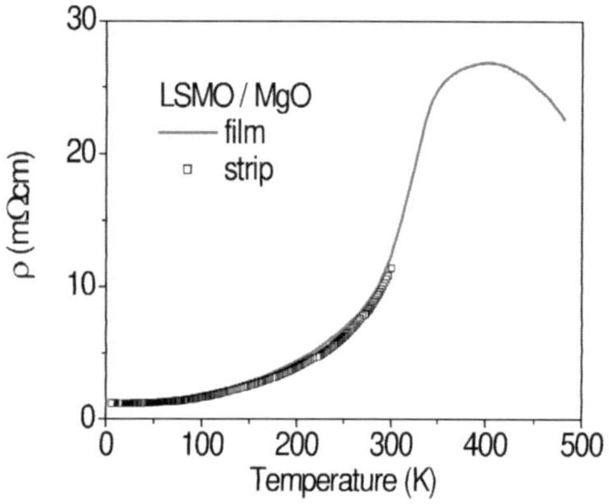

Figure 2. ρ–T dependencies of LSMO films on a MgO substrate before (full line) and after patterning (open squares).

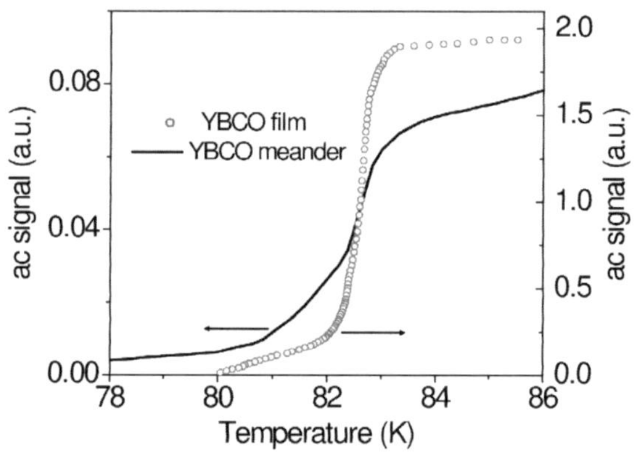

Figure 3. Superconducting transitions of YBCO film (open circles) and patterned YBCO meander (full line) deposited on MgO with patterned LSMO.

Figure 4. Realized HTS/FM heterostructure.

The LSMO prepared films exhibited also a high T_{Curie} (Curie temperature) above 400 K [11]; at room temperature about 76 % and at 77 K (expected working temperature of HTS/FM heterostructures) even 97-98 % spin polarization was reached [11,12]. So, the LSMO films are suitable for spin-injection into the HTS film in the HTS/FM heterostructures.

The transition to a superconducting state of the deposited YBCO films exhibiting standard properties is shown in figure 3. The midpoint of the transition, measured by a contactless method, is the same for the film and the patterned YBCO.

The onset of this transition corresponds to the zero resistance critical temperature as measured by the 4-point dc method. The realized HTS/FM heterostructure for the spin injection is shown in figure 4. The preliminary results indicate that the resistance of the HTS/FM interface is low ($R < 10^{-4}\ \Omega$) despite the *ex-situ* process of YBCO deposition. Investigations of the spin-polarized current injection effect on the electrical and microwave characteristics of the YBCO strip are in progress.

Usually, LSMO and YBCO materials are used in HTS/FM heterostructures or, alternatively, a $La_{0.67}Ca_{0.33}MnO_3$ as FM material. Now, we present the application of a more exotic HTS material, namely, a Tl-based cuprate superconductor. The Tl-based HTS superconductors of several phases exhibit a higher T_C (i.e., a higher operating temperature), a lower surface resistance R_s at microwave frequencies and, at a given temperature, lower $1/f$ noise caused by flux motion compared to the YBCO film. The XRD analyses (θ-2θ scan) revealed that the c-axis-oriented grown Tl-based superconductor was single-phase with peaks belonging to the Tl-2212 phase ($Tl_2Ba_2CaCu_2O_8$-TBCCO) (figure 5). Besides the Tl-2212 phase, LSMO and MgO substrate peaks could be seen. We note that the LSMO in this case was prepared by dc magnetron sputtering [9], so that T_{MI} of only about 300 K was achieved. The resistance vs temperature (R-T) dependence of the TBCCO/LSMO bilayer is shown in figure 6 (full line). This dependence is compared with the R-T of the LSMO film (open circles) measured before the growth of the TBCCO layer in the two-step technological process. The resulting R-T dependence can be interpreted in terms of the relation $R_\Sigma^{-1}(T) = R_{HTS}^{-1}(T) + R_{FM}^{-1}(T)$, where R_{HTS} is the HTS resistance and R_{FM}, the resistance of the ferromagnetic films [13]. The zero resistance critical temperature of the TBCCO layer on LSMO reached about 95 K; therefore, improving the TBCCO layer is necessary.

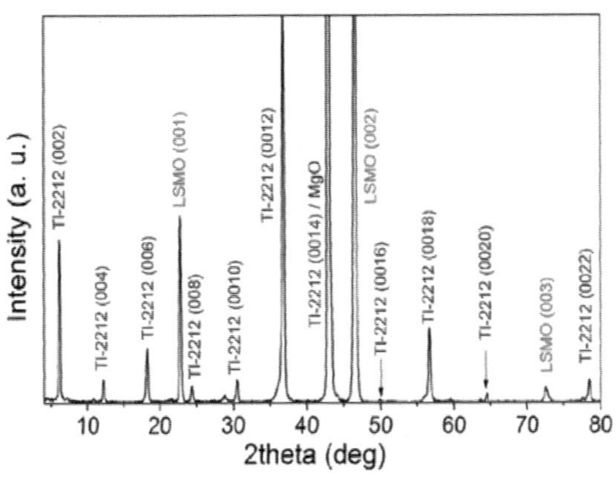

Figure 5. θ-2θ scan of a TBCCO/LSMO bilayer on a MgO substrate.

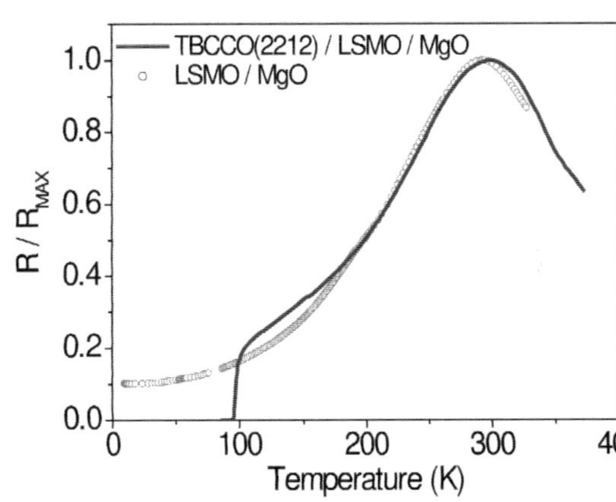

Figure 6. Dependences of a TBCCO/LSMO bilayer (full line) and a LSMO film before deposition of the TBCCO layer (open circles).

4. Conclusions

We prepared two types of bilayers from ferromagnetic manganite and high-temperature superconductors. In the first case, we present the bilayer YBCO/LSMO combination, where LSMO exhibits enhanced parameters – Curie temperature T_{Curie}, as well as temperature of metal-insulator transition T_{MI}, both above 400 K, and, thus, very high spin polarization (about 97-98 %) at 77 K. The wet-etching processes used (developed separately for each layer) allow one to pattern the LSMO and YBCO layers without degradation of their properties. We were able to fabricate YBCO/LSMO heterostructures to investigate the spin-polarized current injection effect on the electrical properties of the YBCO strip in the dc and low frequency regimes and on the microwave characteristics of the strip.

In the latter case, we prepared the bilayer with TBCCO/LSMO composition. We demonstrated the possibility to prepare also another HTS material with a complicated structure and technology of fabrication (the two-step process) on the LSMO film. The use of TBCCO films is convenient for HTS/FM bilayer applications where the TBCCO has better parameters than the YBCO.

Acknowledgement
This work was supported by the Slovak Research and Development Agency, APVV-0494-11, APVV LPP-0078-07, Slovak Grant Agency for Science, VEGA 2/0120/14, VEGA 2/0173/13, the project CENTE II, R&D Operational Program funded by the ERDF, ITMS code 26240120019.

References
[1] Terakago M, Mine S, Sakatani T, Hontsu S, Nishikawa H, Nakamori M, Tabata H and Kawai T 2001 *Supercond. Sci. Technol.* **14** 1140
[2] Nurgaliev T, Miteva S, Jenkins A and Dew-Hughes D 2001 *IEEE Trans. Appl. Supercond.* **11** 446
[3] Nurgaliev T, Miteva S, Jenkins A and Dew-Hughes D 2003 *IEEE Trans. Microwave Theory and Technique* **51** 33
[4] Nurgaliev T, Blagoev B, Mateev E, Štrbík V, Beňačka Š, Šmatko V, Gaži Š and Chromik Š 2012 *J. Phys.: Conf. Series* **356** 012020
[5] Blagoev B, Gostev I, Nurgaliev T, Štrbík V, Bineva I, Uspenskaya L, Mateev E, Neshkov L, Dobročka E and Chromik Š 2014 *J. Phys.: Conf. Series* **514** 012041
[6] Štrbík V, Beňačka Š, Šmatko V, Gaži Š, Chromik Š, Mateev E, Blagoev B and Nurgaliev T 2012 *J. Phys.: Conf. Series* **356** 012021
[7] Sojková M, Štrbík V, Chromik Š, Liday J, Vogrinčič P, Dobročka E, Španková M, Talacko M and Gaži Š 2015 *Vacuum* **119** 250
[8] Bramley A P, Ơconnor J D and Grovenor C R M 1999 *Supercond. Sci. Technol.* **12** R 57
[9] Španková M, Chromik Š, Vávra I, Sedláčková K, Lobotka P, Lucas S and Stanček S 2007 *Appl. Surf. Sci.* **253** 7599
[10] Španková M, Rosová A, Dobročka E, Chromik Š, Vávra I, Štrbík V, Machajdík D, Kobzev A and Sojková M 2015 *Thin Solid Films* **583** 19
[11] Štrbík V, Reiffers M, Dobročka E, Šoltýs J, Španková M and Chromik Š 2014 *Appl. Surf. Sci.* **312** 212
[12] Chromik Š, Štrbík V, Dobročka E, Laurenčíková A, Reiffers M, Liday J and Španková M 2013 *Appl. Surf. Sci.* **269** 98
[13] Nurgaliev T, Blagoev B, Štrbík V, Chromik Š and Sojková M 2016 *Proc. VEIT 2015, this volume of J. Phys.: Conf. Series* (21-25 Sept. 2015 Sozopol Bulgaria)

Properties of ns-laser processed polydimethylsiloxane (PDMS)

P A Atanasov[1,4], N E Stankova[1], N N Nedyalkov[1], T R Stoyanchov[1], Ru G Nikov[1], N Fukata[2], J W Gerlach[3], D Hirsch[3] and B Rauschenbach[3]

[1]Emil Djakov Institute of Electronics, Bulgarian Academy of Sciences, 72 Tsarigradsko Chaussee, 1784 Sofia, Bulgaria
[2]International Center for Materials for Nano-Architectonics (MANA), National Institute for Materials Science (NIMS), 1-1Namiki, Tsukuba 305-0044, Japan
[3]Leibniz Institute of Surface Modification (IOM), Permoserstrasse 15 D-04318 Leipzig, Germany

E-mail: paatanas@ie.bas.bg

Abstract. The medical-grade polydimethylsiloxane (PDMS) elastomer is a widely used biomaterial in medicine and for preparation of high-tech devices because of its remarkable properties. In this work, we present the experimental results on drilling holes on the PDMS surface by using ultraviolet, visible or near-infrared ns-laser pulses and on studying the changes of the chemical composition and structure inside the processed areas. The material in the zone of the holes is studied by XRD, μ-Raman analyses and 3D laser microscopy in order to obtain information on the influence of different processing laser parameters, as wavelength, fluence and number of consecutive pulses on the material transformation and its modification.

1. Introduction

Polydimethilsiloxane (PDMS) belongs to the group of polymeric materials called silicones possessing one common characteristic – the polymer backbone is composed by an alternate succession of Si and O atoms joined together via strong, covalent inter-atomic bonds. The Si atoms are coupled to two adjacent O atoms and two organic radicals. PDMS is among the most popular technical polymeric materials due to its advantageous properties: mechanical flexibility and stability; non-toxicity; permeability to gasses; chemical and biological inertness; high dielectric constant and breakdown field; optical transparency in the ultraviolet (UV) ÷ visible (VIS) regions of the spectra; high biocompatibility and biostability; very hydrophobic; simple and inexpensive fabrication.

PDMS has been widely used in the fabrication of various micro-electro-mechanical systems (MEMS) [1-6] and flexible microelectrode arrays for monitoring and/or stimulation of neural activity [8-10]. The hydrophilic, hydrophobic recovery and surface energy of the material have been studied [11]. Furthermore, the cytotoxicity and antibacterial activity of the material has also been assessed; the results obtained revealed that the PDMS surface modifications performed do not affect the material's biocompatibility, but decrease their hydrophobic character and bacterial adhesion. Thus, the modification or structure formation on the surface of the PDMS is of great importance and growing research interest.

[4] To whom any correspondence should be addressed.

The use of laser radiation offers a useful approach to processing and selective surface functionalization of the PDMS material. In particular, ns-laser pulses (e.g., KrF, ArF and F_2) have been applied to the fabrication of optical elements [12], modification of optical properties [13] and surface processing or functionalization [2, 8, 10]. KrF excimer laser processing of flexible silicone rubber has been used to photo-decompose its surface or to change the surface relief [2, 8, 14, 15]. Information on the changes of the PDMS structures fabricated by ns-laser pulses has been typically obtained by μ-Raman measurements [8]. Laser irradiation can obviously induce breaking of the chemical bonds, but it can also be used in the ablation mode to micro- or nano-structuring the polymer surface, thus enhancing the adhesion of a metal [14, 15].

Nowadays, UV and VIS fs- or ns-laser irradiation of the PDMS-elastomer under ambient conditions has been proven to be an easy and powerful method for surface micromachining in order to fabricate tracks and structures with surface activation without altering its bulk properties [16-18]. Samples of the PDMS elastomer processed by UV, VIS and NIR fs- and ns-laser pulses have been investigated [14-18]. A limited number of laser processed areas has been viewed by SEM and optical microscopy. Selective Pt or Ni metallization of the laser-processed traces have been produced successfully via electro-less plating. The metallization process is not sensitive to the time interval after the laser treatment. The DC resistance has been measured to be as low as 0.5 Ω/mm [17].The present paper reports the results on the changes in the PDMS elastomer properties inside the holes produced as a result of multi-pulse ns-laser ablation at the wavelengths of 266, 355, 532 and 1064 nm. The laser fluence and number of consecutive pulses were varied. The laser-processed areas were studied by laser microscopy, XRD and μ-Raman spectroscopy. The as-processed areas were successfully metallized by Ni via electro-less plating.

2. Experimental

180 μm thick medical grade PDMS-elastomer sheets (MED 4860) were irradiated by the fundamental harmonic (NIR) (λ = 1064 nm), and the 2nd (λ = 532 nm), 3rd (λ = 355 nm) and 4th (λ = 266 nm) harmonics of a Q-switched Nd:YAG laser (pulse duration of 15 ns and repetition rate of 10 Hz) in air at ambient temperature. The samples were mounted on a microcontroller-driven, stepper-motor operated X–Y table. The laser beam was focused by a lens with a focal length of 22 cm. The micro-holes were produced without perforating the foil by up to 110 consecutive laser pulses. The laser fluence was varied at the different wavelengths. The corresponding data of the laser processing parameters are presented in table 1.

Table 1. Summary of the laser processing parameters.

Wavelength (nm)	266	355	532	1064
Laser Fluence (J cm^{-2})	0.1 ÷ 1.0	1.0 ÷ 2.5	3.5 ÷ 5.0	8.0 ÷ 13.0

The areas of the holes produced were viewed and their physical dimensions measured by a VK-9700 K Color 3D Laser Microscope (KEYENCE, Japan). The crystalline properties were studied by an XRD – ULTIMA IV X-Ray Diffractometer (RIGAKU, Japan), while the chemical modifications were analyzed at different points of the holes by a μ-Raman Spectrometer – RMS-310 (Photon Design, Japan) with laser excitation at 532 nm. The spectrometer has a resolution of 0.2 cm^{-1} and image dimension <1 μm^2. Additionally, continuous trenches with dimension of 12×3 mm^2 and different depths were prepared at various laser fluences in order to investigate the laser-processed area by 1° grazing incidence XRD.

3. Results and discussions

3.1. XRD analyses

The results from the XRD analyses are presented in figure 1. Figure 1 a) depicts the XRD spectrum of the non-processed PDMS. The spectrum is similar to that reported in reference [11].

As it is illustrated in figure 1 b), all diffraction patterns are similar, presenting two diffraction halos: a first and larger one located at around 12.3° and a second, smaller and broader one, at 22.2°. These results indicate that the microstructure of the samples is amorphous and that the ns-laser processing does not change the bulk properties of the base material. Additionally, the intensity of the material treated by UV (λ = 355 nm) ns-laser pulses has about twice as high intensity compared to the other wavelengths applied, which is probably related to the better ordering of the material processed.

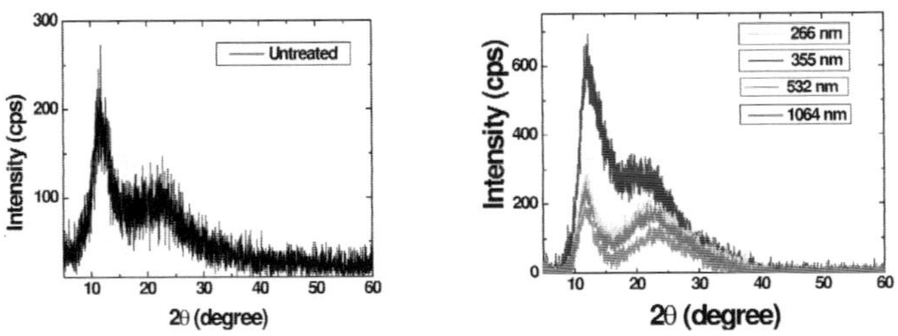

Figure 1. XRD spectra of the unprocessed – a) and ns-laser processed area of PDMS – b).

3.2. Color microscope views of the holes

The laser produced micro holes were viewed by a 3D color laser microscope – figure 2 a) – c).

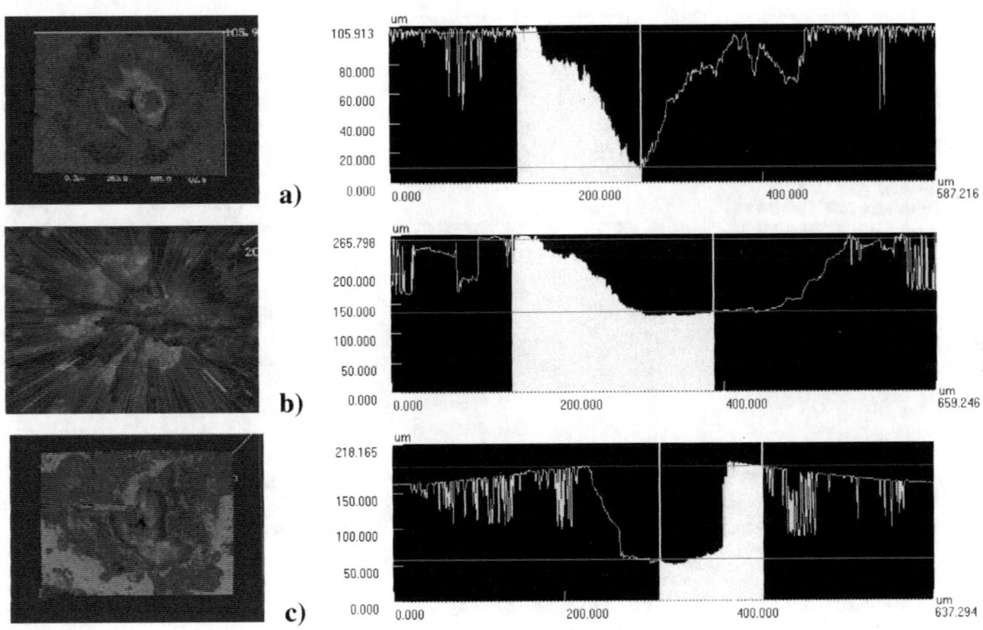

Figure 2. a), b) and c) show photographs and profiles of some of the holes produced by a ns-laser: a) 266 nm, 110 pulses; b) 532 nm, 33 pulses; c) 1064 nm, 33 pulses.

Due to the multimode structure of the laser beam, the holes' profiles are not uniform. This leads to the production of holes with an inhomogeneous structure, especially in the cases of a higher number of consecutive incident pulses. The peculiarities of the properties at different points of the areas processed resulted from the different energy density inside the laser beam, i.e. the existence of some "hot" spots. These differences were clearly expressed in the Raman studies. It is worth noting that when producing tracks by a ns-laser, the quality of the tracks was better due to the equalization of the processing after overlapping of a large number of the pulses (> 30) during the slow movement of the X-Y table.

However, a further effort was needed in order to improve the quality of the laser-beam profile. In this respect, a laser beam equalizer was introduced and the results will be the subject of further investigations.

3.3. Raman spectroscopy inside the holes

The μ-Raman spectra taken from the center or periphery of the holes are presented in figure 3 a) - d). Figure 3 a) – c) correspond, respectively, to the pictures given in figure 2 a) - c). In addition, figure 3 e) depicts the spectrum of the native PDMS given as a comparison. It clearly shows that the appearance of a crystalline Si peak is observed for all laser wavelengths applied. The intensity of the crystalline Si peak increases with the laser fluence and the number of consecutive pulses. The spectra in figure 3 a) – d) illustrate a well-defined change of the chemical composition regardless of the laser wavelength used for processing. An extra peak at ~512 ÷ 520 cm^{-1}, assigned to crystalline silicon (c-Si), is observed after ns-laser processing. Simultaneously, the intensity of the Si-O-Si peak at 488 cm^{-1}, typical for the native PDMS, decreases. Probably, this is a consequence of the reduction of the Si-O-Si bonds by breaking of the Si-O bond and, as a result, Si crystallites are formed. Also, the intensities of the Si-O-Si symmetric stretching at 488 cm^{-1}, the Si-CH_3 symmetric rocking at 685 cm^{-1}, the Si-C symmetric stretching at 709 cm^{-1}, the CH_3 asymmetric rocking, the +Si-C asymmetric stretching at 787 cm^{-1}, and the CH_3 symmetric rocking at 859 cm^{-1} modes decrease, which could contribute to the c-Si formation as well.

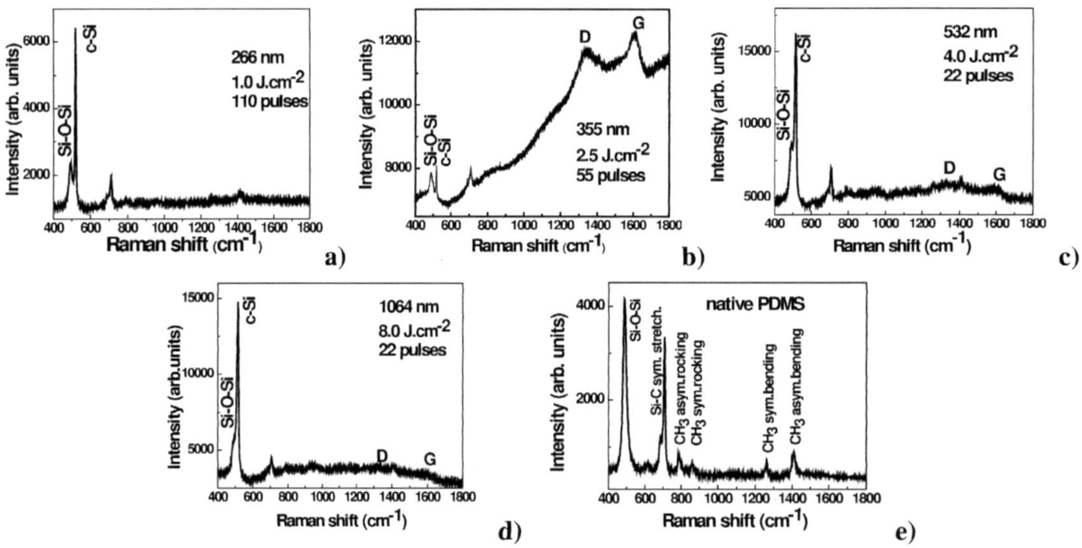

Figure 3. μ-Raman spectra of the holes as in Fig. 2: a) 266 nm – at a point inside the hole produced; b) 355 nm – at a point inside the hole produced; c) 532 nm – at the center of the hole; d) 1064 nm – at the center of the hole. e) the spectrum of the native (non-processed) PDMS.

Two wide peaks in the ranges of 1320 ÷ 1336 cm^{-1} and 1590 ÷ 1615 cm^{-1} are observed in the Raman spectra of some points of the areas treated with both UV lights, which are well expressed at number of pulses > 33. These two peaks are much weaker when the PDMS is irradiated by VIS and NIR wavelengths. In some cases, the high signal of these peaks' tails hinders the expression of the other Raman features. When the peak of crystalline Si is well expressed, these two peaks are almost hidden (figure 3 a)).

As it is known, the amorphous carbon contains a mixture of sp^3, sp^2 and sp^1 sites. The Raman spectrum taken by VIS excitation contains two peaks – D and G modes, which appear at ~ 1360 and 1560 cm^{-1} [19]. These peaks dominate, because the sp^2 sites have cross-sections much higher than that of the sp^3. However, in the case of a high number of sp^3 sites (which are related to the diamond phase) the D peak appears at 1332 cm^{-1}. It is worth noting that the position of the D peak could be extended in the range between 1300 and 1350 cm^{-1} if the phase is mixed with the hexagonal and cubic diamond [20]. Usually, this peak is broadened and lowered because of the phonon confinement at VIS excitation. It has also been shown that carbon materials, which contain simultaneously sp^2 and sp^3 phases, reveal a shift of the G band over the 1600 cm^{-1} [19, 21]. Hence, it could be concluded that the laser treated areas of PDMS, especially by both UV wavelengths, contain a significant fraction of sp^3 carbon bonds.

The micro-Raman analysis also showed that local chemical modifications (i.e. silicone decomposition) occurred after the laser processing of the PDMS, irrespective of the corresponding wavelengths used. This behavior is due to the rather significant influence of the laser fluence and/or the number of the pulses rather than that of the wavelength. The formation of inorganic products like Si and C is responsible for the chemical activation of the surface, which is expected to influence significantly the nucleation and the growing rate of the metal layer (Ni or Pt) during the following process of metallization.

6. Conclusions

Results on processing of a medical grade PDMS elastomer by UV, VIS and NIR ns-laser pulses are presented. A change of the chemical composition of the areas processed (holes) is observed and compared to the non-processed material. An XRD study of the laser-processed PDMS material at NIR, VIS and UV lights is conducted for the first time. The results obtained can be summarized as follows: i) no crystalline phases caused by laser treatment are observed in the XRD spectra; very broad "peaks" are related to the amorphous phases of the material; ii) for all wavelengths, the appearance of a crystalline Si peak is observed in the μ-Raman spectra; its intensity increases with the laser fluence and the number of pulses; iii) amorphous carbon containing sp^2 and sp^3 sites is formed during the laser activation of the surface; iv) the local chemical decomposition is a complex function of the laser fluence, the number of the pulses (incubation) and the wavelength; v) the quality of the laser beam is found to be very important for the processing and functionalization of the PDMS surface for further metallization.

Acknowledgements

The financial support is acknowledged of the Bulgarian National Science Fund under project T02/24 "New advanced method for processing nanocomposite materials for creation of microsystems for medical and high-tech applications" and of the German-Bulgarian bilateral project 01/1 "Ion beam and laser techniques for surface nanostructuring of different materials and application to high resolution analyses (SERS)".

References

[1] Editorial Overview 2009 Lasers in surface science *Current Opinion in Solid State and Materials Science* 1-3
[2] Laude L, Rowley A, Humayun M and Weiland J 2008 *US Patent* US 2008/10305320 A1
[3] McDonald J C, Duffy D C, Anderson J R, Chiu D T, Wu H, Schueller O J and Whitesides G M 2000 *Electrophoresis* **21**/1 27
[4] Makamba H, Kim J H, Lim K, Park N and Hahn J H 2003 *Electrophoresis* **24**/21 3607
[5] 2004 *Silicon Compounds: Silanes and Silicones* (Gelest, Inc. Morrisville PA) 560
[6] Hermanson G T, Mallia A K and Smith P K. 1992 *Immobilized Affinity Ligand Techniques* (Academic Press San Diego CA) 454
[7] Okoshi M, Li J and Herman P R 2005 *Opt. Lett.* **30**/20 2730
[8] Dicara C, Robert T, Kolev K, Dupas-Bruzek C and Laude L D 2003 *Proc. SPIE* **5147** 255
[9] Laude L D 1997 *US Patent* No 5599592
[10] Dupas-Bruzek C, Robbe O, Addad A, Turrell S and Derozier D 2009 *Appl. Surf. Sci.* **255**

8715

[11] Ferreira P, Carvalho A, Correia T R, Bernardo, Antunes P, Correia I J and Alves P 2013 *Sci. Technol. Adv. Mater.* **14** 055006

[12] Okoshi M, Li J and Herman P R 2005 *Opt. Lett.* **30** 2730

[13] Okoshi M, Sekine D, Inoue N and Yamashita T 2007 *Jpn. J. Appl. Phys. Part 2* **46** L356

[14] Laude L D 1997 *U.S. Patent* N 5.599.592

[15] Dini J W 1997 *Met. Finish.* **95** 10

[16] Atanasov P A, Nedyalkov N N, Valova E I, Georgieva Zh S, Armyanov S A, Kolev K N, Amoruso S, Wang X, Bruzzese R, Sawczak M and Śliwiński G. 2014 *J. Appl. Phys.* **116**/2 023104

[17] Stankova N E, Atanasov P A, Nedyalkov N N, Stoyanchov T R, Kolev K N, Valova E I, Georgieva J S, Armyanov St A, Amoruso S, Wang X, Bruzzese R, Grochowska K, Śliwiński G, Baert K, Hubin A, Delplancke M P and Dille J 2015 *Appl. Surf. Sci.* **336** 321

[18] Armyanov S, Stankova N E, Atanasov P A, Valova E, Kolev K, Georgieva J, Steenhaut O, Baert K and Hubin A 2015 *Nucl. Instr. Meth. Phys. Res.* B **360** 30

[19] Ferrari A C 2007 *Solid State Commun.* **143** 47

[20] Ona S, Nakamoto Y, Kagayama T, Shimizu K, Nishikawa Y, Murakami M, Kusakabe K, Watanuki T and Ohishi Y 2008 *J. Phys.: Conf. Series* **121** 062006

[21] Yadav V S, Sahu D K, Singh M and Kumar K 2009 *Proc. WCECS* **I** (San Francisco USA)

Thin nanocrystalline zirconia films prepared by pulsed laser deposition

A Og Dikovska[1,5], G B Atanasova[2], G V Avdeev[3] and V Y Strijkova[4]

[1]Emil Djakov Institute of Electronics, Bulgarian Academy of Sciences, 72 Tsarigradsko Chaussee, 1784 Sofia, Bulgaria
[2]Institute of General and Inorganic Chemistry, Bulgarian Academy of Sciences, Acad. G. Bonchev str., bl. 11, 1113 Sofia, Bulgaria
[3]R. Kaischew Institute of Physical Chemistry, Bulgarian Academy of Sciences, Acad. G. Bonchev str., bl.11, 1113 Sofia, Bulgaria
[4]Acad. J. Malinowski Institute of Optical Materials and Technologies, Bulgarian Academy of Sciences, Acad. G. Bontchev str., bl. 109,1113 Sofia, Bulgaria

E-mail: dikovska@ie.bas.bg

Abstract. In the present work, thin zirconia films were prepared by pulsed laser deposition at different substrate temperatures and oxygen partial pressures. The substrate temperature was varied from 400 °C to 600 °C, and the oxygen pressure, from 0.01 to 0.05 mbar. The effect was investigated of the substrate temperature and oxygen pressure on the formation of m-zirconia and t-zirconia phases.The formation of a cubic phase of ZrO_2 by using targets doped with 3 and 8 mol % content Y_2O_3 was also investigated. The variation in the optical properties was studied and discussed in relation with the zirconia films' microstructure.

1. Introduction

Zirconium oxide (ZrO_2) is a transition metal oxide offering several interesting technological applications due to its properties, such as a high melting point, a high refractive index, good thermal, chemical and mechanical stabilities, a wide band gap, and high dielectric constant and electrical resistivity. These include optical coatings, protective coatings and insulating layers, catalyst for alcohol synthesis, fuel cells, thermal barrier coatings and application in the field of biomedicine [1-4]. ZrO_2 has three well-defined polymorphs depending on the temperature conditions of preparation: monoclinic, tetragonal and cubic [5]. The phase transformations cause extensive cracking in the sintered ZrO_2 material; if this aspect is not controlled or suppressed, it would preclude the use of ZrO_2 in the field of advanced ceramics. On the one hand, the stability of zirconia could be appropriately controlled through carefully tailoring its chemical composition and grain size [5]. On the other hand, the high temperature phases can be retained at room temperature by doping with some suitable oxides, such as aluminium oxide, yttrium oxide, magnesium oxide, and calcium oxide [5].

ZrO_2 thin films have been prepared by several methods, such as chemical vapor deposition [6], electron beam evaporation [7], sputtering [8], pulsed laser deposition (PLD) [9-12], etc. Among these methods, the PLD stands out as a simple technique enabling one to prepare high-quality thin films

[5] To whom any correspondence should be addressed.

from a wide variety of materials and compounds. Mishra et al. [9] reported fabrication of zirconia thin films by PLD. The authors focused their attention on the phase evolution of zirconia produced in a wide range of substrate temperatures and oxygen pressures. Balakrishnan and co-workers [10] investigated the influence of the oxygen partial pressure on the microstructure, morphology, and optical properties of pulsed-laser-deposited ZrO_2 films. However, the process parameters for stable zirconia films deposition were not defined there.

In this work, thin ZrO_2 films were deposited by PLD using pure and Y_2O_3 doped zirconia targets. The influence was investigated of the substrate temperature and oxygen partial pressure on the structure, composition, morphology and optical properties of the films. The optimal process parameters for deposition of monophase zirconia films were determined.

2. Experimental

Thin ZrO_2 films were produced by PLD in a standard on-axis configuration. The third harmonic of a Nd:YAG laser (λ = 355 nm, pulse duration τ = 18 ns, and repetition rate of 10 Hz) with a laser fluence of 3 J cm^{-2} was used for ablation of a ceramic ZrO_2 target. The target was prepared by using commercially available zirconia powder (99.99 % purity) compressed at a pressure of 10 MPa using a uniaxial press and sintered at 1200 °C for 4 h. Both (100)-oriented Si and quartz substrates were used for thin-film depositions. The target-substrate distance was kept at 4.5 cm. The experiments were carried out at different substrate temperatures in the range of 400 - 600 °C. All experiments were performed in an oxygen atmosphere with the pressure being varied from 0.01 to 0.05 mbar. The duration of samples depositions was fixed at 60 or 90 min. A set of experiments were made by using Y_2O_3-doped ZrO_2 targets, namely, 3 and 8 mol% yttria stabilised zirconia (YSZ). The YSZ films were produced under conditions of growth optimized for deposition of pure ZrO_2 films.

The phase identification and phase evolution of the as-deposited films were studied by a Philips PW 1050 X-ray diffractometer (XRD) using $CuK_{\alpha 1}$ (0.15406 nm) radiation. Using the standard software of the XRD apparatus, the crystallites size, volume fraction and strain of both the monoclinic and tetragonal phases were calculated. The surface morphology of the films was examined by AC mode atomic force microscopy (AFM) (MFP-3D^{TM}, Oxford Instruments). The chemical composition and the oxidation state of the elements on the film surface were investigated by X-ray photoelectron spectroscopy (XPS). The measurements were carried out on an AXIS Supra electron-spectrometer (Kratos Analitycal Ltd.) using monochromatic AlK_{α} radiation with a photon energy of 1486.6 eV and a charge neutralization system. The binding energies (BE) were determined with an accuracy of $\pm$ 0.1 eV. The thickness of the films was measured using a Talystep profilometer. The optical properties of the films were determined by measuring the normal-incidence transmittance (T) of the films over the wavelength range 200 - 1200 nm using an Ocean Optics spectrophotometer.

3. Results and discussion

3.1. ZrO_2 films

The chemical composition of the ZrO_2 surface was investigated by XPS. The XPS spectra of ZrO_2 films deposited at a substrate temperature of 600 °C and under different oxygen pressures are presented in figure 1(a) and (b). Figure1(a) shows the doublet peaks centered at 182.1 and 184.5 eV, corresponding to $Zr3d_{5/2}$ and $Zr3d_{3/2}$, respectively. The binding energies of $Zr3d_{5/2}$ in ZrO_2 (oxidation state Zr^{4+}) were reported to be in the range 182.2-182.6 eV [13,14]. At a substrate temperature of 600 °C, the increase of the oxygen pressure during deposition did not change the position of the Zr3d peaks (figure 1(a)). In the O1s region (figure 1(b)), the low binding energy component was attributed to oxygen in zirconium oxide (Zr^{4+}). The values reported for the binding energy of this component are in the range 529.5-530.2 eV [13,14]. The high binding energy component reported in the range of 531.0-532.1 eV probably corresponds to oxygen in the form of a hydroxide or other adsorbed species [13,14]. As it was expected, the high binding energy component increased with the increase of the oxygen pressure during deposition (figure 1(b)). Similar behaviour was observed with the samples produced at 400 °C and 500 °C. At a fixed oxygen pressure, no significant changes in the chemical states of the films were observed as the substrate temperature was increased from 400 °C to 600 °C. All deposited films could be classified as stoichiometric ZrO_2. However, at the highest oxygen pressure used in this work, the estimated oxidation state of the films was in the range 2.10-2.29.

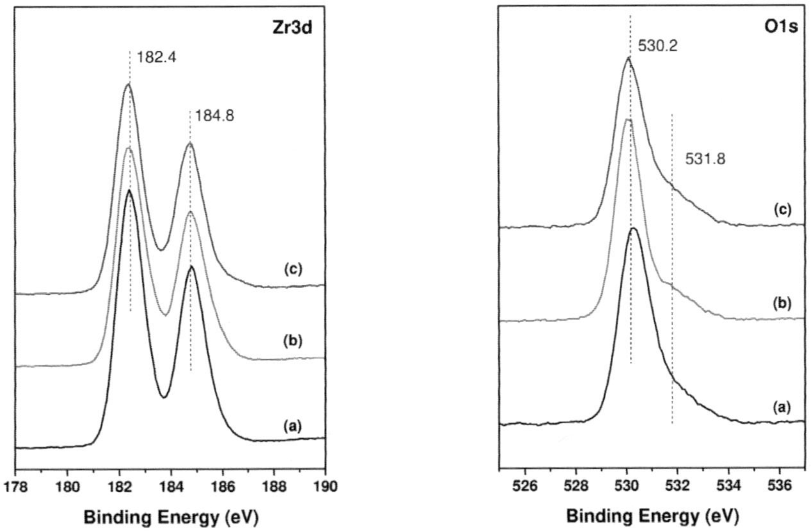

Figure 1. XPS spectra of thin ZrO_2 films deposited at 600 °C substrate temperature and (a) 0.01, (b) 0.03, and 0.05 mbar oxygen pressure.

The crystalline structure of the zirconia thin films deposited at different substrate temperatures and oxygen pressures was investigated by XRD. The XRD patterns were indexed using the Reference Code 98-009-4931 and 98-008-6692. All the films were polycrystalline in nature having both monoclinic and tetragonal phases. In general, the XRD study indicated that the tetragonal phase content increased with the increase of the substrate temperature at a fixed oxygen pressure. The XRD patterns of the ZrO_2 films deposited at an oxygen pressure of 0.05 mbar and different substrate temperatures are presented in figure 2. As it is seen, the intensity of the peaks corresponding to t-zirconia increased with the increase in the substrate temperature, which indicated that the formation of a tetragonal phase is favoured at higher substrate temperatures [9,11]. The XRD patterns of ZrO_2 films deposited at a substrate temperature of 600 °C and different oxygen pressures are presented in figure 3. The XRD results indicated a decrease in the intensity of the peak of m-zirconia with the increase in the oxygen pressure. At the oxygen pressure of 0.05 mbar, all the peaks corresponding to m-zirconia disappeared pointing to the fact that the higher oxygen partial pressure favours the formation of t-zirconia [9,11]. The volume fraction of the phases and the crystallites size were calculated from the intensity of the (11-1) and (011) reflections corresponding to the m- and t-zirconia phases, respectively. The volume fraction of the tetragonal phase increased with the increase in the substrate temperature and the oxygen pressure. The crystallites size for both m- and t-phases of zirconia was evaluated as being in the range of 10-20 nm. In summary, the results of the XRD analysis showed that a higher substrate temperature and oxygen partial pressure favour the formation of the tetragonal phase, which is in agreement with results reported previously [9].

The surface morphology and roughness of the zirconia films were determined by AFM measurements. The AFM images of the ZrO_2 films produced at different deposition times are presented in figure 4. In both figures (a and b), one can see that the structure of the films is nanocrystalline. The films have well-defined grains with a narrow distribution and a surface roughness in the range of 1-3 nm. However, the increase of the deposition time led to the fabrication of films with a higher peak-to-value ratio (figure 4(a) and (b)).

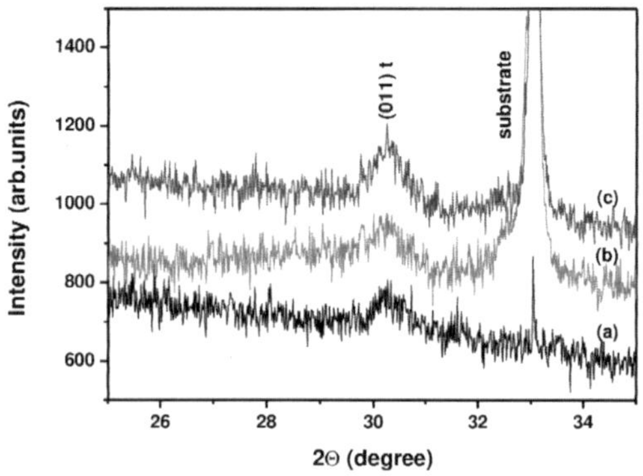

Figure 2. XRD patterns of ZrO_2 films deposited at substrate temperature of (a) 400, (b) 500, and (c) 600 °C and 0.05 mbar oxygen pressure.

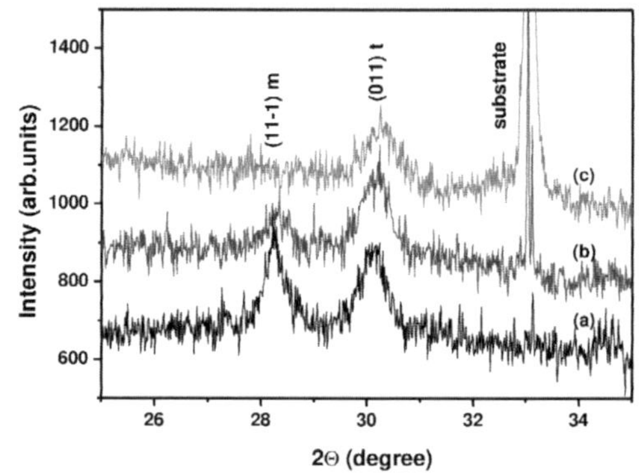

Figure 3. XRD patterns of ZrO_2 films deposited at substrate temperature of 600 °C and (a) 0.01, (b) 0.03, and (c) 0.05 mbar oxygen pressure.

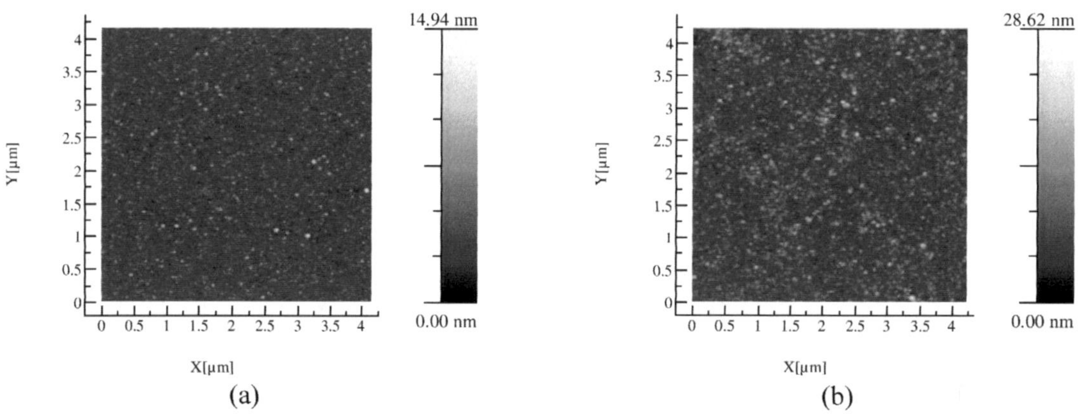

Figure 4. AFM images of thin zirconia films produced at (a) 60 and (b) 90 min deposition time.

As it was mentioned it the Introduction part, the application of zirconia films strongly depends on the possibility to suppress the phase transformation during the fabrication procedure. The results reported above gave us a reason to define the optimal process parameters (substrate temperature of 600 °C and oxygen pressure of 0.05 mbar) for obtaining monophase nanocrystalline zirconia films with a slightly over-stoichiometric composition.

However, the thicknesses of the films, as measured by the profilometer, were found, as it was expected, to decrease with the increasing oxygen pressure [10]. At a pressure of 0.05 mbar, the deposition rate was estimated to be 0.006 nm/s. The increase of the deposition time led to a change in the crystalline structure of the samples and until the appearance of m-zirconia. This means that the optimal process parameters were determined for defined film thickness.

In order to overcome the phase transformation in zirconia, a set of experiments were performed by using 3 and 8 mol% Y_2O_3 doped zirconia targets.

3.2. YSZ films

The use of doped zirconia targets changed the crystalline structure of the films (not shown). A 3 mol % content of Y_2O_3 in the target led to the deposition of tetragonal zirconia, which corresponds well to the previous reports [5]. Increasing the doping content to 8 mol % Y_2O_3 in the target resulted in the fabrication of a cubic phase of zirconia [5,15]. It should be mentioned that doping with Y_2O_3 (for the concentrations used in this work) did not change the estimated value of the crystallites size.

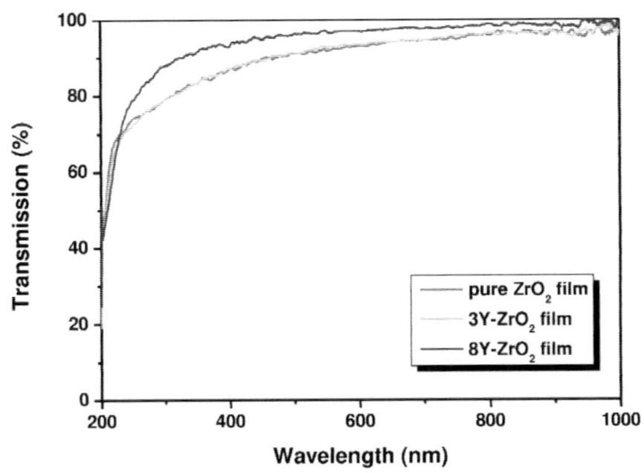

Figure 5. Optical transmission spectra of pure, 3 and 8mol% yttrium stabilized zirconia films.

A doping content of 3 mol % Y_2O_3 did not change significantly the surface morphology and film roughness (not shown). However, at the same film roughness, the films with 8 mol % YSZ had a higher peak-to-trough ratio (45 nm). Isolated droplets on the surface were also observed.

The influence of the dopants concentration was also revealed in the optical properties of the zirconia films. The optical transmission spectra of pure and yttrium stabilized zirconia films are presented in figure 5. All the films are transparent in a wide range of wavelengths. No significant difference was observed in the transmission spectra taken from pure and 3 mol% yttrium stabilized ZrO_2. The result is in agreement with the structural and morphology analysis discussed above. When the surface morphology was similar, a small volume f raction of m-zirconia did not change significantly the optical properties of the films. A major increase of the optical transmission was observed in the 8 mol% yttrium stabilized ZrO_2 films, its value reaching 94-97% in the visible range; we related this to the presence of cubic zirconia.

4. Conclusions

Thin nanocrysttaline zirconia films were prepared by PLD. It was confirmed that the phase formation in zirconia is a complex process that strongly depends on the method of deposition, the process conditions, such as substrate temperature and oxygen pressure, and the thickness of the films. The optimal process parameters for preparing t-phase nanocrystalline zirconia films with a slightly over-stoichiometric composition were determined, namely, a substrate temperature of 600 °C and an oxygen pressure of 0.05 mbar. The cubic phase of zirconia films was prepared under the same process parameters by using a 8 mol% YSZ target.

Acknowledgments

The authors acknowledge the financial support of Operational Program "Development of the Competitiveness of the Bulgarian Economy", project BG16 1P0003-1.2.04-0034-C0001 for the AFM measurements.

References

[1] Zhu J and Liu Z G 2004 *Appl. Phys.* A **78** 741
[2] Baltrus J P, Ohodnicki P R, Joy N A and Carpenter M A 2014 *Appl. Surf. Sci.* **313** 19
[3] Leclercq J, Giraud F, Bianchi D and Gaillard F 2015 *Catalysis Commun.* **67** 35
[4] Ito H, Sasaki H, Saito K, Honma S, Yajima Y and Yoshinari M 2013 *Dental Mater. J.* **32** 122
[5] Patil K C, Hegde M S, Rattan T and Aruna S T 2008 *Chemistry of Nanocrystalline Oxide Materials Combustion Synthesis, Properties and Applications* (Singapore: World Scientific)

[6] Harasex S, Wanzenboeck H D, Basnar B, Smoliner J, Brenner J, Stoeri H S, Gornik E and Bertagnoll E 2002 *Thin Solid Films* **414** 199
[7] Patel K J, Desai M S and Panchal C J 2012 *Adv. Mat. Lett.* **3** 410
[8] Ma C Y, Lapostolle F, Briois P and Zhang Q Y 2007 *Appl. Surf. Sci.* **253** 8718
[9] Mishra M, Kuppusami P., Singh A, Ramya S, Sivasubramanian V and Mohandas E 2012 *Appl. Surf. Sci.* **258** 5157
[10] Balakrishnan G, Sairam T N, Kuppusami P, Thiumurugesan R, Mohandas E, Ganesan V and Sastikumar D 2011 *Appl. Surf. Sci.* **257** 8506
[11] Cancea V N, Birjega R, Ion V, Filipescu M and Dinescu M 2012 *Physics AUC* **22** 50
[12] Al-Kuhaili M F and Durrani S M A 2011 *J. Alloys & Compounds* **509** 9536
[13] Martin P J and Bendavid A 2010 *Thin Solid Films* **518** 5078
[14] Chang S and Doong R 2005 *Chem. Mater.* **17** 4837
[15] Frison R, Heiroth S, Rupp J L M, Conder K, Barthazy E J, Müller E, Horisberger M, Döbeli M and Gauckler L J 2013 *Solid State Ionics* **232** 29

Comparison of different substrates for laser-induced electron transfer desorption/ionization of metal complexes

A A Grechnikov[1,5], V B Georgieva[2], N Donkov[3], A S Borodkov[1], A V Pento[4], Z G Raicheva[2] and Tc A Yordanov[2]

[1]Vernadsky Institute of Geochemistry and Analytical Chemistry, Russian Academy of Sciences, 19 Kosygin str., 119991 Moscow, Russia
[2]Georgi Nadjakov Institute of Solid State Physics, Bulgarian Academy of Sciences, 72 Tzarigradsko Chaussee, 1784 Sofia, Bulgaria
[3]Emil Djakov Institute of Electronics, Bulgarian Academy of Sciences, 72 Tzarigradsko Chaussee, 1784 Sofia, Bulgaria
[4]Energomashtekhnika Ltd, 123 Vernadskogo pr., bl. 2, 119571 Moscow, Russia

E-mail: grechnikov@geokhi.ru

Abstract. Four different substrates, namely, graphite, tungsten, amorphous silicon (α-Si) and titanium dioxide (TiO_2) films, were compared in view of the laser-induced electron transfer desorption/ionization (LETDI) of metal coordination complexes. A rhenium complex with 8-mercaptoquinoline, a copper complex with diphenylthiocarbazone and chlorophyll A were studied as the test analytes. The dependencies of the ion yield and the surface temperature on the incident radiation fluence were investigated experimentally and theoretically. The temperature was estimated using the numerical solution of a one-dimensional heat conduction problem with a heat source distributed in time and space. It was found that at the same temperature, the ion yield from the different substrates varies in the range of three orders of magnitude. The direct comparison of all studied substrates revealed that LETDI from the TiO_2 and α-Si films offer a better choice for producing molecular ions of metal coordination complexes.

1. Introduction

Matrix-free laser desorption/ionization of organic and bioorganic molecules has been a subject of extensive studies in the last decades. This technique is based on using specially prepared substrates absorbing laser radiation as emitters of ions of chemical compounds. The substrate assists the laser-induced desorption/ionization of the analytes preliminary deposited onto the substrate surface, so that this technique is usually referred to as surface-assisted laser desorption/ionization (SALDI) [1]. A variety of inorganic materials in the form of powders, monolithic plates or thin films have been studied as possible substrates for SALDI; it has been found that only a few selected substrates have the ability to promote strongly the ion formation. Among them, graphite and silicon materials have been and still are the most commonly used as ion emitters.

[5] To whom any correspondence should be addressed.

Most results reported on SALDI have been obtained with basic compounds, which are detected as protonated molecules. The ionization efficiency of protonation strongly depends on the analyte molecules gas-phase basicity of [2]. SALDI ensures the detection of highly basic compounds at an attomolar level (10^{-18} mol) [2], but chemical compounds with low basicity could not be ionized by SALDI. Recently, we demonstrated that laser-induced electron transfer desorption/ionization (LETDI), a novel variant of SALDI [3,4], is applicable for determination of chemical compounds with low basicity. In LETDI, the ionization is based on electron transfer from the excited adsorbed analyte molecule to the substrate. However, understanding of the detailed LETDI mechanism still requires further studies.

In this report, we compare the analytical performance of various substrates, including tungsten, graphite, amorphous silicon (α-Si) and titanium dioxide (TiO_2) films, for the laser-induced electron transfer desorption/ionization of metal coordination complexes. Previous studies recognized the important role of laser-induced surface heating for ion emission in traditional variants of SALDI [5,6].The focus of the present work is on the investigation of the thermal effects on the ionization efficiency in LETDI.

2. Experimental

The scheme of the experimental setup was described in detail in [7]. The setup consisted of a mass spectrometer, a Nd:YAG laser oscillating at higher harmonics and recording equipment.

The mass spectrometer was a laboratory-built, linear, turbopumped time-of-flight mass spectrometer with a 0.70 m flight tube. All measurements were performed in the positive ion detection mode. A diode-pumped Nd:YAG laser (RL02.355, ELS-94, Moscow) with near-diffraction limited beam quality, 0.35 ns pulse duration, was operated at 300 Hz. A frequency-tripling crystal yielded a 355 nm laser pulse of about 50 μJ. The laser pulse energy was varied using an attenuator.

Amorphous silicon (α-Si) films were deposited on monocrystalline silicon by standard RF sputtering of Si in a low-pressure (8×10^{-2} Pa) Ar atmosphere. The thickness of the deposited film was approximately 0.5 mm.

The titanium dioxide TiO_2 films were deposited on monocrystalline silicon by e-beam evaporation of TiO_2 powder (working pressure of 3×10^{-3} Pa). The thickness of the deposited film was approximately 0.35 mm.

The pyrolytic graphite plate and tungsten foil were obtained commercially (Baum-Lux, Moscow).

A rhenium complex with 8-mercaptoquinoline and a copper complex with diphenylthiocarbazone were synthesized according to the well-known procedures [8,9]. Chlorophyll A (magnesium coordination complex with chlorine) was obtained by acetone-based extraction from leaf samples according to the procedure in [10]. The complexes obtained were dissolved in the appropriate solvents and deposited onto the substrates using electrospray deposition.

3. Results and discussion

The four studied substrates have different UV absorptivity, thermal conductivity and specific heat capacity. Thus, the first step of the thermal effects investigations was to determine the dependencies of the surface temperature on the fluence of the incident laser radiation for each substrate. Because of the small size of the laser-induced heated area and the short lifetime of the state with high temperature, the experimental determination of the surface temperature under laser irradiation is a complex and laborious task. An easier approach for estimating the surface temperature consists in numerically solving the problem of heat conductivity with a space and time distributed heat source. This approach is well known to yield good agreement with the experiments [11-13].

The parameters necessary for calculations were determined as follows. The absorption of radiation by the material is described by Bouguer's law:

$$I(z)=I_0\, exp(-\alpha z)\,, \qquad (1)$$

where $I(z)$ describes the change in the incident radiation intensity with the depth z; z is directed normally to the irradiated surface; α is the absorbance. The temporal dependence of the laser pulse power W is described as:

$$W(t,z) = W_0 exp\left(-\left(\frac{t-2\tau}{\tau/2}\right)^2\right)\exp(-\alpha z). \quad (2)$$

The normalization condition for a given surface energy density F_0 is specified as

$$F_0 = \iint_0^{\infty} W(t,z)dt\, dz = W_0 \frac{\tau}{2}\frac{\sqrt{\pi}}{2} erfc(-2)\frac{1}{\alpha}. \quad (3)$$

So, the normalization factor is:

$$W_0 = F_0 \frac{4\alpha}{\tau\sqrt{\pi} erfc\,(-2)} \quad . \quad (4)$$

Because the depth of laser radiation absorption by the materials studied is much smaller than the linear cross-sectional dimensions of the laser spot focused on the surface, in calculating the surface temperature we used the approach of a one dimensional model of heat fluxes. The equation of non-steady-state heat conduction in this model takes the form

$$c_p \rho \frac{\partial T}{\partial t} = \frac{\partial}{\partial z}\left(\lambda \frac{\partial T}{\partial z}\right) + f \,, \quad (5)$$

where c_p is the heat capacity; λ is the heat conductivity; ρ is the density of the material; $f = f(z,t)$ describes the heat source, which is determined by equation (2), the normalization condition (3) and by taking into account the amplitude reflection coefficient:

$$f = (1 - R)W(t,z)\,, \quad (6)$$

where R is the reflection coefficient, which can be found using Fresnel equations.

The optical and thermophysical properties, based on which one can formulate the conditions for solving a non-steady state problem of heat conductivity, are known for the substrates studied [11-17]. The FlexPDE V5.0.2 software package was used for calculations.

Figure 1 shows the calculated dependencies of the maximum surface temperature on the laser radiation fluence. It is seen that the maximum temperature grows linearly as the fluence is increased. The slopes of the curves are quite different for the different materials. This means that different laser energies are needed to achieve the same temperature at the surfaces of the α-Si, graphite, tungsten and TiO_2 substrates.

Figure 2 shows the time dependencies of the surface temperature, calculated for fluences corresponding to a maximum temperature of 1000 K.

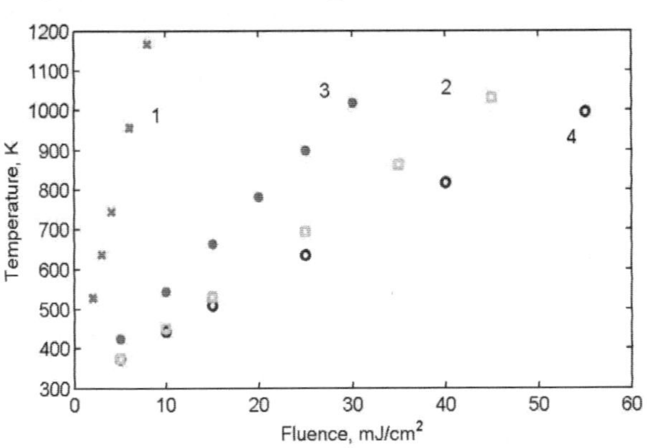

Figure 1. Calculated maximum temperature of the surface versus the laser radiation fluence: 1 – α-Si film, 2 – graphite, 3 – tungsten, 4 – TiO_2 film.

It is seen that the temperature of the surface reaches its maximum nearly at the end of the laser pulse, but the characteristic cooling time is different for the materials studied.

To compare the ionization properties of LETDI substrates, a Re complex with 8-mercaptoquinoline, a Cu complex with diphenylthiocarbazone and chlorophyll A were used as test analytes. Quinolines and thiocarbazones, common organic reagents, are widely used for highly efficient separation and preconcentration of metal element traces, while chlorophyll A is one of the most important bioorganic compound. For all

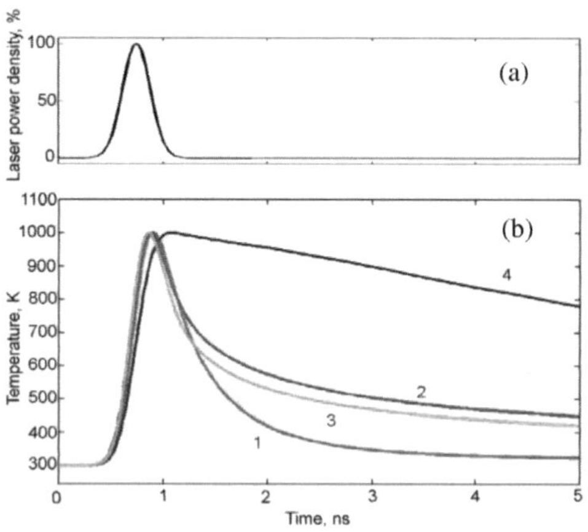

Figure 2. (a) time dependence of the laser pulse power density; (b) calculated time dependencies of the surface temperature at laser radiation fluences corresponding to a maximum temperature of 1000 K: 1 – α-Si film, 2 – graphite, 3 – tungsten, 4 – TiO_2 film.

these metal complexes, LETDI mass spectra were recorded and identified. It was found that each of the substrates studied produces ions of analytes. In general, mass spectra obtained from different substrates are similar to each other. All mass spectra are dominated by the molecular ions for the most abundant peaks. Additionally, mass spectra of chlorophyll A and a Cu complex with diphenylthiocarbazone include one peak of metal-containing fragment ions. The LETDI mass spectra are relatively "clean" with low chemical background, except for the peaks corresponding to the ions of sodium and potassium. As examples, figure 3 and figure 4 show mass spectra of the Cu complex with diphenylthiocarbazone obtained from α-Si and graphite.

In order to reflect better the LETDI performance of the four substrates, the ionization efficiencies of the LETDI process from different

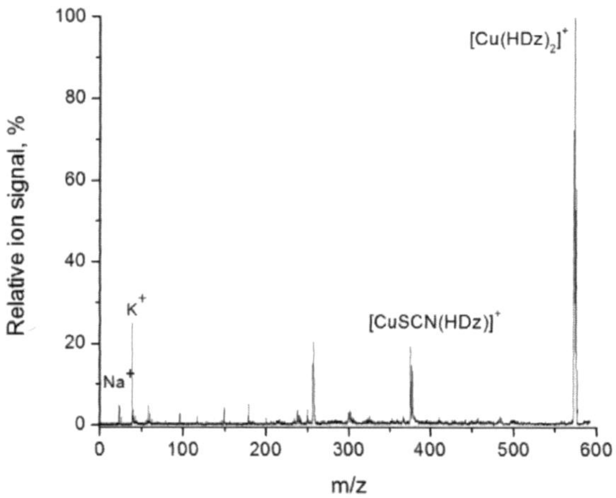

Figure 3. LETDI mass spectrum of Cu complex with diphenylthiocarbazone (H_2Dz) obtained from α-Si.

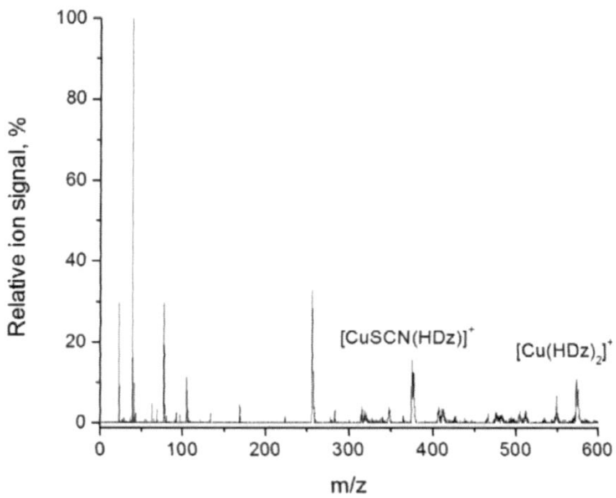

Figure 4. LETDI mass spectrum of Cu complex with diphenylthiocarbazone (H_2Dz) obtained from graphite.

Table 1. Comparison of the ionization efficiencies of the LETDI process from different substrates.

Substrate	Analyte		
	Re complex with 8-mercaptoquinoline	Cu complex with diphenylthiocarbazone	Chlorophyll A
tungsten	1	1	1
graphite	8	7	5
α-Si	5×10^2	5×10^2	4×10^2
TiO_2	10^3	5×10^2	7×10^2

substrates, but at the same maximum temperature, were compared. The results obtained for the temperature of 1000 K are presented in the table 1. The ionization efficiency from tungsten was taken to be 1. It is seen that at the same temperature, the ion yield from different substrates varies in the range of about three orders of magnitude. Tungsten exhibits the lowest ionization efficiency, while TiO_2 film provides the highest efficiency. Also, α-Si films can be successfully used as an ion emitter.

Obviously, the efficiency of LETDI is the product of the probabilities of two processes: the probability of ionization of the analyte molecule and the probability of desorption in the form of molecular ion under the action of the laser pulse. All materials studied absorb efficiently UV radiation and transfer the laser energy to the thermal energy. The observed great difference between the materials in the ability to produce the molecular ions of the investigated metal complexes suggests that the efficiency of LETDI is determined by the ionization stage; more specifically, by electron transfer from the molecule. However, the thermal desorption step can limit the yield of ions with high molecular weights.

3. Conclusions

Based on the results of surface temperature calculations and the experimental study of the efficiencies of ionization of metal complexes from various substrates, it was concluded that the efficiency of the LETDI process is determined mainly by the ionization stage.

The direct comparison of the four common SALDI substrates, namely, graphite, tungsten, amorphous silicon and titanium dioxide films, for laser-induced electron transfer desorption/ionization of metal complexes reveals that titanium dioxide and amorphous silicon offer a better choice for producing molecular ions of metal complexes.

Acknowledgement

The present work was supported by the Ministry of Education and Science of the Russian Federation (Contract No. 14.579.21.0016, unique identifier RFMEFI57914X0016).

References

[1] Murray K K, Boyd R K, Eberlin M N, Langley G J, Li L and Naito Ya 2013 *Pure Appl. Chem.* **85** 1515
[2] Grechnikov A A, Borodkov A S, Alimpiev S S, Nikiforov S M and Simanovsky Ya O 2013 *J. Anal. Chem.* **68** 19
[3] Grechnikov A, Nikiforov S, Strupat K and Makarov A 2013 *Anal. Bioanal. Chem.* **406** 3019
[4] Borisova L V, Borodkov A S, Grechnikov A A, Ugolkova E A and Minin V V 2013 *Russ. J. Inorg. Chem.* **58** 940
[5] Grechnikov A A, Borodkov A S, Zhabin S N and Alimpiev S S 2014 *J. Anal. Chem.* **69** 1361
[6] Zhabin S N, Pento A V, Grechnikov A A, Borodkov A S, Sartakov B G, Simanovskii Ya O, Nikiforov S M and Alimpiev S S 2011 *Quantum Electron.* **41** 835
[7] Alimpiev S S, Grechnikov A A, Sunner J, Karavanskii V A, Simanovsky Ya O and Nikiforov S M 2011 *Rapid Commun. Mass Spectrom.* **25** 140
[8] Bankovskii Yu A 1978 *Chemistry of chelate compounds of mercaptoquinoline and its derivatives* (Riga: Zinatne)
[9] Iwantscheff G 1972 *Das Dithizone und seine anwendung in der mikro- und spurenanalyse.* (Weinheim: Verlag Chemie)
[10] Hu X, Tanaka A and Tanaka R 2013 *Plant Methods* **9** 1
[11] Ong C K, Tan H S and Sin E H 1986 *Mater. Sci. Eng.* **79** 79
[12] Webber H C, Cullis A G and Chew N G 1983 *Appl. Phys. Lett.* **43** 669
[13] Moon S., Hatano M, Lee M and Grigoropoulos C P 2002 *Int. J. Heat Mass Transfer* **45** 2439
[14] Lee S M, Cahill D G and Allen T H 1995 *Phys Rev.* B **52** 253
[15] Chirkin V S 1967 *Thermophysical properties of materials for nuclear technology* (Moscow: Atomizdat)

[16] Djurišić A B and Li E H 1999 *J. Appl. Phys.* **85** 7404
[17] Lee S M, Cahill D G and Allen T H 1995 *Phys. Rev. B.* **52** 253

Corrosion properties of zirconium-based ceramic coatings for micro-bearing and biomedical applications

J Walkowicz[1], V Zavaleyev[1], E Dobruchowska[1], D Murzynski[1], N Donkov[2,5], A Zykova[3], V Safonov[3] and S Yakovin[4]

[1]Institute of Technology and Education, Koszalin University of Technology, 2 Sniadeckich str., 75453 Koszalin, Poland
[2]Emil Djakov Institute of Electronics, Bulgarian Academy of Sciences, 72 Tzarigradsko Chaussee, 1784 Sofia, Bulgaria
[3]National Science Center, Kharkov Institute of Physics and Technology, 1 Academicheskaya str., 61108 Kharkov, Ukraine
[4]Department of Physical Technologies, Kharkov National University, 4 Svobody Sq., 61077 Kharkov, Ukraine

E-mail: nikolaidd@abv.bg

Abstract. Ceramic oxide ZrO_2 and oxynitride ZrON coatings are widely used as protective coatings against diffusion and corrosion. The enhancement of the coatings' mechanical properties, as well as their wear and corrosion resistance, is very important for their tribological performance. In this work, ZrO_2 and ZrON coatings were deposited by magnetron sputtering on stainless steel (AISI 316) substrates. The adhesion, hardness and elastic properties were evaluated by standard methods. The surface structure of the deposited coatings was observed by electron scanning microscopy (SEM) and atomic force microscopy (AFM). The composition of the coatings was analyzed by X-ray photoelectron spectroscopy (XPS) and energy dispersive X-ray spectroscopy (EDS). The corrosion resistance properties were evaluated using the potentiodynamic method. The results show that the corrosion parameters are significantly increased in the cases of both oxynitride and oxide coatings in comparison with the stainless steel (AISI 316) substrates.

1. Introduction

Novel functional coatings are widely used in different industrial areas for tribological applications [1]. The modification of metal materials by means of ceramic coatings deposition is an effective way of forming alternative bearing surfaces. Moreover, many such applications require a high stability in an aggressive and corrosive environment.

Wear and corrosion are the major processes that cause problems with metals and alloys [2]. Corrosion in the aqueous medium of human body fluids takes place via electrochemical reactions [3]. The body fluid environment may decrease the fatigue strength of the metal implant and enhance the release of iron, chromium, nickel, titanium ions, which are found to be powerful allergens and carcinogens [4]. Failure of metal products have been reported due to stress corrosion cracking and poor wear resistance [5,6]. The release of wear and corrosion debris and further accumulation of

[5] To whom any correspondence should be addressed.

harmful products in internal organs of animals were observed in *in vivo* tests [7]. The corrosion and wear resistance properties of modern coatings will play a key role in future effective applications for micro-bearing and biomedical needs.

The aim of the present work was to study the effect of the compositional, structural and morphological properties of oxide ZrO_2 and oxynitride ZrON films on the mechanical and corrosion characteristics of the ceramic coatings prepared.

2. Materials and methods

The process of ZrO_2 and ZrON coatings deposition by magnetron sputtering was performed in a high-vacuum pumping system with a base pressure of about 10^{-3} Pa.

Stainless steel (AISI 316) discs with a diameter of 32 mm and a thickness of 3 mm were used as substrates. The substrates were successively ultrasonically cleaned in acetone, ethanol and deionized water and dried.

The magnetron discharge power was 4 – 5 kW. An ICP source for oxygen activation with RF power up to 1 kW was used. The target consisted of pure zirconium. The main parameters of the deposition process were the following: Ar pressure P_{ar} = 2.3×10^{-1}Pa, oxygen mass flow rate $q = 35$ sccm for ZrO_2 deposition and nitrogen mass flow rate q = 27 sccm for ZrON deposition, magnetron voltage U_m = 500 – 520 V, magnetron current I_m = 7.0 – 7.6 A, total pressure P = 2.8 – 3.0×10^{-1} Pa, coating deposition rate 6 – 8 μm/hour. An ion source was used for cleaning the samples' surface before deposition. More details on the magnetron and the ion source in the sputtering chamber were given previously in [8,9].

The thickness of the as-deposited coatings was measured by a Calotest device. The adhesion properties, hardness and elastic modulus were evaluated by standard methods by using a Revetest (CSM Instrumets) device; the Rockwell test was carried out by a type C diamond Rockwell indenter. To determine the micro-hardness, a Fischerscope® HM2000 tester was used. The surface roughness parameters, such as the arithmetic average roughness (R_a) and mean roughness depth (R_z) were measured by a Hommel Werke T8000 profilometer. The coatings' thickness and roughness parameters are presented in table 1.

Table 1. Characteristics of deposited coatings.

Material/Coating type	Roughness parameters		
	R_a [μm]	R_z [μm]	Thickness [μm]
(AISI 316) / ZrO_2	0.145	0.849	1.0
(AISI 316) / ZrON	0.129	0.807	0.8

The surface morphology and topography were observed by a JSM 5500 LV electron scanning microscope and by atomic force microscopy (AFM). X-ray photoelectron spectroscopy (XPS) was carried out using an ESCALAB MkII (VG Scientific) electron spectrometer at a base pressure in the analysis chamber of 5×10^{-8} Pa (1×10^{-6} Pa during the measurement) with an Al Kα X-ray source (excitation energy hv = 1486.6 eV). The instrumental resolution measured as the full width at half maximum (FWHM) of the Ag3d5/2 photoelectron peak was 1 eV [9]. The composition of the coatings was analyzed by energy dispersive X-ray (EDX) spectroscopy (Oxford Link ISIS 300).

For the potentiometric study, all samples were washed in ethanol in an ultrasonic bath before the measurements. The active area of the sample was 0.287 cm^2. All samples were kept in contact with the electrolyte (0.9% isotonic NaCl solution, volume of the electrolyte 50 ml) for 1 hour before the measurements. The potential was in the range 0.7 – 0.8 V for the reference sample and 0.5 – 1.0 V for all samples with deposited coatings, the sweep rate was 1 mV/s, and the reading density, 3 mV. The temperature during the measurements was kept at 25 ± 1 °C. The values of the corrosion potential and the corrosion current density was estimated using the Tafel slopes extrapolation method.

3. Results and discussion

The composition of the ZrO_2 and ZrON coatings was analyzed by XPS and the photoelectron spectra of Zr3d, O1s, N1s coatings were observed. Two peaks can be seen in the Zr3d XPS spectrum, the first one being observed at the binding energy position E = 182.4 eV, and the second one exhibiting an energy shift of 2,43 eV away from the 3d level due to the spin-orbit coupling (figure 1a). The N1s peak was detected at a binding energy of E = 396.2 eV, associated with the Zr-N chemical bond. In addition, a weak N1s peak at a binding energy of E = 398.0 eV was recorded, corresponding to Zr-N-O chemical bonds. The O1s high-resolution spectra demonstrated a peak at the binding energy position E = 530.2 eV, associated with the Zr-O chemical bond.

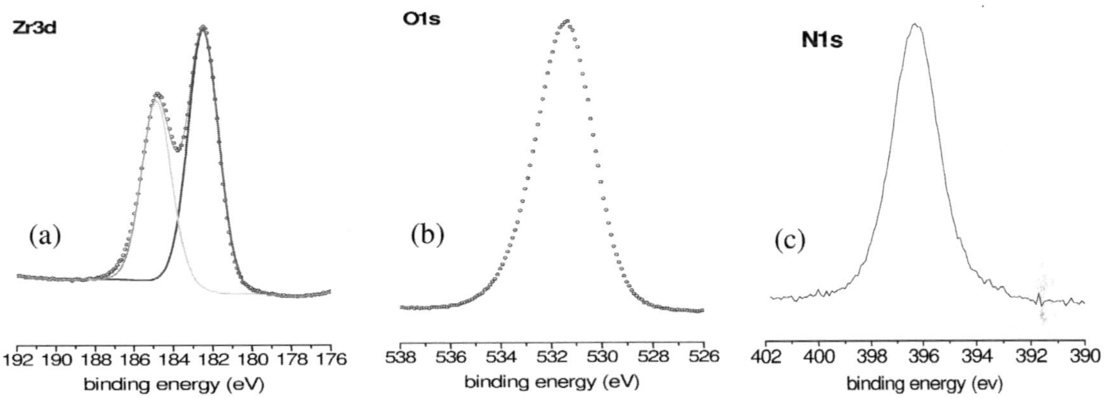

Figure 1. XPS spectra of Zr based coatings: (a) Zr3d, (b) O1s, (c) N1s.

The elemental compositions of the magnetron-sputtered coatings were investigated by energy dispersive X-ray (EDX) spectroscopy. The EDX spectra of the as-deposited coatings were in a good agreement with the XPS data. The elemental compositions of the ZrO_2 and ZrON coatings are presented in table 2.

Table 2. Elemental composition of the deposited coatings.

Element	Zr	Nb	Cr	Fe	O	N	Ar	Total
ZrO_2								
Weight %	64.11	3.08	1.25	4.14	27.05		0.37	100
Atomic %	27.74	1.31	0.95	2.93	66.71		0.37	100
ZrON								
Weight %	68.82	3.89	3.97	8.82	4.43	9.75	0.32	100
Atomic %	31.12	2.95	3.71	7.57	15.82	38.46	0.37	100

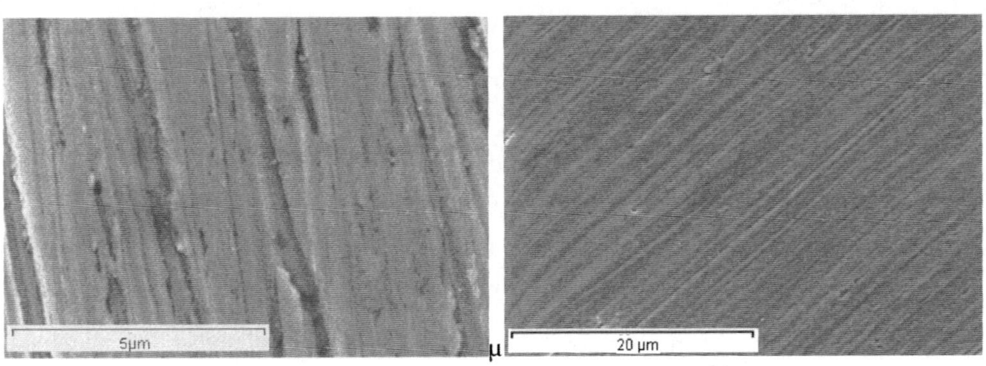

a) b)

Figure 2. SEM micrographs of ZrO_2 (a) and ZrON (b).

The surface topography and morphology of ZrO_2 and ZrON coatings were investigated by electron scanning microscopy (figure 2). The surfaces of the deposited coatings have smooth relief with a uniform cross-section structure.

The mechanical properties of the coatings are the hardness *H* and effective Young's modulus *E**. The films' mechanical behavior is characterized by the ratio *H/E** [10,11], which is proportional to the film's fracture toughness and to the material's resistance to plastic deformation. There is a correlation between the coatings' mechanical properties and structure. The formation of oxynitride coatings improves the mechanical properties of the coated materials. The hardness parameters increase in comparison with the oxide films. The mechanical parameters of ZrO_2 and ZrON coatings deposited on the stainless steel samples (AISI 316) substrates are presented in the table 3.

Table 3. Mechanical characteristics of oxide and oxynitride coatings deposited on stainless steel (AISI 316) substrates.

Material/Coating	Mechanical parameters (average results 10 tests)			
	Hardness *HV* [GPa]	Young Modulus [GPa]	*H/E**	Adhesion [N]
AISI 316 / ZrO_2	14.5	133.7	0.1	17.1
AISI 316 / ZrON	15.7	132.8	0.11	19.6

The wear resistance properties of the obtained coatings are very important for applications as bearings [12,13]. Tribological tests were conducted previously [14]. The abrasive wear exhibited a minimum value of about 1.2×10^{-4} mm^3/Nm in the case of oxynitride coatings.

We further present corrosion tests of anodic polarization carried out by a potentiodynamic method in the potential range 0.5 – 1.0 V at a scanning rate of 1 mV/s. The corrosion behavior was analyzed of stainless steel samples (AISI 316), stainless steel substrates with ZrO_2 (figure 3a) and ZrON (figure 3b) coatings after corrosion tests at a temperature 25 ± 1 °C in an isotonic NaCl (0.9%) solution. The values of the corrosion potential (E_{cor}) and the corrosion current density (J_{cor}) are presented in table 4.

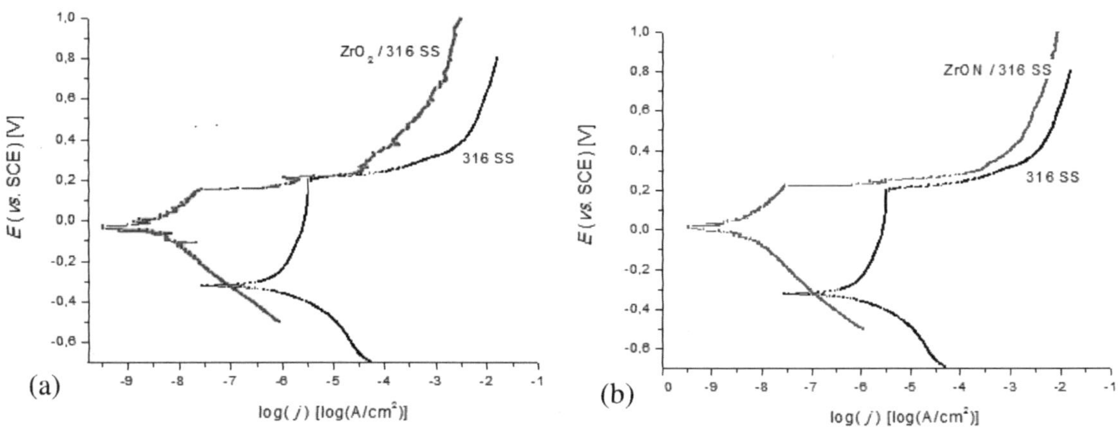

Figure 3. Corrosion resistance of stainless steel samples (AISI 316), stainless steel substrates with ZrO_2 (a) and ZrON (b) coatings after corrosion tests in an isotonic NaCl (0.9%) solution.

The results demonstrate the improvement of the corrosion characteristics of stainless steel samples coated by ceramic coatings. The corrosion resistance parameters increased in the cases of both oxynitride and oxide coatings.

The surface topography and morphology of stainless steel samples (AISI 316), stainless steel substrates with ZrO_2 and ZrON coatings after corrosion tests were analyzed by atomic force microscopy (figure 4).

The surface failure after a corrosion test in an isotonic NaCl (0.9%) solution further confirmed the main results as obtained from the polarization curves.

Table 4. Corrosion characteristics of oxide and oxynitride coatings deposited on stainless steel (AISI 316) substrates.

Material/ Coating	Corrosion potential E_{cor} (V)	E_v (V)	Corrosion current density J_{cor} (A/cm^2)
AISI 316	-0.312	0.208	8×10^{-7}
AISI 316 / ZrO_2	-0.023	0.150	3×10^{-9}
AISI 316 / ZrON	0.015	0.224	4×10^{-9}

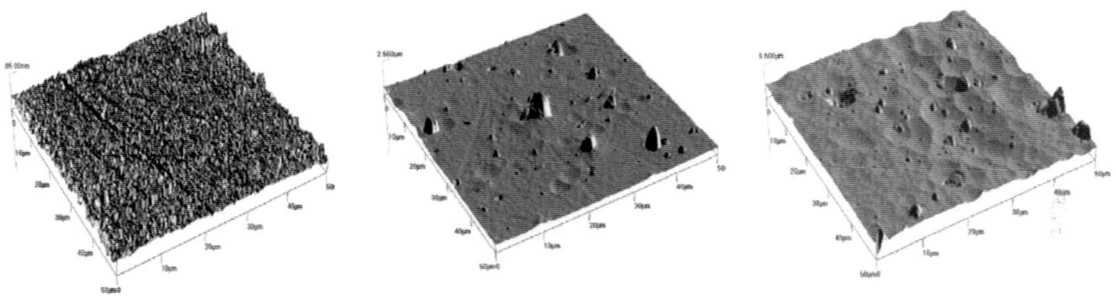

(a) (b) (c)

Figure 4. AFM images of stainless steel samples (AISI 316) (a), stainless steel substrates with ZrO_2 (b) and ZrON (c) coatings after corrosion tests in an isotonic NaCl (0.9%) solution.

4. Conclusions

A technology was developed allowing one to deposit oxide and oxynitride coatings with high hardness parameters up to 15 GPa and enhanced corrosion resistance characteristics on AISI 316 stainless steel substrates. The formation of oxynitride coatings improves the mechanical properties in comparison with the oxide films. The results show that both the oxide and oxynitride ceramic coatings possess excellent corrosion resistance characteristics towards an isotonic NaCl (0.9%) solution. The development of new low-cost and innovative bearing surfaces with advanced corrosion characteristics is very promising in view of biomedical applications, such as micro-bearings and joints replacement.

Acknowledgment

The study was supported by project IMBeing- FP7-PEOPLE-2013-IRSES-612593 within the 7th FP of the European Commission and the Program for Cooperation between Bulgarian Academy of Sciences and the National Ukrainian Academy of Sciences

References

[1] Serro A P, Completo C, Colaco R, dos Santos F, Lobato da Silva C, Cabral J M S, Araujo H, Pires E and Saramago B 2009 *Surf. Coat. Technol.* **203** 3701
[2] Eisenbarth E, Velten D, Schenk-Meuser K, Linez P, Biehl V, Duschner H, Breme J and Hildebrand H F 2002 *Biomol. Eng.* **19** 243
[3] Dion L, Bordenave F, Lefebvre R and Boreille C V 1994 *J. Mat. Sci.- Mat. Med.* **18** 115
[4] Kim D J, Lee M H, Lee D Y and Han J S 2000 *J. Biomed. Mat. Res.* **53** 438
[5] Ratner B D 1993 *J. Biomed. Mat. Res.* **27** 837
[6] Velten D, Biehl V, Aubertin F, Valeske B, Possart W and Breme J 2002 *J. Biomed. Mat. Res.* **59** 18
[7] Zykova A V, Luk'yanchenko V V and Safonov V I 2005 *Surf. Coat. Technol.* **200** 90

[8] Zykova A, Safonov V, Walkowicz J, Rogovska R and Yakovin S 2010 *J. Phys.: Conf. Series* **223** 012024
[9] Safonov V, Zykova A, Smolik J, Rogovska R, Donkov N, Goltsev A, Dubrava T, Rassokha I and Georgieva V 2012 *J. Phys. Conf. Series* **356** 012046
[10] Leyland A and Mathews A 2004 *Surf. Coat. Technol.* **177/178** 317
[11] Musil J, Sklenka J and Čerstvý R 2012 *Surf. Coat. Technol.* **206** 2105
[12] Kusano Y, Van Acker K and Hutchings I M 2004 *Surf. Coat. Technol.* **183** 312
[13] Camara Cozza R 2013 *Surf. Coat. Technol.* **215** 224
[14] Warcholinski B, Gilewicz A, Lupicka O, Rochowicz J, Zykova A, Safonov V and Yakovin S 2014 *Prob. At. Sci. Technol.* **94** 219

Surface modification of tantalum pentoxide coatings deposited by magnetron sputtering and correlation with cell adhesion and proliferation in *in vitro* tests

A Zykova[1,7], V Safonov[1], A Goltsev[2], T Dubrava[2], I Rossokha[2], N Donkov[3], S Yakovin[4], D Kolesnikov[5], I Goncharov[5] and V Georgieva[6]

[1]National Science Center "Kharkov Institute of Physics and Technology",
1 Academicheskaya Str., 61108 Kharkov, Ukraine
[2]Institute for Problems of Cryobiology and Cryomedicine NASU, Kharkov, Ukraine
[3]Emil Djakov Institute of Electronics, Bulgarian Academy of Sciences,
72 Tsarigradsko Chaussee, 1784 Sofia, Bulgaria
[4]Department of Physical Technologies, Kharkov National University,
4 Svobody Sq., 61077 Kharkov, Ukraine
[5]Belgorod State National Research University,
85 Pobedy Str., 308015 Belgorod, Russia
[6] Georgi Nadjakov Institute of Solid State Physics, Bulgarian Academy of Sciences,
72 Tsarigradsko chaussee blvd., 1784 Sofia, Bulgaria

E-mail: zykova.anya@gmail.com

Abstract. The effect was analyzed of surface treatment by argon ions on the surface properties of tantalum pentoxide coatings deposited by reactive magnetron sputtering. The structural parameters of the as-deposited coatings were investigated by means of transmission electron microscopy, atomic force microscopy and scanning electron microscopy. X-ray diffraction profiles and X-ray photoelectron spectra were also acquired. The total surface free energy (SFE), the polar, dispersion parts and fractional polarities, were estimated by the Owens-Wendt-Rabel-Kaeble method. The adhesive and proliferative potentials of bone marrow cells were evaluated for both Ta_2O_5 coatings and Ta_2O_5 coatings deposited by simultaneous bombardment by argon ions in *in vitro* tests.

1. Introduction

In recent years, a new branch of medicine has undergone rapid development based on the use of nanocomposite coatings that promote positive biological processes in a living organism. Medical products with ceramic coatings form normal biopotentials in damaged areas of tissue and organs, prevent the necrosis, tissue atrophy, significantly accelerate the healing process and the post-operative rehabilitation [1-5]. The novel implant and stent types coated by oxide and oxynitride coatings have demonstrated excellent biocompatibility. The positive effect results in lower rates of postoperative infections, inflammatory reactions and other complications [6-8].

[7] To whom any correspondence should be addressed.

Tantalum and tantalum-based compounds are promising for biomedical applications due to their good dielectric properties. The results of tests on implanting Ta in both soft and hard tissue of rats showed the good biocompatibility and osteogenesis of this metal [9]. TaC and TaN materials possess relatively high hardness due to the covalent nature of their bond [10] and demonstrate high corrosion resistance [11]. The blood compatibility of TaN films was shown to be better than that of Ta [12]. The aim of the present work was to study the effect of surface treatment by argon ions on the structure and surface properties of tantalum pentoxide (Ta_2O_5) coatings deposited by reactive magnetron sputtering and, further, the correlation of the surface characteristics with the cell's adhesion and proliferation.

2. Materials and methods

Ta_2O_5 coatings were deposited on Petri dishes glass substrates in a high vacuum pumping system with a base pressure of about 10^{-2} Pa by ion-source-assisted magnetron sputtering. For the reactive deposition, oxygen was delivered through the ICP plasma source, q = 60 sccm, the magnetron voltage was U_m = 700 V, and the magnetron current was about I_m = 5,7 A. The sputtering process was conducted in modes far from the target poisoning one for further oxide coatings deposition with highly stoichiometric composition. Also, such deposition conditions allowed us to avoid micro-arcs and micro-drops formation. The optimum conditions were implemented in the upper part of the volt-ampere characteristic curves of a magnetron discharge in argon with oxygen (figure 1). At sufficiently high mass-flow rates of oxygen, a hysteresis effect was observed.

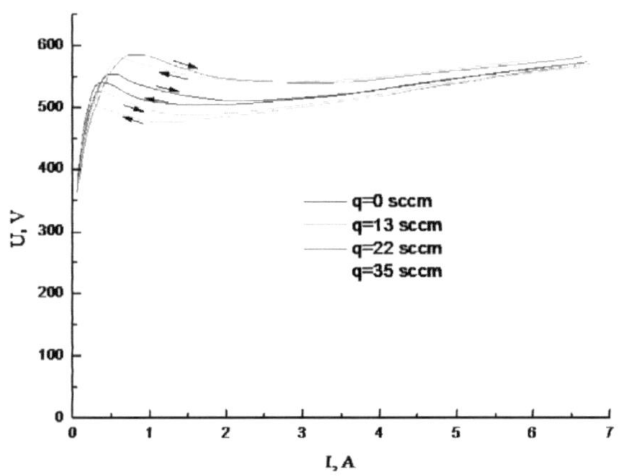

Figure 1. Volt-ampere characteristics of the magnetron discharge with a tantalum target in argon for different values of the oxygen flow: q = 0 sccm, q = 13 sccm, q = 22 sccm, q = 35 sccm.

For other substrates, the deposition process was carried out by a simultaneous bombardment of the growing film by argon ions using an ion source. The ion source parameters were: ion acceleration voltage of 2.5 kV, and ion current source of 30 mA.

The structural parameters of the as-deposited coatings were investigated by means of transmission electron microscopy (TEM) using a JEOL JEM 2100 instrument. The surface topography of the tantalum pentoxide ceramic coatings were evaluated by scanning electron microscopy (QUANTA 600 FEG) (SEM) and by atomic force microscopy (NTEGRA-AURA) (AFM). The coating thickness and cross-section structure were obtained using SEM (Nova NanoSEM, FEI) cross-section measurements. The X-ray diffraction profiles were acquired by means of a DRON-3 diffraction device. X-ray photoelectron spectroscopy (XPS) was carried out by an ESCALAB MkII (VG Scientific) electron spectrometer using an AlKalpha X-ray source (excitation energy hν = 1486.6 eV).

The advancing contact angles were measured by Wilhelm's method (Kruss K12) at a temperature of 20 °C. Standard liquids with well-known values of the surface tension, dispersion component and polar interaction were used. The estimation of the surface free-energy SFE and its polar and dispersion parts were performed by means of the Owens-Wendt-Rabel-Kaeble' method for thee liquid system α-bromonaphthalene-formamide-ethylene glycol-diidomethane-glycerol-water.

Culturing of bone marrow (BM) cells (CBA/H) was performed with densities of 0.5–1×10^4 cells/cm^2. After passage, the cells were removed by a trypsin-EDTA solution («Sigma-Aldrich», USA). The structural and functional characteristics of the BM cells were then evaluated. Visual control of the cell cultures was carried out using a Primo Star light phase-contrast microscope (Zeiss,

Germany) and an Axiovert 40C inverted microscope (Zeiss, Germany). The adhesive potential of the BM cells was also evaluated.

A statistical analysis of test results was performed using the non-parametric Mann-Whitney method. The difference was considered statistically significant at $P < 0.05$.

3. Results and discussion

The difference of the structural parameters of the deposited coatings was investigated by means of TEM (figure 2). The coatings had a nano-globular structure with characteristic sizes of the structural elements about 15 nm in the case of as-deposited tantalum pentoxide film and 20 nm after bombardment of the growing film. The deposition process carried out by simultaneous bombardment by argon ions resulted in surface structure changes of the growing films.

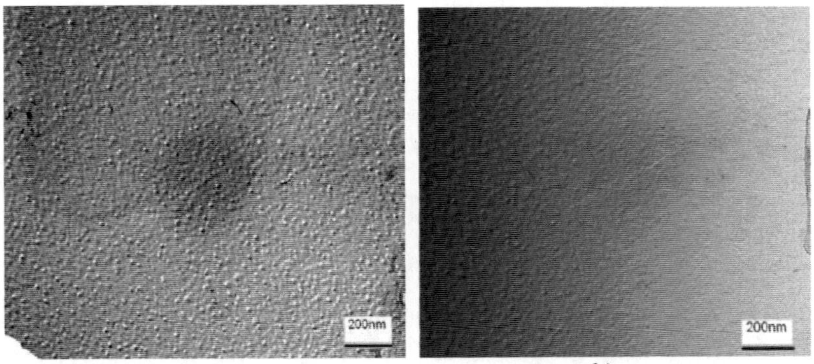

a) b)

Figure 2. Brightfield TEM images of as-deposited oxide ceramic coationgs: a) Ta_2O_5 coatings, b) Ta_2O_5 coatings deposited with simultaneous bombardment by argon ions.

X-ray diffraction profiles and XPS spectra of ceramic Ta_2O_5 coatings have been analyzed previously [13]. The as-deposited magnetron sputtered Ta_2O_5 coatings demonstrated an amorphous nature, i.e, no peaks were observed. All XPS spectra consisted of well-defined XPS lines of Ta 4f, 4d, 4p and 4s; O 1s; C 1s. Ta 4f doublets are typical for Ta_2O_5 coatings and have two peaks: Ta 4f7/2 at ~ 26.5 eV and Ta 4f5/2 whose binding energy is higher by 1.9 eV. The Ta 4f lines of the deposited films were in a good agreement with the Ta 4f doublet that is representative for the Ta-0 bond in Ta_2O_5. The Ta/O ratio estimated from the spectra was about 0.4 for all coatings investigated. The O 1s peaks were centered at binding energies of 530.6 eV for the deposited coatings.

The scanning electron microscopy images of the magnetro-sputtered Ta_2O_5 ceramic coatings revealed a relatively flat surface without cracks. The coating thickness and cross-section structure were obtained from the SEM cross-section measurements (figure 3 a, b). The surface topography and roughness parameters were studied by AFM (figure 3 c, d). The ions bombardment resulted in a change of the surface roughness from $S_a = 0.9$ nm, $S_q = 1.2$ nm to $S_a = 0.6$ nm, $S_q = 0.7$ nm. The difference in the structural parameters of the Ta_2O_5 coatings and the Ta_2O_5 coatings deposited by simultaneous bombardment by argon ions was in a good agreement with the TEM results.

The changes in the surface structure parameters may lead to changes in the surface characteristics and functional properties of deposited coatings. An analysis has been previously performed of the capacity, dielectric constant, and dielectric loss tangent depending on the frequency of the electromagnetic field for tantalum pentoxide ceramic coatings deposited in different sputtering modes [13,14]. The magnetron-sputtered Ta_2O_5 ceramic coatings demonstrated high dielectric parameters. The effect of the deposition process conditions on the electrical behavior of the magnetron-sputtered Ta_2O_5 coatings was analyzed.

Journal of Physics: Conference Series **700** (2016) 012027 doi:10.1088/1742-6596/700/1/012027

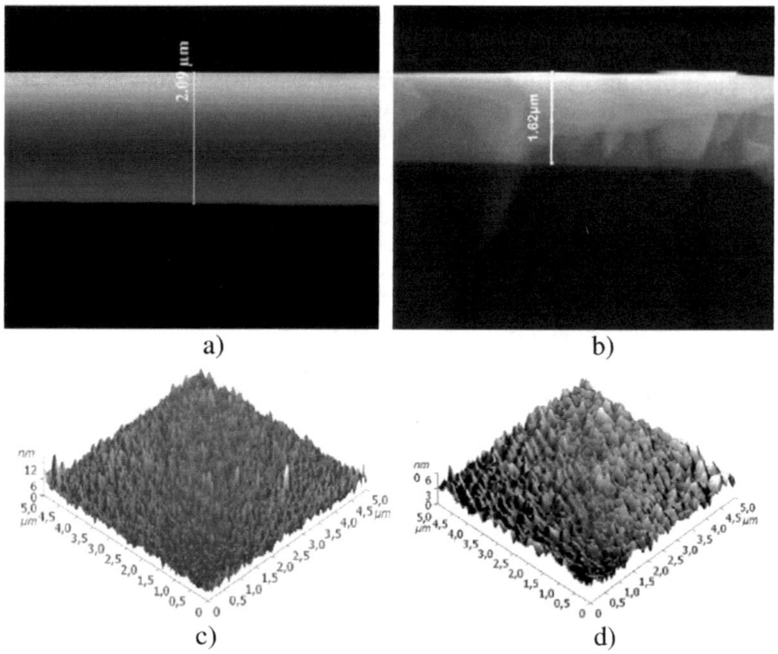

Figure 3. SEM cross-section images (a, b) and AFM surface images (c, d) of as-deposited oxide ceramic coatings: (a, c) Ta_2O_5 coatings, (b, d) Ta_2O_5 coatings deposited by simultaneous bombardment by argon ions.

The surface free energy (SFE) plays an important role in the mechanism of cell/biomaterial response [15]. The total SFE and its polar and dispersion parts and fractional polarities were estimated by the Owens-Wendt-Rabel-Kaeble method for the liquid system α-bromonaphthalene-formamide-ethylene glycol-diidomethane-glycerol-water (table 1) at a temperature of 20 °C.

The data demonstrated that the surface free energy of Ta_2O_5 coatings thus calculated was in the range 41 – 44 mN/m, the SFE polar parts were in the range 11 – 13 mN/m, and the fractional polarity values were in the range 0.26 – 0.29.

Table 1. Surface free energy components by the Owens-Wendt-Rabel-Kaeble method.

Substrate /coating	Total γ	Components of SFE mN/m		
		Dispersion part γ^d	Polar part γ^p	Fractional polarity $\gamma^p / (\gamma^d + \gamma^p)$
Glass	56.28	30.31	25.97	0.46
Glass/ Ta_2O_5	43,96	31.06	12.90	0.29
Glass/ Ta_2O_5 (argon ions bombardment)	41,11	30,03	11,08	0.26

The tantalum pentoxide ceramic coatings deposited by a simultaneous argon ion bombardment showed the minimal SFE and fractional polarity values. The properties of the oxide ceramic coatings shifted to the more hydrophobic region after the surface treatment by argon ions.

The BM cells were cultured during 18 – 20 days. The medium was changed every three days to achieve a sub-confluent monolayer, and then the culture was prepared for one passage. At the end of each passage, the cells were removed by a solution of trysin-EDTA (Sigma-Aldrich) using a standard method, inactivated by a medium containing 10% FCS, washed out and counted. The effect of the conditions of depositing oxide ceramic coatings on the adhesive potential of the BM cells was registered (table 2).

During the culturing on the Ta_2O_5 coated substrates, a reduction of the BM cells adhesion potential was observed (table 2).

Table 2. Adhesive potential of the BM cells on glass and Ta_2O_5 coated substrates.

Substrate/coatings	Adhesive potential of BM cells, %	Percent of attached cells,%
Glass (control)	$7,12\pm0,64$	100
Glass/ Ta_2O_5	$5,73\pm0,53^*$	$80,46^a$
Glass/ Ta_2O_5 (argon ions bombardment)	$3,00\pm0,24^*$	$42,22^*$

[a]Statistically significant difference to control, $P < 0,05$.

The minimal values of cell adhesive potential were found in the case of Ta_2O_5 coatings deposited by a simultaneous bombardment by argon ions (figure 4). Therefore, the low adhesive and proliferative activity suggest the possibility of applying the Ta_2O_5 coatings investigated to angioplasty and vascular surgery in view of reducing the likelihood of thrombosis.

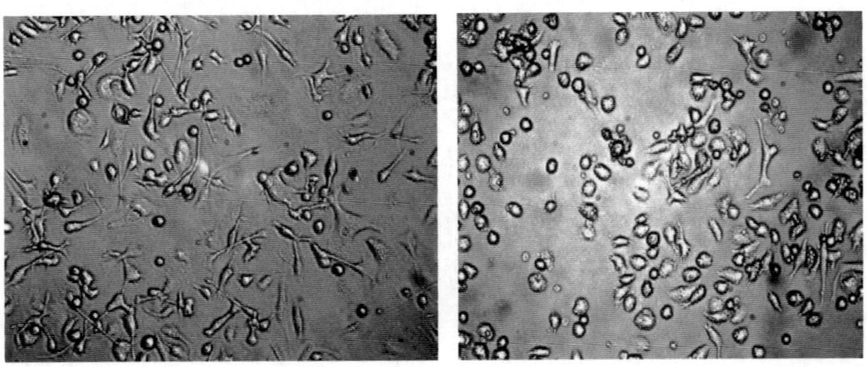

Figure 4. Morphology of adhesive BM cells by means of inverted microscopy: a) Ta_2O_5 coatings, b) Ta_2O_5 coatings deposited by simultaneous bombardment by argon ions (phase contrast, magnification 1000).

4. Conclusions

The deposition process carried out by a simultaneous bombardment by argon ions resulted in a surface modification of the growing films. The surface roughness parameters decreased from $S_a = 0.9$ nm to $S_a = 0.6$ nm following the ion bombardment. The changes in the surface structure parameters led to changes in the surface characteristics. The minimal values of SFE and the fractional polarity were exhibited by the tantalum pentoxide ceramic coatings deposited by a simultaneous bombardment. The properties of the oxide ceramic coatings shifted to the more hydrophobic region after a surface treatment by argon ions.

It was shown that the argon ions bombardment during tantalum pentoxide coating deposition process reduces the adhesive and proliferative potential of BM cells. Therefore, the Ta_2O_5 coatings are promising for application to angioplasty and vascular surgery due to the low adhesive and proliferative activity.

Acknowledgment
This work was supported under a program for international scientific collaboration between the Bulgarian Academy of Science and the National Academy of Science of Ukraine, research program of NASU No 24-04-14 and program of RFBR research project No №14-02-90457.

References

[1] Dalby M J, Riehle M O, Sutherland D S, Agheli H and Curtis A S G 2004 *Biomaterials* **25** 5415
[2] Mendonca G, Mendonca D B S, Simoes L G P, Araujo A L, Leite E R, Duarte W R, Aragao F J L and Cooper L F 2009 *Biomaterials* **30** 4053
[3] Goltsev A N, Rassokha I V, Dubrava T G, Ostankova L V, Ostankov M V, Gordienko E A, Safonov V I and Zykova A V 2013 *Cell Transplant. Tissue Eng.* **8** 46
[4] Georges P C and Janmey P A 2005 *J. Appl. Physiol.* **98** 1547
[5] Safonov V, Zykova A, Smolik J, Rogowska R, Luk'yanchenko V and Kolesnikov D 2014 *J. Appl. Surf. Sci.* **310** 174
[6] Karjalainen P P, Ylitalo A, Airaksinen J K and Nammas W 2011 *J. Interven Cardiol.* **24** 1
[7] Zhang F, Zheng Z, Chen Y, Liu X, Chen A and Jiang Z 1998 *J. Biomed. Mater. Res.* **42** 128
[8] Tsyganov I, Maitz M F, Wieser E, Richter E and Reuther H 2005 *Surf. Coat. Technol.* **200** 1041
[9] Matsuno H, Yokoyama A, Watari F, Uo M and Kawasaki T 2001 *Biomaterials* **22** 1253
[10] Miyazaki T, Kim H M, Kokubo T, Ohtsuki C and Nakamura T 2002 *Biomaterials* **23** 827
[11] Georgiev G, Fescheschiev N, Popov D and Uzunov Z 1989 *Vacuum* **36** 595
[12] Leng X Y, Sun H, Yang P, Chen J Y, Wang J, Wan G J, Huang N, Tian X B, Wang L P and Chu P K 2001 *Thin Solid Films* **398/399** 471
[13] Donkov N, Mateev E, Zykova A, Safonov V, Kolesnikov D, Goncharov I, Sudzhanskaya I and Yakovin S 2014 *J. Phys.: Conf. Series* **558** 012036
[14] Atanassova E, Kalitzova M, Zollo G, Paskaleva A, Peeva A, Georgieva M and Vitali G 2003 *Thin Solid Films* **426** 191
[15] Hallab N J, Bundy K J, O'Connor K, Moses R L and Jacobs J J 2001 *Tissue Eng.* **7** 55

Surface photovoltage and photoluminescence study of thick Ga(In)AsN layers grown by liquid-phase epitaxy

V Donchev[1,5], M Milanova[2], J Lemieux[3], N Shtinkov[3] and I G Ivanov[4]

[1]Faculty of Physics, St. Kliment Ohridski University of Sofia, 5 J. Bourchier blvd., 1164 Sofia, Bulgaria
[2]Central Laboratory of Applied Physics, Bulgarian Academy of Sciences, 59 St. Petersburg blvd., 4000 Plovdiv, Bulgaria
[3]Department of Physics, University of Ottawa, Ottawa, ON, K1N 6N5, Canada
[4]Department of Physics, Chemistry & Biology, Linköping University, 581 83 Linköping, Sweden

E-mail: vtd@phys.uni-sofia.bg

Abstract. We present an experimental and theoretical study of Ga(In)AsN layers with a thickness of around 1 μm grown by liquid-phase epitaxy (LPE) on n-type GaAs substrates. The samples are studied by surface photovoltage (SPV) spectroscopy and by photoluminescence spectroscopy. Theoretical calculations of the electronic structure and the spectral dependence of the dielectric function are carried out for different nitrogen concentrations using a full-band tight-binding approach in the $sp^3d^5s^*s_N$ parameterisation. The SPV spectra measured at room temperature clearly show a red shift of the absorption edge with respect to the absorption of the GaAs substrate. This shift, combined with the results of the theoretical calculations, allows assessing the nitrogen concentration in different samples. The latter increases with increasing the In content. The analysis of the SPV phase spectra provides information about the alignment of the energy bands across the structures. The photoluminescence measurements performed at 2 K show a red shift of the emission energy with respect to GaAs, in agreement with the SPV results.

1. Introduction

The addition of nitrogen in small concentrations is known to shift the absorption edge of III-V materials to longer wavelengths, which makes dilute III-V nitrides promising materials for advanced optoelectronic devices [1]. The commonly used fabrication methods for dilute nitrides are molecular beam epitaxy (MBE) and metalorganic chemical vapor deposition (MOCVD). Another technique, liquid-phase epitaxy (LPE), has recently shown promise in obtaining high-quality dilute nitride layers with low defect concentrations for optoelectronic applications [2,3]. In contrast to other epitaxial methods, the crystallization at LPE growth is carried out under near equilibrium conditions, which provide a high-quality material in terms of lifetime, mobility and absence of defects in the epitaxial layers. However, since LPE has rarely been used for growing this type of materials, the data for the

[5] To whom any correspondence should be addressed.

structural, optical and electronic properties of LPE-grown Ga(In)AsN alloys are still scarce. Therefore, these materials need to be extensively studied in comparison with materials fabricated by more conventional techniques like MBE and MOCVD.

Among the experimental techniques applied to dilute nitrides, surface photovoltage (SPV) spectroscopy has been relatively rarely used [4,5] although it provides information on the optical absorption spectrum and carrier transport in the sample [6]. Besides, the few reported SPV studies on dilute nitrides only concern MBE or MOCVD grown samples. The present contribution partially fills this gap by presenting an original study of the electronic and optical properties of LPE-grown Ga(In)AsN layers based on SPV spectroscopy, combined with photoluminescence (PL) and semi-empirical tight-binding calculations.

2. Experimental

A series of GaAsN and GaInAsN epitaxial layers were grown by the horizontal graphite slide-boat technique for LPE on (100) n-type GaAs:Si ($\sim 10^{18}$cm^{-3}) substrates. A flux of Pd-membrane purified hydrogen at atmospheric pressure was used for the experiments. No special baking of the system was carried out before epitaxy. The starting materials for the solutions consisted of 6N pure solvent metals Ga, In and of polycrystalline GaAs and GaN with a purity of 5N. Ternary GaAsN compounds were grown from Ga-rich solution, and quaternary GaInAsN with low In content were fabricated from mixed In+Ga melt. The N content in the melt was 0.5 at % for all grown layers. The 1-2 μm thick epitaxial layers were grown from initial epitaxy temperature varying in the range 620-570 °C at a cooling rate of 0.8 °C/min for 3-4 minutes. The layer thicknesses and composition were determined using a LYRA I XMU (Tescan) scanning electron microscope equipped with an energy dispersive X-ray (EDX) microanalyzer (Quantax, Bruker) in the Laboratory of Material Technology, Department of Solid State Physics and Microelectronics, Sofia University. The In content in the GaInAsN layers varied in the range 0.01-0.02 depending on the melt composition and epitaxy temperatures. Hall measurements revealed n-type doping of the layers in the $6\text{-}7\times10^{17}$cm^{-3} range.

The SPV spectra were recorded at room temperature using the metal-insulator-semiconductor operation mode of the SPV technique [6]. A 250 W halogen lamp along with a SPEX grating monochromator (f = 0.25 m, 600 gr/mm) and an optical chopper (94 Hz) were used for illumination. The probe electrode was a semitransparent SnO_2 film evaporated on the bottom surface of a quartz glass. A sheet of mica (15μm) was placed between the probe and the sample. The probe signal with respect to the ground was fed to a high-impedance unity gain buffer and then measured by an SR830 lock-in amplifier. The photon flux was kept constant ($\approx 1.5\times10^{14}$ cm^{-2} s^{-1}) within ± 0.5 % for all the wavelengths by positioning a neutral density filter with a variable optical density. The reference signal from the optical chopper defined the zero value of the SPV phase. More details about the SPV experimental set-up and measurement procedure can be found in [7]. The spectral resolution of the measurements was limited to 10 meV, which determined the precision of the energy positions presented below.

The photoluminescence was measured at 2 K. The sample was excited by a CW Ar-ion laser at the wavelength of 351 nm (3.53 eV) and a power of ~3 kW/cm^2. A CCD camera was used for detection.

3. Theoretical calculations

The electronic structure of $Ga_{1-y}In_yAs_{1-x}N_x$ was calculated using the semi-empirical tight-binding method for dilute nitrides in the $sp^3d^5s^*s_N$ tight-binding parameterization, including the spin-orbit coupling [8,9]. The imaginary part of the dielectric function was computed using the Fermi Golden Rule and the dipole approximation for the transition probabilities [10]. The integration over the Brillouin zone was carried out over 1.25×10^5 k-points and the spectra were smoothed out using a convolution with a Gaussian with half-width at half maximum equal to 15 meV. As the tight-binding parameters were calculated for 0 K, a temperature correction of 96.5 meV [11] was applied to compare the theoretical dielectric function to the experimental spectra. The absorption edge offset of $Ga_{1-y}In_yAs_{1-x}N_x$ with respect to pure GaAs was calculated using non-linear bandgap dependences (bowing parameters) and band offsets from Ref. [11] for N and In concentrations in the expected experimental range ($x \leq 0.01$; $y \leq 0.1$, respectively).

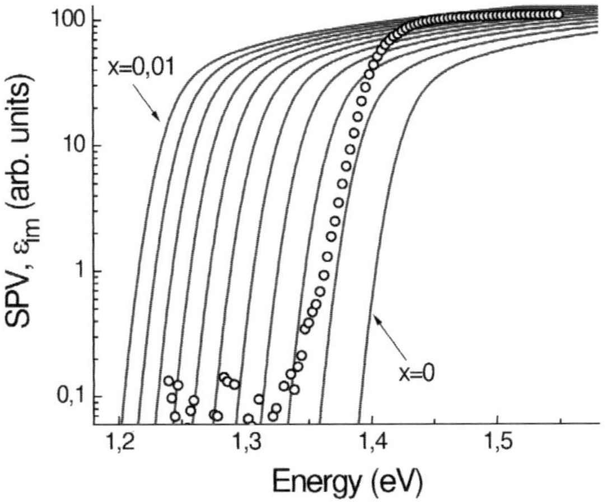

Figure 1. SPV spectrum of a GaAsN sample (symbols) compared with the calculated imaginary part of the dielectric function of $GaAs_{1-x}N_x$ (lines).

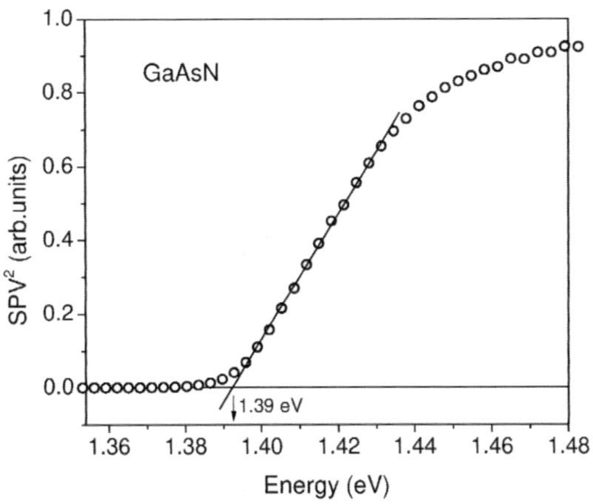

Figure 2. Determining the bandgap of the GaAsN sample from figure 1 using the squared SPV spectrum.

4. Results and discussions

The symbols in figure 1 represent the SPV spectrum of a GaAsN sample. It is compared with the calculated imaginary part of the dielectric function of $GaAs_{1-x}N_x$ for x ranging from 0 (right curve) to 0.01 (left curve) with a step of 0.001. The comparison shows that the N content in the experimental sample is likely to be $x = 0.001$ (0.1%). This is confirmed by the following considerations. To assess the bandgap of the material we plot the square of the SPV vs photon energy, which for the ideal case of a bulk semiconductor should give a straight line resulting from the shape of the 3D combined density of states. In the present case we consider the linear part of the obtained curve and extrapolate the straight line to find its intersection with the abscissa (see figure 2), which is assumed to be the bandgap, E_g. For the current sample E_g = 1.39 eV and therefore the bandgap shift ΔE_g with respect to GaAs (E_g=1.42 eV) is 30 meV. Figure 3 shows calculated bandgap shifts as a function of the In content (y) in $Ga_{1-y}In_yAs_{1-x}N_x$ for several N content values as in figure 1. For $y = 0$ (i.e. GaAsN) and $x = 0.001$, ΔE_g = 31 meV in accordance with the experimental finding.

Figure 4 represents the normalized SPV amplitude spectrum of a GaInAsN sample with In content $y \cong 0.015$ (~ 1.5 %) compared with that of the GaAs substrate. The small undulations in the spectra are due to the interference effects caused by the mica sheet. Plotting the square of the SPV vs photon energy for both curves we obtain E_g = 1.35 eV and E_g = 1.42 eV for the GaInAsN layer and the GaAs substrate, respectively, which gives ΔE_g = 70 meV. Comparing the values of y and ΔE_g with the curves from figure 3 we obtain $x \cong 0.002$ (~0.2%) for the N content in this sample. Similar considerations applied to samples with In content about 2% give $\Delta E_g \cong 90$ meV and therefore $x \cong 0.003$ (0.3%). In samples with 1% In content we obtain $\Delta E_g \cong 40$ meV and $x \cong 0.001$ (0.1%). The above described results lead to the conclusion that the N content in the layers increases from 0.1% to 0.3% with increasing the In content from 1% to 2%.

Figure 5 represents with stars the bandgap reduction in $Ga_{1-y}In_yAs$ due to the addition of In [11]. As can be seen, it is only 19 meV when the In content increases from 0 to 2%. The above discussed experimental bandgap shifts ΔE_g are represented with circles on the same figure. They are quite larger as compared to the shifts only due to In. Therefore, they are mainly due to the presence of N in the samples. On the other hand, the difference between ΔE_g of the $GaAs_{0.999}N_{0.001}$ (30 meV) and

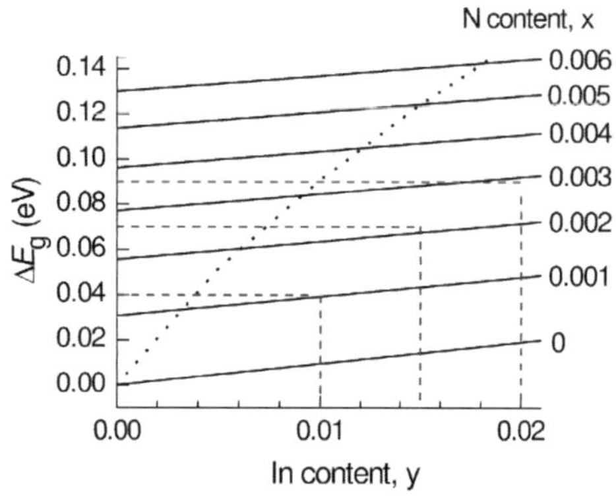

Figure 3. Calculated bandgap shifts with respect to pure GaAs as a function of the In content (y) in $Ga_{1-y}In_yAs_{1-x}N_x$. The dotted line corresponds to the lattice-matched condition $y = 3x$. The dashed lines mark the experimentally obtained bandgap shifts and the corresponding In contents.

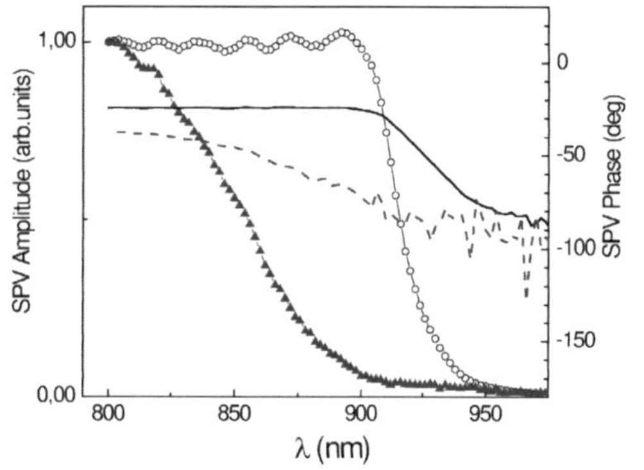

Figure 4. SPV amplitude (symbols) and phase (lines) spectra of n-GaAs substrate (triangles and dashed line) and a GaInAsN sample (circles and solid line), measured at 300K.

$Ga_{0.99}In_{0.01}As_{0.999}N_{0.001}$ (40 meV) samples is due to the presence of In in the second case and is consistent with the value of In induced bandgap reduction [11].

Figure 4 also shows the SPV phase spectra of the GaInAsN and substrate samples. Both phases are in the IV quadrant and their behavior is typical for the case of upwards (with respect the surface) energy band bending [7], which is in accordance with the n-type doping. The phase spectra suggest also that the energy band bending at the interface layer/substrate is also upwards. This is expected from the lower doping of the layer as compared to the substrate.

Finally we present in figure 6 the PL spectrum of a GaInAsN sample measured at 2 K. The spectrum reveals a strong asymmetric peak at 1.457 eV with a low-energy tail containing smaller peaks and shoulders. The bandgap shift ΔE_g with respect to GaAs, determined from the peak position, in this case is 62 meV. From the calculations shown in figure 3, this corresponds to N content x close to 0.002 (0.2%). Bearing in mind the large absorption coefficient of GaAs and InAs at the excitation wavelength used (> 5×10^5 cm^{-1}) we have to assume that the information provided by the PL is only taken from the upper 20-40 nm of the sample. Some gradient of the N content along the growth direction is not excluded.

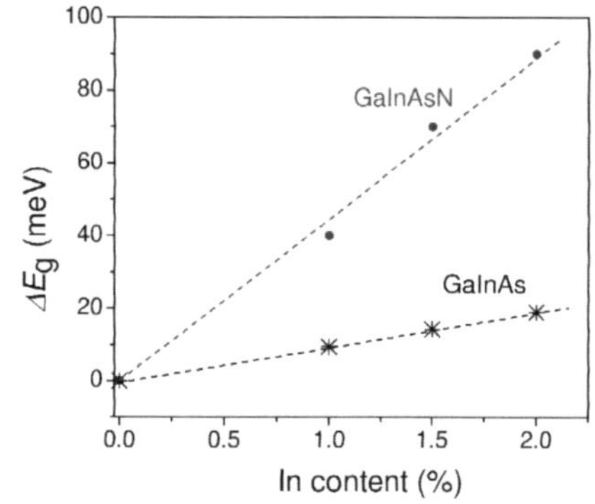

Figure 5. Bandgap shifts as a function of In content in GaInAs [11] (stars) and GaInAsN (circles). The lines are guides for the eyes.

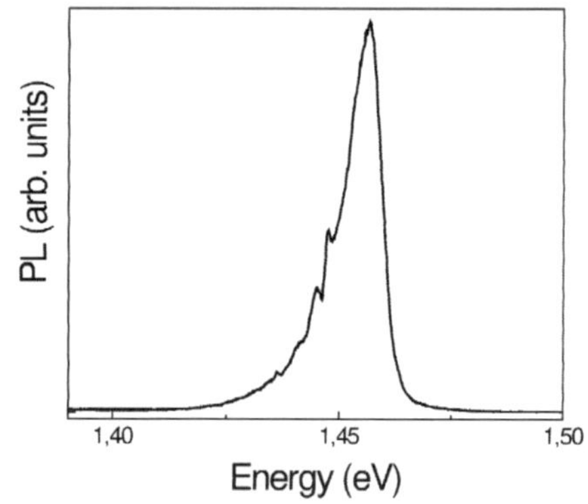

Figure 6. Photoluminescence spectrum of a GaInAsN sample, measured at 2 K.

5. Conclusions

An original study is presented of the optical properties of thick LPE grown Ga(In)AsN layers. The absorption features, measured by SPV spectroscopy, are compared with results obtained by tight-binding calculations. In this way, an indirect assessment of the N content in the layers is achieved. For the GaInN sample the N content is found to be 0.1%. On the other hand, with increasing the In content in the layers from 1% to 2% the incorporation of N increases from 0.1 % to 0.3%. The bending of the energy bands determined by the SPV phase spectra is found to be upward both at the surface and at the interface layer/substrate. This work demonstrates the applicability of the SPV technique as an alternative of the optical absorption spectroscopy for studying dilute nitride materials. The results obtained contribute to the better understanding of the physical properties of dilute nitrides grown by LPE and their potential for optoelectronic applications. Work is under progress for achieving GaInAsN samples with larger N contents, respectively with larger bandgap shifts.

Acknowledgements

This work was supported by the research fund of the Sofia University (contract 71/2015) and COST action 1406 MutiscaleSolar. Prof. B. Arnaudov is acknowledged for the Hall measurements and K. Genkov for the EDX measurements.

References

[1] Erol E A ed 2008 *Dilute III-V Nitride Semiconductors and Material Systems. Physics and Technology* (Berlin: Springer)
[2] Dhar S, Halder N, Kumar J and Arora B M 2004 Observation of a 0.7 eV electron trap in dilute GaAsN layers grown by liquid-phase epitaxy *Appl. Phys. Lett.* **85** 964–6
[3] Milanova M, Vitanov P, Terziyska P, Popov G and Koleva G 2012 Structural and electrical characteristics of InGaAsN layers grown by LPE *J. Crystal Growth* **346** 79-82
[4] Kudrawiec R et al. 2014 Surface photovoltage and modulation spectroscopy of E_- and E_+ transitions in GaNAs layers *Thin Solid Films* **567** 101–4
[5] Bansal B, Kadir A, Bhattacharya A, Arora B M and Bhat R B 2006 Alloy disorder effects on the room temperature optical properties of GaInNAs quantum wells *Appl. Phys. Lett.* **89** 032110
[6] Kronik L and Shapira Y 1999 Surface photovoltage phenomena: Theory, experiment, and applications *Surf. Sci. Rep.* **37** 1–206
[7] Donchev V, Ivanov T, Germanova K and Kirilov K 2010 Surface photovoltage spectroscopy – an advanced method for characterization of semiconductor nanostructures *Trends in Appl. Spectr.* **8** 27–66
[8] Shtinkov N, Desjardins P and Masut R 2003 Empirical tight-binding model for the electronic structure of dilute GaNAs alloys *Phys. Rev.* B **67** 081202
[9] Turcotte S, Shtinkov N, Desjardins P, Masut R and Leonelli R 2004 Empirical tight-binding calculations of the electronic structure of dilute III–V–N semiconductor alloys *J. Vac. Sci. Technol.* A **22** 776
[10] Voon L C L Y and Ram-Mohan L R 1993 *Phys. Rev.* B **47** 15500
[11] Vurgaftman I, Meyer J R and Ram-Mohan L R 2001 *J. Appl. Phys.* **89** 5815

Journal of Physics: Conference Series **700** (2016) 012029 doi:10.1088/1742-6596/700/1/012029

Interface characterization of nanoscale SiO_x layers grown on RF plasma hydrogenated silicon

E Halova[1], N Kojuharova[1,3], S Alexandrova[1] and A Szekeres[2]

[1]Department of Applied Physics, Technical University of Sofia, 8 St. Kl. Ohridski blvd., 1797 Sofia, Bulgaria
[2]Georgi Nadjakov Institute of Solid State Physics, Bulgarian Academy of Sciences, 72 Tzarigradsko Chaussee, 1784 Sofia, Bulgaria

E-mail: nkojuharova@tu-sofia.bg

Abstract. In the present paper, results are presented on electrical characterization of the interface Si/SiO_x, formed by oxidation on Si wafers previously exposed to rf hydrogen plasma. As tools of investigations, multiple frequency C-V and G-V measurements were applied. The data analysis was performed using a two-frequency method to extract a generalized frequency independent C-V characteristic. The interface trap densities were evaluated from the generalized C-V data by comparison with theoretical data for an ideal interface. A set of localized states, acting as interface traps, was found that characterize the interface of Si to substoichiometric SiO_x, layer with $x < 2$. The dielectric constant of the oxides was calculated from the capacitance in accumulation of the generalized C-V curves. The thickness and the refractive index of the oxide layers were obtained from ellipsometric data analysis assuming the oxide-Si substrate as being a single layer system. From the data for the dielectric constant and refractive index, the suggestion is made that the oxides grown on hydrogenated Si contain voids thus reducing the dielectric constant. Correlation with oxide mechanical stress is found.

1. Introduction

Control and characterization of the defects at the interfaces in single or multilayered structures is an important issue in modern semiconductor devices. The incorporation of hydrogen into a semiconductor structure is known for its beneficial effect, namely, stabilization of device characteristics mainly through gettering of impurity defects. In silicon technology, hydrogenation has been widely used in the past decade in lower concentrations for improving the defect and leakage characteristics, and in higher concentrations for future nanoelectronic applications, such as exfoliation of silicon in smart cut process, increase of carrier life time in solar cells, etc.

Contemporary semiconductor electronics is still largely dominated by the silicon technology. Beyond for CMOS digital applications, Si/SiO_2 has been used for applications in various structures from solar cells to graphene films and carbon nanotubes, new 2D geometry, double gate devices, ballistic nanotube transistors etc. A detailed discussion on future developments of nanodevices can be found in the excellent reviews [1, 2, 3]. It can be concluded that at least in the next decade Si/SiO_2 will remain an important building block.

[3] To whom any correspondence should be addressed.

The aim of the present work is characterization of the interface between rf plasma hydrogenated (100) p-Si wafers and the overgrown nanoscale SiO_x by thermal oxidation at 850°C. The interface traps density spectra are assumed to depend on the amount of incorporated hydrogen. For this purpose, plasma hydrogenation of the Si wafers was conducted either without heating of the Si substrate or heating them up to 300 °C; the interface traps density spectra were examined in detail.

2. Experimental details

2.1 Sample preparation

The MOS capacitors investigated in this work were formed on substrates of p-type (100)-oriented single-crystal Si wafers, 5-10 Ohm cm. The Si wafers were initially cleaned using H_2SO_4/H_2O_2 solutions by standard wet RCA pre-gate oxide cleaning followed by a dip in diluted HF and a rinse in deionized H_2O. Parts of the wafers were subsequently hydrogenated by exposure to a rf hydrogen plasma in a planar plasma unit at 13,56 MHz, gas pressure of 133 Pa and input power of 15 W. The substrates were kept on the lower electrode. The duration of the plasma exposure was 15 min. The hydrogenation was performed on Si wafers, either unheated or at 300 °C. Other wafers were left with RCA cleaning only, to serve as a reference, and will be referred to as RCA oxides. The differently cleaned wafers were thermally oxidized at 850 °C by exposing to dry O_2 flow in a conventional atmospheric furnace. The thickness of the SiO_x layers varied depending on the conditions of the pre-oxidation treatment, the hydrogenation level at different temperatures or wet RCA, but was in the nanoscale range below 10 nm, as determined from the ellipsometric measurements. In order to obtain information about the charged states in the structures, the oxides were incorporated in MOS capacitors formed by thermally-evaporated circular aluminum contacts with an area $1{,}96\times10^{-3}$ cm^2 on the SiO_x surface using a mechanical mask, and a continuous Al layer on the back side of the substrates.

2.2 Measurements

Information on the concentration of the charged defects and their location in the Si/SiO_x interface region was obtained by using multiple frequency capacitance-voltage (C-V) and conductance-voltage (G-V) measurement techniques. The measurement unit was a Precision Component Analyzer Wayne-Kerr 6425. Applying the two-frequency method to the C-V and G-V characteristics taken at various frequencies ranging from 500 Hz to 300 kHz and a 30-mV test signal, at room temperature, a generalized C-V curve was generated.

The density of interface trap $D_{it}(E)$ profiles in the Si bandgap was investigated by the standard high-frequency method of obtaining C-V characteristics of the MOS structures and comparing the experimental curve with the ideal theoretical one [4]. For that purpose, a two-frequency method was developed and applied [5] to explore thin and/or leaky oxides. The energy spectra of the interface traps $D_{it}(E)$ were extracted from comparison of the experimental and the generalized C-V curves. The three-element circuit model of a MOS capacitor was used. This technique allowed the evaluation of a generalized frequency independent C-V curve.

From the measured values of C-V and G-V curves for two frequencies in the range 500 Hz - 300 kHz, the dissipation D is calculated from the formula

$$D = \frac{G}{\omega C}, \tag{1}$$

where G is the parallel conductance and C is the capacitance measured at the different frequencies.

Based on the method of Yang and Hu, presented in [5], the generalized capacitance C' is

$$C' = \frac{f_1^2 C_1(1+D_1^2) - f_2^2 C_2(1+D_2^2)}{f_1^2 - f_2^2}, \tag{2}$$

where C_1 and D_1 are the values measured at the frequency f_1 and C_2 and D_2 are the values measured at the frequency f_2.

Additional information about the oxide and the interface region to the Si was acquired using a Rudolf Research spectroscopic ellipsometer, in the wavelength range 280-820 nm and at an incidence angle of 70°. The accuracy of the thickness determination was ± 0,2 nm. The ellipsometric data and the capacitance in strong accumulation of the C-V curves used used for calculation of the thickness and the dielectric constants of the oxide layers. The refractive index was calculated assuming the oxide-Si as a single oxide layer system.

3. Results and discussion

By applying the two-frequency method to the C-V and G-V characteristics taken at various frequencies [5], generalized C-V curves for all samples were generated. In these curves, the series resistance and the leakage through the oxide layer is accounted for.v Figure 1 presents the generalized C-V curves. The study of the electrical characteristics showed that the MOS structures with SiO_x layer grown on hydrogenated Si exhibit shifts and shapes of the measured curves that refer to a small oxide charge and low leakage currents, i.e., a high-quality dielectric layer.

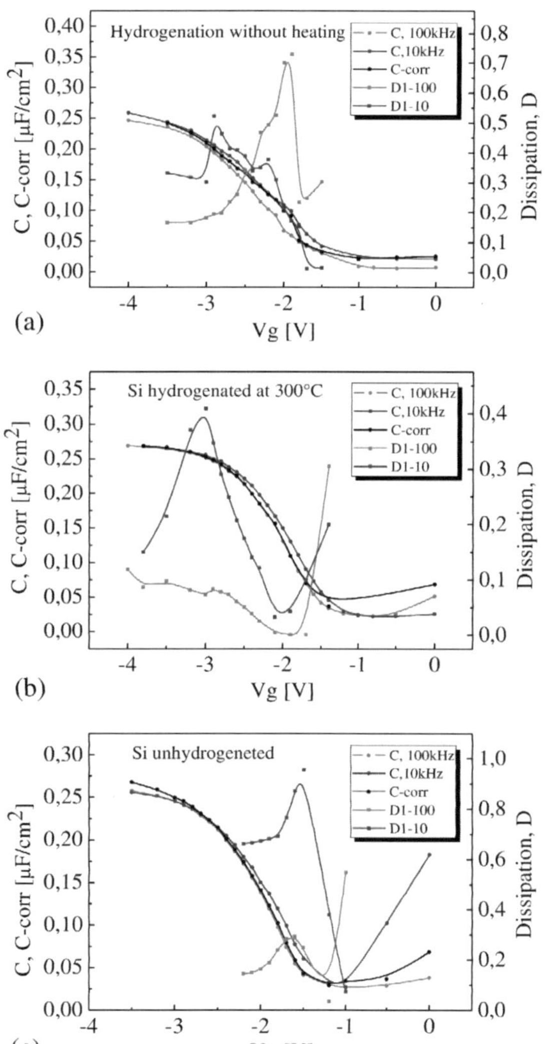

Figure 1. C-V and D-V curves at different frequencies and generalized C-V curve (C-corr) for oxides grown on a) Si hydrogenated without heating; b) Si hydrogenated at 300 °C; c) unhydrogenated Si.

The shape of the curves of plasma-treated Si without heating shows variations typical of a high density of interface traps. In the shape of curves of Si oxides hydrogenated without heating (figure 1a), one observes deviations from the ideal shape, typical for a high density of interface traps, even higher than the oxide of the unhydrogenated Si (figure 1c). Hydrogenation of the Si wafer at 300 °C results in an oxide with C-V and generalized C-V curves with a regular shape, as seen in figure 1b. The C-V curves of this sample are also steeper in comparison to the C-V curves of the oxides on Si unheated during the plasma hydrogenation, which indicates an overall lower density of traps, which can be attributed to the higher degree of an annealing or passivation effect.

The interface traps density was calculated over the Si bandgap using the distributions D_{it} obtained from comparison of the generalized C-V curves with the theoretical C-V curves, calculated for the idealized case without the presence of interface traps. The results for all oxides grown on differently treated Si wafers prior to oxidation are presented in figure 2. The densities around midgap are below 10^{12} eV^{-1} cm^{-2} for all oxides. The interface density spectra show the presence of localized interface traps energetically distributed in the Si bandgap. Localized interface levels appear approximately at the same energy positions in the bandgap for the oxides grown

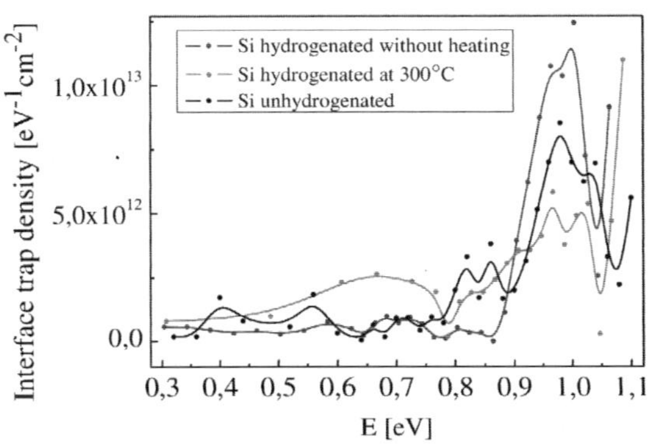

Figure 2. Interface trap density spectra.

on hydrogenated and unhydrogenated Si wafers. The positions and the density of the trap levels depend on the hydrogenation temperature of the substrate and are assumed to be related to particular structural defects at the interface and in a thin region inside the oxide.

It is known that the dangling Si bonds form a peak at 0,82 eV above the valence band edge E_v [6]. A double peak at about this position is seen most clearly in our oxides on unhydrogenated Si wafers. Generation of defects is often related to intrinsic interface stress of Si = SiO_2 [7]. For our oxides, the stress levels as obtained by ellipsometric measurement [8] are presented in the table. Another double peak found in all oxides around 0,97 eV near E_v can be attributed to vacancy-oxygen (V/O) complex defects. A localized interface trap level was found near this energy position in irradiated SiO_2 [9]. It can be seen that in the oxide grown on a Si substrate that has been heated up to 300°C during plasma hydrogenation, the density of the traps is reduced. A higher trap density in this region has an important contribution to the flatband and threshold voltage values so that the reduction of the density of traps is an important result. Over the rest of the Si bandgap, the density is close for all studied oxides. Moreover, the oxides grown on unhydrogenated Si exhibit a set of levels around midgap which disappear in oxides on hydrogenated Si and the spectra flatten. The explanation of the behavior of the interface traps can be found in the combination of annealing effect at the increased temperature and hydrogen incorporated during the plasma hydrogenation, which serve for the reconstruction of chemical bonds in the region of the interface.

The results of ellipsometry measurements, summarized in the table 1, also indicates that the grown oxides must be regarded as SiO_x, where $x < 2$. The values of the refractive index over 1,46 are typical of non-stoichiometric SiO_x. For our oxides, x was estimated from simulated data for the refractive index for SiO_x when varying the oxygen content [10]. The results are given in the table 1. Similar values for x were also found for SiO_x films deposited by reactive sputtering [11].

Table 1. Parameters of oxides grown on Si with different pre-oxidation treatment.

Si wafer pre-oxidation treatment	Refractive index n	Oxide thickness d_{ox} (nm)	Dielectric constant ε	$\sigma\times10^8$ (N/m^2)	x
Plasma hydrogenation: unheated	1,606	13,27	3,63	3,1	1,26
Plasma hydrogenation: 300 °C	1,678	12,57	3,81	2,0	1,16
Without hydrogenation	1,660	12,02	3,64	4,0	1,20
SiO_2	1,460	-	3,90	-	2,00
SiO	1,960	-	5,00	-	1,00

The dielectric constants of the oxides were calculated from the capacitance in strong accumulation of the generalized C-V curves using the thickness of the oxide layers from the ellipsometric data. The results for the dielectric constant of the oxide are also summarized in the table. For comparison, additionally in the table the data for ε and n for stoichiometric SiO_2 and silicon oxide SiO are given [12, 13, 14]. The parameters for oxides on hydrogenated Si can differ also from those of oxides grown on unhydrogenated Si before oxidation, which is evident from the table. The refractive index data reveal higher values as compared to stoichiometric SiO_2 and substantially lower values than for SiO monoxide. This implies that the oxide composition can be characterized as rather substoichiometric silicon dioxide with $x < 2$. From the data for the dielectric constant, one can also draw conclusions about the structure of oxides. It can be supposed that the lower dielectric constant are due to the presence of voids ($\varepsilon = 1$) in the upper Si-layer formed during the hydrogenation plasma, which persist even after oxidation [15].

4. Conclusions

Electrical characterization of silicon oxides grown on rf plasma hydrogenated Si was studied. The densities of interface traps were evaluated from C-V and G-V measurements taken at different frequencies, using a two-frequency technique to generate independent frequency and leakage curves. The C-V and G-V analysis demonstrated the generation of localized interface traps due to dangling Si bonds, and a vacancy-oxygen (V/O) complex. A series of localized states acting as interface traps was found that characterize the interface of Si as substoichiometric SiO_x layer. Most of the traps were annealed for oxide grown on Si hydrogenated at increased wafer temperature. The dielectric constants were determined from the generalized C-V curves. The suggestion was made that grown oxides on hydrogenated Si contain voids that reduce the dielectric constant. The mechanical stress of the oxide, estimated by ellipsometry, indicated a decrease due to the presence of voids in the oxide. It was concluded that the concentration of interface defects can be optimized by heating the Si to 300 °C during plasma hydrogenation.

References

[1] Saraswat K *How far can we push Si CMOS?* http://www.ohio.edu/people/starzykj/network/class/ee516/slides/Future%20Devices.pdf
[2] Iwai H 2013 Ultimate CMOS scaling *Proc. Korean Int. Summer School on Nanoelectronics* (2–5 July 2013 Daejon Korea) http://www.iwailab.ep.titech.ac.jp/pdf/iwaironbun/201307nanokiss.pdf
[3] Balestra F *Beyond CMOS Nanodevices* (New York: Wiley & Sons - Interscience)
[4] Nicollian E H and Brews J R 1982 *MOS Physics and Technology* (New York: Wiley)
[5] Yang K J and Hu Ch 1999 *IEEE Trans. Electron Dev.* **46** 1500
[6] Nelson S A, Hallen H D and Buhrman R A 1988 *J. Appl. Phys.* **63** 5027
[7] Stesmans A 1993 *Phys. Rev.* B **48** 2418
[8] Szekeres A, Paneva A, Alexandrova S, Lisovskyy I, Litovchenko V and Mazunov D 2003 *Vacuum* **69** 355
[9] Kaschieva S, Halova E, Vlaikova E, Alexandrova S, Valcheva E and Dmitriev S 2006 *Plasma Process. Polym.* **3** 237
[10] Tomozeiu N 2011 *Silicon Oxide (SiO_x, $0 < x < 2$): a Challenging Material for Optoelectronics, Optoelectronics - Materials and Techniques* ed P. Predeep
[11] Miyazaki H 2010 *Phys. Chem. of Glasses - European J. Glass Sci. Technol.* B **51** 136
[12] Hass G and Salzberg C D 1954 *J. Opt. Soc. Am.* **44** 181
[13] Philipp H R 1971 *J. Phys. Chem. Solids* **32** 1935
[14] Gupta T 2009 *Dielectric Materials in Copper Interconnect Technology* (New York: Springer Verlag) chapter 2 p 67
[15] Alexandrova S, Szekeres A and Halova E 2010 *IOP Conf. Ser.: Mater. Sci. Eng.* **15** 012037

Journal of Physics: Conference Series **700** (2016) 012030 doi:10.1088/1742-6596/700/1/012030

Study of the new CSAR62 positive tone electron-beam resist at 40 keV electron energy

R Andok[1], A Bencurova[1], K Vutova[2], E Koleva[2], P Nemec[1], P Hrkut[1], I Kostic[1,3] and G Mladenov[2]

[1] Institute of Informatics, Slovak Academy of Sciences, 9 Dubravska, 84507 Bratislava, Slovakia
[2] Emil Djakov Institute of Electronics, Bulgarian Academy of Sciences, 72 Tsarigradsko Chaussee, 1784 Sofia, Bulgaria

E-mail: ivan.kostic@savba.sk

Abstract. One of the few "top-down" methods for nano-device fabrication is the electron-beam lithography, which allows flexible patterning of various structures with a nanoscale resolution down to less than 10 nm. Thinner, more etching durable, and more sensitive e-beam resists are required for the better control, linearity, and uniformity of critical dimensions of structures for nano-device fabrication. Within the last decade, researchers have made significant efforts to improve the resolution of the nanoscale e-beam lithography. The resist material properties are an important factor governing the resolution. Only the e-beam resist ZEP 520 of the Japanese manufacturer ZEON is characterized by relatively good properties and thus meets most users' expectations. This paper deals with the investigation and simulation of the characteristics of the new less-expensive AR-P 6200 (CSAR 62) positive e-beam resist (available since May 2013, manufactured by Allresist GmbH company).

1. Introduction

The SX AR-P 6200/2 electron-beam resist is a positive-tone electron beam resist which provides, depending on the respective developer, a high to a very high sensitivity and allows realizing resolutions down to the sub-10-nm range. The resist is, furthermore, characterized by high process and plasma etch stability and, in addition, is very suitable for lift-off processes up to structure sizes of 10 nm.

Due to the very high contrast of the order of 15, a resolution of 10 nm can be achieved at a film thickness of 180 nm. Realization of an aspect ratio of 18 is thus possible.

By selecting an appropriate developer (X AR 600-54/8), the sensitivity of the resist can be increased to 10 μC/cm². An even higher resolution power is obtained with the X AR 600-54/6 developer.

All customers who employed the good qualities of the popular ZEP 520 can now look forward to an equivalent e-beam resist with a clear price advantage. After an intensive development work, Allresist was able to reach and partly even exceed the application benefits of the ZEP 520, e.g. with respect to the easy fabrication of lift-off structures.

The main components of the resist are poly(α-methylstyrene-co-methyl chloroacrylate), an acid

[3] To whom any correspondence should be addressed.

generator and the safer solvent anisole. The acronym CSAR is deducted from the mechanism used: **C**hemical **S**emi **A**mplified **R**esist.

A particularly advantageous feature is the easy fabrication of lift-off structures. The plasma etch resistance of the CSAR62 and ZEP 520 is comparably high and twice as good as the plasma etch resistance of the PMMA resist.

In comparison with the PMMA resists, the CSAR 62 is characterized by a higher sensitivity and a substantially improved plasma etch resistance. The higher sensitivity results from the introduction of halogen atoms into the polymer chain. Generally, chlorine is used for this purpose (as well as for the CSAR 62 – methyl chloroacrylate), but bromine or iodine are also possible alternatives. The chlorine atom supports breaking of the polymer chain during irradiation with electrons. In addition, a halogen-containing acid generator enhances this effect. The introduction of further reactive halogens accelerates the influence on the polymer chain even more. As a result, less energy (a lower dose) is required to break the high-molecular polymer into smaller fragments. These fragments quickly dissolve in the developer, while the unexposed, still high-molecular resist, remains intact.

Due to the temperature-stable acid generator, the fractions of the chain are activated during the process of electron irradiation, and the resist layer does not require any tempering after exposure.

The improved plasma etch resistance results from aromatic substitutions of activating groups, such as e.g. phenyl, naphthyl or anthracyl groups into the polymer. The CSAR furthermore contains α-methylstyrene to provide etch stability. Due to their π-electrons, aromatic compounds are considerably more stable in the presence of different plasmas as compared to aliphatic polymers such as PMMA and thus reach the same level of stability like photoresists composed of novolacs (cresol mixtures condensed with formaldehyde).

2. Experiments and calculations

2.1. Experimental results

The investigations were performed on a 400-nm layer of CSAR 62 positive e-beam resist prepared by a spin coating technique onto a Si substrate for the case of 40 keV electron energy. The contrast and sensitivity characteristics were obtained experimentally for various development times.

The process conditions for the CSAR 62 (AR-P 6200.13) e-beam resist are summarized in table 1.

Table 1. Process conditions for the positive e-beam resist CSAR 62 (AR-P 6200.13).

Coating	AR-P 6200.13, 4000 rpm, 0.4 μm film thickness
Pre-bake (± 1 °C)	150 °C, 1 min hot plate or 150 °C, 30 min convection oven
E-beam exposure	40 keV Dose-to-clear E_o: 30 μC/cm²
Development (21-23 °C ± 0,5 °C) puddle	AR 600-548, 60 sec
Stopping / Rinse	AR 600-60, 30 s / DI-H_2O, 30 s
Post-bake (optional)	130 °C, 1 min hot plate or 130 °C, 25 min convection oven for slightly enhanced plasma etching resistance
Customer-specific technologies	Generation of semiconductor properties
Removal	AR 600-71 or O_2 plasma ashing

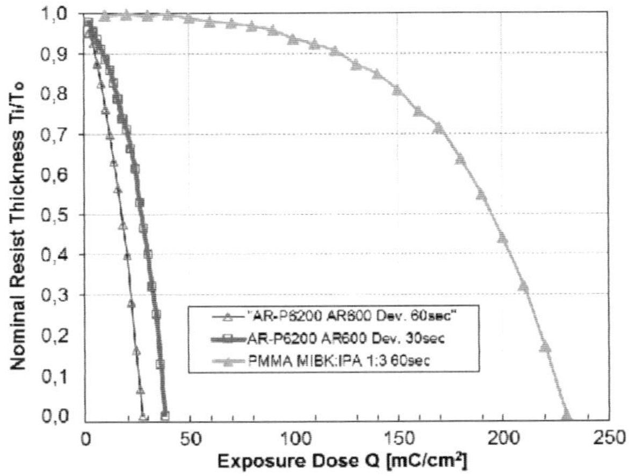

Figure 1. Contrast curves: for the CSAR62 resist developed in AR600 for 30 sec and 60 sec, and for the PMMA resist developed in MBIK:IPA 1:3 for 60 sec (irradiated by electrons accelerated to an energy of 40 keV), respectively.

A comparison between the contrast curves for the CSAR62 and PMMA is shown in figure 1. The advantages of the sensitivity and the contrast of the CSAR62 resist in comparison with the PMMA resist are clearly seen. A higher sensitivity and a better contrast for the CSAR62 resist are observed.

2.2. Simulation of the characteristics of the CSAR and PMMA resists

The proximity effect parameters were simulated for 400-nm thicknesses of the CSAR resist and of the PMMA resist layer on Si substrates. All exposures were performed at 40-keV e-beam energy (point or Gaussian sources). Discrete data for the energy deposition function (EDF) at the interface (at a depth of 400 nm) and near the resist surface (at a depth of 10 nm) were obtained by Monte Carlo simulation [1,2] using a point source for 10 000 particles (figures 2 and 3). An analytical approximation of the discrete data obtained for the EDF was performed using a sum of two Gaussians.

Table 2. Proximity effect parameters at the resist surface (*d* = 10 nm) and at the CSAR/Si interface (*d* = 400 nm).

Parameter	d = 10 nm	d = 400 nm
β_f	0.3208712000E-01 [μm]	0.6138025000E-01 [μm]
β_b	0.3918174000E+01 [μm]	0.5028665000E+01 [μm]
η_E	0.1249845000E+00	0.3074930000E+00

Table 3. Proximity effect parameters at the resist surface (*d* =10 nm) and at the PMMA/Si interface (*d* = 400 nm).

Parameter	d = 10 nm	d = 400 nm
β_f	0.2211289000E-01 [μm]	0.6308963000E-01 [μm]
β_b	0.3219354000E+01 [μm]	0.5082692000E+01 [μm]
η_E	0.3177008000E+00	0.3497990000E+00

Afterwards, the proximity effect parameters (β_f, β_b, η_E) along the resist depth were investigated and their variations were evaluated by the Monte Carlo methodology [1,2]. Tables 2 and 3 represent values of the parameters (β_f, β_b, η_E) at two resist depths – at the resists surface and at the resists/substrate interface for the CSAR and PMMA, respectively, for a point e-beam source. The values obtained of the proximity effect parameters are similar for both CSAR and PMMA resists.

The results of the Monte Carlo simulation and comparisons between the obtained EDF discrete data (▪) and the corresponding analytical fit (solid curves) at different resist depths are presented in figures 2 and 3. Figure 2 represents the discrete data and analytical fits for the EDF at the resist/substrate interface, while figure 3 shows the results at the resist surface (at a depth of 10 nm) for a point source

(10 000 particles) for both resists. A comparison between the EDF obtained at the interface for both investigated positive resists CSAR and PMMA is shown in figure 4.

2.3. Modeling and calculation of dissolution rates for the CSAR and PMMA resists

Table 4 shows estimated regression models for the dissolution rates (V, nm/s) depending on the change of the exposure dose (D, $\mu C/cm^2$) for the investigated positive CSAR and PMMA resists.

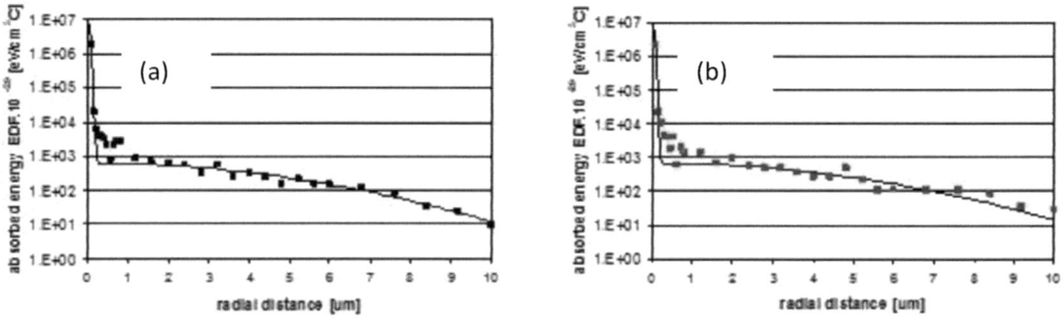

Figure 2. Comparison between the discrete data obtained (▪) and the analytical fit (solid curve) at the resist/substrate interface in the case of: (a) CSAR/Si; (b) PMMA/Si.

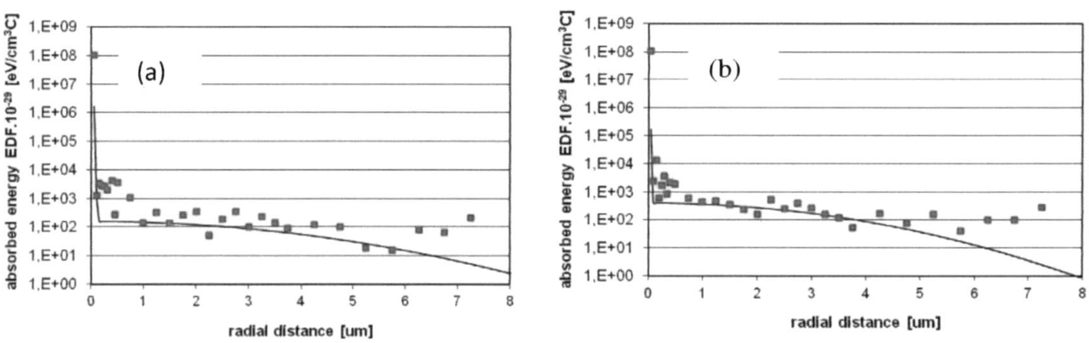

Figure 3. Comparison between the discrete data obtained (▪) and the analytical fit (solid curve) at the resist surface (d = 100 Å) for: (a) CSAR/Si; (b) PMMA/Si.

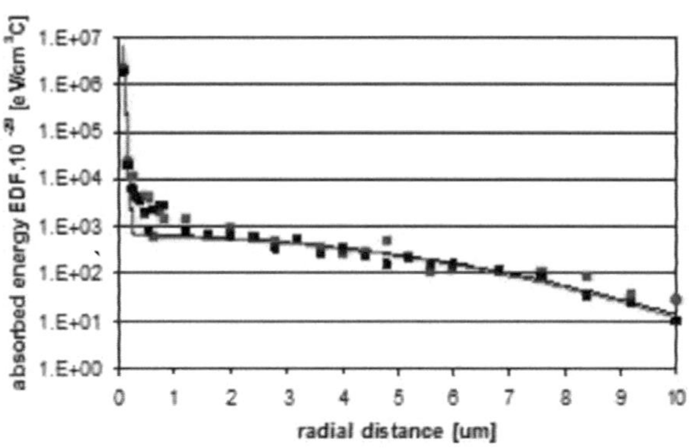

Figure 4. Obtained EDF discrete data (▪) and an analytical fit (solid curves) at the interface for the PMMA/Si (▪ and black line, respectively) and for the CSAR/Si (▪ and pink line).

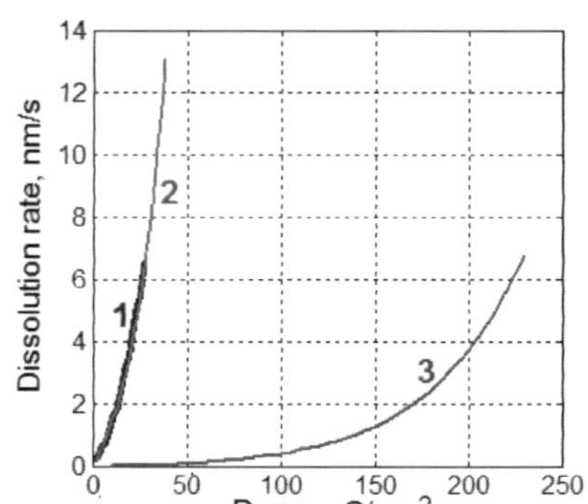

Figure 5. Dissolution rates *vs.* the exposure dose for different e-beam resists and development times: 1 - CSAR 62 resist, 60 s; 2 - CSAR 62 resist, 30 s; 3 – PMMA resist, 60 s.

The values of the squared multiple correlation coefficient, which is a measure for the models accuracy (the closer the coefficient R^2 value is to 100%, the better is the model) are also presented in table 4. The root mean square prediction error RSME is calculated by:

$$RSME = \left(\frac{1}{n}\sum_{i=1}^{n}(V_i - \hat{V}_i)^2\right)^{1/2}$$

where n is the number of experimental data for the dissolution rates V_i, $\hat{V}_i$ represent the dissolution rates estimated by the equations in table 4. As can be seen, the estimated models are adequate and good enough for prediction of the dissolution rates for the investigated e-beam resists.

The dependencies obtained of the dissolution rates on the exposure doses for both CSAR and PMMA resists at different development conditions are presented in figure 5.

It can be seen (figure 5) that for 60 s development time the dissolution rate for the CSAR 62 resist (curve 1) reaches the same values as for the PMMA (curve 3), but at much smaller doses. At one and the same exposure dose the dissolution rate in the case of the CSAR 62 resist is much higher than the one for the PMMA resist. The increase of the development time for the CSAR 62 resist leads to a decrease of both the dissolution rate and the dose needed to clear the resist (figures 1 and 5).

Table 4. Regression models for the dissolution rates *V* depending on the exposure dose *D*.

Resists and conditions	**Regression equations**	**R^2**	**RSME**
AR-P6200.13_4 (CSAR 62), development time - 60 s, resist thickness - 393 nm, exposure dose range is 2 - 27 μC/cm^2.	$\mathbf{V} = 0.006034*\mathbf{D}^2 + 0.07813*\mathbf{D} + 0.02899$	99.94%	0.0491
AR-P6200.13_1 (CSAR 62), development time - 30 s, resist thickness - 397 nm, exposure dose range is 2 - 38 μC/cm^2.	$\mathbf{V} = 0.0001701*\mathbf{D}^3 - 0.001539*\mathbf{D}^2 + 0.1604*\mathbf{D} - 0.1668$	99.90%	0.1190
PMMA, development time - 60 s, resist thickness - 404 nm, exposure dose range is 10 - 230 μC/cm^2.	$\mathbf{V} = 5.871*10^{-9}*\mathbf{D}^4 - 1.524*10^{-6}*\mathbf{D}^3 + 0.0001972*\mathbf{D}^2 - 0.007249*\mathbf{D} + 0.09337$	99.98%	0.0248

3. Application

The Strausberg-based company Allresist won the Brandenburg Innovation Award 2014 for the development of the innovative product CSAR 62, which we consider as an appropriate replacement for the very expensive ZEP520A resist. The performance of the CSAR 62 being similar to that of ZEP520A in resolution, speed, and etch resistance, is confirmed in references [3,4].

A difference in the dependencies of the relative resist thickness (and solubility rates) *vs.* the doses of exposure at two development durations – 30 s and 60 s, respectively, can be seen on the experimental contrast curves obtained by us and shown in figure 1. Similar nonlinear behavior has been observed for the CAR CAMP6 resist [5-8]. In these references, the multiciphered functions for the solubility rates provide different development kinetics behavior of the images developed, which can be predicted by the new computer algorithm. For the CSAR 62 resist, the equations in table 4 are implemented in the numerical code to be used for simulation of the profiles developed in this new positive-tone chemical self-amplified resist.

In the current R&D of nanoscale devices, patterning in various combinations of thin films and substrates is required, e.g. in the III-V semiconductors (GaN, GaInN, AlGaN, AlGaInN, GaInN), SOI wafers (Silicon on Insulator), photonic crystals, etc. The simulation methodology of the characteristics of e-beam resists, described in this paper, is suitable for an effective and precise control of structure dimensions in an e-beam resist applied as a masking layer for patterning onto various thin films. As an example of such applications, we present a dimension control of photonic crystals structures. In this case, hexagonal gratings are required with various diameters and periods of the holes. Exposures were performed with a Gaussian beam to obtain optimal circular shapes of the holes. Thin film of CSAR 62 positive resist was prepared by spin coating onto 220 nm silicon nitride/1900 nm silicon dioxide thin films on a silicon substrate. The thickness of 600 nm was chosen as appropriate for dry etching of 220 nm thin silicon nitride. The diameter of the holes was controlled by variyng the dose. In figure 6, hexagonal gratings with hole diameters of 80 nm (figure 6a), 130 nm (figure 6b) and 200 nm (figure 6c) are demonstrated. The distance between the holes forming a hexagonal pattern was 380 nm for all the cases.

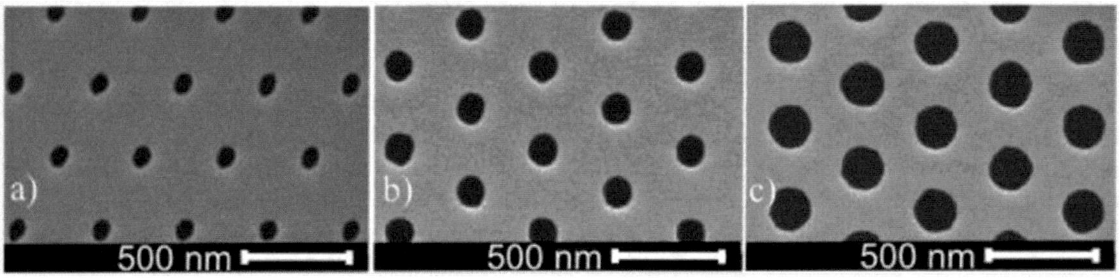

Figure 6. Hexagonal gratings in CSAR 62 with various hole diameters: a) 80 nm, b) 130 nm, and c) 200 nm. The distance between the holes is 380 nm for all cases.

4. Conclusions

In this paper, we present and discuss the results of an investigation of the characteristics of a new positive tone chemical self-amplified resist, namely, CSAR 62. It is shown that they are appropriate for various nano-lithography applications. Additionally, this resist is a cheaper alternative to the ZEP type nano resist. It was also observed that its development rate exhibited a nonlinear behavior, i.e. it depended on the time and/or the resist depth. Simulations were carried out in order to study the development of the CSAR 62 resist for hexagonal gratings with different hole diameters. The results demonstrated an efficient dimension control of photonic crystals structures.

Acknowledgments

The work was financially supported by the Joint Research Project SAS-BAS 2015-2017, the projects ITMS 26240120011 (50%) and ITMS 26240120019 (20%) co-funded by the European Regional Development Fund, VEGA Agency Grant 2/0134/15, and the contract BIn-5/2009 funded by the Bulgarian National Fund for Scientific Research.

References

[1] Vutova K and Mladenov G 1994 Modelling of exposure and development processes in electron and ion lithography *Modelling and Simulation in Mater. Sci. and Engin.* **2** 239-54
[2] Vutova K and Mladenov G 2010 Computer simulation of processes at electron and ion beam lithography *Lithography* ed M. Wang (Vukovar InTech) part 1 chapter 17 319-50
[3] Thoms S and Macintyre DS 2014 Investigation of CSAR 62, a new resist for electron beam lithography *J. Vacuum Sci. Technol.* B **32** 06FJ01 dx.doi.org/10.1116/1.4899239
[4] http://www.tnw.tudelft.nl/en/about-faculty/departments/quantum-nanoscience/kavli-nanolab-delft/news/kavli-nanolab-news-no-9/new-resist-as-alternative-for-zep520a/
[5] Mladenov G, Koleva E, Vutova K, Kostic I, Spivak V, Bencurova A and Ritomsky A 2009 Resists for electron beam nanolithography *J. Electrotechnica and Electronica* **5-6** 13-20
[6] Vutova K, Koleva E, Mladenov G, Kostic I and Tanaka T 2010 Computer simulation of resist profiles at electron beam nanolithography *Microelectr. Engin.* **87** 1108-11
[7] Vutova K, Koleva E, Mladenov G and Kostic I 2009 Some peculiarities of resist-profile

simulation for positive-tone chemically amplified resists in electron-beam lithography *J. Vacuum Sci. Technol.* B – *Microelectr. and Nanometer Struct.* **27** 52-7

[8] Vutova K, Mladenov G, Koleva E, Kostic I, Bencurova A, Nemec P and Tanaka T 2011 Nonlinear solubility behavior of polymer and oligomer resists at electron beam modification *J. Mater. Sci. Engin.* B **1** 523-9

Enhanced photoanisotropic response in azopolymer doped with elongated goethite nanoparticles

L Nedelchev[1], D Nazarova[1], N Berberova[1,4], G Mateev[1], D Kostadinova[1], R Mariño-Fernández[2], V Salgueiriño[2] and D Schmool[3]

[1]Institute of Optical Materials and Technology, 1113 Sofia, Bulgaria
[2]Departamento de Física Aplicada, Universidade de Vigo, 36310, Vigo, Spain
[3]Laboratory PROMES CNRS, University of Perpignan, 66100 Perpignan, France

E-mail: natali.berberova@gmail.com

Abstract: We present a study of the photoinduced birefringence in nanocomposite films of an azopolymer (PAZO) doped with goethite (α-FeOOH, a characteristic antiferromagnetic material) nanoparticles (NPs). The NPs had an elongated shape with a size 15×150 nm, i.e., a ratio of 1:10. Samples were prepared with different concentrations of the NPs in the azopolymer varying from 0% (undoped azopolymer film) to 15 wt %. An unusual dependence of the birefringence on the concentration was observed – two peaks of enhancement at 1 % and at 10 % concentration. Our previous studies on ZnO and SiO_2 NP have indicated only one peak of increase at low concentrations – 0.5 wt % and 2 wt %, respectively. This effect could be related to the elongated shape of the nanoparticles and the presence of two characteristic NPs sizes – 15 and 150 nm. Moreover, the birefringence increase for the samples with 10 wt % NPs concentration (compared with the non-doped samples) was rather significant – nearly 70%.

1. Introduction

Azobenzene and its numerous derivatives, including azopolymers, are well-known as one of the most efficient media for polarization recording [1,2]. One of the most significant optical characteristics of the azopolymers is the birefringence Δn that can be induced in them upon illumination. This is defined as the difference between the component of the refractive index perpendicular to the polarization of the recording beam and the component parallel to it – $\Delta n = n_{\perp} - n_{\parallel}$.

Recently it has been reported that nanocomposites based on an azopolymer and inorganic dopants – silver nanoparticles [3], ZnO NP with different sizes [4,5], as well as silica NPs [6] – show an increased birefringence compared with the non-doped azopolymer. In all of these studies, however, nanoparticles with spherical shape have been used which possess only one characteristic dimension – their diameter.

In order to establish the influence of the shape of the nanoparticles on the birefringence enhancement, we studied the photoinduced birefringence in nanocomposite films of an azopolymer doped with elongated nanoparticles (nanorods) of the magnetic material goethite (α-FeOOH) in

[4] To whom any correspondence should be addressed.

concentrations varying from 0 wt % (pure azopolymer) to 15 wt %. The nanorods had dimensions 15×150 nm, i.e. a ratio of 1:10.

2. Experimental results and discussion

2.1. Samples preparation

In the present study we used the commercially available azopolymer poly[1-4-(3-carboxy-4-hydrophenylazo) benzensulfonamido]-1,2-ethanediyl, sodium salt] (Sigma Aldrich) or PAZO, for short. Its chemical structure is shown in figure 1(a). An aqueous dispersion of goethite nanorods was synthesized by co-precipitation of ferrous and ferric ions solutions at acid pH (pH<4) [7]. 20 mL of aqueous 1 M $FeCl_3$ and 5 mL of 2 M $FeSO_4 \cdot 7H_2O$ in 1.3 M HCl were simultaneously injected into 250 mL of 0.4 M NH_4OH under rapid mechanical stirring. Stirring was allowed to continue for 30 min, and then the yellow-brown solid product was allowed to precipitate. The sediment was redispersed in 50 mL distilled water, and subsequently three 30 mL aliquots of tetramethylammonium hydroxide solution (1 M) in 50 mL of water were added, again with rapid stirring. Finally, water was added to the dispersion up to a total volume of 250 mL. Figure 1(b) shows a transmission electron microscope (TEM) image of the goethite NPs used in the experiments. Their dimensions were 150±40 nm by 15±4 nm. Finally, three thin films from each concentration were prepared.

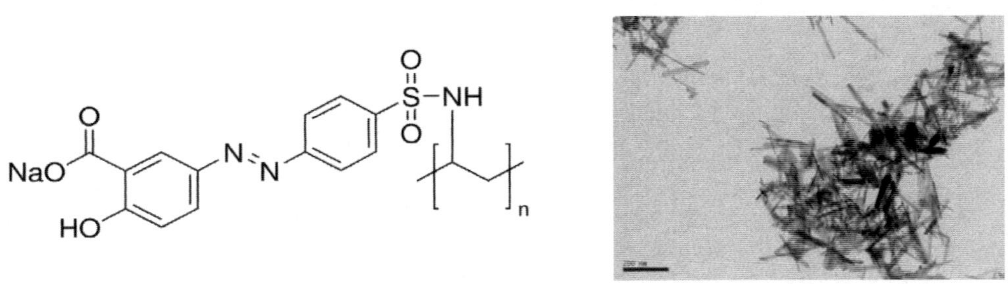

(a) (b)

Figure 1. (a) Chemical structure of the azopolymer matrix. (b) TEM image of the goethite NPs.

In order to produce samples with various concentrations of goethite in the azopolymer, we used the following procedure: we dissolved PAZO in water and, to ensure complete dissolution of the polymer, we heated the solutions to 60 °C. The nanoparticles were separately dispersed in water by sonication. Then, the polymer solutions and the nanoparticles dispersions were combined to achieve the desired concentration (C) of the goethite nanoparticles in the azopolymer. Seven different concentrations were obtained in this way: $C = 0$, 0.5, 1, 2, 5, 7.5, 10 and 15 wt % . Finally, thin films were prepared by depositing the solutions on a glass substrate (BK7) and spin-coating at 1500 rpm for 30 seconds.

2.2. Characterisation of the films

As it will be seen, to calculate both the absorbance and birefringence, it is necessary to know the precise value of each sample's thickness. For this purpose, we used a Talystep profiler with 1 nm vertical and 0,1 μm horizontal resolution, additionally upgraded with a digital recording device connected to a PC. A groove was made on the surface, with the distance between the "plateau" and the bottom of the groove corresponding to the film's thickness. Figure 2 shows a typical profile obtained from such measiurements.

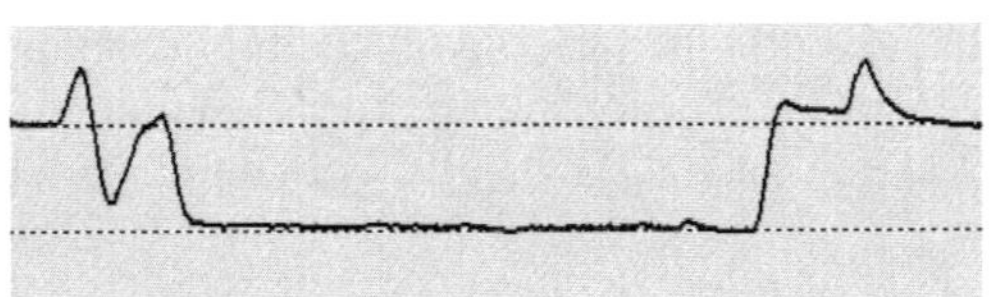

Figure 2. Profile for a sample with 0.5 wt % NP concentration. The thickness measured is 460 nm.

Besides the thickness, we also measured the absorbance spectra of the samples. This is essential for choosing correctly the wavelengths of the pump and the probe lasers for measuring the birefringence. The wavelength of the pump laser has to be within the absorbance band, but not at the peak, because then the recording light will not penetrate into the depth of the film. On the other hand, the probe laser's wavelength has to be outside the absorbance band, so that it only measures the birefringence but does not destroy it.

The absorbance spectra were determined using a Varian Cary 5E spectrophotometer. The data directly obtained from the measurement were the transmittance spectra $T(\lambda)$, which were then transformed to absorbance using $\alpha = -(1/d)\log(T)$. Typical spectra of the absorbance coefficient are shown in figure 3.

As seen, the peak of absorbance is at 360 nm for all samples. The arrows in the figure indicate the selected wavelengths for the pump and probe lasers.

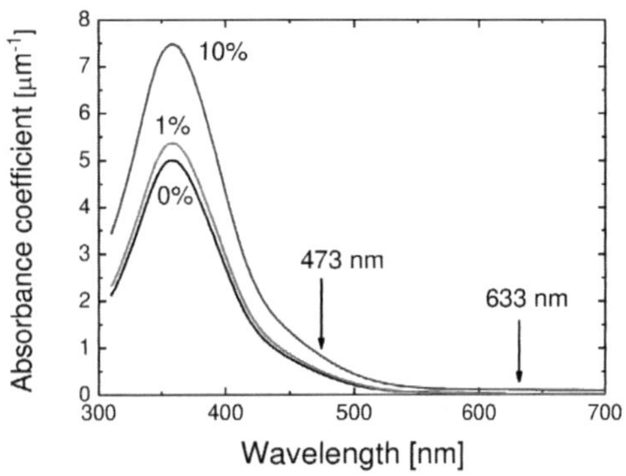

Figure 3. Absorbance coefficient spectra for the samples with concentrations $C = 0$, 1 and 10 wt %.

2.3. Measurement of the photoinduced birefringence
Figure 4 shows the experimental setup used to measure the birefringence.

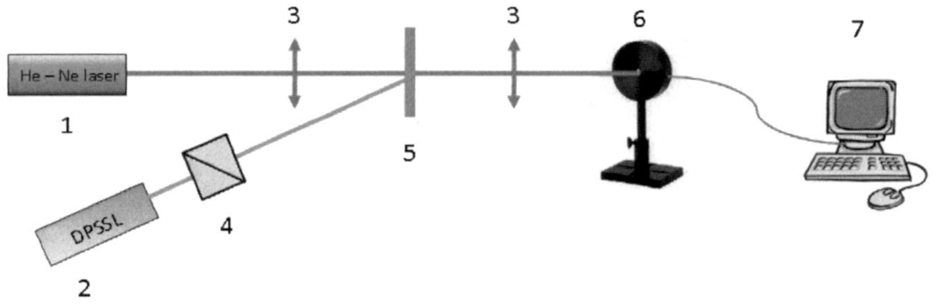

Figure 4. Setup for birefringence measurement. 1 – probe He-Ne laser (λ = 633 nm), 2 – DPSS laser (λ = 473 nm), 3 – polarizers, 4 – Glan prism, 5 – sample, 6 – photodetector and 7 – PC.

Birefringence was induced by a vertically polarized beam from a diode-pumped solid-state laser (DPSSL) with an intensity of 400 mW/cm^2. The sample was placed between two crossed polarizers oriented at ±45° with respect to the vertical. The probe beam passed through the system polarizer-sample-polarizer and was measured by a digital optical power meter (Thorlabs). The birefringence was calculated from the intensity using the following expression [1]:

$$\Delta n(t) = \frac{\lambda}{\pi d} \arcsin \sqrt{\frac{I_{\perp}(t)}{I_{\parallel}}}, \tag{1}$$

where λ is the wavelength of the probe laser (633 nm), $I_{\perp}$ is the intensity of the probe beam transmitted through the system "crossed polarizers – sample", and $I_{\parallel}$ is the intensity of the probe beam transmitted through the parallel polarizers and the sample before the start of the recording. We thus obtained the time dependence of the photoinduced birefringence $\Delta n(t)$, graphically presented for three of the samples in figure 5(a).

Each data point on figure 5(b) was averaged over three samples with the same NP concentration. Each measurement consisted of three stages. Initially, we measured the probe laser signal for 60 seconds before starting the pump beam in order to take into account the background light noise.

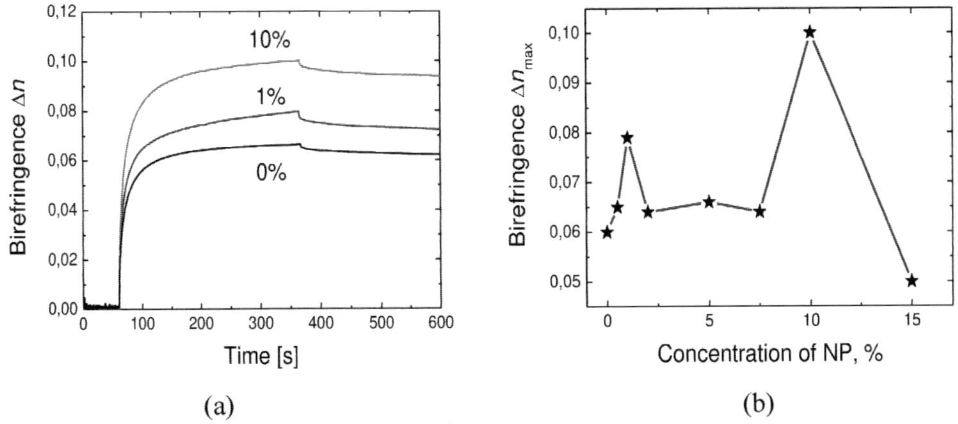

Figure 5. (a) Time dependence of Δn for the samples with 0, 1 and 10 % concentration. (b) Dependence of Δn on the goethite NP concentration C.

Then the pump laser was started and the birefringence began to increase. When it approached saturation after 300 s of irradiation, the pump laser was stopped and the birefringence relaxation was measured for another 240 seconds. Thus, it was observed that more than 90 % of the birefringence was retained after the relaxation period.

The samples thicknesses and the maximal values of the photoinduced birefringence (Δn_{max}) for all samples are summarized in table 1. We studied three samples of each concentration.

Table 1. Samples thicknesses and the maximal values of the photoinduced birefringence (Δn_{max}).

Concentration, wt %	d, µm	Δn_{max}
0	0.45	0.060
0.5	0.46	0.064
1	0.43	0.079
2	0.47	0.065
5	0.42	0.071
7.5	0.41	0.064
10	0.40	0.10
15	0.45	0.046

The dependence of Δn_{max} on the concentration of goethite nanoparticles in the PAZO azopolymer is plotted in figure 5b. Two peaks of birefringence enhancement can be clearly observed. The first one is at $C = 1$ wt %, which corresponds to a 32 % increase of the birefringence compared to the undoped azopolymer. The second peak is at 10 wt % concentration, where the birefringence is enhanced by nearly 70 %. A possible reason for the presence of two peaks in the Δn_{max} (C) dependence may be the geometry of the nanorods and the two characteristic dimensions – 15 and 150 nm. In order to confirm this, further experiments will be focused on determining how the size of the nanoparticles used for doping the azopolymer influences the birefringence.

3. Conclusions

Photoinduced birefringence was measured in nanocomposite films of an azopolymer doped with elongated nanoparticles (nanorods) of the magnetic material goethite (α-FeOOH) in concentrations varying from 0 wt % (pure azopolymer) to 15 wt %. An unusual dependence of the birefringence on the concentration was observed – two peaks of enhancement were seen at 1 wt % and at 10 wt % concentration. Our previous studies with ZnO and SiO_2 NP have indicated only one peak of increase at low concentrations – 0,5 wt % and 2 wt % respectively. This effect could be related to the elongated shape of the nanoparticles and the presence of two characteristic NP sizes – 15 and 150 nm. The birefringence increase for the samples with 10 wt % NP concentration compared with the undoped samples was about 70%.

Acknowledgments

Authors are grateful for the support provided by FNI T02/26, COST Action MP1205, Taiwan and Bulgarian-Taiwanese international project under contract 103-2911-I-009-516. R. M.-F. and V. S. acknowledge the financial support from the Xunta de Galicia (Regional Government, Spain) under projects EM2014/035 and Inbiomed.

References

[1] Nikolova L and Ramanujam P S 2009 *Polarization Holography* (Cambridge University Press, Cambridge)
[2] Natansohn A and Rochon P 2002 *Chem. Rev.* **102** 4139-75
[3] Zhou J, Yang J, Sun Y, Zhang D, Shen J, Zhang Q and Wang K 2007 Effect of silver nanoparticles on photo-induced reorientation of azo groups in polymer films *Thin Solid Films* **515** 7242–6
[4] Nedelchev L, Nazarova D, Dragostinova V and Karashanova D 2012 Increase of photoinduced birefringence in a new type of anisotropic nanocomposite: azopolymer doped with ZnO nanoparticles *Opt. Lett.* **37** 2676-8
[5] Nazarova D, Nedelchev L, Dragostinova V, Berberova N 2013 Influence of the size of nanoparticles doped in series of azopolymers of the photoinduced birefringence *Proc. SPIE* **8770** 877009
[6] Nazarova D, Nedelchev L, Sharlandjiev P and Dragostinova V 2013 Anisotropic hybrid organic/inorganic (azopolymer/SiO_2 NPs) materials with enhanced photoinduced birefringence *Appl. Opt.* **52** E28-E33
[7] Mariño-Fernández R, Masunaga S H, Fontaíña-Troitiño N, Morales M P, Rivas J and Salgueiriño V 2011 Goethite (α-FeOOH) nanorods as suitable antiferromagnetic substrates *J. Phys. Chem.* C **115** 13991-99

Journal of Physics: Conference Series **700** (2016) 012032 doi:10.1088/1742-6596/700/1/012032

Photoinduced variation of the Stokes parameters of light passing through thin films of azopolymer-based hybrid organic/inorganic materials

Nataliya Berberova[1], Dimana Nazarova, Lian Nedelchev, Blaga Blagoeva, Desislava Kostadinova, Vera Marinova and Elena Stoykova

Acad. J. Malinowski Institute for Optical Materials and Technologies, Bulgarian Academy of Sciences, Acad. G. Bonchev str., bl. 109, 1113 Sofia, Bulgaria

E-mail: natali.berberova@gmail.com

Abstract. Hybrid materials synthesized by combining polymers and inorganic nanoparticles (NPs) have received considerable attention recently due to their advantageous electrical, optical, or mechanical properties. Here we present a polarimetric study of thin films prepared from photoanisotropic hybrid materials: azopolymer (poly[1-[4-(3-carboxy-4-hydroxyphenylazo)benzenesulfonamido]-1,2-ethanediyl, sodium salt]) doped with different concentrations of ZnO NPs with sizes below 50 nm. To obtain full information about the polarization response of these thin films, we determined the kinetics of the Stokes parameters of light passing through the films on illumination by linearly polarized light within the absorption band of the material, namely 473 nm. Using the Stokes parameters, the birefringence values were calculated as a function of time and compared for the films with different NP concentrations.

1. Introduction

The research on azobenzenes during the last decades revealed their remarkable potential for a broad spectrum of applications: polarization holography [1,2], optical data storage [3-6], formation of surface relief gratings [7,8] etc. Using polarization holography makes it theoretically possible to achieve 100 % diffraction efficiency even in thin films of photoanisotropic media [2]. Being able to record the polarization state of light, azobenzenes also enable one to apply polarization multiplexing and increase the density of recording [9].

The most important characteristic of any photoanisotropic media, including all azopolymer based materials, is the maximal value of the photoinduced birefringence Δn_{max}. Different methods have been exploited to determine this parameter for a given material. One of the most informative and reliable is to measure the Stokes parameters of light passing through the spot on the sample illuminated by the recording beam [5,10]. Therefore, this method was applied in our present study.

Recent reports have indicated that doping nanoparticles in azopolymers can increase the value of the photoinduced birefringence Δn_{max} [11-14]. Thus, we synthesized hybrid organic/inorganic materials based on an amorphous azopolymer (PAZO) doped with different concentrations of ZnO

[1] To whom any correspondence should be addressed.

nanoparticles. We calculated the kinetics of birefringence by measuring the photoinduced variation of the Stokes parameters of light passing through thin films of our composite materials.

2. Experimental procedures

2.1. Azopolymer matrix and preparation of the thin films

The azopolymer used as organic host in our hybrid samples was poly[1-[4-(3-carboxy-4-hydroxyphenylazo)benzenesulfonamido]-1,2-ethanediyl, sodium salt] or shortly PAZO. The azopolymer is commercially available from Sigma Aldrich. Its chemical structure is shown in figure 1a. Water solubility is an important advantage of this polymer. The nanoparticles from ZnO (Sigma Aldrich) have an average diameter <50 nm (figure 1b).

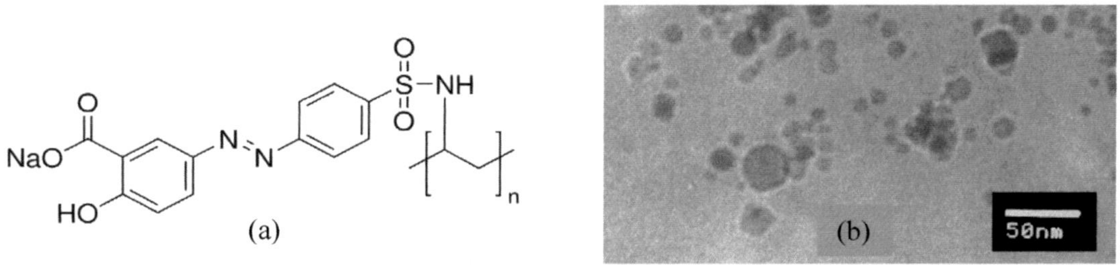

Figure 1. (a) Chemical structure of the azopolymer matrix. (b) TEM image showing the NPs size.

For this study, thin film samples were produced with different concentrations of the ZnO NPs in the azopolymer. Initially the azopolymer was dissolved in water and the nanoparticles were separately dispersed in water by sonication. The polymer solutions and the nanoparticles dispersions were then combined to achieve the desired concentration (C) of the ZnO nanoparticles in the azopolymer, namely $C = 0$ % (or non-doped azopolymer), 0.5, 1, 2, 5 and 10 %. The resulting dispersions were deposited on glass substrates and spin coated for 30 seconds at 1500 rpm. The substrates had been cleaned and polished in advance.

2.2. Thickness measurement

The precise value of the samples thickness, denoted as d, is a crucial parameter for accurate determination of the thin films optical properties, namely, the absorbance spectra and the birefringence. The thickness was measured by a high-precision Talystep profilometer with 10 nm accuracy. A groove was made on the surface of each sample, inside which the nanocomposite film was removed, with the distance between the "plateau" and the bottom of the groove corresponding to the layer's thickness, as shown in figure 2. In order to ensure that the glass substrate was not affected, we used a tip with hardness lower than the hardness of glass.

Although all the samples contained the same amount of azopolymer, their thicknesses varied slightly and were in the range 410 – 450 nm.

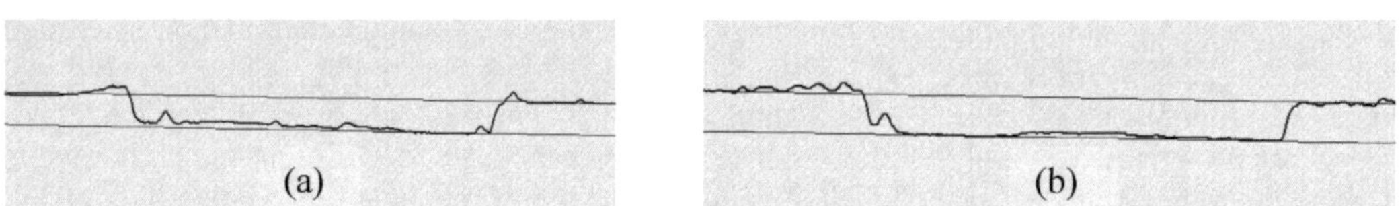

Figure 2. Typical thickness profiles: (a) $C = 2$ %, $d = 410$ nm, (b) $C = 10$ %, $d = 440$ nm.

2.3. Spectra of the absorption coefficient

The absorbance spectra were determined by a UV-Vis-NIR Cary 5E spectro-photometer (Varian). The normalized spectra of the transmittance $T_{norm}(\lambda)$ were directly measured and then the absorption coefficient spectra $\alpha(\lambda)$ were calculated using the Beer-Lambert law i.e. $\alpha = -(1/d)\log(T_{norm})$ (figure 3). The reflections from the front and back side of the samples were compensated by placing

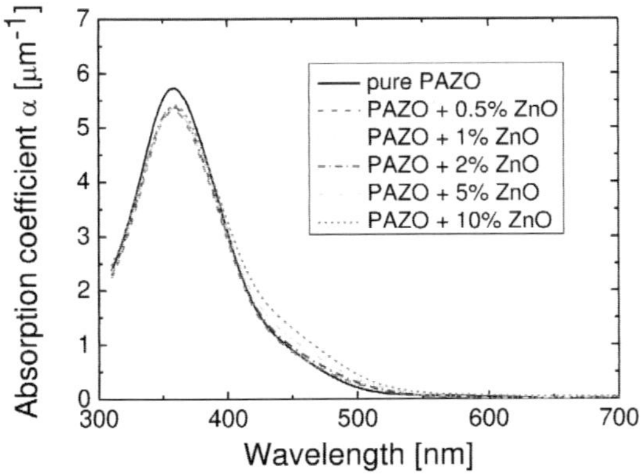

Figure 3. Absorption coefficient spectra for all samples with $C = 0, 0.5, 1, 2, 5$ and 10 %.

a glass substrate in the reference arm of the spectrophotometer. Thus, the absorption coefficient spectra characterized the properties of the material and were not influenced by the film thickness, as was the absorbance A.

Doping the azopolymer with ZnO NPs did not shift the peak of the spectra which was at 359 nm for all samples. On the other hand, there was some increase of the absorbance with the increase of the concentration in the 400-500 nm wavelength range. It can be related to an interaction between the NPs and the azopolymer matrix. Still, the scattering was low even for the sample with the highest NPs concentration (10 %), which was confirmed by the low absorbance in the range 600-700 nm.

The spectra also showed that the pump laser wavelength (473 nm) was absorbed by the samples, but at the same time, being far from the absorbance peak, the laser light could penetrate and influence the entire film. In contrast, the light at the probe beam wavelength (633 nm) was virtually not absorbed by the samples.

2.4. Kinetics of the Stokes parameters

The Stokes parameters contain all the information about the polarization state of light. Knowing the state of polarization of the incident light and measuring the Stokes parameters of the beam after the sample, we can determine its photoinduced birefringence [2]:

$$\Delta n = \frac{\lambda}{2\pi d} \arctan\left(\frac{S_3}{S_2}\right), \tag{1}$$

where λ is the wavelength of the probe laser, d is the film thickness, and S_2 and S_3 are components of the Stokes vector (S_0, S_1, S_2, S_3).

Birefringence was induced in the sample by a vertically polarized beam from a diode-pumped solid-state (DPSS) laser (λ_{rec} = 473 nm, intensity 400 mW/cm^2). The probe beam from a He-Ne laser (λ_{probe} = 633 nm) was linearly polarized at 45°. It passed through the spot on the sample illuminated by the pump laser; the Stokes parameters of the output beam were measured by a PAX5710 Polarization Analyzing System (Thorlabs). A screenshot illustrating the device interface is given in figure 4a, while the kinetics of the Stokes parameters for the sample with $C = 10$ % is presented in figure 4b.

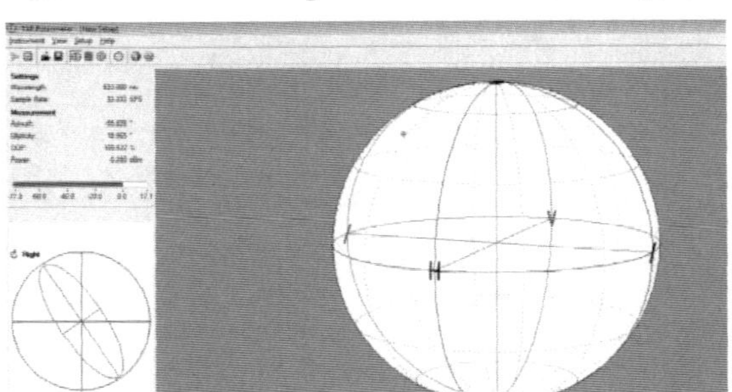

(a)

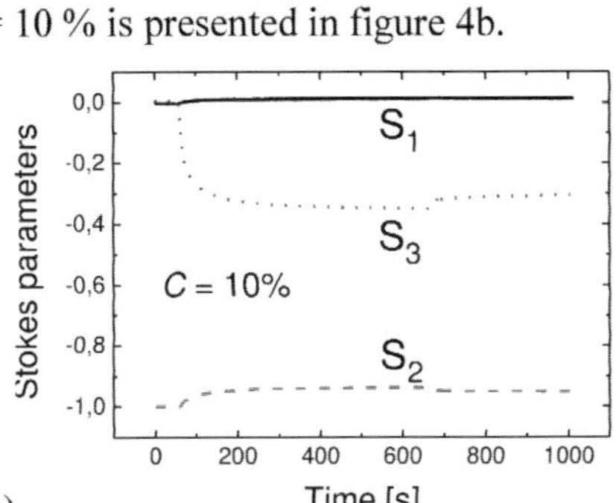

(b)

Figure 4. (a) Polarimeter interface. (b) Example of the Stokes parameters kinetics.

3. Results and discussion

Using equation (1), we can obtain the kinetics of birefringence shown in figure 5a.

As seen in figure 5b, doping the azopolymer with ZnO nanoparticles led to an increase of the maximal value of the photoinduced birefringence. In order to quantify this increase or enhancement, we can use the "Increase Ratio" parameter, defined as follows [14]:

$$IR(C) = \frac{\Delta n_{\max}(C) - \Delta n_{\max}(0)}{\Delta n_{\max}(0)} \cdot 100\%, \qquad (2)$$

where $\Delta n_{\max}(C)$ is the maximal saturated value of the photoinduced birefringence for nanocomposite film with concentration of the NPs relative to the azopolymer C, and $\Delta n_{\max}(0)$ is the maximal saturated value of the photoinduced birefringence for the non-doped (or "pure") film.

In the present study, the maximal $IR = 34$ % was obtained for the sample with 10 % concentration of the ZnO NPs in the PAZO polymer. It is interesting to compare this with our earlier results for nanocomposites of P_1 azopolymer and ZnO NPs [12]. Then, we observed a peak at 0.5 % followed by a decrease of the birefringence at higher dopant concentrations. The reason for the continued increase of the birefringence in the PAZO + ZnO NPs even at high concentrations of the dopant is probably the better refractive index matching between the PAZO polymer used here and the nanoparticles, leading to lower scattering of the hybrid samples.

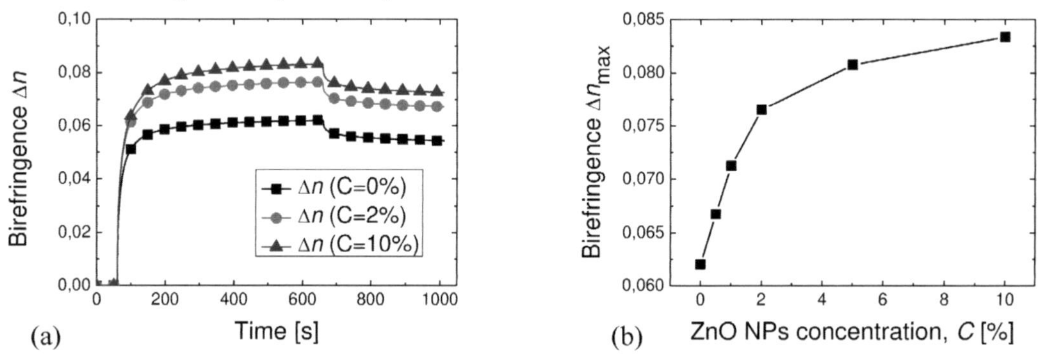

Figure 5. (a) Birefringence kinetics for samples with ZnO NP concentrations $C = 0$, 2 and 10 %. (b) Dependence of the maximal value of the birefringence on the concentration for all samples.

4. Conclusions

In this article we presented a polarimetric study of thin films prepared from photoanisotropic hybrid materials: an azopolymer denoted as PAZO doped with different concentrations of ZnO NPs: $C = 0$ % (or non-doped azopolymer), 0.5, 1, 2, 5 and 10 %. Measuring the photoinduced variation of the Stokes parameters of light, we calculated the kinetics of birefringence and its maximal value for each NPs concentration. The highest increase ratio of 34 % was obtained at 10 % concentration of the ZnO nanoparticles.

Acknowledgement

The authors acknowledge the support provided by FNI T02/26, COST Action MP1205, Taiwan and Bulgarian-Taiwanese international project under contract 103-2911-I-009-516.

References

[1] Todorov T, Nikolova L and Tomova N 1984 Polarization Holography *Appl. Opt.* **23** 4309-12
[2] Nikolova L and Ramanujam P S 2009 *Polarization Holography* (Cambridge University Press)
[3] Natansohn A, Rochon P, Gosselin J and Xie S 1992 *Macromolecules* **25** 2268-73
[4] Berg R, Hvilsted S and Ramanujam P S 1996 *Nature* **383** 505-8

[5] Nedelchev L, Matharu A S, Hvilsted S and Ramanujam P S 2003 *Appl. Opt.* **42** 5918-27
[6] Matharu A S, Jeeva S and Ramanujam P S 2007 *Chem. Soc. Rev.* **36** 1868-80
[7] Kim D, Li L, Jiang X, Shivshankar V, Kumar J and Tripathy S 1995 *Macromolecules* **28** 8835-9
[8] Natansohn A and Rochon P 2002 *Chemical Rev.* **102** 4139-75
[9] Ilieva D, Nedelchev L, Petrova Ts, Tomova N, Dragostinova V and Nikolova L 2005 *J. Opt.* A*: Pure Appl. Opt.* **7** 35–9
[10] Todorov T and Nikolova L 1992 *Opt. Lett.* **17**/5 358-9
[11] Zhou J, Yang J, Sun Y, Zhang D, Shen J, Zhang Q and Wang K 2007 *Thin Solid Films* **515** 7242–6
[12] Nedelchev L, Nazarova D, Dragostinova V and Karashanova D 2012 *Opt. Lett.* **37** 2676-8
[13] Nazarova D, Nedelchev L, Sharlandjiev P and Dragostinova V 2013 *Appl. Opt.* **52** E28-E33
[14] Nazarova D, Nedelchev L and Mintova S 2014 *Optofluid, Microfluid, Nanofluid* **1** 43–48

Journal of Physics: Conference Series **700** (2016) 012033 doi:10.1088/1742-6596/700/1/012033

Investigation of the thermal processes in electron-beam surface modification by means of a scanning electron beam

M Ormanova[1,3], Vl Angelov[2] and P Petrov[1]

[1]Emil Djakov Institute of Electronics, Bulgarian Academy of Sciences, 72 Tzarigradsko Chaussee, 1784 Sofia, Bulgaria
[2]Institute for Nuclear Research and Nuclear Energy, Bulgarian Academy of Sciences, 72 Tzarigradsko Chaussee, 1784 Sofia, Bulgaria

E-mail: maria_mecheva@abv.bg

Abstract. In this work we present a study of the thermal processes taking place during surface modification of steels performed by a scanning electron beam. The model is based on solving the heat transfer equation by means of Green functions. The thermal field was calculated, together with the size of the zone of structural changes in tool steel samples. The comparison of the zones of thermal treatment as experimentally obtained and theoretically calculated and the corresponding structural changes show a very good agreement.

1. Introduction

The use of high-intensity electron beams for modification of metals and alloys leads to enhancement of the techniques for improvement of their physical and mechanical properties.

When a beam of fast electrons interacts with the metal surface, the electron's kinetic energy is transformed into heat, which spreads from the surface to the inner core of the metallic sample. The rates of heating and cooling with this processing method are significantly higher as compared to the oven heating and bath cooling. This leads to specifics in the phase and structural transitions of the metals and their alloys [1-3].

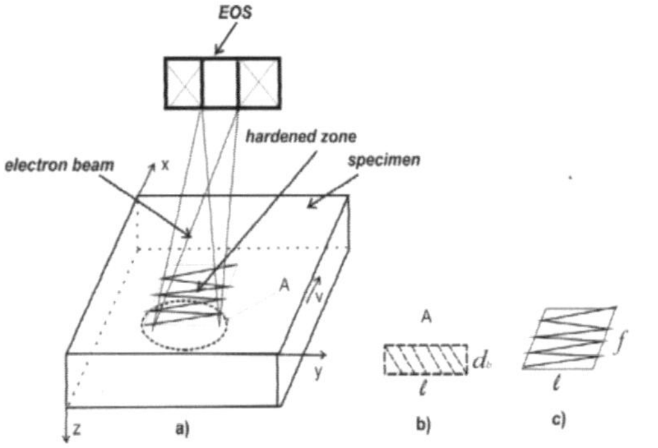

Figure 1. Schematic diagram of a scanning electron-beam treatment setup.

Experimental and numerical methods are used for determination of the temperature field, and thus, of the heating and cooling rates, the final micro-structure and the mechanical properties of the processed samples. The experimental methods are extremely time- and effort-consuming and the obtained results are not always reliable [4,5]. Therefore, the numerical methods are more promising and are widely used for modelling the thermal processes described above [6]. Works [7,8] present results from the investigation of the thermal field distribution

[3] To whom any correspondence should be addressed.

Journal of Physics: Conference Series **700** (2016) 012033 doi:10.1088/1742-6596/700/1/012033

during processing of steels by means of a scanning electron beam (figure 1a). The method of finite elements has been used for solving the equation of heat transfer, with the assumption that the electron beam energy is evenly distributed over a rectangular with area $S = d_b l$ (figure 1b). Here d_b is the electron beam diameter, l is the beam deviation amplitude. In actual experiments, however, the electron beam scans the sample surface following a certain trajectory, as shown in figure 1c, with given scanning frequency and amplitude.

In this work we present a thermal model, whereby the electron-beam energy distribution is obtained by taking into account the frequency and amplitude of beam scanning over the trajectory according to figure 1c. The model is based on solving the heat transfer equation by means of Green functions. The thermal field and the size of the structural changes zone were calculated for instrumental steel samples.

2. Thermal model

The heat transfer equation is solved by means of Green functions, presented in [9] for the case of high-intensity scanning electron beams. The equation has the form:

$$\frac{1}{\alpha}\frac{\partial T}{\partial t} - \nabla^2 T = \frac{f(r,t)}{\lambda}, \tag{1}$$

where $\alpha = \frac{\lambda}{c\rho}$ is the heat exchange coefficient.

$T(r,t)$ is expressed in terms of the three-dimensional Green's function as follows:

$$T(r,t) = \frac{k}{\lambda}\int_{\tau=0}^{t} dt \int_R G(r,t|r',t)f(r',\tau)dv' + \int_R G(r,t|r',\tau)|_{\tau=0}F(r')dv', \tag{2}$$

where $F(r')$ is the initial temperature distribution.

In orthogonal coordinates, the Green's function has the form:

$$G(x,y,z,t|x',y',z',\tau) = G1(x,t|x',\tau) \times G2(y,t|y',\tau) \times G3(z,t|z',\tau), \tag{3}$$

where each of the Green functions G1, G2 and G3 depends on the type of the region (i.e. finite, semi-infinite, or infinite) and on the boundary conditions. The infinite medium Green function is obtained as:

$$G(x,t|x',\tau) = [4\pi\alpha(t-\tau)]^{-1/2} exp\left(-\frac{(x-x')^2}{4\alpha(t-\tau)}\right) \tag{4}$$

and the semi-infinite medium Green function when the boundary at $z' = 0$ is:

$$G(z,t|z',\tau) = [4\pi\alpha(t-\tau)]^{-1/2} \times \left[exp\left(-\frac{(z-z')^2}{4\alpha(t-\tau)}\right) + exp\left(-\frac{(z+z')^2}{4\alpha(t-\tau)}\right)\right], \tag{5}$$

From equations (4, 5) and equation (3) we obtain:

$$G(x,y,z,t|x',y',z'=0,\tau) = 2[4\pi\alpha(t-\tau)]^{-1.5} \times exp\left(-\frac{(x-x')^2+(y-y')^2+z^2}{4\alpha(t-\tau)}\right). \tag{6}$$

Substituting $x' = x - r\cos\theta$, $y' = y - r\sin\theta$ and $G(r,t|r',\tau)$ from equation (6) in equation (2), we obtain the temperature distribution:

$$T(x,y,z,t) = T_0 + \frac{2\alpha}{\lambda}\int_0^t \int_{-\pi}^{\pi} \int_0^{\infty} \Big[f(x-r\cos\theta, y-r\sin\theta) \times [4\pi\alpha(t-\tau)]^{-1.5} \times exp\left(-\frac{r^2+z^2}{4\alpha(t-\tau)}\right)\Big] r dr d\theta d\tau \;, \tag{7}$$

where α is the heat exchange coefficient, λ is the heat conduction coefficient, x, y, z are the coordinates of the point of calculating the temperature, t is the time of calculating the temperature, r and θ are the polar coordinates of the points of the integration.

For a dithering beam $f_c(x - r\cos\theta\,,\ y - r\sin\theta,\ \tau)$ we have:

$$f_c(\xi,\eta,\tau) = \frac{3Q}{\pi r_0^2} exp\left(-3\frac{(\xi - v\tau)^2 + \left(\eta - \iota_{max}\cos\left(\frac{2\pi\tau}{\tau^*}\right)\right)^2}{r_0^2}\right), \quad (8)$$

where $(\xi = x - r\cos\theta, \eta = y - r\sin\theta)$, Q is the transferred power, r_0 is the characteristic beam radius, v is the speed of specimen motion, ι_{max} is the scanning amplitude, τ^* is the scanning period.

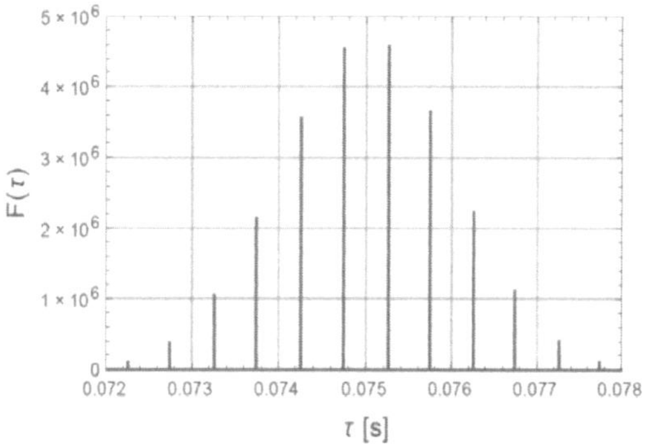

Figure 2. Integrand peaks positions for cosine scanning electron beam along the x axis with velocity 40 mm/s, frequency 1 kHz, amplitude 20 mm, beam size 0.1 mm for $\theta = \pi/2$ and $r = 1$ mm.

The integrand in (7) consists of $2t/\tau^*$ very well separated parts with very small time dimensions and corresponding sharp peaks, as shown in figures 2–4. If the point where we calculate the temperature rise lies on the x axis (y=0), then the time difference of the peaks for two consecutive integrand parts is $\tau^*/4$.

For the calculations we used the Mathematica numerical integration tool, which works adaptively, reaching a pre-set accuracy. Knowing the exact positions of the integrand parts, the integration was transformed into $2t/\tau^*$ integrations of each integrand part. Finally, the total temperature increase at a given point was calculated as the sum of the contributions of all integrands. Separating the numerical integration into many independent numerical integrations allows the use of parallel computing implemented in Mathematica for multicore computers, achieving a significant acceleration of the calculation process.

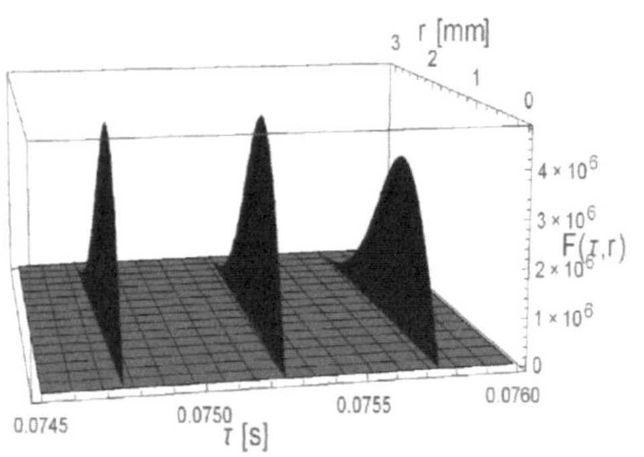

Figure 3. Integrand peaks positions for cosine scanning electron beam along the x axis with velocity 40 mm/s, frequency 1 kHz, amplitude 20 mm, beam size 0.1 mm for $\theta = \pi/2$.

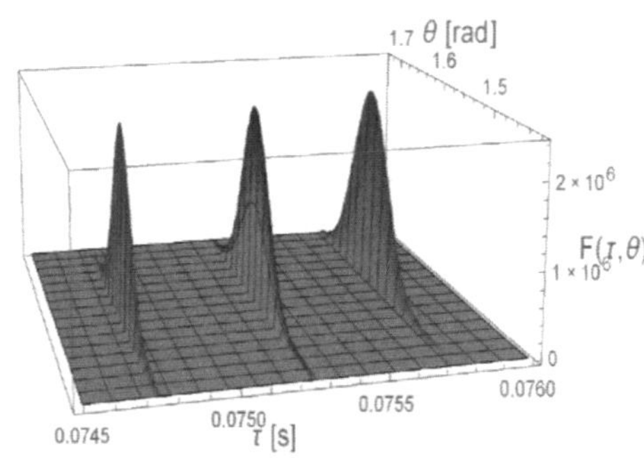

Figure 4. Integrand peaks positions for cosine scanning electron beam along the x axis with velocity 40 mm/s, frequency 1 kHz, amplitude 20 mm, beam size 0.1 mm for r = 1 mm.

The thermal model used in this work is based on the exact solution of the heat transfer equation by means of the Green functions method and is realized without any approximations. The model is implemented with Mathematica package and all calculations are performed with high accuracy settings. Excellent agreement between results obtained with the Green functions method and the FEM methods results for different cases is reported in [9], which confirms the high reliability of this method.

3. Results and discussion

The numerical calculations of the temperature field were performed according to equations (7,8) with the following technological parameters: accelerating voltage U = 55 kV, electron-beam scanning frequency f = 1 kHz, electron-beam current I = 40 mA, samples motion speed V = 40 mm/s, deviation amplitude l = 20 mm, electron-beam diameter d_l =0,01 mm (q = 2200 W). The specimen was tool steel with the following thermo-physical parameters $\alpha = 7$ mm^2/s, $\lambda = 0.033$ W/mm.K [10, 11].

Figure 5 shows the distribution of the thermal field in the sample depth from $z = 0$ to $z = 1$mm, and figure 6, the temperature distribution on the surface, $z = 0$.

It can be seen that the thermal field is highly inhomogeneous and has a non-stationary nature. The maximal temperature obtained on the surface of the sample (figure 6) is T = 784,625 °C. The calculated value of the treated zone depth (in the temperature interval 500 – 800 °C) is $h_{calc} \approx 100\mu$m (figure 5). The layers in the sample depth reach a temperature maximum later than the layers close to the surface, the latter having higher heating and cooling rates.

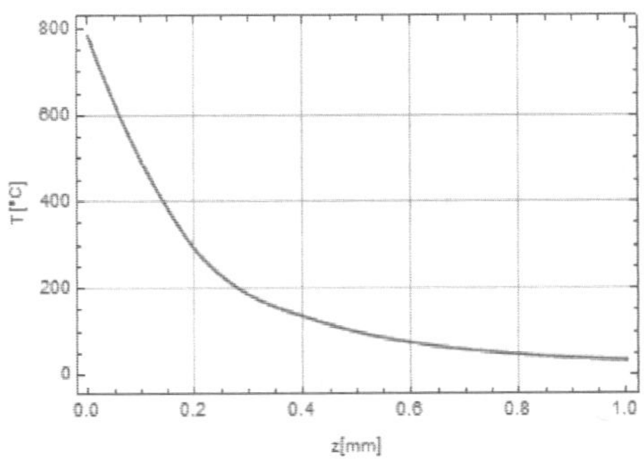

Figure 5. Temperature distribution in depth, from $z = 0$ to $z = 1$ mm.

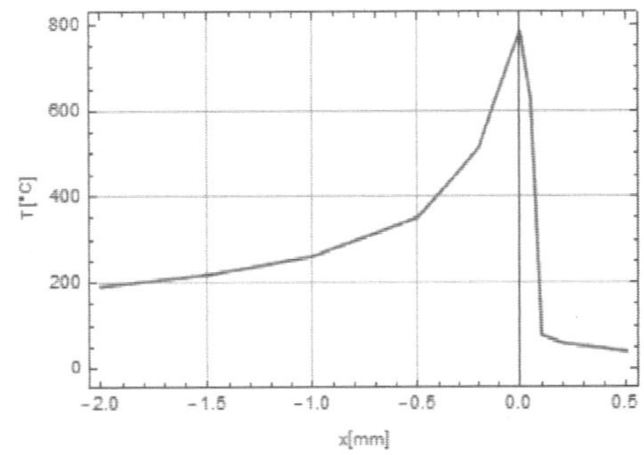

Figure 6. Temperature distribution at the surface, $z = 0$.

In order to evaluate the applicability of our thermal model, we carried out experiments on samples of tool steel W 320 (0.31% C, 0.30% Si, 0.35% Mn, 2.9% Cr, 2.8% Mo, 0.5% V). Samples with size 20×20×4 mm were subjected to hybrid processing, consisting of preliminary electron-beam hardening (EBH), followed by plasma nitriding (PN) and subsequent electron-beam treatment by means of a scanning electron beam (EBH).

The plasma nitriding was performed under the following technological parameters: T_N = 620 °C; t_N = 24 h; processing gas composition – 70% N_2 + 30% H_2. After plasma nitriding, a layer with a thickness of 10 μm saturated with nitrogen was observed.

The experiments were performed on an electron-beam installation (Leybold Heraus EWS 300/15-60) using a scanning electron beam for the sample processing (figure 1). The technological parameters were the same as those used for the numerical calculations described above.

Figure 7 shows a cross-section of the sample after hybrid processing (EBH+PN+EBH). In-depth grain refinement of the sample structure can be seen, which is due to the overlapping (appr. 35÷40 μm) of the plasma nitriding and the subsequent electron beam treatment. In the plasma nitriding, the nitrogen atoms are embedded in the steel crystal lattice, forming nitride phases.

Figure 8 shows the results on the distribution of the hardness in the depth of the treated zone. As a result of the structural changes, the hardness increases up to a distance d=23,76 μm (625,4 kg/mm^2)

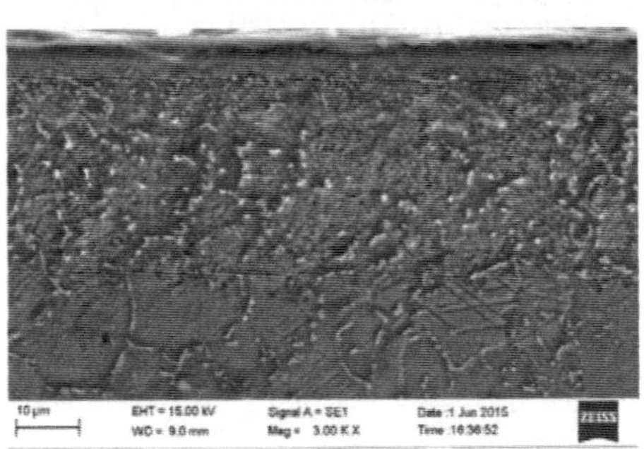

Figure 7. SEM metallographic cross-section after EBH+PN+EBH treatment.

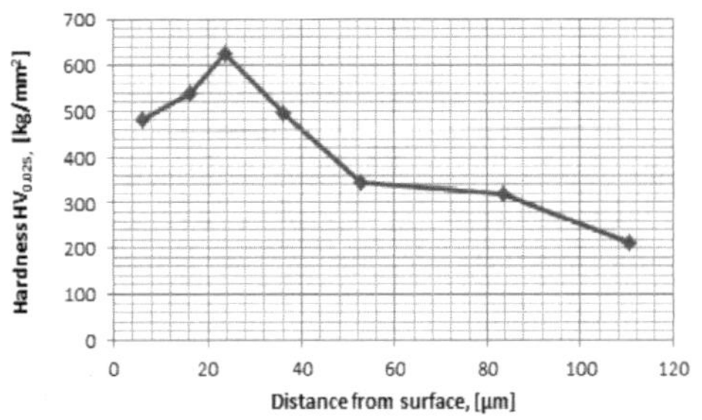

Figure 8. Hardness distribution in depth of the treated zone.

and then gradually decreases. At a distance h_{exp}=110,46 µm, the hardness reaches a value of 210,6 kg/mm^2, which is the hardness of the initial untreated steel. This value can be explained by means of X-ray structural analysis (XRD), which reveals the phases of $FeN_{0.076}$ and Fe_3O_4, having face-centered crystal lattices with lattice parameters a = 0,3626 nm and 0,8379 nm. The presence of $FeN_{0.076}$ in the sample is favorable for the final hardness – it increases by a factor of three.

The comparison of the experimentally obtained (figure 8) and the theoretically calculated (figure 5) treated zone depths shows very good agreement (h_{calc} = 100 µm; h_{exp} = 110,46 µm).

4. Conclusions

A method based on analytical solution of the heat transfer equation by means of Green functions is proposed. The model describes the thermal processes during processing by a scanning electron beam, taking into account the scanning frequency, amplitude and trajectory of the electron beam. The comparison of the experimentally obtained with the theoretically calculated zones of thermal treatment and the corresponding structural changes show a good agreement.

References

[1] Shiller S, Heisig U and Panzer S 1972 *Elektronenstrahltechnologie* (Forschungsinstitut Manfred von Ardene Dresden VEB Verlag Technik) 528
[2] Zenker, R, Sscher G, Buchwalder A, Liebich J, Reiter A and Häßler R 2007 *Surf. Coat. Technol.* **202** 804
[3] Arata Y 1984 *Trans. JWRI* **13** 121 -45
[4] Basile G and Moisan A 1978 *Proc. 2nd Int. Colloq. Welding and Melting by Electron and Laser Beams* (Avignon) 85
[5] Blakeley P 1983 *Proc .3rd Int. Colloq. Welding and Melting by Electron and Laser beams,* (Lyon) 379
[6] Dowden J 2001 *The Mathematics of Thermal Modelling - an Introduction to the Theory of Laser Materials Processing* (Chapman&Hall/CRC) 291
[7] Petrov P and Dimitroff D 1998 *Proc. 6th Int. Conf. Welding and Melting by Electron and Laser Beams* (Toulon) 101
[8] Petrov P 2010 *J. Phys.: Conf. Series* **223** 012029
[9] Farrahi G H and Sistaninia M 2009 *IJE Trans.* A: *Basics* **22**/2 169
[10] Mihailov V, Karhin V and Petrov P 2012 *Fundamentals of Welding* (Tehnika) (In Bulgarian)
[11] Goldak J and Akhlaghi M 2005 *Computational Welding Mechanics* (Springer: USA)

Neutron diffraction of titanium aluminides formed by continuous electron-beam treatment

S Valkov[1,4], D Neov[2], D Luytov[3] and P Petrov[1]

[1]Emil Djakov Institute of Electronics, Bulgarian Academy of Sciences, 72 Tzarigradsko Chaussee, 1784 Sofia, Bulgaria
[2]Frank Laboratory of Neutron Physics, Joint Institute for Nuclear Research, 141980 Dubna, Moscow Region, Russia
[3]Department of Solid State Physics and Microelectronics, Faculty of Physics, St. Kl. Ohridsky University of Sofia, 5 J. Bourchier blvd., 1164 Sofia, Bulgaria

E-mail: stsvalkov@gmail.com

Abstract. Ti-Al-based alloys were produced by hybrid electron-beam technologies. A composite Ti-Al film was deposited on a Ti substrate by electron-beam evaporation (EBE), followed by electron-beam treatment (EBT) by a continuously scanned electron beam. The speed of the specimens motion during the EBT were V_1 = 1 cm/sec and V_2 = 5 cm/sec, in order to realize two different alloying mechanisms – by surface melting and by electron-beam irradiation without melting the surface. The samples prepared were characterized by XRD and neutron diffraction to study the crystal structure on the surface and in depth. SEM/EDX analysis was conducted to explore the surface structure and analyze the chemical composition. Nanoindentation measurements were also carried out. No intermetallic phases were registered in the sample treated at velocity V_1, while the sample treated at V_2 exhibited a Ti_3Al/TiAl structure on the surface, transformed to Ti/TiAl in depth. The nanoindentation test demonstrated a significant negative hardness gradient from the surface to the depth of the sample.

1. Introduction

Ti-Al-based alloys combine attractive mechanical properties, light weight and excellent oxidation behavior at high temperatures. Titanium aluminides are widely used as a protective material, e.g., in the field of automotive and aircraft industry. Some aircraft engines manufacturers have successfully introduced γ-TiAl for turbine blades [1-3]. It has been found that the mechanical properties are greatly improved when the alloys consist of a mixture of γ-TiAl and α_2-Ti_3Al phases, rather than the single phase structure of γ-TiAl [4]. Due to the very good oxidation behavior of aluminium, the presence of a $TiAl_3$ phase is capable of causing the formation of a thin oxide film, such as Al_2O_3, on the surface, which acts as a protective barrier to organic environmental pollutants [5].

There exist several technologies for production of the alloys discussed above, including thin-films deposition [6], ion implantation [7], etc. It should be noted that many studies have been published describing the formation of Ti-Al alloys induced by pulsed electron-beam surface treatment, [8,9], but

[4] To whom any correspondence should be addressed.

the formation of Ti-Al alloys by irradiation with continuous electron beams has been less well investigated.

In this study, a composite Ti-Al film was deposited on an α-Ti substrate by electron-beam evaporation followed by electron-beam treatment with a continuously scanned electron beam, thus realizing two different alloying mechanisms – by surface melting and by surface irradiation without melting. The crystal structure was determined by XRD on the surface and by neutron diffraction in depth. SEM/EDX analyses were carried out in order to explore the surface structure and to analyze the chemical composition; a nanoindentation test was also conducted to examine the mechanical properties.

2. Experimental part

A composite Ti-Al film with a thickness a 10 μm was deposited on an α-Ti substrate by electron-beam evaporation followed by electron-beam treatment with a continuously scanned electron beam. When applying this method, the accelerated electrons provoke interactions of the atoms from the deposited layer and the substrate, forming intermetallic alloys. The experiments were conducted on a Leybold Heraeus (EWS 300/ 15 - 60) installation with the following technological parameters: accelerating voltage $U = 55 - 60$ kV, beam current $I = 40$ mA, focusing lens current $I_f = 512$ mA. The alloying process was controlled by varying the speed of motion of the specimens. A low speed of specimen motion corresponds to a long interaction time of the electron beam and the treated area, increasing the temperature and, thus, melting the surface. In order to realize two different alloying mechanisms (by melting the surface, and by surface irradiation without melting), two different speeds were chosen: $V_1 = 1$ cm/sec and $V_2 = 5$ cm/sec.

The phase composition on the surface was characterized by X-ray diffraction. The experiments were performed on a Bruker D8 Advance diffractometer with Cu Kα radiation and a LynxEye detector. The XRD patterns were registered within the range from 20° to 80° at 2θ scale with a step of 0.02°.

Scanning electron microscopy (SEM) analysis was carried out in order to explore the surface structure. The experiment was performed on a dual-beam scanning electron/focused system (SEM/FIB LYRA I XMU, TESCAN), equipped with an EDX detector (Quantax 200, Bruker). In this study, secondary electrons were used with high voltage $HV = 30$ kV and a work distance of 8.7 mm.

Neutron diffraction experiments were performed in order to study the phase composition in the volume of the samples. The measurements were conducted on a DN-2 diffractometer of the IBR-2 fast pulsed reactor at the Frank Laboratory of Neutron Physics (FLNP) JINR (Dubna, Russia). IBR-2 is a high-flux pulsed reactor with a peak thermal neutron flux of 10^{16} neutrons/cm^2 s at the moderator. The reactor operates in a time-of-flight mode, which implies time-dispersive measurements instead of the most common angle-dispersive diffraction. The initial velocity of each neutron is determined by the time of arrival of the neutron into the detector. Since the neutron flux is formed after passing through the moderator, it has a velocity (and hence, wavelength) distribution defined by the Maxwell thermal distribution and the resultant neutron spectra are a blend of this distribution and the sample diffraction pattern. The scattering angle and direction are fixed by the position of the detector during the experiment. Diffraction maxima in the scattered intensity appear when the Bragg conditions are fulfilled. The relation between the time of the flight and the interplanar distance d_{hkl} is as follows (modified Bragg law):

$$d = \frac{ht}{2mL\sin\theta}, \tag{1}$$

where d – interplanar distance; t – time of neutron flight; L –distance from the neutron source to the detector; θ –angle position of the detector; h – Plank constant; m – mass of the neutron.

Figure 1 represents schematically the DN-2 diffractometer. The neutrons are transported from the neutron source to the sample position in a bent neutron guide constructed from straight mirror sections with a length of 0.5 m. The guide together with the specially designed neutron moderator enables the use of the long-wavelength neutrons in the range 1.2 – 20 Å. The time-of-flight neutron experiments were performed for 13.5 hours with an SNM17 neutron detector filled with ^{3}He. The detector was positioned at 63.8° and the neutron flight path from the moderator to the detector was 24.27 m. The neutron flux at the sample was ~ 5×10^6.

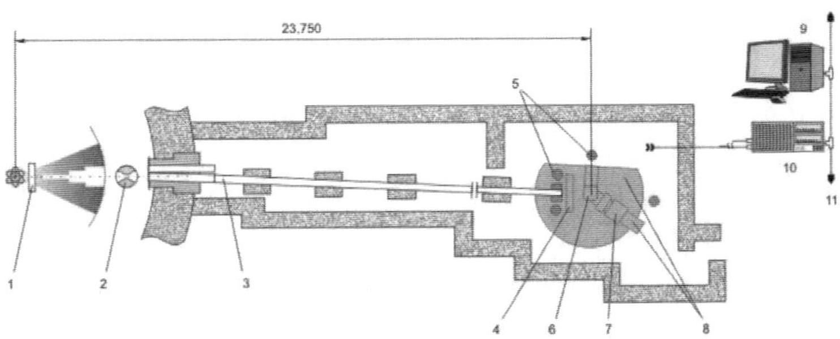

Figure 1. DN-2 diffractometer at the IBR-2 pulsed reactor.

In figure 1: 1 – moderator; 2 – background chopper; 3 – neutron guide tube; 4 – ring detector; 5 – ^{3}He detectors; 6 – goniometer; 7 – two-coordinate position-sensitive detector; 8 – mechanical part with a moveable arm; 9 – control and operative visualization/analysis; 10 – data acquisition; 11 – data transfer.

The nanoindentation test was performed to study the mechanical properties of the samples prepared. The experiment was conducted on a nanomechanical tester (Bruker, USA). Prior to the indentation, the surface of the samples was polished; the force applied was 200 mN.

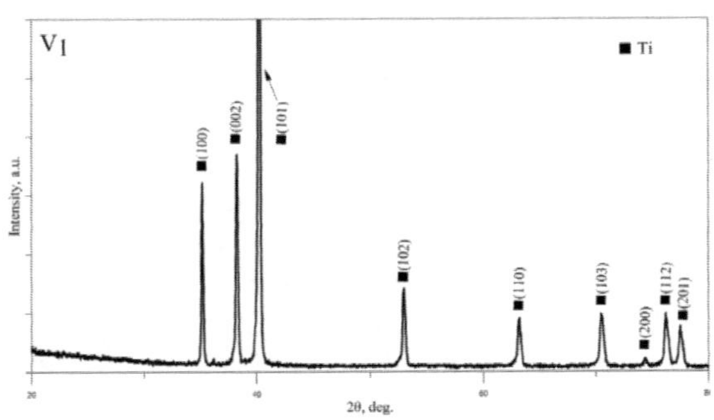

Figure 2. XRD pattern of the sample treated at V_1 speed of specimen motion.

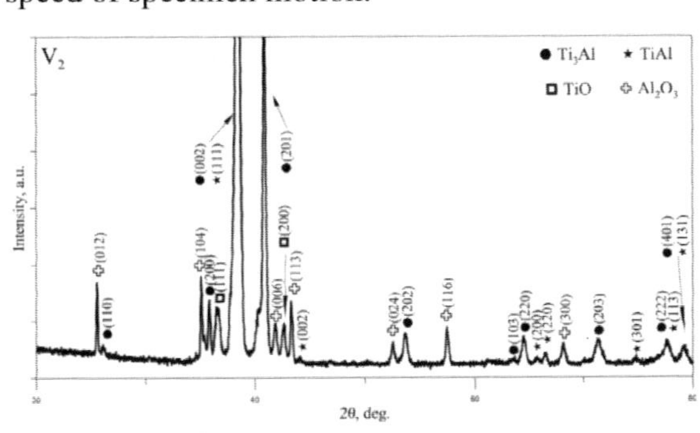

Figure 3. XRD pattern of the sample treated at V_2 speed of specimen motion.

3. Results and discussion

Figures 2 and 3 represent the XRD patterns of the samples. The diffractogram of the specimen treated at speed V_1 does not contain any peaks corresponding to intermetallic Ti-Al phases, while at speed V_2, an intermetallic Ti-Al structure is visible with a main phase α_2-Ti_3Al. In addition, the pattern exhibits several peaks corresponding to γ-TiAl and some peaks of Al_2O_3 and TiO phases, because the sample was retrieved from the vacuum chamber directly after the EBT process. The latter phases demonstrate the good oxidation behavior of the sample.

A SEM image taken from the surface of the sample treated at speed V_2 is shown in figure 4. The EDX analysis showed that an intermetallic Ti-Al structure has been formed with a significant degree of oxidation. The mean ratio of Ti:Al was about 3:1, which corresponds to the main Ti_3Al phase, according to the XRD analysis. The dark precipitations, marked by asterisks, correspond to Al_2O_3 with a negligible amount of Ti (~4%). Thus, the SEM/EDX results confirmed the phase composition as evaluated by XRD.

Figure 4. SEM image of the surface of a Ti-Al alloy.

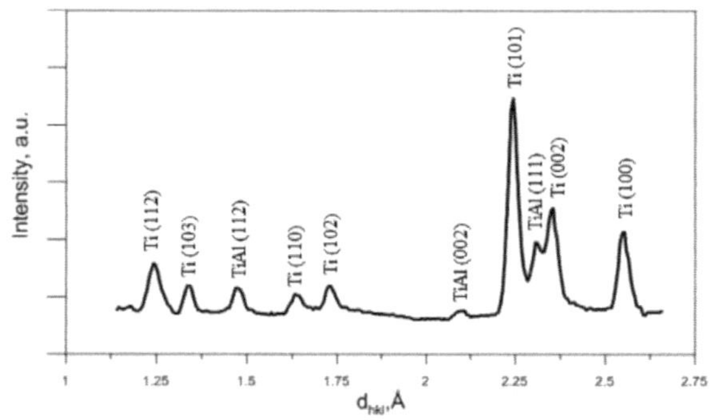

Figure 5. Time-of-flight neutron diffraction pattern of a TiAl alloy.

Figure 5 shows the time-of-flight neutron diffraction spectrum. It should be noted that, while for X-rays the penetration depth is limited by photo-electric absorption, the material is essentially transparent for neutrons [10]. Thus, neutron diffraction is an appropriate technique for in-depth crystal structure investigations. The pattern is presented by the interplanar distance distribution, contrary to the XRD spectra presented as an angle distribution. One can see several diffraction peaks corresponding to polycrystalline Ti, and some peaks of γ-TiAl. Thus, besides on the surface, an intermetallic Ti-Al structure has been formed in depth. The results obtained from neutron diffraction Measurements showed that the Ti_3Al/TiAl surface structure has been transformed to Ti/TiAl in the volume. It was thus shown that the combination of X-ray and neutron diffraction is an appropriate method of studying the in-depth phase composition distribution. Therefore, we can predict a transformation of the mechanical properties from the surface to the volume.

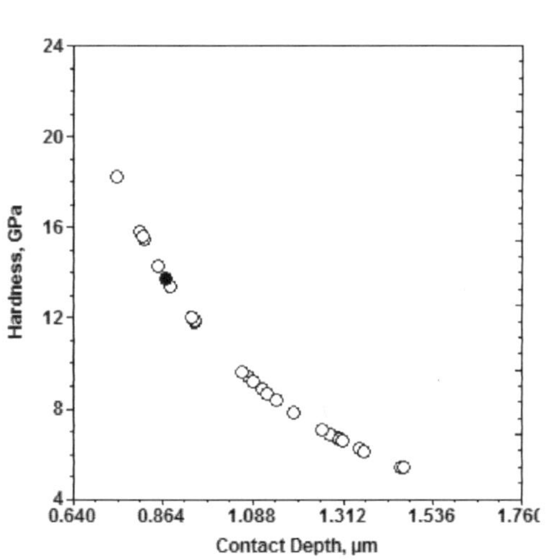

Figure 6. Nanohardness depth profile of a Ti-Al alloy.

According to [4], the hardness on the surface should be much greater than in the volume of the sample produced. In order to study the mechanical properties, a nanoindentation test was carried out whose results are presented in figure 6, where a significant decrease from the surface to the volume is visible. The hardness measured at a depth of 0.7 µm was about 18.5 GPa, while at 1.5 µm it decreased to about 5 GPa.

4. Conclusions

Ti-Al alloys were successfully produced by continuous electron-beam treatment without surface melting of the Ti substrate/Ti-Al coating system with the technological parameters discussed above, in contrast with the case of electron-beam surface melting. We demonstrated that the combination of the XRD and neutron diffraction techniques is an effective method of evaluating the in-depth phase composition distribution. The results obtained showed that the surface structure of Ti_3Al/TiAl was transformed to Ti/TiAl in the volume, which means that the mechanical properties on the surface are superior to those in the volume. In addition, a nanoindentation test demonstrated a decrease in the nanohardness in the depth, with the values being ~18.5 GPa at 0.7µm and ~5 GPa at 1.5 µm.

Acknowledgments
The authors would like to thank the Bulgarian Nuclear Regulatory Agency for providing financial support. The SEM/EDX experiments were conducted at the Materials Science Laboratory of the Department of Solid State Physics and Microelectronics, Faculty of Physics, St. Kliment Ohridsky University of Sofia.

References

[1] Wu X 2006 *Intermetallics* **14** 1114 - 22
[2] Dimiduk M 1999 *Mater. Sci. Eng.* A **263**/2 281 – 8
[3] Clements H and Mayer S 2013 *Adv. Eng. Mater.* **15** 191 – 215
[4] Appel F, Paul J and Oehring M 2011 *Gamma Titanium Aluminide Alloys: Science and Technology* (Wiley-VCH Verlag & Co. KGaA)
[5] Eckert J, Gao W, Golovin I, Imai Y, Li Z, Rennhofer M, Roth S, Russel A, Stoica M, Vanghan G, Zavrazhnov A and Zlomanov V 2008 *Intermetallic Research Progress* (Nova Science Publ. New York)
[6] Hampshire J, Kelly P and Teer D 2002 *Thin Solid Films* **420 – 421** 386 – 91
[7] Tsyganov I, Wieser E, Matz W, Mucklich A, Reuther H, Pham M and Richter E 2000 *Thin Solid Films* **376** 188 - 97
[8] Mei X, Fu J, Li X, Rotshtein V, Koval N and Ma T 2014 *Rare Meter.* **33**/2 155 – 60
[9] Allian-Bonasso N, Rotshtein V, Bouzy E, Germain L, Ivanov Y and Grosdidier T 2013 *J. Phys,: Conf. Series* **416**
[10] Zabel H 1990 *Adv. Solid State Phys.* **30** 197 – 217

Structural modification of Ga^+ and N^+ ion implanted ta-C films

M Berova[1], M Sandulov[1], T Tsvetkova[1,4], D Karashanova[2], R Boettger[3] and L Bischoff[3]

[1]Georgi Nadjakov Institute of Solid State Physics, Bulgarian Academy of Sciences, 72 Tzarigradsko Chaussee, 1784 Sofia, Bulgaria
[2]Acad. J. Malinovski Institute of Optical Materials and Technologies, Bulgarian Academy of Sciences, Acad. G. Bontchev Str., Bl. 109, 1113 Sofia, Bulgaria
[3]Institute of Ion Beam Physics and Materials Research, Helmholtz-Zentrum Dresden-Rossendorf, POB 510119, 01314 Dresden, Germany

E-mail: tsvet@issp.bas.bg

Abstract. Thin-film samples (d ~ 40 nm) of tetrahedral amorphous carbon (ta-C) deposited by filtered cathodic vacuum arc (FCVA) were implanted with Ga^+ at ion energy E = 20 keV and ion fluences $D = 3\times10^{14}\div3\times10^{15}$ cm^{-2} and N^+ with the same energy and a dose $D = 3\times10^{14}$ cm^{-2}. The Ga^+ ion beam induced a structural modification of the implanted material. This resulted in a considerable change of its structural properties, manifested as the formation of a new phase under non-equilibrium conditions, which could be accompanied by considerable changes in the ta-C films optical properties. The N^+ implantation also resulted in a modification of the surface structure. These effects were explored using transmission (TEM) and scanning (SEM) electron microscopy.

1. Introduction

In recent years, carbon-based materials, and in particular tetrahedral amorphous carbon (ta-C), have attracted great interest from both scientific and industrial points of view. The term tetrahedral is used to describe amorphous carbon films with a large percentage of sp3 bonding. The high sp3 content in the films results in unique properties [1-3]. These properties also offer advantages as compared to another wide-optical-band gap material – silicon carbide (SiC) – for uses in nano-scale optical data storage using focused ion beams techniques, where SiC films have recently found useful applications [4-9].

In both polycrystalline and amorphous SiC film materials, a considerable role in the creation of a useful optical contrast between irradiated and non-irradiated areas of the films is played by the transformation of a substantial part of the present diamond-like (sp3) carbon bonds, before the irradiation, into graphite-like (sp2) carbon bonds, as a result of it [6]. It is expected that a similar mechanism of carbon bonds transformation would result when applying ion bombardment with different ions, e.g. nitrogen (N^+) and gallium (Ga^+) ions, in the case of ta-C films, so that a useful optical contrast would be achieved between irradiated and non-irradiated areas of the films. The use of

[4] To whom any correspondence should be addressed.

gallium as an ion-implanted species is particularly attractive since it is available in standard focused ion-beam (FIB) machines, and, in addition, has been shown to be capable of generating large optical contrasts [10,11]. The implantation of N^+ leads to the formation of crystals on the sample surface. The underlying structural modification, induced by the N^+ and Ga^+ ion bombardment, was investigated by transmission (TEM) and scanning (SEM) electron microscopy measurements.

2. Experimental

Thin ta-C films ($d \sim 40$ nm) were deposited on Corning glass substrates using a commercial FCVA system (Commonwealth Scientific Corporation). Carbon plasma is produced from the arc spot on the cathode, 99.999% pure graphite in high vacuum. The cathodic arcs are prolific generators of highly-ionized carbon plasmas. With the FCVA technique, the plasma stream is steered through a magnetic filter to eliminate neutral particles generated at the cathode. At the filter exit, the fully ionized plasma, consisting of carbon ions and electrons, streams towards the substrate. The films were deposited at room temperature with an arc current of 120 A under floating conditions.

Ion implantation of N^+ and Ga^+ was carried out at room temperature (RT) using a commercial broad-beam ion implanter. The ion-beam intensity was $I \sim 2$ μA/cm^2, the ion energy was $E = 20$ keV, and the ion fluencies used were $D = 3\times10^{14} \div 3\times10^{15}$ cm^{-2} for the Ga^+-implanted, and $D = 3\times10^{14}$ cm^{-2} for the N^+-implanted samples. The SRIM simulation program [12] was used to determine the projected range Rp ~ 29 nm and the struggle ΔRp ~ 10 nm, for the N^+, and Rp ~ 17 nm and ΔRp ~ 4 nm, for the Ga^+ implanted ions into the ta-C film samples ($d = 40$ nm).

The chemical composition and state of the elements in the ta-C films surfaces with and without N^+ ions, as well as in the case of Ga^+ implantation, were studied using transmission electron microscopy (TEM) and scanning electron microscopy (SEM) measurements. The TEM experiment was carried out on a JEOL JEM 2100 high-resolution scanning transmission electron microscope with 200 keV energy and a height of 25 cm between the objective and projector lenses. The SEM experiment was conducted using a Philips 515 scanning electron microscope.

3. Results and discussion

3.1. Analysis and treatment of TEM images

Below, TEM images are presented of ta-C films before (figure 1) and after implantation with different elements and doses of ions (N^+ ions for figure 2 and Ga^+ ions for figure 3).

Figure 1 presents a TEM image of a clear ta-C film without any implantation impact. On the surface, one can see a few little dark spots, which are identified as a graphite-like fraction in an amorphous carbon matrix.

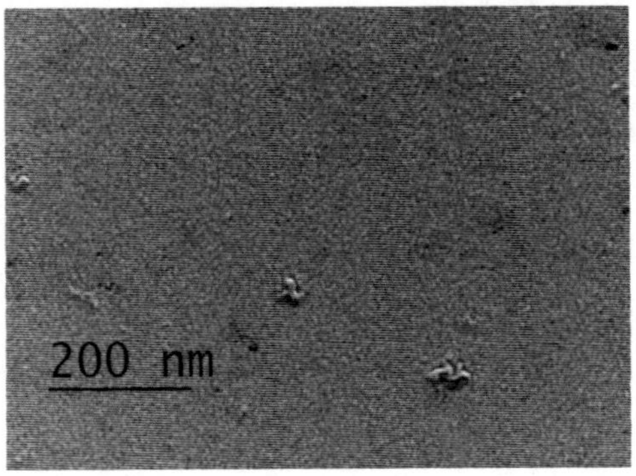

Figure 1. TEM image presenting area of a clear ta-C film without any implantation effects.

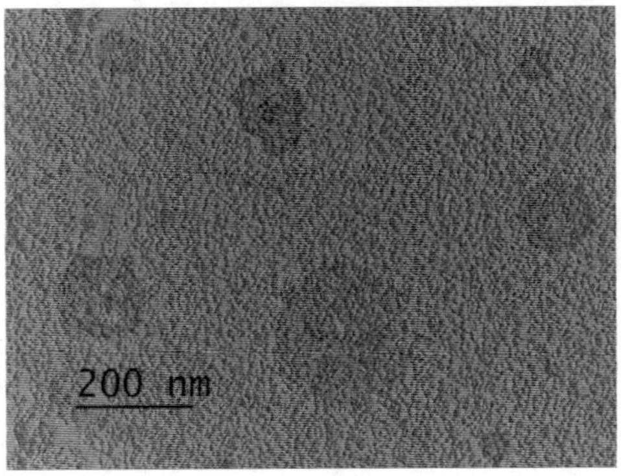

Figure 2. TEM image of area of a N^+-implanted ta-C film with a dose $D1 = 3\times10^{14}$ cm^{-2}.

Figure 2 shows a TEM image of a ta-C film after N^+ ion implantation effect of a dose $D1 = 3\times10^{14}cm^{-2}$. It is noticed that the dark spots are larger compared with those in the film without any implantation effect. It could be presumed that the implantation effects contribute to clustering of the graphite fraction.

Using selected-area electron-diffraction (SAED) microscopy for the N^+ implanted sample, electron-beam diffraction images were obtained from these darker spots. The result showed that they were crystals in the amorphous carbon.

In the case of Ga^+ implanted ta-C films, the TEM images revealed gallium precipitates in the form of approximately spherical colloids (figure 3). The SAED patterns indicated that they were amorphous gallium.

The size of these colloids clearly increases with the Ga^+ ion dose (figure 3a and figure 3b). These results are in good agreement with previous results of TEM analysis of Ga^+-implanted float glass samples [13]. Figure 3c is a zoomed area from Figure 3b, where one can see that the inclusions look like droplets in the layer. In the next section we will discuss further the nature of these colloids.

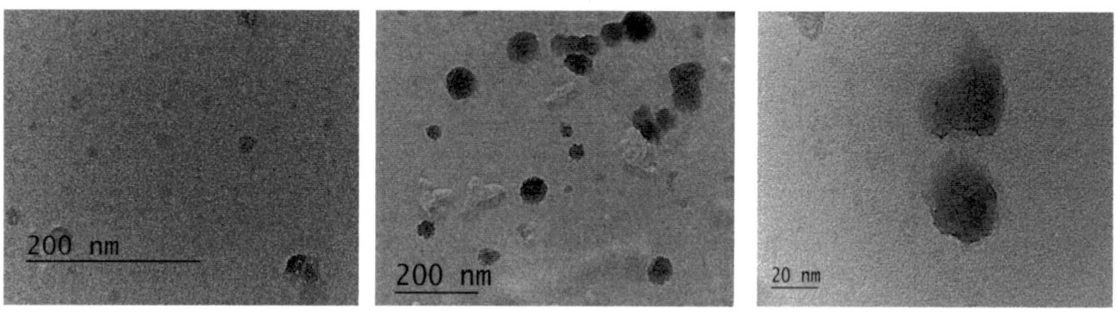

(a) (b) (c)

Figure 3. TEM image of an area of Ga^+-implanted ta-C film: a) dose $D2 = 3\times10^{14}cm^{-2}$; b) dose $D2 = 3\times10^{15}cm^{-2}$; c) dose $D2 = 3\times10^{15}cm^{-2}$ at higher magnification.

3.2. Analysis and treatment of the SEM images

A SEM image taken from the surface of the unimplanted ta-C film is given in figure 4. In comparison with the reference sample, the N^+ implanted ta-C layer is highly damaged with randomly located formations on the surface (figure 5). In the magnified image in figure 6, it is seen that the formations are small crystals (figure 6).

Figure 4. SEM image of an unimplanted ta-C film, electron-beam energy 25 kV, magnification 5000.

Figure 5. SEM image of a ta-C film implanted with N^+, $D1 = 3\times10^{14}$ cm^{-2}, electron-beam energy 25 kV, magnification 10000.

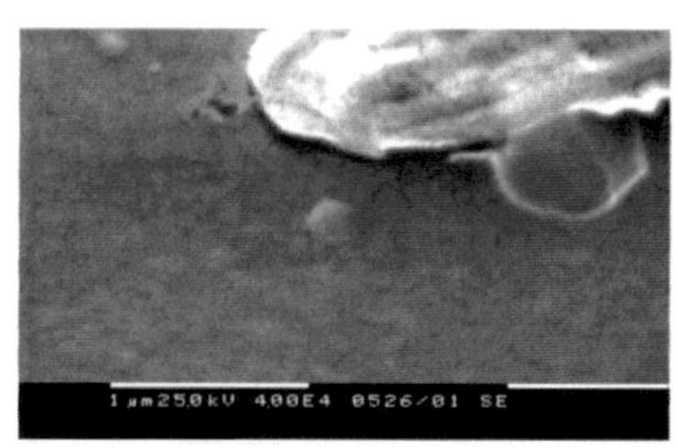

Figure 6. SEM image of a ta-C film implanted with N^+, $D1 = 3\times10^{14}$ cm^{-2}, electron-beam energy 25 kV, magnification 40000.

We also compared two samples implanted with Ga^+ with different doses: $D_1 = 3\times10^{14}$ cm^{-2} and $D_2 = 3\times10^{15}$ cm^{-2} with the same reference sample as above (figure 4). In figure 7, the image of sample D_1 shows that there are a few convexities. In figure 8 one can see that the sample implanted with the second dose shows an increase of the inclusions.

Figure 7. ta-C film implanted with Ga^+, $D_1 = 3\times10^{14}$ cm^{-2}.

Figure 8. ta-C film implanted with Ga^+, $D_2 = 3\times10^{15}$ cm^{-2}.

Figure 9. ta-C film implanted with Ga^+, $D_1 = 3\times10^{14}$ cm^{-2}, electron beam energy 25 kV, magnification 20000.

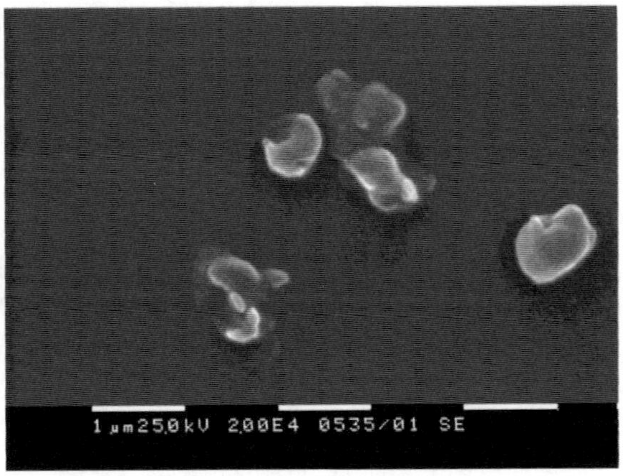

Figure 10. ta-C film implanted with Ga^+, $D_2 = 3\times10^{15}$ cm^{-2}, electron beam energy 25 kV, magnification 20000.

Figure 9 and figure 10 both show in greater detail at higher magnification the colloid nature of the Ga^+ inclusions at the two doses. This phenomenon is in good agreement with those observed by D.E. Hole, P.D. Townsend, J.D. Barton, L.C. Nistor, J. Van Landuyt [13] for Ga^+ implantation effects in float glass samples. The results of their experiment were obtained by TEM and RBS.

From the analysis of the distribution of the dark spots in the images of the TEM investigation, we can conclude that the number of dark areas is increasing with the dose of implantation with Ga^+ ions in a nonlinear way. The contrast of the dark spots (graphite in the case of N^+ and gallium precipitates in the case of Ga^+) rises after the implantation with a higher dose. The change of the structures could also lead to changes in the optical properties, which effect is a subject of others our publications [14, 15]. There is a difference between the sizes of the spots in the case of the N^+ and the Ga^+ implantation, caused by the different masses of the two ions (N^+ produces a larger size).

The TEM and SEM investigations of the N^+-implantation-induced structural modification in ta-C films are also very helpful in elucidating the role of this implantation in enhancing the resistive switching properties of ta-C films.

4. Conclusions

The present study of Ga^+- and N^+-implanted ta-C films revealed that there can be significant differences between the impact of various elements and their doses to the size and distribution of particles created by ion implantation. It was concluded that the TEM and SEM analyses conducted indicated the formation of new phases in both cases of implantation. Comparing the two types of elements, we can conclude that Ga^+ ions formed colloids with a more spherical shape and distinct boundaries than the clusters in the N^+ implanted samples. As the dose is increased, the Ga^+ colloids became bigger and their amount grew. In consequence, it is expected that these formations contribute to the changes observed in the optical properties of ta-C films, which can be used for optical data recording applications.

Acknowledgements

The authors would like to thank the group of Prof. Daniela Karashanova for the samples preparation and characterization. The authors would also like to thank the staff of the ion-beam center at the Helmholtz-Zentrum Dresden-Rossendorf e.V., a member of the Helmholtz Association, for performing the ion implantation.

References

[1] Fallon P J, Veerasamy V S, Davis C A, Robertson J, Amaratunga G A J, Milne W I and Koskinen J 1993 *Phys. Rev.* B **48** 4777

[2] Silva S R P, Xu S, Tay B K, Tan H S, Scheibe H J, Chhowalla M and Milne W I 1996 *Thin Solid Films* **290–291** 317

[3] Veerasamy V S, Yuan J, Amaratunga G, Milne W I, Gilkes K W R, Weiler M and Brown L M 1993 *Phys. Rev.* B **48** 17954

[4] Müller G 1993 *Nucl. Instrum. Methods* B **80-81** 957

[5] Kalbitzer S 2004 *Nucl. Instrum. Methods* B **218** 343

[6] Tsvetkova T 1996 in: *Beam Processing of Advanced Materials*, Eds. J Singh, S Copley, J. Mazumder, ASM International, Metals Park p. 207.

[7] Tsvetkova T, Takahashi S, Zayats A, Dawson P, Turner R, Bischoff L, Angelov O and Dimova-Malinovska D 2005 *Vacuum* **79** 94

[8] Tsvetkova T, Takahashi S, Zayats A, Dawson P, Turner R, Bischoff L, Angelov O and Dimova-Malinovska D 2005 *Vacuum* **79** 100

[9] Takahashi S, Dawson P, Zayats A V, Bischoff L, Angelov O, Dimova-Malinovska D, Tsvetkova T and Townsend P D 2007 *J. Phys.* D: *Appl. Phys.* **40** 7492

[10] Bischoff L, Teichert J, Kitova S and Tsvetkova T 2003 *Vacuum* **69** 73

[11] Tsvetkova T, Angelov O, Sendova-Vassileva M, Dimova-Malinovska D, Bischoff L, Adriaenssens G J, Grudzinski W and Zuk J 2003 *Vacuum* **70** 467

[12] Ziegler J F, Biersack J P and Littmark U 1985 *The Stopping and Range of Ions in Matter* **1** (Pergamon New York)

[13] Hole D E, Townsend P D, Barton J D, Nistor L C and Van Landuyt J 1995 *J. Non-Crystalline Solids* **180** 266-74

[14] Sandulov M, Berova M and Tsvetkova T 2014 *J. Phys.: Conf. Ser.* **558** 012044

[15] Sandulov M, Berova M, Tsvetkova T and Zuk J *Acta Physica Polonica* A in press

MeV electron irradiation of Si-SiO_2 structures with magnetron sputtered oxide

S Kaschieva[1,4], Ch Angelov[2] and S N Dmitriev[3]

[1]Georgi Nadjakov Institute of Solid State Physics, Bulgarian Academy of Sciences, 72 Tsarigradsko Chaussee, 1784 Sofia, Bulgaria
[2]Institute for Nuclear Research and Nuclear Energy, Bulgarian Academy of Sciences, 72 Tsarigradsko Chaussee, 1784 Sofia, Bulgaria
[3]Flerov Laboratory of Nuclear Reactions, Joint Institute of Nuclear Research, 141980 Dubna, Moscow Region, Russia

E-mail: kaschiev@issp.bas.bg

Abstract. MeV electrons influence on the characteristics of Si-SiO_2 structure with magnetron sputtered oxide was studied by ellipsometry and the thermally stimulated current (TSC) method. The MOS structures used in this study were fabricated on <100> oriented p-Si wafers of 12.75-17,25 Ω.cm resistivity. Magnetron sputtered oxides with different thicknesses of 20 and 100 nm were deposited on p-Si substrates. Both groups of samples were irradiated by 23 MeV electrons. The oxide thicknesses and TSC characteristics of the MOS samples were measured before and after MeV electron irradiation with doses of 4.8×10^{15} and 4.8×10^{16} el.cm^{-2}. The oxide thicknesses of both groups of samples increased after irradiation. The main defects generated by the MeV electrons were evaluated. It was shown that the trap concentration increases with the electron irradiation dose. The main peak in the TSC characteristics gives information about the main radiation defects at the Si-SiO_2 interface of the MOS structures. These defects can be related to the vacancy-boron complexes which are associated with the main impurities in the p-Si substrate. These results correspond to our results reported earlier for MeV electron irradiated Si-SiO_2 structures with thermally grown oxide. But (in this case) the effects observed are more pronounced for the magnetron sputtered oxide. A possible reason is the higher defect concentration generated in the magnetron sputtered oxide during its deposition on Si-substrates.

1. Introduction

The silicon MOS structure is the basis of electronic and microelectronic devices, although novel configurations and compositions are constantly being introduced. The long-term application of Si-SiO_2 structures in the modern semiconductor device industry is based on the fact that an extremely stable Si-SiO_2 interface with a low interface state density can be achieved. It is well known that the Si-SiO_2 interface is the part of the MOS structures that is most sensitive to different kinds of irradiation. One solution of the problem of reducing the radiation effects on the MOS structure is to modify the oxide materials in such a way as the required parameter can be achieved. We report on a study of MOS structures characteristics with magnetron sputtered SiO_2 before and after MeV electron irradiation.

[4] To whom any correspondence should be addressed.

MeV electron irradiation is a very useful tool for studying the operation of MOS devices in a radiation environment. The MeV electrons penetrate through the entire MOS structures, their energy being sufficient to produce hard radiation defects in the whole system: metal, oxide, Si-SiO_2 interface and silicon substrate. Our earlier studies showed that MeV electrons increase the thermally grown oxide thickness of MOS structures [1]. The phase analysis of n- and p-type SiO_2/Si samples irradiated by MeV electrons carried out by soft X-ray emission spectroscopy have shown that the effect of radiation-stimulated oxidation depends on the type of Si substrate. The increase of the SiO_2 contribution to the spectra is more pronounced for the specimens prepared on n-type silicon wafers. This means that MeV electron irradiation leads to the oxidation of the Si/SO_2 samples, and this oxidation effect for the thermally grown oxide is more prominent for the samples prepared on n-Si substrates [2].

We present a study on MeV electrons irradiation effect on the characteristics of p-Si-SiO_2 structures with magnetron sputtered SiO_2. The reactive magnetron sputtering relies on the principle of compound formation by a chemical reaction between the sputtered material, a metal, and a reactive gas added to the sputtering gas, generally Ar [3]. Magnetron sputtering is a quite difficult process to control, mainly because during the deposition the reactive gas reacts not only with the material sputtered to form the compound on the substrate (or on the walls) but also with the target surface [4]. The deposition rate of dielectric compounds is an important parameter, whose control is essential in what concerns the characteristics of the deposited layer [5-6].

In the present work the characteristics of p-Si-SiO_2 structure with magnetron sputtered oxide are studied by ellipsometry and the thermally stimulated current technique before and after different doses of 23 MeV electron irradiation.The oxide thickness and the electrical properties of the Si-SiO_2 structure were investigated as a function of the MeV electron dose irradiation.

2. Experimental details

In this experiment, 12.75 - 17.25 Ω cm p-type <100> oriented single-crystal Si wafers were used to prepare Si-SiO_2 structures. The oxide was deposed on Si wafers by radio frequency magnetron sputtering using a 13.56 MHz planar Leybold-Heraeus system. The deposition was carried out in argon atmosphere at a pressure of 1 Pa and a cathode voltage of 2.0 kV. The oxide thickness of each sample was determined by ellipsometry. The Si-SiO_2 samples were divided into two groups after the oxide deposition - with oxide thicknesses of 20 nm and of 100 nm. Al gate electrodes were photolithographically formed onto the oxide of some of these samples so that MOS capacitors (with oxide thicknesses of 20 and 100 nm) were produced. The Si-SiO_2 and MOS samples were irradiated by 23 MeV electrons produced by a Microtron MT-25 installation at the Flerov Laboratory of Nuclear Reactions of the Joint Institute for Nuclear Research (FLNR, JINR) Dubna, Russia. The beam current during irradiation was about 9 μA. The MeV electrons' effect on the oxide thickness of the Si-SiO_2 structures and the thermally stimulated current characteristics of the MOS structures were evaluated after two doses (4.8×10^{15} and 4.8×10^{16} el cm^{-2}) of high-energy electron irradiation.

3. Results and discussion

We measured the oxide thickness of non-irradiated and 23-MeV electron-irradiated Si-SiO_2 structures covered by with 20-nm and 100-nm thick magnetron sputtered oxides studied. The oxide thickness changes as the electron irradiation dose was raised were measured by ellipsometry. We found that the oxide film thickness of both groups of the samples increased after irradiation. Table 1 presents the oxide thickness of the Si-SiO2 structures with magnetron sputtered SiO_2 before and after the doses of 23 MeV electron irradiation. It is seen that the oxide thickness increases slowly. The oxide increase for the Si-SiO_2 structures with the thinner film (20 nm oxide) was 2 nm (a value within the experimental error). The oxide thickness of the structures with a thicker (100 nm) magnetron sputtered oxide film increased more pronouncedly after electron irradiation. This could be connected with a higher defect concentration (in the thicker oxide) generated during the magnetron sputtering deposition. We assume that radiation-stimulated oxidation takes also place as a result of the high-energy electron irradiation.

The radiation-stimulated oxidation was more effective in the case of MeV-electron-irradiated p-type Si-SiO_2 structures covered with magnetron-sputtered oxide. The thickness of thermally-grown oxide on p-type Si-SiO_2 structures remained the same after MeV electron irradiation [7, 8].

Table 1. Oxide thickness before and after electron irradiation.

SiO_2 oxide thickness [nm]	Energy 23 MeV, electron irradiation dose [el.cm^{-2}]	
	4.8×10^{15}	4.8×10^{16}
20	22	24
100	112	125

The MOS samples were studied using the thermally stimulated current (TSC) technique before and after irradiation by the two doses of MeV electrons.

Figures 1-2 show typical TSC spectra of p-type MOS structures with magnetron-sputtered silicon dioxide of a thickness of 20 nm and 100 nm before and after irradiation by 23 MeV electrons. The MeV electrons create different kinds of defects in the Si-SiO_2 structures (which depend on the oxide thickness) with concentrations depending on the irradiation dose. The thermo-stimulated current is shown as a function of the sample temperature during the measurement for the two electron doses (curves 1 and 2). The TSC spectra obtained of the irradiated samples were composed of several overlapping peaks, so that a thermal and differential field "cleaning" of the spectra were applied to reveal the main peaks [9, 10].

No TSC characteristics of the p-type MOS structures with magnetron-sputtered oxide thickness of 20 nm could be registered before electron irradiation. Figure 1 presents TSC curves after 23 MeV electron irradiation of the MOS samples with doses of 4.8×10^{15} and 4.8×10^{16} el cm^{-2} (curves 1 and 2), respectively. It is seen that the 23 MeV electrons create new states at the Si-SiO_2 interface, i.e., the high-energy electron irradiation creates new discrete energy levels in the silicon forbidden gap. MeV electrons create several kinds of defects at the Si-SiO_2 interface whose concentration increases as the irradiation dose is increased.

Figure 2 presents TSC characteristics of p-type MOS structures with 100-nm magnetron-sputtered oxide thickness – before (curve 0) and after (curves 1 and 2) 23 MeV e lectron irradiation. Curve 0

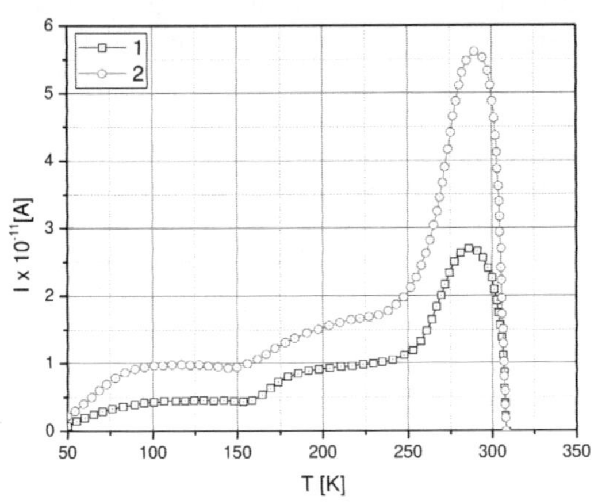

Figure 1. TSC curves of p-type MOS structures with 20 nm magnetron sputtered oxide thickness after 23 MeV electron irradiation with doses of 4.8×10^{15} and 4.8×10^{16} el cm^{-2} (curves 1 and 2) respectively.

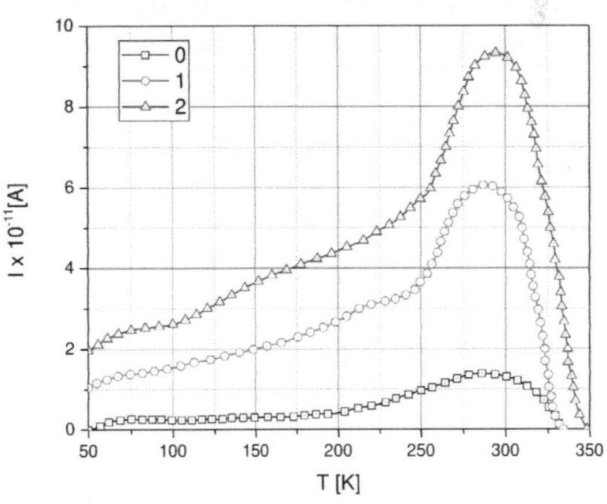

Figure 2. TSC characteristics of p-type MOS structures with 100 nm magnetron sputtered oxide thickness – before (curve 0) and after 23 MeV electron irradiation with doses of 4.8×10^{15} and 4.8×10^{16} el cm^{-2} (curves 1 and 2) respectively.

demonstrates that the radiation defects are created at the Si-SiO_2 interface of the thicker magnetron-sputtered oxide before MeV electron irradiation. This could be connected with the higher defect concentration generated in the magnetron-sputtered oxide during its deposition. It is seen that the irradiation by MeV electrons leads to an increase in the concentration of defects generated in the magnetron-sputtered oxide and, obviously, high energy electron irradiation introduces new radiation defects in the MOS structures. Curve 1 shows an increase in the height and in the areas enclosed by TSC curve after MeV electron irradiation. This is an indication of an increase in the density of all kinds of radiation defects induced in the oxide during its preparation. A further increase in the irradiation dose furthers this tendency, as it is shown by curve 2.

We used two techniques to obtain the activation energy of the traps - the initial rise plot method and the Grossweiner' techniques were applied to the TSC curves of the two MOS structures (with 20 nm and 100 nm tick oxide). The last peak of the TSC curves (in figure 1 and figure 2) is attributed to the level with energy position Ev = +0.45 eV and is associated with boron-vacancy complex defects. The generation of this kind of defects increases as the boron concentration in the Si substrate is increased, and, in contrast, the generation decreases as the oxygen concentration in the Si wafer is increased [11]. Curves 1 and 2 in both figures also confirm the fact that the main defects created by the high-energy electron irradiation correspond to the V-B defects and their concentration increase with the increasing of the irradiation dose. The area enclosed by the higher temperature peaks increases effectively with the increase of the electron dose. These results correspond to our earlier studies that showed that most of the radiation defects induced by high-energy electrons at the Si-SiO_2 interface of MOS structures (with thermally grown oxide) have to do with the main impurities in the Si substrate [6,7]. This result can be explained by taking into account that the high-energy electron irradiation generates vacancies, which tend to form complexes with the basic impurities of the substrate (such as boron in p-type Si wafers).

We, therefore, assume that the higher defects concentration generated in the magnetron-sputtered oxide during its deposition creates conditions for a higher degree of defects concentration following the MeV electrons irradiation.

4. Conclusions

The oxide thickness of Si-SiO2 and MOS structures with magnetron-sputtered oxide deposited on p-type Si substrates increases after irradiation by different doses of 23 MeV electrons. The thickness of structures with a thicker (100 nm) magnetron-sputtered oxide film increases to a larger extent after electron irradiation. Our results also showed that radiation defects are created at the Si-SiO_2 interface of the thicker magnetron-sputtered oxide (100 nm) during its deposition. The concentration of the radiation defects induced in MOS structures by MeV electrons depends also on the oxide thickness. The defects concentration at the Si-SiO_2 interface of these structures is higher than in the MOS structure with a thinner oxide (20 nm) film irradiated under the same conditions. A possible reason is the higher defects concentration generated in the thicker magnetron sputtered oxide during its deposition.

Acknowledgement

This study was supported by Grant № 41-17 (30.01.2013) from JINR, Dubna, Russia.

References

[1] Kurmaev E, Shamin S, Galakhov V, Kirilova M, Kurennykh T, Viykhodets V and Kaschieva S, 1997 *J .Phys.Condens.Matter.* **9** 6969

[2] Shamin S N, Galakhov V R, Kaschieva S, Dmitriev S N and Belov A G 2003 *J. Mater. Sci.-Mater. Electr.* **14** 809

[3] Westwood W D 1998 Reactive sputtering: introduction and general discussion *Handbook of Thin Solid Film Process Technology* (IOP Bristol)

[4] Snyders R, Dauchot J-P and Hecq M 2007 *Plasma Processes and Polymers* **4**/2 113-26

[5] Bartzsch H, Glöß D, Böcher B, Frach P and Goedicke K 2003 *Surf. Coat. Technol.* **174–175** 774–8
[6] Le-Nian He and Jin Xu 2002 *Vacuum* **68**/2 197–202
[7] Kaschieva S, Todorova Zh and Dmitriev S N 2004 *Vacuum* **76** 307-10
[8] Halova E, Alexandrova S, Kaschieva S and Dmitriev S N 2010 *J. Phys.: Conf. Series* **253** 012047
[9] Nicolas K H and Woods J Br 1964 *J. Appl. Phys.* **15** 783
[10] Saunders A F and Wright GT 1970 *Electron. Lett.* **6** 207
[11] Vavilov V, Plotnikov A and Tkachev V 1982 *FTT* **4** 3446-9

Influence of the thickness on the morphology and sensing ability of thermally-deposited tellurium films

T Hristova-Vasileva[1], I Bineva[1,4], A Dinescu[2], D Nesheva[1], D Arsova[1] and B Pejova[3]

[1]Georgi Nadjakov Institute of Solid State Physics, Bulgarian Academy of Sciences, 72 Tsarigradsko Chaussee, 1784 Sofia, Bulgaria
[2]National Institute for R&D in Microtechnologies – IMT Bucharest, 126 Erou Iancu Nicolae Str., 077190 Bucharest, Romania
[3]Faculty of Natural Sciences and Mathematics, Sts. Cyril and Methodius University, P.O. Box 162, 1001 Skopje, Macedonia

E-mail: irka@issp.bas.bg

Abstract. Tellurium films with nominal thicknesses of 30, 90 and 300 nm were prepared by thermal evaporation in vacuum at a low deposition rate of 0.3 nm/s. The morphology evolution with the increase of the film thickness was observed by scanning electron microscopy and atomic force microscopy.

Nanorods with a width of about 40 nm were observed on the thinnest films surface. On the 90 nm thick films, the formations grew in priority in the *z*-direction to nanoblades with the same width, but a length of about 100 nm. The further increase of the thickness led to an increase of the 2D nanoobjects' width and length and formation of a stacked nanosheet structure. The surface root-mean-square roughness (*S*q) increased with the thickness of the films.

Preliminary investigations of the sensing ability of the as-deposited tellurium films with different thicknesses towards water (H_2O), ethanol (C_2H_5OH), acetone (C_3H_5OH), and ammonia (NH_3) vapors were performed by measuring the vapor-induced changes in the film dark current. The films showed appreciable response only to ammonia vapors; their sensitivity was almost equal for the 30 and 90 nm thick films, and decreased significantly for the film tkness of 300 nm.

1. Introduction

Tellurium is a crystalline semiconductor, whose atoms tend to form polymeric, covalently bonded helical chains, readily packed into a hexagonal lattice through van der Waals forces [1]. Due to this specific nature, many authors reported the formation of various nanoobjects (nanorods, nanotubes, nanobelts, nanoblades, etc.) on the films, using complicated physical or soft chemistry methods [2-4]. It can be assumed that when using rather low deposition rates during the Te film preparation by conventional thermal vacuum evaporation, which is a relatively cheap and accessible method, the condensing film will also form 1D and 2D nanoobjects, thanks to its typical crystalline structure.

[4] To whom any correspondence should be addressed.

Recently the researchers in almost all scientific areas have been working on fighting the air, water and soil pollution by development of approaches for purification or reutilization of contaminated media, or by elaboration of equipment for quantitative measurements of the pollution extent. The tellurium films have been found appropriate and, thus, have been widely studied in view of resistive sensing of CO, NO_2 and NH_3 pollution [5-9].

Taking into account these two objectives, in this work we explored the thickness-morphology and thickness-sensing behavior of slowly deposited Te films.

2. Experimental procedures

Thin films of Te (purity of 5N) were prepared on Si (100) wafers by thermal vacuum evaporation at room substrate temperature, deposition rate of 0.3 nm/s, residual pressure of 7×10^{-3} Pa, and distance between the crucible and the substrate of 20 cm. The deposition rate and the thickness of the films was controlled using a quartz-crystal microbalance (QCM) Miki MSV 1841/A device, and verified by cross-section scanning electron microscopy.

The morphology of the deposited films was investigated by atomic force microscopy (AFM) and scanning electron microscopy (SEM). The AFM measurements were performed on a Multimode V scanning probe microscope (Bruker, ex. Veeco, Santa Barbara, CA). The images were taken in tapping mode, and each sample was investigated on multiple points of the surface. Measurements at scales of 1, 3 and 10 μm were carried out at scanning rates in the interval 0.5-2.0 Hz, and image resolution of 512 lines per scan direction (l/s.d.) for scales of 1 and 3 μm and 256 l/s.d. for 10×10 μm^2. Aluminum coated silicon TAP150-Al-G cantilevers (Budget Sensors Innovative Solutions Bulgaria Ltd.) with a resonant frequency of 150 kHz and a spring constant of 5 N/m were used. The radius of the cantilever's tip is smaller than 10 nm. The root-mean-square roughness (*Sq*) of the samples surface was determined in a scale of 3 and 10 μm. Before the analysis, the images were only flattened. They were further processed using the SPIP^{tm} 6.1.0 program.

The SEM observations were performed using an e-Line EBL equipment (Raith GmbH., Germany) in a SEM mode with an accelerating voltage of 10 kV. The films were observed on top and cross-section with magnification from 50 to 300 000 times.

The structure of the samples was investigated by X-ray diffraction (XRD) using a Bruker D8 diffractometer with a LynxEye solid-state detector at CuK_α irradiation (Ni-filter) and a Bragg angle (2θ) ranging from 10 to 60°.

For preliminary investigation of the films' potential application as resistive sensors for gas detection, their electrical current was measured in the presence of pollutant's vapors at room temperature and atmospheric air pressure. The film sensitivity is defined as the ratio of the current change with and without presence of active vapors $(I_{initial}-I_{final})/I_{initial}$). The measurements were performed on the as-prepared Si/Te structures provided with planar aquadag contacts on the top of the Te surface. The reagents used (deionized water (H_2O), 99.7 % absolute alcohol (C_2H_5OH), p.a. acetone (C_3H_5OH) and p.a. ammonia water (25 % NH_3 in H_2O)) in a quantity of 400 μl were introduced in a heated crucible (placed in the investigation chamber), immediately vaporized and distributed into the chamber volume (1.4 l) using a propeller. The measurement of the dark current change was performed up to saturation of the signal.

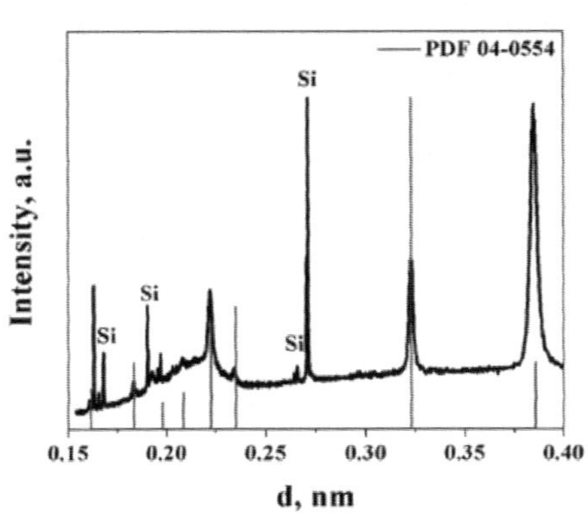

Figure 1. Typical X-ray diffraction pattern of as-deposited Te films. The silicon peaks belong to the substrate.

3. Results and discussion

A typical XRD pattern of the as-deposited Te films is presented in figure 1. The films consisted of hexagonal tellurium, corresponding to IJPCS powder diffraction file No. 04-0554. Some peaks of the silicon substrate appeared, and are marked in the figure. No phase or structural changes were observed with changing the film thickness.

Cross-section SEM microphotographs of as-deposited Te films are presented in figure 2, and the results for the film thickness are given in table 1. The films can be considered as bilayered structures, composed by a loose near-surface part and a relatively dense bottom part. This specificity of the samples causes the observed significant difference between the nominal thickness, defined by the QCM technique at set tellurium density of 6.24 g/cm^3 [10], and the actual one d_{actual}, roughly calculated as the sum of the bottom layer thickness d_{bottom} (determined by SEM) and the sample's root-mean-square roughness *Sq* (determined by AFM).

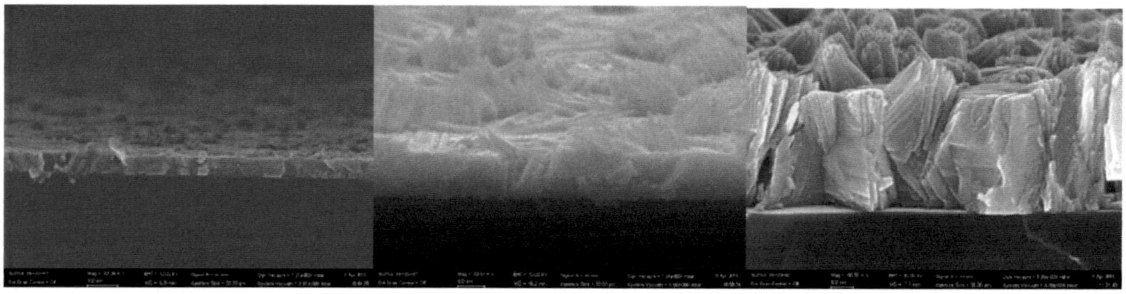

a) b) c)

Figure 2. Cross-section SEM of Te films with nominal thickness of a) 30, b) 90 and c) 300 nm – magnification of 80 000 times.

Table 1. Nominal (d_{nom}), bottom layer (d_{bottom}) and actual (d_{actual}) thicknesses, and root-mean-square roughness (*Sq*) of thermally deposited Te films.

d_{nom}, nm	d_{bottom}, nm	*Sq*, nm (for 3 μm scan)	*Sq*, nm (for 10 μm scan)	d_{actual}, nm
30	57	7.32	7.25	64
90	142	33.9	34.3	176
300	530	55.7	56.1	586

The results from the surface morphology observations by SPM and SEM are presented in figure 3. The images, obtained by these techniques, are in a good agreement.

As seen in table 1, *Sq* is practically the same for scanned areas of 3×3 and 10×10 μm^2, which shows that the surface region of the deposited films is extremely homogeneous. The films become rougher with the increase of their thickness.

The thinnest films are mainly formed by laterally distributed fibers, as some of their ends rise on the surface, appearing as randomly distributed nanorods (figure 3a). In the films with nominal thickness of 90 nm, these nanoobjects grow predominantly in the *z*-direction by forming highly disoriented bundled nanoblades having width of ~40 nm and length of ~150 nm (figure 3b). The further increase of the thickness causes agglomeration of the bundles and formation of nanocrystalline conglomerates with a columnar structure (figure 3c).

The well-developed surface observed of the as-obtained Te-films makes them interesting in view of sensing applications. The results from the investigations on the films' dark current change during introduction of different vapors are presented in figure 4. All films show a weak response towards water, ethanol and acetone.

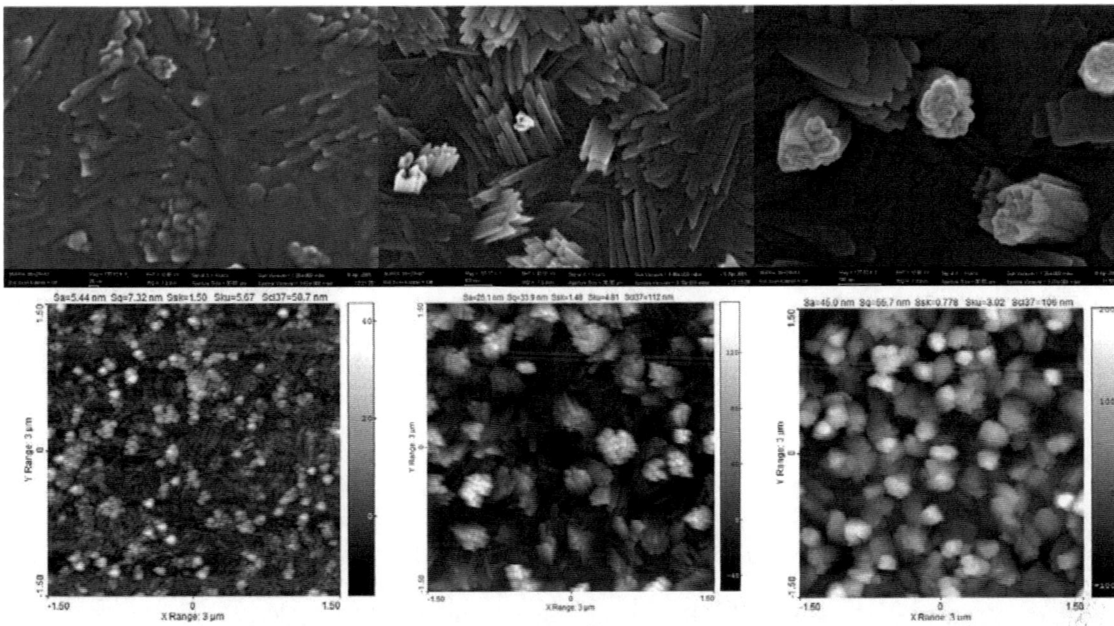

a) b) c)

Figure 3. SEM microphotographs (magnification of 100 000 times) and 2D AFM images (3 μm ^{scan}) of Te films with nominal thickness of a) 30, b) 90 and c) 300 nm.

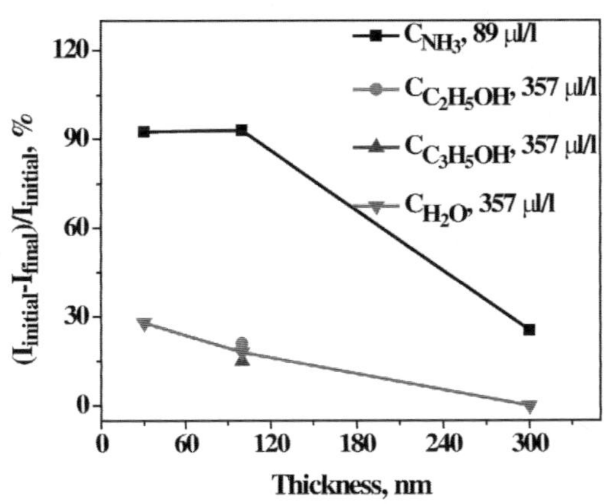

Figure 4. Thickness dependence of room temperature sensitivity to ammonia (NH_3) and water vapors of Te films on Si substrates. The room temperature sensitivity of 90 nm films towards ethanol (C_2H_5OH) and acetone (C_3H_5OH) is also presented.

The weak sensitivity to water allowed us to study the samples' sensitivity towards ammonia vapors by using a standard 25 % water solution in a quantity of 89 μl/l NH_3. A change of the sensitivity with the thickness was observed – it decreased with the increase of the thickness (figure 4).

The highest, almost equal, sensitivity to ammonia was observed for samples with a nominal thicknesses of 30 and 90 nm. This is probably due to their bilayered structure: the highly active nanostructured surface layer allows easy attachment of the ammonia molecules, while the relatively dense bottom part ensures good electrical transport conditions.

The sensing mechanism towards ammonia of the films is based on an adsorption-desorption physical reaction. The desorption process is very slow – it takes more than six hours for the films' initial current values to be restored.

4. Conclusions

Thin films with nominal thicknesses of 30, 90 and 300 nm were deposited on Si (100) wafers by physical vapor deposition at a low deposition rate of 0.3 nm/s.

The influence of the film thickness on the surface morphology was investigated. Formation of nanorods was observed on the 30 nm films, which grew predominantly in the *z*-direction forming 2D nanoblades on the 90 nm samples. The further thickness increase to 300 nm led to consolidation of the objects and formation of a stacked nanosheets structure.

Preliminary resistive sensing investigations were performed on the prepared Si/Te structures at different thicknesses of the active tellurium film. The investigation of the dark current change in the presence of water, ethanol, acetone and ammonia vapors showed significant sensitivity towards NH_3 only. The highest sensitivity values were obtained for the samples with nominal thicknesses of 30 and 90 nm, while the sensitivity decreased with the increase of the film thickness.

Acknowledgements
The authors acknowledge the support of the Bulgarian Academy of Sciences (contract ISSP-VK 03/14 and the bilateral projects with the Romanian Academy of Sciences, and with the Macedonian Academy of Sciences and Arts, as well as with the Bulgarian National Science Fund (contract DMU 0391/2011).

References
[1] Xia Y, Yang P, Sun Y, Wu Y, Mayers B, Gates B, Yin Y, Kim F and Yan H 2003 *Adv. Mater.* **15** 353-89
[2] Rao C N R, Deepak F L, Gundiah G and Govindaraj A 2003 *Progress in Solid State Chem.* **31** 5-147
[3] Ujjal K, Gundiah G and Rao C N R 2004 *J. Mater. Chem.* **14** 2530–5
[4] Métraux C and Grobéty B 2004 *J. Mater. Res.* **19** 2159-64
[5] Her Y C and Huang S L 2013 *Nanotechnology* **24** 215603
[6] Tsiulyanu D, Marian S and Liess H-D 2002 *Sensors and Actuators* B **85** 232-8
[7] Tsiulyanu D, Tsiulyanu A, Liess H-D and Eisele I 2005 *Thin Solid Films* **485** 252-6
[8] Tsiulyanu D, Stratan I, Tsiulyanu A, Liess H-D and Eisele I 2007 *Sensors and Actuators* B **121** 406-13
[9] Bianchetti M F, Heredia E, Oviedo C and Walsoe de Reca N E 2005 *J. Argentine Chem. Soc.* **93** 27-34
[10] Lide D R ed 2005 Physical constants of inorganic compounds *CRC Handbook of Chemistry and Physics, 85th Edition* (CRC Press, Boca Raton, FL)

Detection of powder formation in SiH_4/H_2 glow discharges

G Alexiou, G Tsigaras[1], E Amanatides and D Mataras

Plasma Technology Laboratory, Department of Chemical Engineering,
University of Patras, Greece

E-mail: lef@plasmatech.gr

Abstract. Time-resolved self-bias voltage measurements and spatiotemporal resolved optical emission spectroscopy were used in order to monitor particle formation and instabilities in the initial stage of 13.56 MHz SiH_4/H_2. The effect of total gas pressure on the particles formation and the time required for the plasma to reach steady state was investigated. Both techniques were very sensitive in monitoring the nucleation and coagulation phase of particles formation. The characteristic times required for reaching steady state were similar for both techniques indicating that electrical properties and chemical kinetics are affected in a similar way from particle formation. The increase of pressure resulted in a significant shortening of the instability period and this can be a significant advantage in the deposition of device grade silicon thin films.

1. Introduction

Plasma enhanced chemical vapour deposition (PECVD) using radio frequency (RF) capacitively coupled silane/hydrogen (SiH_4/H_2) discharges, is the most common technique, that is used to produce hydrogenated microcrystalline (μc-Si:H) and amorphous (a-Si:H) silicon. These materials are currently applied on single, tandem and multi-junction solar cells and modules for more effective use of solar spectrum [1]. There are still several issues and technological aspects related to these devices that are under investigation for their efficiency improvement and the reduction of cost [2]. More precisely, the increase of the μc-Si:H growth rate while maintain or even improve the film quality is crucial. The increase of the SiH_4/H_2 gas pressure and the so called high-pressure depletion method is the most promising technique, which can reach the goal of high deposition rates and device grade materials [3]. On the other hand, pressure raise will enhance the powder formation in the discharge which in turn can affect the film quality or the process costs due to the necessity of larger reactor cleaning times between depositions [4]. Gases residence time in the plasma and electron/ion-molecule, molecule-molecule collision frequencies will increase with pressure favoring secondary gas phase reactions, nucleation of clusters and dust formation.

Particles formation in SiH_4 plasmas has been the subject of investigation for several studies that can be found in the literature [5], [6], [7], [8]. The mechanism of clusters nucleation, coagulation and agglomeration has been monitored by different plasma diagnostics [9], [10], [11], [12], [13], theory and simulations [14], [15], [16]. However, all these studies are focused on low pressure, pure SiH_4 or SiH_4 – Ar discharges, which are not typically used for μc-Si:H films deposition. Molecular hydrogen

[1] To whom any correspondence should be addressed.

that is used in mixtures with SiH_4 for the formation μc-Si:H is a reactive gas and can significantly alter the mechanism of particle formation.

In the present work, we focused on the effect of the total SiH_4/H_2 pressure on the initial electrical and chemical instabilities that are induced by the nucleation and coagulation of silicon clusters. The investigation was carried out at typical conditions of μc-Si:H film growth. Time resolved electrical and optical plasma diagnostics were implemented and the characteristic times for instabilities ignition and steady state conditions were monitored. Spatio-temporal optical measurements were also used to follow changes on the electron heating mechanisms during the nucleation phase. The increase of pressure resulted in significant drop of the time that is required to reach steady state plasma conditions and this can be an important advantage for the material properties as it will minimize the undesirable incubation layer [17].

2. Experimental

A capacitively coupled, ultrahigh vacuum (10^{-8} Torr base vacuum) reactor, with parallel plate electrodes was used in this work [18]. The gas was fed into the plasma through a showerhead electrode. The process pressure was monitored by a capacitance manometer and was controlled via a throttle valve. All experiments were performed at room temperature with an excitation frequency of 13.56 MHz and interelectrode space of 15 mm. The total gas flow rate was 150 sccm, with SiH_4 input concentration of 4%, while the chamber pressure was varied from 0.5 to 4 Torr (table 1).

Table 1. Experimental conditions.

	Set 1	Set 2
Electrode gap	15 mm	15 mm
Total gas flow	150 sccm	150 sccm
SiH_4 concentration	4%	0%
Discharge Power	11 W	11 W
Substrate Temperature	30 °C	30 °C
Chamber Pressure	0.5 – 4 Torr	0.5 – 4 Torr

The amount of RF power dissipated in the discharge was calculated by using the transmission ABCD matrix method [19], after measuring the voltage and current waveforms and their relative phase difference outside the reactor. For all the experimental conditions that are presented here, the real power dissipated to the plasma was fixed to 11 W.

Plasma imaging and spatiotemporal emission profiles of α-balmer, b-balmer and SiH ($A^2\Delta$-$X^2\Pi$) lines were measured according to the method described in [20], [21] by using an iCCD detector (Andor, iStar734) and the proper interference filters. Self-bias voltage was measured by a handmade low-pass filter attached to the RF electrode.

3. Results and discussion

Figure 1 shows the time evolution of the self-bias voltage (V_{dc}) as a function of time for (a) the pristine Hydrogen and (b) SiH_4/H_2 discharges. Different gas pressures (P) from 0.5 to 4 Torr are also plotted for each case and as we can observe V_{dc} values become less negative with pressure for both H_2 and SiH_4/H_2 plasmas. This is typical behaviour in RF capacitively coupled asymmetric sheaths, where previous studies have shown that V_{dc} will scale linearly with $P^{-1/2}$ due to the plasma confinement and the drop of the electron temperature [22], [23]. In addition, V_{dc} is less negative in SiH_4/H_2 compared to pure H_2 discharges for the same pressure again due to the lower electron temperature that is expected in the SiH_4-containing mixture.

As for the time evolution of the V_{dc}, in pure Hydrogen it reaches a steady state value few ms after plasma ignition. On the other, hand, for the SiH_4/H_2 mixture V_{dc} gets the most negative value few ms

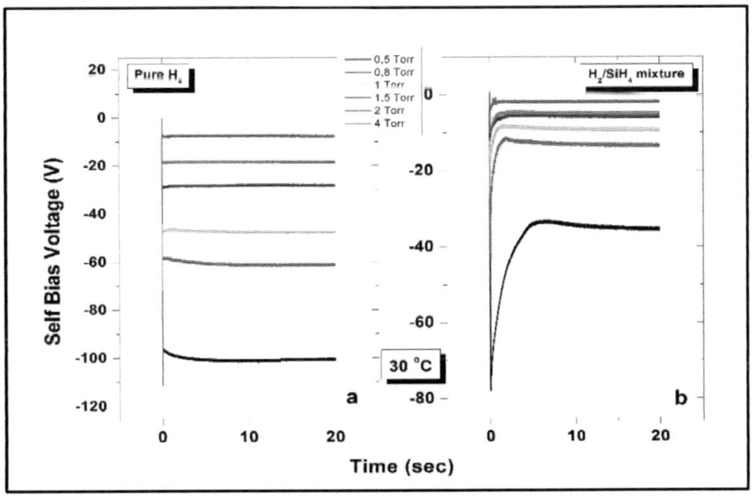

Figure 1. Time evolution of the self-bias voltage for several values of chamber pressure for the cases of pristine H_2 (a) and SiH_4/H_2 (b) discharges.

after plasma is on and then decreases in absolute value asymptotically. It is worth noticing that the time required reaching steady state is of the order of seconds and becomes shorter with pressure increase. Similar variation of V_{dc} values have been reported [8], [12], [24] in SiH_4-Ar and SiH_4-H_2 plasmas and are related to the first two stages of dust formation i.e the nucleation and coagulation of clusters.

Figure 2 presents the time evolution of the H_α line intensity in the case of pristine Hydrogen (a) and SiH_4/H_2 mixture (b) for different values of total pressure. The increase of pressure results in a drop of Hα emission intensity for both H_2 and SiH_4/H_2 mixture and this is related to the drop of electron temperature with pressure, which in turn affects all the high energy electron – molecule collision processes. Moreover, as in the case of V_{dc} the emission intensity in pure hydrogen plasma increases

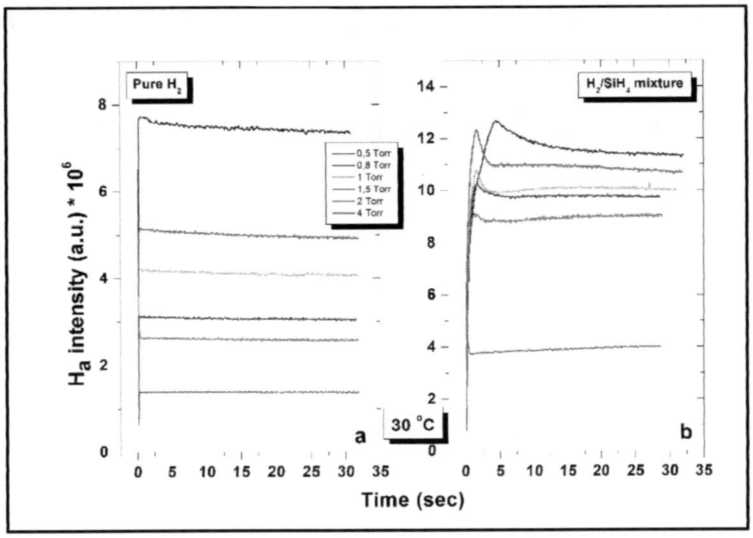

Figure 2. Emission intensity of H_a line for several values of chamber pressure for the cases of pristine H_2 (a) and SiH_4/H_2 (b) discharges.

very fast and stabilizes few msec after the plasma ignition. In SiH_4-H_2 discharges, H_α emission intensity increases also rapidly immediately after plasma ignition and continues to rise in a slower rate for the next few sec. The duration of this second stage becomes shorter with pressure and is followed by a slow drop of H_α emission intensity until steady state conditions are reached. The slower rate rise of emission intensity in the second stage should be attributed to an increase in electron temperature during the particle nucleation phase in order to compensate the charge loss due to the electron attachment [7]. On the other hand, the slight drop of H_α emission intensity and the subsequent stabilization reflects the coagulation phase which is followed by decrease of both electron density and temperature until steady state is reached [14].

Figure 3 shows the time required for the plasma to reach state conditions (t_{st}) as a function of the SiH_4/H_2 total gas pressure. The characteristic times were calculated from the temporally resolved V_{dc} and H_α emission intensity measurements and there is almost no difference between the estimated times for stabilization. Finally, the increase of pressure results in significant drop of the t_{st}. This can be attributed to the enhancement of chemical kinetics and can be an important advantage for the material properties as it will minimize undesirable incubation layer, which is caused by process instabilities during the film initial growth [17].

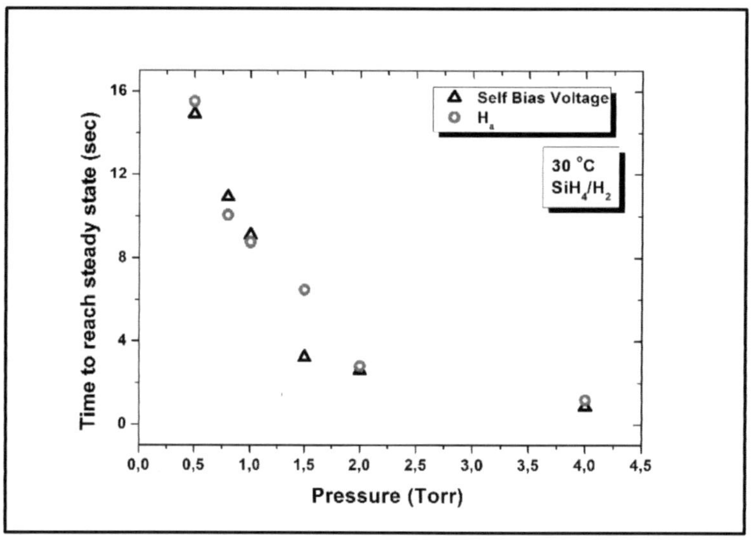

Figure 3. Time to reach steady state in self-bias voltage (3a) and H_a emission line (3b).

4. Conclusions

Self-bias voltage and time-resolved optical emission spectroscopy (OES) were used to detect particles formation in SiH_4/H_2 discharges and instabilities during the initial plasma stage. It was shown that both techniques are very sensitive to particle detection and can be used to monitor nucleation and coagulation stage of particles formation. In dust free conditions (pure H_2) we observe that self-bias voltage as well as H_a intensity stabilized few msec after plasma ignition. In the SiH_4/H_2 discharges, the time required for plasma to reach steady state conditions is of the order of seconds. Both techniques lead to similar values of t_{st} indicating that plasma electrical properties and chemical kinetics are affected in a similar way from the formation of particles. The increase of pressure results in a significant drop of the time required for process steady state conditions and this can be considered as an important advantage that favors the growth of good quality silicon thin films under high pressure conditions.

References

[1] Schuttauf J W, Bugnon G, Stuckelberger M, Hänni S, Boccard M, Despeisse M, Haug F J, Meillaud F and Ballif C 2014 *IEEE J. Photovoltaics* **4**/3 757
[2] Shah A, Moulin E and Ballif C 2013 *Solar Energy Materials and Solar Cells* **119** 311
[3] Rech B, Roschek T, Repmann T, Muller J, Schmitz R and Appenzeller W 2003 *Thin Solid Films* **427** 157
[4] Selwyn G S, McKillop J S, Haller K L and Wu J J 1990 *J. Vac. Sci. & Technol.* A **8** 1726
[5] Howling A A, Sansonnens L, Dorier J L and Hollenstein Ch 1993 *J. Phys.* D*: Appl. Phys.* **26** 1003
[6] Stoffels W W, Stoffels E, Kroesen G M W and de Hoog F J 1995 *J. Appl. Phys.* **78** 4867
[7] Boufendi L, Bouchoule A and Hbid T 1996 *J. Vac. Sci. Technol.* A **14**/2 572
[8] Bouchoule A 1999 *Dusty plasma physics chemistry and technological impacts in plasma processing* (NY: Wiley)
[9] Howling A A, Dorier J L and Hollenstein Ch 1993 *Appl. Phys. Lett.* **62**/12 1341
[10] Thompson S, Perrey C R, Carter C B, Belich T J, Kakalios J and Kortshagen 2005 U *J. Appl. Phys.* **97** 034310
[11] Fukuzawa T, Obata K, Kawasaki H, Shiratani M and Watanabe Y 1996 *J. Appl. Phys.* **80**/6 3202
[12] Wattieaux G, Mezeghrane A and Boufendi L 2011 *Phys. Plasmas* **18**/9 093701
[13] Mikikian M, Cavarroc M, Couëdel L and Boufendi L 2006 *Phys. Plasmas* **13**/9 092103
[14] Bhandarkar U V, Swihart M T, Girshick S L and Kortshagen U R 2000 *J. Phys.* D*: Appl. Phys.* **33** 2731
[15] Bogaerts A, De Bleecker K, Kolev I and Madani M 2005 *Surf. & Coat. Technol.* **200** 62
[16] Vladimirov S V and Ostrikov K 2004 *Phys. Rep.* **393** 175
[17] Rezek B, Stuchlik J, Fejfar A and Kocka J 1999 *Appl. Phys. Lett.* **74**/101475
[18] Amanatides E, Mataras D and Rapakoulias D E 2002 *J. Vac. Sci. Technol.* A **20**/1 68
[19] Sobolewski M A 1992 *J. Vac. Sci. Technol.* A **10**/6 3550
[20] Amanatides E, Gkotsis P, Syndrevelis Ch and Mataras D 2006 *Diamond and Related Materials* **15**/4-8 904
[21] Zhang X D, Zhang F R, Amanatides E, Mataras D, Xiong S Z and Zhao Y 2008 *J. Non-Crystalline Solids* **354** 2208
[22] Catherine Y and Couderc P 1986 *Thin Solid Films* **144**/2 265
[23] Lieberman M A and Lichtenberg A J 2005 *Principles of Plasma Discharges and Materials Processing* (New Jersey: John Wiley & Sons Inc.)
[24] Grootoonk B, Woerdenweber J, Gordijn A, Gabriel O and Meier M 2014 *Can. J. Phys.* **92** 736

Comparative study of RF reactive magnetron sputtering and sol-gel deposition of UV induced superhydrophilic TiOx thin films

V E Vrakatseli[1], E Amanatides and D Mataras

Plasma Technology Laboratory, Department of Chemical Engineering, University of Patras, Greece

E-mail: lef@plasmatech.gr

Abstract. TiO_x and TiO_x-like thin films were deposited on PEEK (Polyether ether ketone) substrates by low-temperature RF reactive magnetron sputtering and the sol-gel method. The resulting films were compared in terms of their properties and photoinduced hydrophilicity. Both techniques resulted in uniform films with good adhesion that can be switched to superhydrophilic after exposure to UVA radiation for similar time periods. In addition, the sputtered films can also be activated and switched to superhydrophilic by natural sunlight due to the higher absorption in the visible spectrum compared to the sol-gel films. On the other hand, the as deposited sol-films remain relatively hydrophilic for a longer time in dark compared to the sputtered film due to the differences in the morphology and the porosity of the two materials. Thus, depending on the application, either method can be used in order to achieve the desirable TiO_x properties.

1. Introduction

It has been almost two decades since Wang et al. [1] revealed the unique photohydrophilic property of crystalline TiO_2, i.e. its ability to become superhydrophilic upon UV irradiation [2], [3]. Nevertheless, low-temperature deposition methods are required for the deposition of photoinduced hydrophilic TiO_2 films on thermosensitive substrates aiming to advanced applications, such as microfluidic and biomedical devices [4], [5]. The RF reactive magnetron sputtering is considered to be a promising low temperature method for deposition of high-quality TiO_2 thin films [6], [7]. The sol-gel route of metal alkoxides is a widespread method due to the fact that it is a cost-effective, simple and equipment-free procedure for deposition of oxide films. However, dense and crystalline pure oxide thin films can be obtained only after intense post annealing. Thus, the choice of the deposition technique of functional coatings is of great significance, since it has a major impact on the properties and quality of the deposited film.

In this work, we deposited TiO_x thin films on PEEK substrates by low-temperature RF reactive magnetron sputtering and the sol-gel method. The deposited films were compared in terms of their UV induced hydrophilicity, hydrophilicity persistence in dark, as well as their optical and structural properties. The differences in the latter properties were directly reflected on the radiatively induced

[1] To whom any correspondence should be addressed.

superhydrophilicity and the dark storage recovery indicating that each of these techniques can be implemented, depending on the desired wettability properties of the TiO_x films.

2. Experimental

2.1. Sol-gel spin coating deposition

For the deposition of TiO_x-like thin films, an isopropanol (iPrOH, PENTA 99.8%) solution of titanium tetraisopropoxide (TTiP, Sigma Aldrich 99.999%) and distilled water were used. The detailed procedure for the preparation of the deposition solution has been previously described in Ref. [8]. The molar ratios of TTiP and H_2O to iPrOH was 0.22 M and 1.38 M respectively, which corresponds to a hydrolysis ratio H = 6.3. The films deposited on PEEK were allowed to dry at room temperature, in dark storage for at least 24 hours resulting in TiOx-like xerogel thin films.

2.2. RF reactive sputtering deposition

A magnetron source (ST-20, AJA International) with a 99.999% purity Ti metal target of two inches in diameter was used for the deposition of TiO_2 thin films. The total process pressure was kept at 3 mTorr by regulating the gas flows of argon and oxygen through two MKS mass flow controllers, and the partial pressure of oxygen was $P_{O2}/(P_{Ar}+P_{O2}) = 0.35$. The TiO_2 films were deposited onto unheated PEEK and microscope glass slides with dimensions 15mm × 15mm which were lying on a cyclical substrate holder at 100 mm above the Ti target. The experiments were carried out at different RF (13.56 MHz) power values (150 W, 250 W, 400 W, 500 W).

2.3. Thin film characterization

A DektakXT stylus profilometer was used for the film thickness measurements. Their structure was determined by X-ray diffractometry (Siemens D-5000) using CuKa radiation (1.54 Å) at 40 kV and 30 mA at the 2θ range of 20° to 70°. The surface wettability was evaluated by carefully placing 2 µl distilled water drops on the surface and calculating the apparent contact angle using a Kruss DSA100S goniometer. The transmittance (%) spectra of the films were recorded by a UV-VIS spectrophotometer (Perkin Elmer 1-35). The films morphology was monitored by using FE-ESEM (FEI QUANTA 250 FEG).

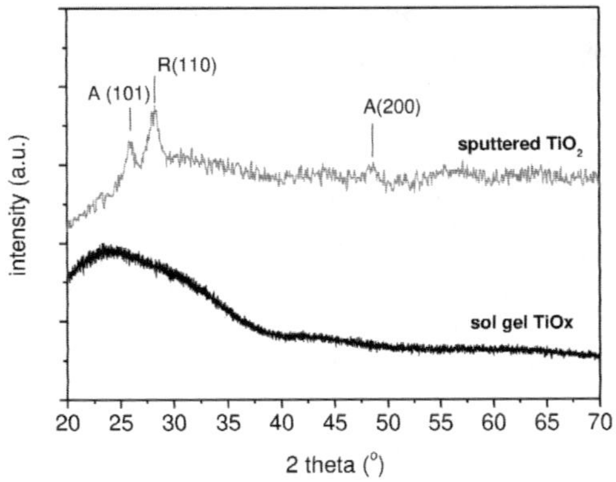

Figure 1. XRD patterns of the sol-gel derived TiOx-like film (SG) and the sputtered TiO_2 film (SP-400W) deposited on glass slides.

3. Results

Both methods resulted in transparent TiOx thin films. A 450-nm-thick xerogel TiOx-like film (SG) and the 400-W sputtered, 30-nm-thick TiO_2 film (SP-400W), which was deposited at a rate of 3.5 nm/min, were used for the comparison of the two deposition methods

3.1. Structure, morphology and optical transparency

The sol-gel films were amorphous due to the room temperature process and the absence of further annealing. In addition, a high content of organic groups in its network or as terminal groups will remain in the film [9]. On the other hand, the high-vacuum sputtering process resulted in crystalline oxide films.

The XRD pattern of the SP-400W TiO_2 film (figure 1) show mixed phases of anatase (101), (200) and rutile (110). It is also worth noticing that during the sputtering process at a 400-W input power, the substrate reached 100 °C due to plasma heating, surface reactions and ion bombardment. This temperature rise will favor the formation of nanocrystals [10]. The mean crystallite sizes (table 1) were estimated by applying Scherrer's equation at the main peaks of the XRD pattern [11].

Table 1. Structural and optical characteristics of the sol-gel deposited xerogel TiOx film and the 400 W RF M sputtered TiO_2 film.

Method	**Approximate mean crystallite size**	**Refractive index**	**Optical E_g**	**Porosity**
RF M Sputtering (400 W)	22.3 (anatase 101) 18.6 (anatase 200) 17.3 (rutile 110)	2.24 (450 nm) 2.1 (632.8 nm)	~3.2 eV	0.3
Sol gel	Amorphous	1.24 (450 nm) 1.15 (632.8 nm)	~3.65 eV	0.9

Moreover, the SG and SP-400W films exhibit very different surface morphology. The SEM image of the SG film shows a uniform smooth surface with nano-sized porosity, whereas the SP-400W film exhibits a denser nanocomposite structure (figure 2). In figure 3 we present the % transmittance spectra of the two films deposited on microscope glass slides. Both films are transparent throughout the visible range 400 nm – 700 nm, although the sputtered film exhibits interference fringes indicating differences in the refractive indices of the materials. A significant difference of the adsorption edge between the sputtered and the sol-gel film is also observed. The absorption edge of the SP-400W film is approximately at 380 nm, while the SG film is blue-shifted and absorbs light only at wavelengths below ~340 nm. The absorption shift to the lower wavenumbers for the SG film can be attributed to the reduction of the grains size (figure 2, SEM images) and can be interpreted as a quantum size effect [12].

(a)

(b)

Figure 2. Top view SEM images of the (a) xerogel TiOx film's surface (250000 × magnification) and (b) RF magnetron sputtered TiO_2 film deposited at 400 W (120000 × magnification).

The optical constants of the films were estimated using the Pointwise Unconstrained Minimization Approach (PUMA) for the fitting of the transmission data [13]. The SG film exhibits a low refractive index n= 1.15 (632.8 nm) which is closer to a polymer rather than an oxide film. Moreover, the porosity of the two films was calculated by equation [14]:

$$p = \frac{n_d - n_{meas}}{n_d - 1} , \quad (1)$$

where n_d is the refractive index of the dense anatase [15] and n_{meas} is the calculated index of the films. The energy band gap (E_g) of the materials was calculated from Tauc's relation [16]. The SP-400W E_g is estimated at 3.2 eV, which is very similar to that of the pure anatase, while the SG film's E_g is much larger at 3.65 eV and confirms the inefficiency of this film to absorb the near visible range radiance.

3.2. *Hydrophilicity*

The contact angles of water with the as-deposited films were measured right after the sputtering deposition and after drying for 24 hours for the SG films. The as-deposited sputtered film is superhydrophilic with a contact angle <10°, while the contact angle of SG films ranges from 30° to 40°. Nevertheless, both surfaces suffer from aging, which is observed by the rise of

Journal of Physics: Conference Series **700** (2016) 012039 doi:10.1088/1742-6596/700/1/012039

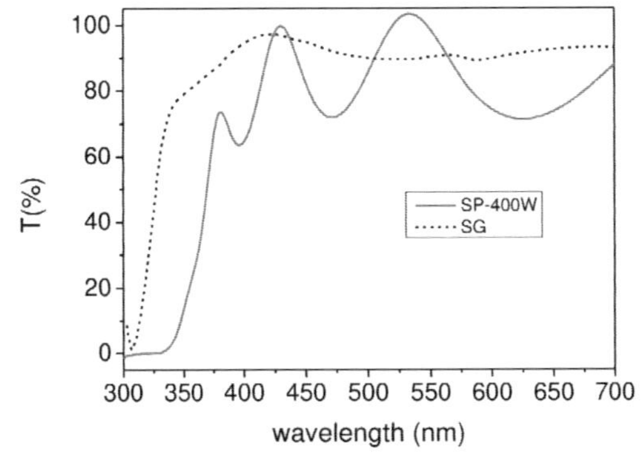

Figure 3. UV-Visible transmission spectra of the sol-gel derived TiOx-like film (SG) and the RF reactive sputtered TiO_2 film deposited on glass slides.

the contact angle mainly during the first 10 days after the deposition. The sputtered film changes to hydrophobic (CA>100°), whereas the SG film contact angle reaches a plateau at ~80°.

The evaluation of the photoinduced hydrophilicity was performed by measurements of the films CA during their illumination with intense UV light (254 nm, 8 mW/cm^2), and also during their exposition to the day's natural sunlight (figure 4a). Upon 15 to 20 minutes of UV illumination, both the SP-400 film and the amorphous SG xerogel film become superhydrophilic exhibiting a 0° contact angle, i.e. total wetting of their surface, no matter their initial wetting state (hydrophobic or relatively hydrophilic). On the other hand, during their exposure to the natural sunlight's spectrum, only the sputtered film reaches superhydrophilicity after 3 h of illumination. The contact angle value of the SG film drops from 88° to 50° after 2 h of exposure, but lower CA cannot be achieved due to the high transmittance of the film in the UV-A and the near visible wavelengths range (figure 3).

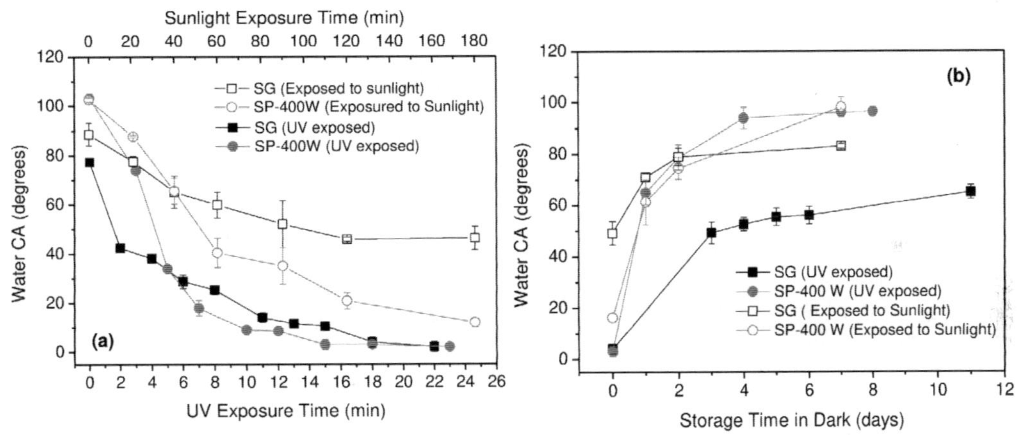

Figure 4. Time-dependence of contact angles of the sputtered TiO_2 film (SP-400W) and the xerogel TiOx film (SG) upon (a) irradiation with UV (8 mW/cm^2) and natural sunlight (shiny spot at ambient conditions of the lab) and (b) during their storage in a dark place (RH = 50 %, T = 23 °C).

Finally, in order to monitor the hydrophobic recovery of the illuminated (UV and sunlight) films, the materials were stored in dark and CA measurements were performed at specific time intervals. The SP-400W film exposed to UV radiation or to sunlight return to its hydrophobic state within 3 days. On the other hand, the SG films exposed to UV light remains hydrophilic for a longer period with CA ~50° for one week. Lee et al. [17] , showed that the slow recovery of CA in dark is closely related to the higher surface roughness. The higher porosity of the SG xerogel film probably provides a larger surface area. Thus, the slow dark recovery rate of the SG film cannot be attributed to the surface roughness, but rather to a higher surface density of chemisorbed –OH after the UV-induced excitation.

4. Conclusions

The sol gel TiOx films deposited at room temperature without annealing exhibit good uniformity and transparency and also UV-induced superhydrophilicity. Despite the amorphous structure, the SG film converts to superhydrophilic after 20 min of irradiation with UV light. The RF reactive sputtering results in crystalline TiO_2 films at low temperature, which also show UV-induced superhydrophilicity in a short exposure time (15 min).

The comparison of the photoinduced performance of the films shows two significant differences: (a) the sputtered film is suitable for outdoor applications, since it can become superhydrophilic also upon exposure to natural sunlight, while the SG film cannot be activated, and (b) the sputtered film's wettability switches between hydrophobic (CA>100°) in dark and superhydrophilic upon UV or sunlight exposure, while the SG film is intrinsically more hydrophilic and remains hydrophilic for a longer time in dark. Depending on the application, both methods can result in good quality films.

Acknowledgements
Part of this work was co-financed by Hellenic Funds and by the European Regional Development Fund (ERDF) under the Hellenic National Strategic Reference Framework (NSRF) 2007-2013, of the Project "THALIS" DESIgn and fabrication of Robust supErhyDROPhobic/philic surfaces and their application in the realization of "smart" microfluidic valves".

References
[1] Wang R, Hashimoto K, Fujishima A, Chikuni M, Kojima E, Kitamura A, Shimohigoshi M and Watanabe T 1997 *Nature* **388** 431-33
[2] Wang R, Sakai N, Fujishima A, Watanabe T and Hashimoto K 1999 *J. Phys. Chem.* B **103**/12 2188
[3] Sakai N, Fujishima A, Watanabe T and Hashimoto K 2003 *J. Phys. Chem.* B **107** 1028
[4] Park H J, Kim I T, Park S H and Kim S K 2011 *J. Micromech. Microeng.* **21**/9
[5] Lai Y, Lin L, Pan F, Huang J, Song R, Huang Y, Lin C, Fuchs H and Chi L 2013 *Small* **9**/17 2945
[6] Okimura K, Maeda N and Shibata A1996 *Thin Solid Films* **281-282** 427-30
[7] Pradhan S S, Pradhan S K, Bhavanasi V, Sahoo S, Sarangi S N, Anwar S and Barhai P K 2012 *Thin Solid Films* **520**/6 1809-13
[8] Vrakatseli V E, Pagonis E, Amanatides E and Mataras D 2014 *J. Phys.: Conf. Series* **550** 012034
[9] Yoldas B E 1986 *J. Materials Sci.* **21** 1087
[10] Mbarek I B, Chaabouni F, Selmi M, Abaab M and Rezig B 2010 *Phys. Status Solidi* **7**/9 2311
[11] Patterson A L 1939 *Phys. Rev.* **56** 978
[12] Akkaya A U and Tepehan F Z 2011 *Surf. Coating Technol.* **206** 37
[13] Birgin E G, Chambouleyron I and Martinez J M 1999 *J. Comput. Phys.* **151** 862
[14] Prusakova V, et al 2015 *Phys. Status Solidi* C **12** 192-6
[15] Kingery W D, Bowen H K and Uhlmann D R 1976 *Introduction to Ceramics*, 2nd ed. (New York: Wiley)
[16] Tauc J, Grigorovici R and Vancu A 1966 *Phys. Status Solidi* **15** 627
[17] Lee H Y, Park Y H and Ko K H 2000 *Langmuir* **16** 7289

Electron transport in Al-doped ZnO nanolayers obtained by atomic layer deposition

B S Blagoev[1,5], D Z Dimitrov[1], V B Mehandzhiev[1], D Kovacheva[2], P Terziyska[1], J Pavlic[1], K Lovchinov[3], E Mateev[4], J Leclercq[1] and P Sveshtarov[1]

[1]Georgi Nadjakov Institute of Solid State Physics, Bulgarian Academy of Sciences, 72 Tzarigradsko Chaussee , 1784 Sofia, Bulgaria
[2]Institute of General and Inorganic Chemistry, Bulgarian Academy of Sciences, Acad. G. Bonchev Str., bl. 10, 1113 Sofia, Bulgaria
[3]Central Laboratory of Solar Energy and New Energy Sources, Bulgarian Academy of Sciences, 72 Tzarigradsko Chaussee, 1784 Sofia, Bulgaria
[4]Emil Djakov Institute of Electronics, Bulgarian Academy of Sciences, 72 Tzarigradsko Chaussee, 1784 Sofia, Bulgaria

E-mail: blago@issp.bas.bg

Abstract. Al-doped ZnO thin films with different Al content were prepared by atomic layer deposition (ALD). To carry out thermal ALD, diethyl zinc (DEZ) and tri-methyl aluminium (TMA) were used as Zn and Al precursors, respectively, and water vapor as oxidant. Various numbers n of DEZ and m TMA cycles was used to obtain different $[ZnO]_n[Al_2O_3]_m$ films, where $n = 100 - 95$, $m = 1 - 5$. The X-ray diffraction analysis showed a predominantly (100) oriented polycrystalline phase for the ZnO:Al films with a low Al content ($m = 1 - 3$) and an amorphous structure for pure Al_2O_3. In ZnO:Al with a higher Al content ($m = 4 - 6$) the (100) reflection disappeared and the (002) peak increased. The resistivity of the films decreased with the increase in the Al content, reaching a minimum of 3.3×10^{-3} Ω cm at about 1.1 % Al_2O_3 for the $[ZnO]_{99}[Al_2O_3]_2$ sample; for higher dopant concentrations, the resistivity increased because of the increased crystal inhomogeneity due to axis reorientation.

1. Introduction

Transparent conducting oxides (TCO) have been investigated extensively in recent years due to their applications in electronic, optoelectronic and photovoltaic devices. In this respect, zinc oxide (ZnO) is among the important TCO semiconductors because of its high transmittance in the visible range and low DC resistivity. Small additions of Al dopant in ZnO films strongly decrease their resistivity without degrading their optical properties [1-4]. Thin ZnO films have been fabricated by various chemical and physical methods, as spray pyrolysis [1], radio-frequency magnetron sputtering [3], sol–gel spin-coating [5], pulsed laser deposition [6], metal-organic chemical-vapor deposition [7], plasma enhanced chemical vapor deposition [8], molecular beam epitaxy [9], atomic layer deposition [2, 10-15], etc. Atomic layer deposition is one of the most suitable techniques for obtaining uniform ultra thin films on large areas. Dopant incorporation with atomic precision could be easily performed

[5] To whom any correspondence should be addressed.

by the ALD technique. One of the most interesting applications of Al-doped ZnO thin films are as a transparent oxide conducting layer in various devices [2, 10-13]. Therefore, the theoretical and experimental studies of the electrical charge transport are of great importance.

In spite of the large number of ZnO:Al publications, no sufficient information about the electron transport is available for samples prepared by ALD. This is why in this paper we present a study of the Al dopant influence on the AC and DC electrical transport in ZnO nanolayers with different low Al dopant concentration grown by ALD. For deposition of the ZnO films and the Al addition we used the liquid precursors DEZ and TMA, respectively, alternating with H_2O.

2. Experimental

Thin ZnO films doped with Al (ZnO:Al) were prepared on a Beneq TFS 200 apparatus by atomic layer deposition (ALD) on p-type Si (100) wafers (15 Ω cm) with a diameter of 10 cm. The deposition process was conducted at 200 °C and a pressure of 2 mbar. Liquid precursors of DEZ with H_2O and TMA with H_2O were used for ZnO and Al_2O_3, respectively, at room temperature vapor pressure. A pure N_2 gas flow of 600 sccm was maintained during the precursor introduction and the purging steps. The pulse and purging durations were the same for all precursors, 200 ms and 1 s, respectively. To obtain different Al concentration in the ZnO films, the number of DEZ/H_2O (n) and TMA/H_2O (m) pulses were varied. Five differently doped ZnO films $[ZnO]_n[Al_2O_3]_m$: $[ZnO]_{100}[Al_2O_3]_1$; $[ZnO]_{99}[Al_2O_3]_2$; $[ZnO]_{98}[Al_2O_3]_3$; $[ZnO]_{97}[Al_2O_3]_4$; $[ZnO]_{95}[Al_2O_3]_6$ were deposited. The lower right indices of the square brackets n and m here indicate the pulse repetition of DEZ/H_2O and TMA/H_2O, respectively. Each subcycle $[ZnO]_n[Al_2O_3]_m$ was repeated ten times to obtain 1010 total cycles for each film.

Usually, a few DEZ cycles followed by only one TMA cycle [2, 11] are used for Al doping of ZnO films. In this work we gradually increased the number of TMA cycles to investigate the change in the resistivity.

The ellipsometric data were taken using a Woollam M2000D rotating compensator spectroscopic ellipsometer with a wavelength range from 193 to 1000 nm. The film thickness was determined by fitting the experimental Psi and Delta data using a three layer model which includes the Si substrate, the ZnO layer and a roughness layer. For the ZnO layer we used the general oscillator model consisting of two Tauc-Laurentz oscillators.

The X-ray diffraction patterns were collected within the 2θ range from 10 to 80° with a constant step of 0.02° on a Bruker D8 Advance diffractometer with Cu Kα radiation and a LynxEye detector. The phase identification was performed with the Diffracplus EVA using ICDD-PDF2 Database.

The sheet resistance was obtained by the standard four-probe technique on a Veeco EPP-100 apparatus at room temperature and the resistivity was calculated using the thickness data taken from the ellipsometric measurements.

An electrochemical impedance spectroscopy (EIS) analyzer with a low-current probe (Bio-Logic SP-200 potentio-/galvano-stat) was used for measuring the frequency response of the samples. The reference voltage range was from 0.1 to 500 mV. The electrochemical impedance spectroscopy measurement was preformed in the frequency range from 1 Hz to 3 MHz. The data were processed with the software package EC-Lab.

3. Results and discussion

Thin ZnO:Al films were prepared by the ALD technique described above. The doped (ZnO:Al) films were with an average thickness of about 190 nm. The films were moderately uniform and smooth across the entire 4″ Si wafers with a standard thickness deviation of 1-2 nm. The thickness reported was taken from the mean value of 54 uniformly distributed points on the Si wafer surface extracted from the ellipsometric data. The roughness was taken from AFM analysis (not presented here) and was between 1 and 2 nm. The volume percentage of the Al_2O_3 layer content α in $[ZnO]_n[Al_2O_3]_m$ was calculated from the formula: $\alpha[\%] = \frac{mA}{mA + nB} \times 100$ [2], where m and n are the TMA and DEZ number of pulses, respectively, *A* and *B* are the growth rates for pure Al_2O_3 (1.14 Å/cycle) and pure ZnO (2.05 Å/cycle) films, respectively. The total growth rates for the Al-doped ZnO films were slower than that for pure ZnO film and decreased with the increase in the Al_2O_3 content (table 1). This result, namely, a decreased ZnO growth rate after TMA exposure, is in accordance with the work of Na et al. [14]. In contrast, Oh et al. [2] and Elam et al. [15] reported an increase in the ZnO growth rate after TMA exposure due to a nucleation process. The sheet resistance deviation across the whole 4″ Si surface for all films was less than 2 %. All films presented possessed a low resistivity in the range of ~10^{-3} Ω cm, which is in agreement with the properties of other ZnO and Al-doped ZnO films obtained by ALD [2, 10-13]. The resistivity of the Al-doped ZnO films was lower than that of pure ZnO films and reached a minimum of ~3.3×10^{-3} Ω cm for $[ZnO]_{99}[Al_2O_3]_2$ sample with 1.1 % Al_2O_3 content (table 1 and figure 2).

The X-ray diffraction analyses showed that the Al_2O_3 film was amorphous, while the ZnO:Al films consisted of a polycrystalline hexagonal phase with wurtzite-type structure (figure 1). The unit cell parameters and the mean crystallite size for films with different Al concentrations are presented in table 2.

In the table one can see that the ZnO films with low Al doping have lower unit cell parameters and a higher mean crystallite size than the undoped ZnO. Moreover, the preferred orientation of <100> is observed for these films. As the level of Al doping increases, the unit cell parameters of the wurtzite phase increase and the mean crystallite size decreases. The preferred orientation changes its direction from <100> to <002> (see figure 1). There is a weak peak of ZnO (110) at 2θ ~ 56.8° (not shown in figure 1), whose intensity also decreases with increasing the Al content. For the highest level Al-doped ZnO film, $[ZnO]_{95}[Al_2O_3]_6$, the (100) and (110) peaks disappear completely and only the (002) peak is seen. It can be concluded that Al doping affects strongly the morphologic parameters (crystallite size and shape) of the ZnO films.

Table 1. Growth rate, volume percentage of Al_2O_3 layer content calculated from the growth rates of pure Al_2O_3 and ZnO and the number of TMA and DEZ cycles, thickness and resistivity of Al-doped ZnO films.

$[ZnO]_n[Al_2O_3]_m$	**Growth rate,** Å/cycle	**Al_2O_3 layer content,** %	**Thickness,** nm	**Resistivity,** Ω.cm
$[ZnO]_{101}[Al_2O_3]_0$	2.05	0	207.00	9.0×10^{-3}
$[ZnO]_{100}[Al_2O_3]_1$	1.94	0.55	196.14	4.2×10^{-3}
$[ZnO]_{99}[Al_2O_3]_2$	1.87	1.11	188.69	3.3×10^{-3}
$[ZnO]_{98}[Al_2O_3]_3$	1.84	1.67	186.33	3.9×10^{-3}
$[ZnO]_{97}[Al_2O_3]_4$	1.81	2.24	182.57	4.2×10^{-3}
$[ZnO]_{95}[Al_2O_3]_6$	1.77	3.39	178.51	4.6×10^{-3}

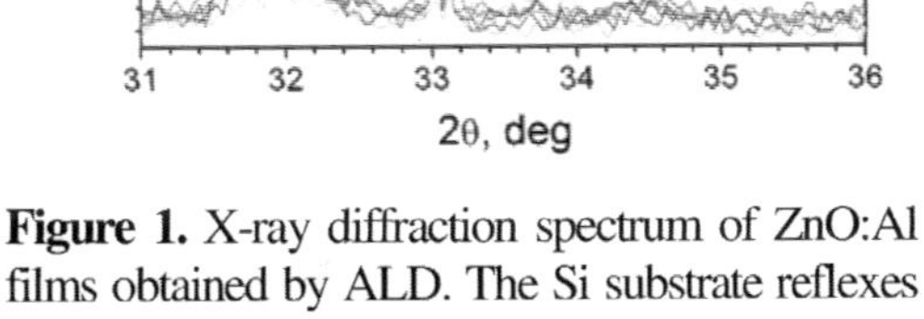

Figure 1. X-ray diffraction spectrum of ZnO:Al films obtained by ALD. The Si substrate reflexes are denoted with an asterisk (*).

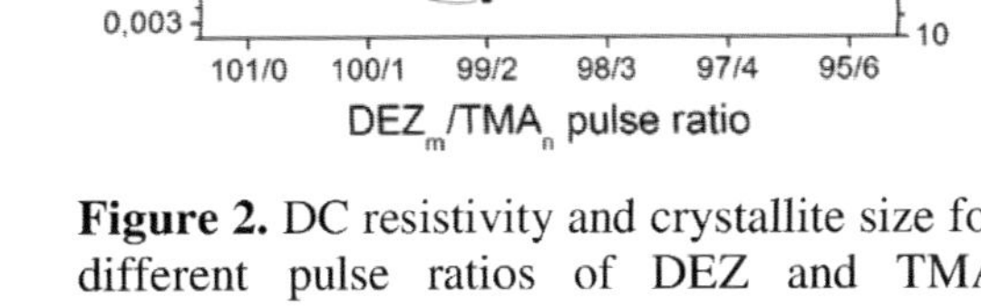

Figure 2. DC resistivity and crystallite size for different pulse ratios of DEZ and TMA precursors.

Table 2. Unit cell parameters and mean crystallite size for the ZnO films with different degree of Al-doping. The space group is SG $P6_3mc$.

Composition	*a* [Å]	*c* [Å]	Crystallite size [nm]
ZnO	3.259(3)	5.192(7)	41±3
$[ZnO]_{100}[Al_2O_3]_1$	3.249(2)	5.181(8)	54±6
$[ZnO]_{99}[Al_2O_3]_2$	3.246(1)	5.184(6)	46±3
$[ZnO]_{98}[Al_2O_3]_3$	3.255(3)	5.21(1)	23±2
$[ZnO]_{97}[Al_2O_3]_4$	3.269(5)	5.24(1)	15±1
$[ZnO]_{95}[Al_2O_3]_6$	3.289(7)	5.28(1)	15±1

The resistivity of Al-doped ZnO films decreased with increasing the Al content, which acts as an electron donor. Above 1.1% Al_2O_3 in the $[ZnO]_{99}[Al_2O_3]_2$ sample, the crystallinity worsened because of the crystal axes reorientation and the formation of new domains and grain boundaries due to the lower grain dimensions (figure 2). The crystal structure degradation suppressed the electron mobility and the resistivity started to increase slightly (figure 2), so that a local minimum of resistivity at 3.3×10^{-3} Ω cm was observed.

Figure 3 presents the real and imaginary parts of the complex impedance of Al-doped ZnO thin films *vs.* the frequency at an amplitude of 50 mV at room temperature. The impedance measurements of pure ZnO, $[ZnO]_{100}[Al_2O_3]_1$, $[ZnO]_{99}[Al_2O_3]_2$ and $[ZnO]_{97}[Al_2O_3]_4$ films at the high-frequency region 0.5 – 3 MHz are also presented. For all samples measured, the impedances decreases as the frequency is increased.

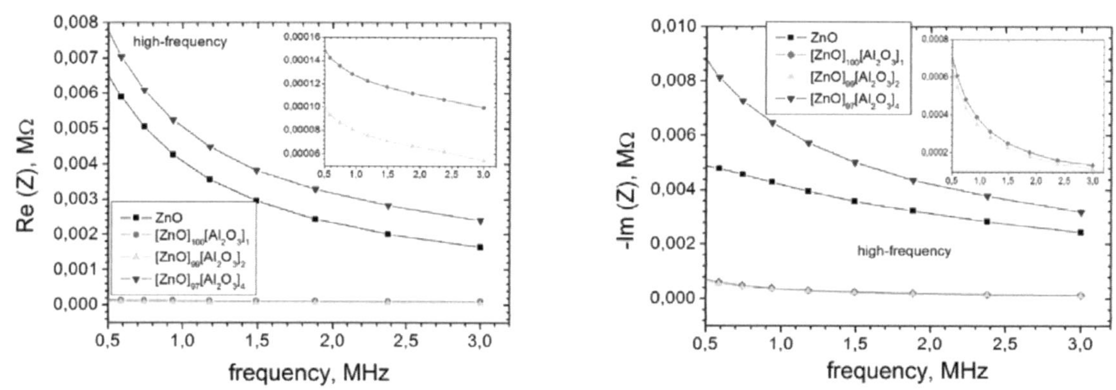

(a) (b)

Figure 3. Impedance measurements of Al-doped ZnO films: the real (in the left figure - a) and the imaginary (in the right figure - b) parts of the impedance versus the frequency. The insets of both figures show an enlarged region of Al-doped ZnO films with a lower Al content, $m = 1, 2$.

The real and imaginary parts of the impedance of the low Al content samples, $m = 1, 2$ (see the insets of figure 3) possessed the lowest values corresponding to the minimum of DC resistivity (see figure 2), while the pure ZnO and $[ZnO]_{97}[Al_2O_3]_4$ films showed values that were several times higher. As previously mentioned, the pure ZnO has a lower concentration of mobile electrons, while the films with a higher Al content ($m > 3$) have strong crystal distortion and inhomogeneity. In spite of the large differences in the impedance values, all samples showed the same behavior in the above frequency range. The real part of the impedance starts at a higher value and decreases smoothly to a lower value. The imaginary part at this frequency range has the same behavior, but with a negative sign. Such behavior is characteristic of a resistance and a capacitance connected in series, where the real part is attributed to the resistance and the imaginary part corresponds to the capacitance. Therefore, the resistance decreases with the increase of the electron donors (the Al content, respectively), and the capacity increases with the increase of crystal inhomogeneity as the Al content increases. In fact, not all of Al^{3+} cations of the deposited Al_2O_3 layer substitute the Zn^{2+}. The Zn substitution by Al depends on the growth temperature and the pulse duration of the TMA precursor

[2]. Moreover, the increased number of TMA cycles ($m > 3$) could lead to a relative decrease of the Al substitution on Zn sites in the neighboring ZnO layers due to the longer distance which will lead to an increase in the resistivity. Therefore, a more detailed investigation, such as cross-section EDAX analysis, is needed to properly define the Al content.

To obtain a deeper understanding on the electrical transport, we plan to perform additional impedance measurements.

4. Conclusions

Thin ZnO:Al films with different content of Al were prepared by atomic layer deposition on 4″ Si. Various number n of DEZ and m of TMA ALD cycles was used to obtain different $[ZnO]_n[Al_2O_3]_m$ films, where $n = 100 - 95$, $m = 1 - 5$. The films were relatively uniform and smooth with a uniform DC resistivity distribution. The films with lower Al content ($m < 3$) showed a predominantly (100) crystal orientation, while increasing the Al content ($m > 3$) caused a reorientation to (002). It was found that the DC and AC resistance decreased with the doping due to an increase of mobile electrons and reached a minimum of 3.3×10^{-3} Ω cm at about 1.1 % Al_2O_3 for $[ZnO]_{99}[Al_2O_3]_2$ and then increased slowly because the crystallinity worsened due to the axis reorientation and because of the relative decrease of Al substitution on Zn sites in the neighboring ZnO layers, which suppressed the electron mobility.

Acknowledgments
This work was supported by the EU FP7 project INERA, Grant No 316309.

References

[1] Gabás M, Gota S, Ramos-Barrado J R, Sánchez M, Barrett N T, Avila J and Sacchi M 2005 *Appl. Phys Lett.* **86** 042104
[2] Oh B-Y, Kim J-H, Han J-W, Seo D-S, Jang H S, Choi H-J, Baek S-H, Kim J H, Heo G-S, Kim T-W and Kim K-Y 2012 *Current Appl. Phys.* **12** 273-9
[3] Kim K-K, Niki S, Oh J-Y, Song J-O, Seong T-Y, Park S-J, Fujita S and Kim S-W 2005 *J. Appl. Phys.* **97** 066103
[4] Noriega R, Rivnay J, Goris L, Kälblein D, Klauk H, Kern K, Thompson L M, Palke A C, Stebbins J F, Jokisaari J R, Kusinski G and Salleo A 2010 *J. Appl. Phys.* **107** 074312
[5] Jun M-C, Park S-U and Koh J-H 2012 *Nanoscale Res. Letters* **7** 639
[6] Tsoutsouva M G, Panagopoulos C N, Papadimitriou D, Fasaki I and Kompitsas M 2011 *Mater. Sci. Eng.* B **176** 480
[7] Tan S T, Chen B J, Sun X W, Hu X, Zhang X H and Chua S J 2005 *J. Crystal Growth* **281** 571
[8] Barankin M D, Gonzalez E, Ladwig A M and Hicks R F 2007 *Solar Energy Mater. & Solar Cells* **91** 924
[9] Heo Y W, Ip K, Pearton S J, Norton D P and Budai J D 2006 *Appl. Surf. Sci.* **252** 7442
[10] Wojcik A, Godlewski M, Guziewicz E, Minikayev R and Paszkowicz W 2008 *J. Crystal Growth* **310** 284
[11] Maeng W J, Lee J-W, Lee J H, Chung K-B and Park J-S 2011 *J. Phys. D: Appl. Phys.* **44** 445305
[12] Saarenpaa H, Niemi T, Tukiainen A, Lemmetyinen H and Tkachenko N 2010 *Solar Energy Mater. & Solar Cells* **94** 1379
[13] Godlewski M, Guziewicz E, Łuka G, Krajewski T, Łukasiewicz M, Wachnicki L, Wachnicka A Kopalko K, Sarem A and Dalati B 2009 *Thin Solid Films* **518** 1145
[14] Na J-S, Peng Q, Scarel G and Parsons G N 2009 *Chem. Mater.* **21** 5585
[15] Elam J W and George S M 2003 *Chem. Mater.* **15** 1020

Stand for coating deposition and coating/materials testing

A A Ayrapetov[1], L B Begrambekov, M Yu Dyachenko, A E Evsin, A V Grunin, A M Kalachev, Ya A Sadovskiy and P A Shigin

National Research Nuclear University MEPhI,
31 Kashirskoe shosse, Moscow, 115409 Russia

E-mail: lexxair@plasma.mephi.ru

Abstract. The paper describes a new laboratory stand constructed for film deposition and for testing of deposited films and materials under pulsed and continuous heat load, ion and electron irradiation. The films are formed on substrates by atoms of target materials as a result of their sputtering by ions of argon plasma. The ion energy and ion flux can be varied independently. This enables the deposition of coatings with variable composition over thickness or of multi-layer coatings. Testing of materials is carried out in plasma under ion or electron irradiation by biasing the tested sample negatively or positively, respectively. The energies of ions or electrons can be varied up to 25 keV. The applied power can reach 4000 W (40 MW/m^2 power density in the case of a 1-cm^2 sample) in both continuous and pulsed regimes. In pulsed regime, pulses of 1 – 99% duty cycle at 0 – 500 Hz can be applied to the sample. The pulsed particle load can be combined with a continuous load. The size of the tested sample must not exceed 100 mm in diameter. The heat flux can irradiate the whole sample or be focused at its center (minimum spot of ~ 4mm^2). Heating of the samples up to 2800 K is possible. At the same time, the backside of the tested sample could be actively cooled. This paper presents the results of deposition and testing of a B_4C coating on tungsten and tungsten testing.

1. Introduction

Coatings with different functional properties are widely used in different branches of technology. They could provide corrosion resistance, could be used as protective coating, could work as diffusion barriers, etc. In particular, a protective coating could be necessary for tungsten tiles of the ITER divertor. A number of studies [1,2] showed that tungsten under high-power density heat and particle loads undergoes various effects, including cracking, erosion, macro-scale particle emission, melting-point lowering. These and other effects could lead to a rapid destruction of the divertor tiles and generation of tungsten dust. The B_4C coating was proposed as a protecting in-situ renewable coating for tungsten tiles of the ITER divertor [3]. Until now, methods of B_4C coating deposition in plasma have not been investigated.

Material testing under intensive plasma and particle irradiation is being performed on a number of special devices and facilities. Quasi-stationary plasma accelerators, such as QSPA-T [4] or QSPA Kh-50 [5], provide sample irradiation by pulsed plasma fluxes with a power density of up to 5 GW/m^2 and a duration of less than 1 ms. The time-gap between the pulses is about several minutes, and this fact

[1] To whom any correspondence should be addressed.

leads to testing series of about a hundred cycles. A high heat-load can be obtained in devices with an electron beam, like JUDITH II [6]. In this machine, an electron beam with a power of up to 200 kW is used, but it scans the surface of the sample being tested and produces an average value of the heating load. The devices with ion beams extracted from a high-current discharge, like Magnum-PSI [7], MARION [8] and PISCES [9] can provide a power density of up to 20 MW/m^2 with a duration of up to hundreds of seconds.

Combining stationary and pulse heat loads is a challenging task and is achieved only in the Magnum-PSI device by periodically ramping up the discharge current to higher values. At present, stationary irradiation is possible only for 100 sec in this device, while in ITER a 1000 sec discharge is expected. All the devices mentioned are of high complexity and require complex maintenance by skilled personnel.

A special device is needed for the development of a coating deposition technique (in particular, B_4C coating deposition on ITER divertor tiles) and for testing of coatings with high particle and heat loads. The paper presents a device, which meets both of these purposes.

2. Experimental device

The stand allows deposition of coatings by means of sputtering of targets by ions of an argon gas discharge with a heated cathode. The advantage of this type of discharge is the possibility of ion irradiation in a wide range of ion energies (from tens of eV to tens of keV) and by various ion fluxes. Therefore, it becomes feasible to try a large amount of various regimes of deposition and choose the best one. By replacing the sputtered target with the tested sample, one can irradiate the latter by ions or electrons of the plasma.

The Stand for COating Deposition and MATerial Testing (CODMATT) is presented schematically in figure 1 ((a)-coating configuration, (b)-testing configuration). The stand has a vacuum chamber (1) with a cylindrical plasma chamber (2) inside it. The plasma discharge is ignited between a heated tungsten cathode (3) and an anode (4).The anode is designed as two concentrically ring-shaped plates with a ring-shaped slit between them. Ions are drawn off from the discharge area through the slit of the anode. The bigger anode plate can be moved by the motion feedthrough (5). The discharge current can reach 20 A with the voltage between anode and cathode being 200 V. The residual pressure is 10^{-6} Pa. The main residual gas components are H_2O (75%) and H_2 (25%), the other components being O_2, N_2 and CO_2 (<< 1%). A quadrupole mass-spectrometer is used to measure the composition of the residual and working gasses. The vacuum chamber is equipped by two lock chambers. The upper lock chamber (6) is used for the substrate manipulator; the bottom lock chamber (7) is used for target installation and the tested sample manipulator. The lock chambers are separated from the main vacuum chamber by gate valves (8). The manipulators move through the sealings (9). The stand control is automated.

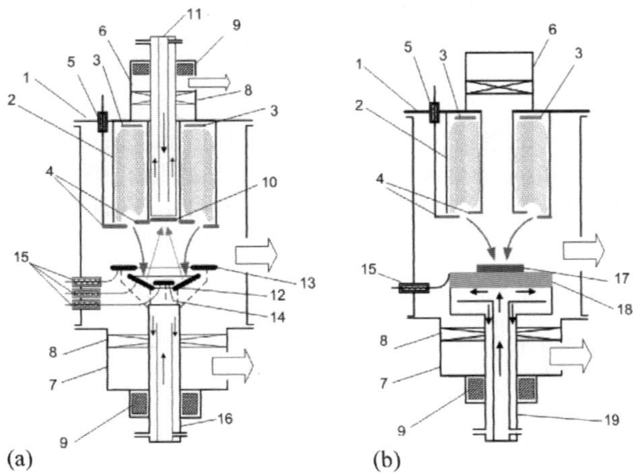

Figure 1. Schematic diagram of the stand: (a) coating configuration; (b) testing configuration. 1 - vacuum vessel; 2 - plasma chamber; 3 - heated tungsten cathode; 4 - anode; 5 - motion feedthrough; 6 - upper lock chamber; 7 - bottom lock chamber; 8 - gate valve; 9 - sealing; 10 - substrate for coating; 11 - substrate manipulator; 12 - target; 13 - outer anti-dynatron electrode; 14 - inner anti-dynatron electrode; 15 - electric feedthrough; 16 - target manipulator; 17 - testing sample; 18 - sample holder; 19 - testing sample manipulator.

2.1. Coating configuration
The coating configuration is presented schematically in figure 1 (a). The substrate for coating (10) with a size of 15×15×1 mm^3 is mounted on a manipulator (11) equipped with a motorized drive (not shown). The motorized drive provides an automatic and simple substrate exchange procedure and transport of the substrate into the vacuum chamber. The manipulator is water-cooled, which allows fixing the substrate temperature in the range of 300 – 900 K during the deposition process. The temperature is measured by a pyrometer. Before and during deposition, the substrate can be irradiated by plasma ions for achieving better adhesion and better structural properties of the film, respectively. The main working gas is argon because it provides a high sputtering rate. For investigation of hydrogen (or some other gas) trapping during coating, the mixture of argon and hydrogen (or other investigated gas) can be used as a working gas.

The target (12) is irradiated by ions extracted from the plasma. Two anti-dynatron electrodes (outer – 13 and inner – 14) are located concentrically around the target. The anti-dynatron electrodes suppress the ion-electron emission from the target. These electrodes are also needed for ion beam formation. The anti-dynatron electrodes and the target are biased by a voltage of ≤ 25 kV by means of the electric feedthroughs (15). The maximum power supplied to the target is 2500 W, to the inner electrode, 500 W, and to the outer one, 1000 W. The target and the electrodes are mounted on the water-cooled target manipulator (16).

The coating is formed on the substrate by atoms sputtered from the target by plasma ions. For better adhesion, creation of an interlayer between the substrate and the coating is needed. It is possible to form this layer with a composition different from that of the target by sputtering the anti-dynatron electrodes. This can be achieved by varying the shape of the ion beam (by moving the electrodes and the anode), as well as by varying the value of the electrode potential. In the case of depositing B_4C onto tungsten, the outer electrode is made of tungsten, the inner one is made of graphite, and the target is a mixture of boron and graphite. The deposition procedure is as follows. First, substrate sputtering is conducted, then the tungsten electrode is sputtered (tungsten atoms are being deposited onto the substrate), after that, the graphite electrode is gradually biased (W-C layers are deposited) and, lastly, sputtering of the target begins, while the sputtering of the electrodes is minimized (boron carbide is deposited).

Due to the automatic control of power supply, one may handle the ion fluxes by a predetermined program and carry out the deposition in an automatic regime, including preliminary ion cleaning of the substrate, deposition of the sublayer and of the coating. The deposition rate of B_4C was measured as ≈10 μm/h.

2.2. Testing configuration
The testing configuration is presented schematically in figure 1 (b). Using the stand, material testing under high heat and particle loads in stationary and pulsed regimes can be conducted. In the testing configuration of the stand, both the target and the anti-dynatron electrodes are replaced by the sample (17) on the sample holder (18). The testing sample manipulator (19) is actively cooled. By using different systems of sample holders, the temperature of the sample during testing can be varied in the range 600 – 2800 K. In thermal cycling tests, ions or electrons irradiate the backside of the tested sample.

The ion irradiation provides a maximum energy release in the surface layer of the sample. Moreover, the possibility exists to reveal the influence of ion irradiation and implantation on the behavior of materials and coatings under thermal cycling. The electron irradiation leads to a widening of the surface area of energy release. At the same time, in this case surface sputtering is excluded and radiation-induced defects formation in the area of energy release is reduced sharply.

During the testing of materials, the sample holder is biased by a voltage of ≤ 25 kV (power of ≤ 4 kW). The irradiation power can be varied during the testing process. The pulse duration of cycling irradiation exceeds 1 ms, while the duty cycle assume any value in the range 1 – 100 %. The diameter of the tested sample is limited to 100 mm. By altering the anode and the sample positions, various ion beam shapes can be obtained and, hence, various surface areas can be irradiated (from 4 mm^2). This regime can also be used for modification of the materials surface by ion or electron irradiation.

3. Results

As part of a pilot program of the CODMATT testing, experiments with boron carbide deposition on tungsten substrates were carried out. Also, thermocycling tests of the deposited coatings were carried out in order to verify their adhesive properties. To test the coatings of B_4C in the T-10 tokamak, it was required that samples with coating thickness of a few microns withstand thermal cycling up to 900 K.

An analysis of the concentrations of the components in the coating and control of the integrity of the coatings after thermal cycling was carried out by the electron microscope TESCAN VEGA 3, which had a console to perform elemental analysis by X-ray energy-dispersive spectroscopy (EDS). Targets formed with 4:1 boron-graphite powder mixture were used for deposition of the coating. The substrate temperature during deposition did not exceed 600 K. The B/C concentration in the coatings deposited was found to be in the range 3,8 – 4,2. The SEM image of a coating after deposition is presented in figure 2 (a). These coatings were tested by thermocycling in the 450 – 1200 K range. After 150 heat cycles, small cracks of a dimension of 0,1 μm with a distance between them around 200 μm appeared. The SEM image of a coating after thermocycling is presented in figure 2 (b). After 100 more cycles, these cracks did not widen and no flake forming was observed. The cross-section of a crack (figure 2 (c)) shows that the crack goes through both the coating and tungsten substrate. One can conclude that the cracking of the coating was caused by substrate cracking in the first place.

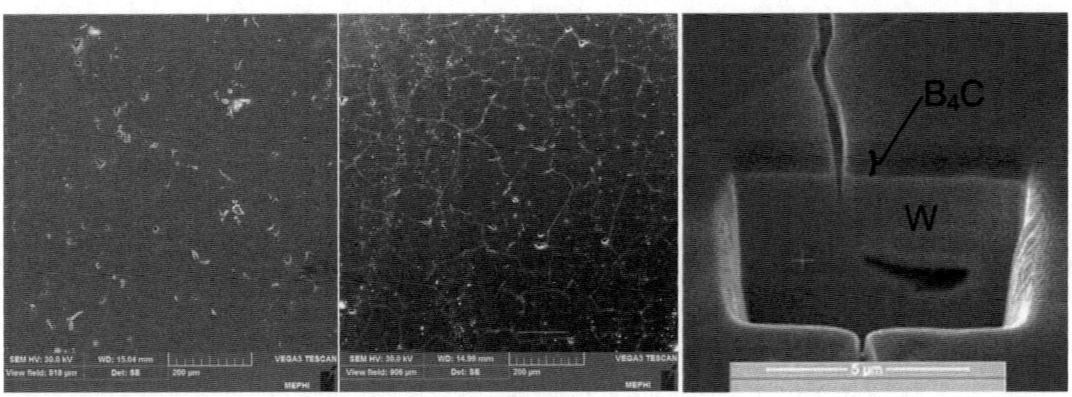

a) b) c)

Figure 2. SEM images of a coating before thermocycling (a), after 150 heat cycles in the range 450 – 1200 K (b) and a FIB crosssection of the crack (c).

In the material test mode, tungsten samples were irradiated. For these experiments, ITER divertor tungsten tiles were used as samples with dimensions of 20 × 20 × 1 mm^3. The samples were prepared by electro-spark cutting, then polished with abrasive polishing paper having a grain size of 2 – 3 microns and washed with ethanol in an ultrasonic bath.

The samples were irradiated by H_2^+ ions (E = 10 keV/at and a flux of $1{,}2\times10^{18}$ at/m^2s). The irradiated zone was in the shape of a ring with a diameter of 10 mm located in the center of the sample. The pulse duration was 10 s and the irradiated zone had a temperature of 1500 K after each pulse. The samples were not actively cooled in the experiment. After the end of the pulse, the sample was cooled down for 20 s to 900 K. After each series of irradiation, the sample surface was studied by SEM. On the irradiated surface, shallow holes formation with 20 μm dimensions and deep cavities were observed. Their position changed and size increased as the number of cycles of irradiation reached 430 pulses.

The sharp edges of the holes, their considerable depth and their shape indicated that their formation was not the result of ion sputtering. On the same side, constant sample heating at the same temperature with a cumulative time of the pulses did not lead to any significant surface relief modification. Interestingly, formation of similar holes on the surface of fused quartz during irradiation by helium ions was observed in [10]. Often, their appearance was related to the growth of whiskers. The paper concludes that their formation, as well as the occurrence of whiskers, are a consequence of the diffusion fluxes and mobility of dislocations under the conditions of ion irradiation. One may suppose that during tungsten irradiation, the same processes lead to surface modification. Cyclic ion bombardment at high temperatures stimulate the processes leading to their formation in the material at a macroscopic depth and the appearance of holes in the irradiated surface.

4. Conclusions

A simple and convenient to operate automated stand was implemented. The stand allows one to investigate the regularities of formation and deposition of single- and multi-component coatings. Also, the stand allows one to study the behavior of materials under thermocycling and high-density ion and electron irradiation.

During pilot testing of the CODMATT stand, a B_4C coating was deposited on tungsten and pulsed ion irradiation of tungsten samples was performed by 20 keV H_2^+ ions. The results of these tests are presented and discussed.

Acknowledgments

This work was supported by Competitiveness Growth Program of the Federal Autonomous Educational Institution of Higher Professional Education National Research Nuclear University MEPhI and by the Ministry of Education and Science of the Russian Federation (agreement contract №14.575.21.0049 of 27.06.14).

References

[1] Krieger K, Geier A, Gong X, Maier H, Neu R and Rohde V 2003 *J. Nucl. Mater.* **313–316** 327–32

[2] Makhlaj V A, Garkusha I E, Aksenov N N, Chuvilo A A, Chebotarev V V, Landman I, Malykhin S V, Pestchanyi S and Pugachov A T 2013 *J. Nucl. Mater.* **438** S233–6

[3] Begrambekov L and Buzhinskiy O 2006 *Problems of Atomic Sci. and Technol. Series: Nuclear Fusion* **4** 14 (in Russian)

[4] Klimov N et al. 2009 *J. Nucl. Mater.* **390-391** 721-6

[5] Garkusha I E et al. 2007 *J. Nucl. Mater.* **363-365** 1021-25

[6] Majerus P, Duwe R, Hirai T, Kühnlein W, Linke J and Rödig M 2005 *Fusion Eng. and Design* **75-79** 365-9

[7] De Temmerman G, van den Berg M A, Scholten J, Lof A, van der Meiden H J, van Eck H J N, Morgan T W, de Kruijf T M, Zeijlmans van Emmichoven P A and Zielinski J J 2013 *Fusion Eng. and Design* **88** 483-7

[8] Nicolai D, Chaumet P, Neubauer O and Uhlemann R 2013 *Fusion Eng. and Design* **88** 2506-9

[9] Goebel D M, Campbell G and Conn R W 1984 *J. Nucl. Mater.* **121** 277-82

[10] Begrambekov L B and Zaharov A M *Ionizing radiation and laser materials* (Moscow: Energoatomizdat) pp 91-100 (in Russian).

Plasma-enhanced CVD preparation of isotopes of group IV and VI elements

P G Sennikov[1] and R A Kornev

Laboratory of Plasma Chemistry, G.G. Devyatykh Institute of Chemistry of High-Purity Substances of the Russian Academy of Sciences, Nizhny Novgorod, 603950, Russia Federation

E-mail: pgsen@rambler.ru

Abstract. The plasma-chemical reduction was studied experimentally of molybdenum, sulfur and germanium fluorides with different isotopic composition by hydrogen in plasmas sustained by inductively coupled (IC) and capacitively-coupled (CC) RF discharges. The emission spectra of plasmas recorded under different experimental conditions are discussed. Reaction mechanisms are proposed. In the case of $^{98}MoF_6$ reduction in CCP, the combined radical-atomic mechanism including intermediate MoF_3 resulting in bulk and powder ^{98}Mo formation was established. In the emission spectrum of the $^{32}SF_6$ + H_2 system in ICP at 0.2 Torr, only lines assigned to SiF_2, SiF and F_2^* originating from the decomposition of $^{32}SF_6$ and a fast reaction of its products with the reactor's quartz walls were observed. Due to this etching process, the yield of ^{32}S was moderate. Increasing the pressure in the reactor above 1 Torr resulted in an increase of the ^{32}S yield. The mechanism of $^{72}GeF_4$ reduction depends on the pressure and plasma type. At 0.25 Torr in ICP, a ^{72}Ge deposit was formed via the ^{72}GeF radical; however, at 3.5 Torr in CCP, the mechanism changed to a molecular one. Results of the study of the structure and the isotopic and chemical purity of isotope samples obtained are briefly presented.

1. Introduction

The RF glow discharges used for plasma-enhanced chemical vapor deposition (PECVD) consist of partially ionized gases containing ions, electrons and a variety of neutral species in both ground and excited states. Plasmas are ignited and sustained by applying an RF voltage at frequencies between 50 kHZ and 13.56 MHz to a low pressure (~10 mTorr – 10 Torr) gas. These conditions produce a highly reactive environment. E.g., the electron (and positive ion) densities are between 10^8 and 10^{12} cm^{-1}, and the average electron energies are several electron volts, while the ion (and neutral) energies are at least two orders of magnitude lower. Because of such differences in the particles' energies, the term *nonequilibrium plasmas* is used to describe these discharges. The electrons gain energy via acceleration by an alternating electric field, which results in elastic collisions between the high-energy electrons and the neutral gas molecules. These impact reactions can be assigned to three general categories: excitation, dissociation, and ionization. The excitation and dissociation processes generally require energies of a few electron volts or less. As a result, the discharge produces a large concentration of free radicals, both by direct dissociation and by electronic excitation into an

[1] To whom any correspondence should be addressed.

antibonding states that may ultimately lead to dissociation. By comparison, the ionization of most molecules or atoms requires electron energies above ~7 eV, so relatively few ions are formed. The ratio of the electron concentration to the neutral species (radicals, molecules) concentration is less than 10^{-3}. This property leads to the fact that neutral species, rather than ions, are the primary contributors to solid phase deposition. That is, neutrals are not more chemically reactive, but are present in much higher concentrations compared to the ions.

Depending on the experimental conditions, the solid deposits have different chemical and phase structures. This is why the PECVD method is widely used for laboratory and industrial production of different chemical substances and elements.

The advantage of this approach is clearly demonstrated in the case of precursors with high-energy chemical bonds. Inorganic fluorides of different elements belong to this group of substances. Besides, fluorine is monoisotopic, so that the volatile fluorides are used in centrifugal processes of isotopic separation (enrichment). The classical illustration of this approach is the isotopic enrichment of uranium hexafluoride.

This report summarizes the main results of recent investigations carried out at the Institute of Chemistry of High-Purity Substances and the Institute of Applied Physics of Russian Academy of Sciences in Nizhny Novgorod in creating the physical-chemical background of plasma-chemical production of chemically and isotopically pure IV and VI group elements (molybdenum, sulfur and germanium) from their enriched fluorides.

Some of the molybdenum isotopes are valuable tools for studies of mineral metabolism in humans [1]. In recent years, a significant interest appeared in uranium-molybdenum (UMo) fuel to be used in Material Test Reactors [2]. This interest was due to the fact that UMo fuel is mechanically stable, even at high uranium concentrations and high U-burnup. For this goal, Mo in a bulk form is required, depleted of ^{95}Mo and with a high thermal neutron cross-section. Of the greatest interest is to use the ^{95}Mo, ^{96}Mo, ^{98}Mo, and ^{100}Mo stable isotopes for production of ^{97}Ru, ^{96}Tc, ^{95m}Tc, and ^{99m}Tc radioisotopes which, in their turn, are widely used in medicine for diagnostic and treatment of cancer and cardiovascular disease. The most popular are the radio-pharmaceutical drugs based on ^{99m}Tc which originate from ^{99}Mo with a half-life of 66.02 hours. The radiative capture method for ^{99}Mo production from ^{98}Mo nano-powder targets is very promising among all other reactor and non-reactor alternatives.

Production of high-purity isotopically enriched sulfur ^{32}S is an important task for determining the isotopic effect influence on the elementary sulfur properties, as well as on the properties of sulfur-based vitreous and semiconductor compounds, such as germanium sulfides or arsenic sulfides glasses and films in terms of their optical and physical characteristics [3].

The isotopes of germanium, both in the form of thin layers and bulk crystals, are used for the development of nano-electronic devices of a new generation, as well as for manufacturing unique instruments in high-energy physics [4].

2. Experimental

The starting fluorides, namely $^{98}MoF_6$ (98.5% enriched), ($^{32}SF_6$ 99.6% enriched), and $^{72}GeF_4$ (52% enriched) were produced by SC "PA Electrochemical Plant", Zelenogorsk, Russian Federation.

The process of their plasma chemical hydrogen reduction was carried out in the set-up shown in figure 1. The power of the RF generator (13.56 MHz) was 1 kW. The power applied to the plasma discharge area was measured by a calibrated coupler in the range 300 – 700 W.

The process of plasma-chemical reduction was studied in capacitively-coupled plasma (CCP) for $^{98}MoF_6$, $^{32}SF_6$ and $^{72}GeF_4$, and inductively-coupled plasma (ICP) for $^{32}SF_6$ and $^{72}GeF_4$). The difference in these types of plasmas is that the plasma-sustaining capacitively-coupled discharge at a pressure P ~ 1 – 10 Torr is characterized by a low current strength (10^{-6} – 10^{-1} A) and a high voltage value (10^2 – 10^3 V). The plasma-sustaining inductively-coupled discharge at a pressure P ~ 0.1 – 1 Torr is characterized by a high current strength (10^2 – 10^3 A) and a low voltage value (1 – 10 V). The different parameters of these discharges result in different mechanisms of electron production and in different plasma-chemical reactions mechanisms. Thus, one can also expect differences in the deposited materials structure.

The flow rate of the plasma-forming gas $H_2 + MF_n$ (M = ^{98}Mo, ^{32}S, ^{72}Ge; n = 4,6) was adjusted by mass-flow controllers and was equal to 0 – 310 ± 5 cm^3/min. The dilution ratio for H_2/MF_n varied from 2 to 25. The plasma chemical reactor was made of quartz with an inductor coil or ring electrodes powered by the RF-generator through a matching device. In the case of CCP, a quartz trap was placed below the quartz reactor to collect the Mo powder. In the case of ICP, an additional quartz tube, the substrate, was placed inside the reactor where the reaction products were deposited. There was a cryogenic trap at the output of substrate tube designed to completely entrap the condensed phase. The total degree of conversion of fluorides to molybdenum, sulfur and germanium was determined by IR spectroscopy (a Bruker Vertex 80v spectrometer) with an accuracy of 0.1%. The condensed phase yield was measured gravimetrically with an accuracy of 1×10^{-3} g. The emission spectrum of the chemically active plasma was examined in the range 350 – 950 nm by an HR4000CJ-UV-NIR emission spectrometer. The characterization of ^{98}Mo, ^{32}S and ^{32}Ge in a solid form (layers, powder, bulk) was carried out by XRD, SEM, and EDX. For isotope and impurities, the controlling methods of SIMS, laser-mass-spectrometry, ICP mass-spectrometry and atomic-emission spectrometry were used.

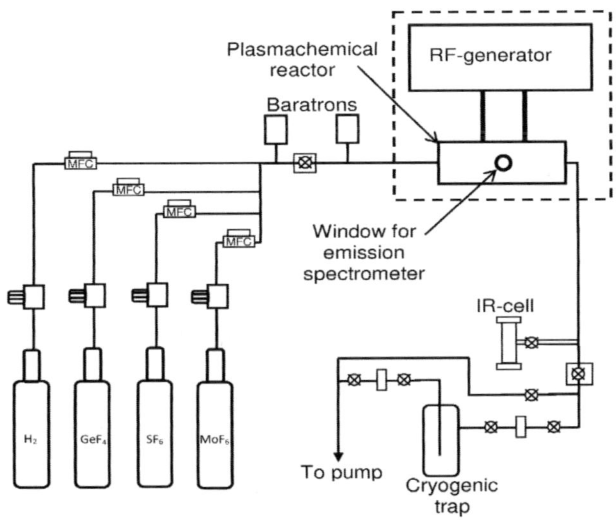

Figure 1. Set-up for the process of plasmachemical hydrogen eduction of $^{98}MoF_6$, $^{32}SF_6$ and $^{72}GeF_4$ in RF CCP and ICP.

3. Results and discussions

3.1. PECVD of molybdenum

In CCP, the molybdenum was mainly deposited on the inside surface of the quartz tube in the form of a thick layer, but a small part (less than 3%) was deposited as powder in the special trap. The absence of $^{98}MoF_6$ vibrational modes in the FTIR spectrum of the exhaust gases is indicative of the practically total conversion of $^{98}MoF_6$ to solid reaction products. To obtain more detailed information about the hydrogen reduction mechanism under the given experimental conditions, spectra of chemically-active plasma of the $H_2 + {}^{98}MoF_6$ mixture were acquired by emission spectroscopy.

The spectrum of CCP of $H_2 + {}^{98}MoF_6$ with ratio $H_2/{}^{98}MoF_6 = 4$ and pressure $p = 3$ Torr contains the lines at 486.13 and 656.2 nm originating from emission of atomic hydrogen H_β and H_α, respectively, while the line at 703 nm was assigned to atomic fluorine (figure 2). We suggest several basic reactions responsible for hydrogen reduction of $^{98}MoF_6$ in CCP:

$$H_2 \rightarrow 2H, \quad (1) \qquad {}^{98}MoF_6 + e \rightarrow {}^{98}MoF_4 + 2F, \quad (2)$$

$$ {}^{98}MoF_4 + e \rightarrow {}^{98}MoF_3 + F, \quad (3) \qquad {}^{98}MoF_3 + e \rightarrow {}^{98}Mo + 3F, \quad (4)$$

$$F + H \rightarrow HF. \quad (5)$$

Journal of Physics: Conference Series **700** (2016) 012042 doi:10.1088/1742-6596/700/1/012042

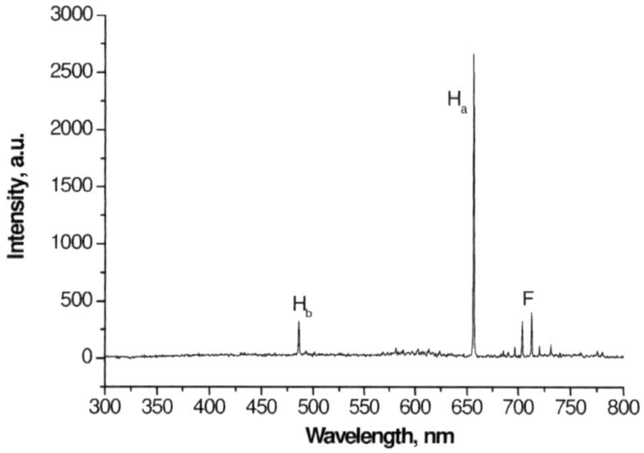

Figure 2. Emission spectrum of H_2/ $^{98}MoF_6$=3,5 in RF CCP at p = 3 Torr.

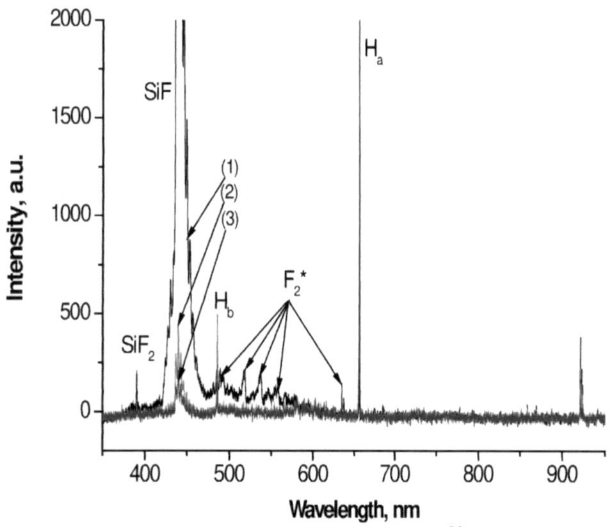

Figure 3. Emission spectra of H_2/ $^{32}SF_6$ = 6 in (1) – ICP at p = 0.2 Torr, (2) – ICP at p = 0.7 Torr, (3) – CCP at p = 1 Torr.

The formation of traces of $^{98}MoF_4$, $^{98}MoF_3$ and the predominant formation of ^{98}Mo accordin to reactions (2), (3) and (4) was confirmed by XRD, and the hydrogen fluoride byproduct according to reaction (5), by the FTIR spectra of the exhaust gases.

The space-centered ^{98}Mo thick layer deposited from CCP contains, according to the EDX data, fluorine (1.3% at) and oxygen (12.0 % at) due to adsorption from the atmosphere during storage. The powder collected in a small amount in the quartz trap, located at the reactor output, also contains the ^{98}Mo powder phase with an average particles size of 25 – 28 nm (XRD data).

The isotopic content did not change as going from the starting gas to solid samples. This means that no isotopic dilution takes place that could be caused due to molybdenum content in the apparatus. The chemical purity of ^{98}Mo samples depends on a) supply of impurities from construction materials, and b) adsorption from atmosphere during storage. The conversion degree of $^{98}MoF_6$ to ^{98}Mo is more than 85%. Concerning the thick layer (bulk) ^{98}Mo, the deposition rate is equal to approximately 1 g/h. The content of main impurities in ^{98}Mo is less than 0,1 ppm wt.

3.2. PECVD of sulfur

Figure 3 shows the emission spectra of ICP at p = 0.2 and 0.7 Torr and of CCP at p = 1 Torr of a H_2 + $^{32}SF_6$ gas mixture (H_2/ $^{32}SF_6$ = 4). We observed intensive bands at 486.13 and 656.2 nm referring to atomic hydrogen H_β and H_α, as well as weaker bands at 390, 440, 490, 517, 538, 552, 636 and 925 nm. The bands at 390, 440 nm can be assigned to SiF_2 and SiF radicals [5,6], and the bands at 490, 517, 538, 552 and 636 nm, to the active particle F_2^* [7,8]. It is a well known fact that $^{32}SF_6$ is an extremely active etchant with respect to Si and SiO_2 [9,10]. This is why one can get indirect information about the mechanisms of its plasma reduction by the appearance of bands referring to SiF_3, SiF_2 and SiF radicals and their intensity, as well as by registration of the SiF_4 band with various intensity in the IR spectrum of the exhaust gases. One can conclude that the band of SiF is quite intensive in the spectrum recorded at p = 0.2 Torr. As the pressure increases up to 0.7 and 1 Torr and as going to CCP, the SiF band intensity decreases sharply and the SiF_2 band disappears. The F_2^* band disappears simultaneously, but at the same time the H_β appears, while it was absent in the spectrum at 0.2 Torr. It was impossible to assign the band at 925 nm. It should only be noted that its intensity decreases while the reactor pressure is increased. Based on these experimental data, as well as on the mechanism of intermediate reactions proposed in [9-11], the $^{32}SF_6$ reduction process in quartz reactor, including the interaction of the corresponding active particles with the surface at 0.2 Torr, could be described by the following reactions:

$H_2 \rightarrow 2H,$ (6) $^{32}SF_6 + e \rightarrow {}^{32}SF_4 + F_2^*,$ (7)
$^{32}SF_4 + e \rightarrow {}^{32}SF_3 + F + e,$ (8) $^{32}SF_3 + SiO_2 \rightarrow {}^{32}SOF_2 + SiOF,$ (9)
$SiOF + {}^{32}SF_3 \rightarrow {}^{32}SOF_2 + SiF_2.$ (10) $^{32}SOF_2 + F_2 \rightarrow {}^{32}SOF_4,$ (11)
$SiF_2 + F_2^* \rightarrow SiF_4,$ (12) $SiF_2 + e \rightarrow SiF + F,$ (13)
$F + H \rightarrow HF.$ (14)

While the pressure increases, the rate of reactions (6-14) essentially decreases, although the SiF band is still observed. Herewith, one observes a reduction of $^{32}SF_6$ with formation of elemental sulfur, as studied in[12]:

$H_2 \rightarrow 2H,$ (15) $^{32}SF_6 + 6H \rightarrow {}^{32}S + 6HF.$ (16)

The conversion degree of $^{32}SF_6$ to ^{32}S under the given conditions is equal to 70%. The deposition rate of ^{32}S is equal to 0,25 g/h. The content of main impurities in ^{32}S prepared by this method is less than 0,1ppm wt.

The high-purity ^{32}S obtained was used for the synthesis of glass of the ^{72}Ge-^{32}S system; its IR transmission spectrum was studied in comparison with the same glass prepared from Ge and S of natural isotope composition.

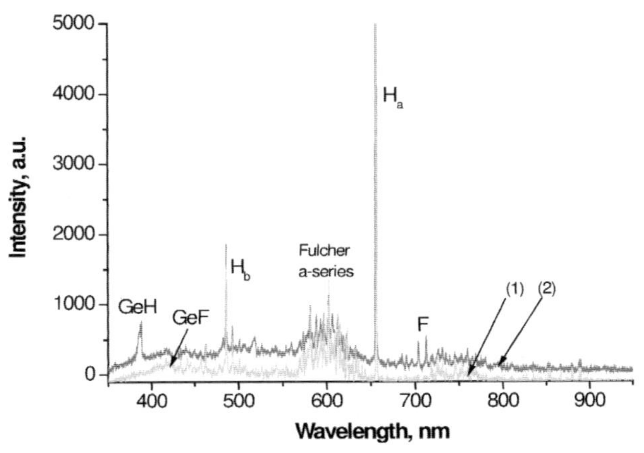

Figure 4. Emission spectra of RF (1) – ICP at $p = 0.25$ Torr, (the ratio of $H_2/^{72}GeF_4 = 4$); and (2) – CCP at 3.5 Torr (the ratio of $H_2/^{72}GeF_4 = 4$).

3.3. PECVD of germanium

The plasma-chemical hydrogen reduction of $^{72}GeF_4$ was studied in ICP and CCP. The analysis of exhaust gases, of the chemically active plasma and of the condensed phase by emission- and FTIR-spectroscopy, as well as by mass spectrometry (MS), made it possible to propose a mechanism accounting for the formation of the intermediate species and the final products observed.

Figure 4 shows the emission spectra of RF plasma of the H_2 + $^{72}GeF_4$ mixture in both types of plasma. The lines at 486.13 and 656.2 nm wavelengths are observed in the spectrum of ICP and correspond to the emission of atomic hydrogen H_β and H_α. The lines of α-Fulcher series in 575 – 625 nm range, the lines connected with the ^{72}GeF radical [13] in the 390 – 450 nm region, and with the $^{72}GeF_2$ radical in the 320 – 390 nm region [14], are present as well. The line at 703.75 nm, which could be assigned to free fluorine [15], was not registered in the spectrum.

As for the spectrum of CCP , the line at 385 nm, assigned to the ^{72}GeH radical [16], as well as the line at 703 nm connected with atomic fluorine [15], are observed, but the lines that could be assigned to the above-mentioned radicals ^{72}GeF and $^{72}GeF_2$ are absent.

Taking into consideration the peculiarities observed in the emission spectra of both types of plasma, as well as the formation of a ^{72}Ge deposit, and ($^{72}GeF_2$)n and HF molecules confirmed by MS and FTIR, the following reaction steps can be proposed to describe the mechanism of reduction of

$^{72}GeF_4$ by hydrogen. The formation of atomic hydrogen takes place in both types of discharge (17). The formation of germanium-containing radicals, as well as of condensed ^{72}Ge and $(^{72}GeF_2)n$ in ICP, can be explained by reactions (18– 21).

$H_2 \rightarrow 2H$,	(17)	$^{72}GeF_4 + e \rightarrow {}^{72}GeF_n^{\cdot} + F_m^{\cdot} + e$,	(18)
$^{72}GeF^{\cdot} + e \rightarrow {}^{72}Ge + F + e$,	(19)	$^{72}GeF^{\cdot} + H \rightarrow {}^{72}Ge + HF$,	(20)
$n^{72}GeF_2^{\cdot} \rightarrow (^{72}GeF_2)n$,	(21)	$^{72}GeF_4 + H \rightarrow {}^{72}GeF_3H + F$,	(22)
$^{72}GeF_3H + H \rightarrow {}^{72}GeF_2H_2 + F$,	(23)	$^{72}GeF_2H_2 + H \rightarrow {}^{72}GeFH_3 + F$,	(24)
$^{72}GeFH_3 + H \rightarrow {}^{72}GeH_4 + F$,	(25)	$^{72}GeH_4 \rightarrow {}^{72}Ge + 2H_2$.	(26)

In CCP, the most probable process is the interaction of atomic hydrogen with the $^{72}GeF_4$ molecule with subsequent substitution of fluorine atoms for hydrogen and formation of fluorogermane and germane molecules in a gas phase with further thermal decomposition of GeH_4, according to 22-26.

One can conclude that, as going from CCP at 3.5 Torr to ICP at 0.25 Torr, a change in the reaction mechanism of germanium formation takes place. In the first case, the radical mechanism with participation of $^{72}GeF^{\cdot}$ is a basic one and formation of metal ^{72}Ge and condensed polymeric $(^{72}GeF_2)n$ product is observed. In the second case, the interaction of atomic hydrogen with germanium tetrafluoride with fluorogermanes, fluorine and germane as intermediate products is responsible for germanium formation. The conversion degree of $^{72}GeF_4$ to ^{72}Ge is 95%. The maximum deposition rate of ^{72}Ge is about 4 g/h. The approach described made it possible to prepare a sample of enriched polycrystalline ^{72}Ge with the following content of the main impurities: B<0.02; P<0.2; Sb~0.003; Al<0.03; As<0.01 ppm wt. The single crystal grown from it has the concentration of electroactive impurities $<10^{13}$ cm^{-3}.

4. Conclusions

We presented some results regarding an experimental study of isotopic hydrogen reduction of isotopic enriched molybdenum, sulfur and germanium fluorides in RF inductively and capacitively coupled plasma. It was shown that formation of a solid deposit in a bulk or powder form, as well as of intermediate products, depends on the pressure in the reactor and the reaction mechanism. Optical emission spectra of the systems studied are helpful for describing this mechanism. In the case of $^{98}MoF_6$ reduction in CCP, a combined radical-atomic mechanism was proposed including intermediate MoF_3 formation resulting in bulk and powder ^{98}Mo deposition. In the case of the $^{32}SF_6 + H_2$ system in ICP, a strong etching ability towards the reactor quartz walls was observed. It can be minimized along with a rise of the ^{32}S yield by increasing the reactor pressure. The deposition of polycrystalline ^{72}Ge was effective both in ICP at low and in CCP at an order of magnitude higher pressure due to the different reaction mechanisms. The main characteristics of prepared materials are briefly discussed.

Acknowledgements

The research was supported by grant No. 16-13-10129 of the Russian Science Foundation and in part by the Program for Basic Research of the Russian Academy of Sciences (Project No. 10).

References

[1] Turnlund J R 2014 *J. Anim. Sci.* **84** E73 408
[2] Mol C A and Bakker K 2011 Method of providing a nuclear fuel and a fuel element provided by such method *Patent 0150166 A1 US*
[3] Churbanov M F, Snopatin G E, Shiryaev V S, Sukhanov M V, Kotereva T V, Koroleva L A, Plotnichenko V G, Koltashev V V, Kutyin A M and Plekhovich A D 2013 *J. Non-Cryst. Solids* **377** 12
[4] Agostini M, Allerdt M and Andreotti E 2015 *Eur. Phys. J.* C **75** 39
[5] Bruno G, Capezzuto P and Cicala G 1992 *Pure and Appl. Chem.* **64** 725
[6] Mutsukura N, Ohuchi M, Satoh S and Machi Y 1983 *Thin Solid Films* **109** 47

[7] Bestwick T D, Oehrlein G S and Angell D 1990 *Appl. Phys . Lett.* **67** 431
[8] Foest R, Olthoff J K, Van Brunt R J, Benck E C and Roberts J R 1996 *Phys. Rev.* **54** 1876
[9] Bruno G, Capezzuto P, d'Agostino R and Molinari E 1980 *J. Fluorine Chem.* **16** 209
[10] Edelson D, Beling C A and Kohman G T 1953 *Ind. Eng. Chem.* **45** 2094
[11] Radovanov S B, Tomcik B, Petrovic Land Jelenkovic B M 1990 *J. Appl. Phys.* **67** 97
[12] Mochalov L A, Kornev R A, Churbanov M F and Sennikov P G 2014 *J. Fluorine Chem.* **160** 48
[13] Tananaev I V and Shpirt M Ya 1967 *Chemistry of Germanium* (Khimiya Moskva) 452 (in Russian)
[14] Saito S and Obi K 1993 *J. Appl. Phys.* **74** 1480
[15] Bruno G, Capezzuto P and Cicala G 1991 *J. Appl. Phys.* **69** 7256
[16] Cicala G, Losurdo M, Capezzuto P and Bruno G 1992 *Plasma Sources Sci. Technol.* **1** 156
[17] Fridman A A 2008 *Plasma Chemistry* (Cambridge University Press New York) 978
[18] Polak L S, Ovsyannikov A A, Slovetskii D I and Vursel' F B 1975 *Teoreticheskaya i prokladnaya plazmokhimiya* (Nauka Moskva) 304 (in Russian)
[19] Gusev A V, Kornev R A and Sukhanov A Yu 2013 *High Energy Chem.* **47** 272

Study of the morphology of Ge quantum dots grown by liquid phase epitaxy

I I Maronchuk[1], D D Sanikovich[1], H Nichev[2,3] and D Dimova-Malinovska[2]

[1] Laboratory of Applied Physics and Nanotechnologies for Energy Applications, Sevastopol State University, 7 Kurtchatova Str., 299033 Sevastopol, Russia
[2] Central Laboratory of Solar Energy and New Energy Sources, Bulgarian Academy of Sciences, 72 Tzarigradsko Chaussee, 1784 Sofia, Bulgaria

E-mail: nitschev@abv.bg

Abstract. This paper reports the results of a study of the morphology of Ge quantum dots (QDs) grown by liquid-phase epitaxy on Si substrate with orientation (111) while applying pulsed cooling of the substrate. The morphology of the structures with QDs is studied by atomic force microscopy. The QDs are grown by applying a different number of cooling pulses – 1, 3, 5 and 10, with a difference in the temperature between the substrate and the heat absorber ΔT_D = 5 °C. The effect is studied of the different technological conditions on the size of the Ge QDs and the density of their distribution.

1. Introduction

Recently, the study of nano-heteroepitaxial structures with quantum size objects, such as quantum dots (QDs), have attracted the interest of many research teams due to their potential for application in optoelectronic devices with improved properties. The Ge quantum dots are regarded as promising materials [1,2]. They could be used as infrared detectors and third-generation solar cells. Based on theoretical calculations, it has been shown that application of Ge QDs in Si based solar cells increases their efficiency up to 53 % [3,4]. It should be noted that the maximal efficiency of the available commercial Si-based solar cells is 22 %.

It has been reported that the application of InAs QDs in the matrix of GaAs grown by MBE and MOCVD did not result in formation of high-efficiency solar cells as has been expected – the application of the QDs resulted in a decrease of the quantum efficiency and of the solar cells efficiency [5]. The authors attributed this result to the presence of stress in the “wetting” layers of the narrow gap material between the QDs where the centers responsible for the carriers’ recombination channels without light emission are created [5].

In this work, liquid phase epitaxy (LPE) in the regime of pulsed cooling (PC) of the Si substrate is applied to deposit Ge QDs arrays without an elastic tense "wetting" layer in between QDs, thus ensuring conditions for their formation close to the equilibrium and growth of structures with reduced carrier recombination [6]. The aim is to study the influence of the technological parameters during the liquid-phase epitaxy (LPE) on the Ge QDs size, namely, the number of the applied cooling pulses, on

[3] To whom any correspondence should be addressed.

the surface morphology of nano-hetero-epitaxial structures with "open" QDs. Atomic force microscopy (AFM) is applied to study the surface morphology.

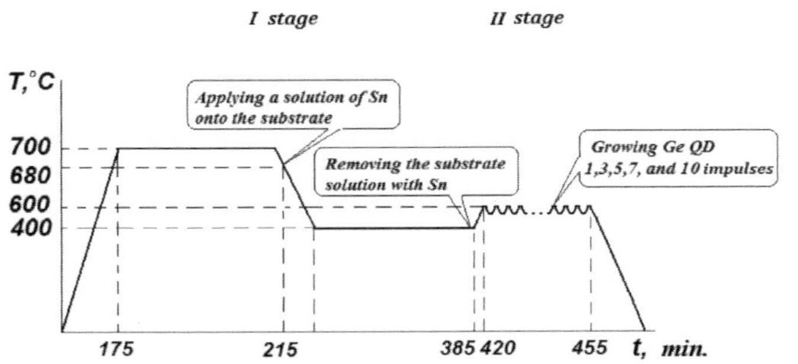

Figure 1. Schematic diagram of the processes of growing a buffer Si layer and Ge "open" QDs on a Si substrate with Sn solution hot fusion.

2. Experimental

Growing of Ge "open" quantum dots on a buffer Si layer formed on a Si substrate with (111) orientation is performed by LPE with application of cooling pulses [6,7] and an original technological scheme. The process of growing presented by the diagram in figure 1 consists of two stages - formation of the buffer layer and growing of Ge QDs.

Special equipment for growing multilayer NHJ structures with a horizontal supplier was constructed.

The formation of the buffer layer on the Si substrate consists of deposition of Sn, homogenization of a saturated melt solution of Si+Sn alloy heated to a temperature of 700^0C and cooling down to 400 °C in order to grow a thin Si layer. The Sn is removed by a specially designed device in the cassette used for growing the QDs. The second stage of the process is growing QDs by applying cooling pulses. A heat sink with temperature, T_{HS}, which is lower than the temperature of the substrate, T, by $\Delta T_D = T - T_{HS}$ is placed on the back surface of the substrate. Growing NHE structures with QDs with a surface density $< 10^{11}$ cm^{-2} of the narrow-gap material Ge on a substrate of the wider band gap material single crystal Si [5] is performed at a temperature of the oven of 600 °C in an atmosphere of highly purified hydrogen. The difference in the temperatures of the substrate and the heat sink is $\Delta T_D = 5$ °C. After a certain time, which depends on the duration of the cooling pulse $\tau \approx 10^{-3} \div 10^{-1}$ sec, the heat sink assumes the temperature of the substrate and a nanolayer with QDs arrays crystallizes on the front surface of the substrate [8]. Due to the different lattice constants of the substrate material (Si) and the QDs material (Ge), in the initial stage a continuous "wetting" layer by the Stranski-Krastanov's mechanism is formed. As the thickness of the growing layer increases, periodic stresses appear at the interface with the substrate. During the growing process, the formation of the Ge QDs takes place in the sites with minimum mechanical stress, i.e. in the positions with a perfect structure [9]. Different structures are prepared by applying different numbers of "cooling" pulses – 1, 3, 5 and 10. The number of pulses determines the different size of the objects grown. According to the theoretical considerations, the diameter and the height of the QDs should increase with the increase of the number of cooling pulses [10, 11]. The density of the crystallized QDs under optimum conditions is limited mainly by the period of the perfect crystalline structure in the interface. The residual Sn on the structures grown is removed by a water solution of HNO_3. The possible presence of Sn after the removal is checked by X-ray fluorescent analysis.

The surface morphology of the samples prepared is studied by a CMM-2000 AFM (Proton—Zelenograd, Russia) with a soft cantilever tip MSCT (Veeco, USA).

3. Results and discussion

Figure 2 presents AFM images of the Ge QDs grown by applying 1 "cooling" pulse. The results show that the QDs have a diameter of about 20 nm and a height of about 5 nm. The deviation in the size is < 20 %. The mean surface roughness is < 1 Å. In this case, the QDs are very small and do not have a well-defined shape. A similar picture is observed in the case of applying 3 cooling pulses (not shown here); however the QDs have grown with a larger size – the diameter is about 60 nm and the height is 10 nm with a deviation < 30%.

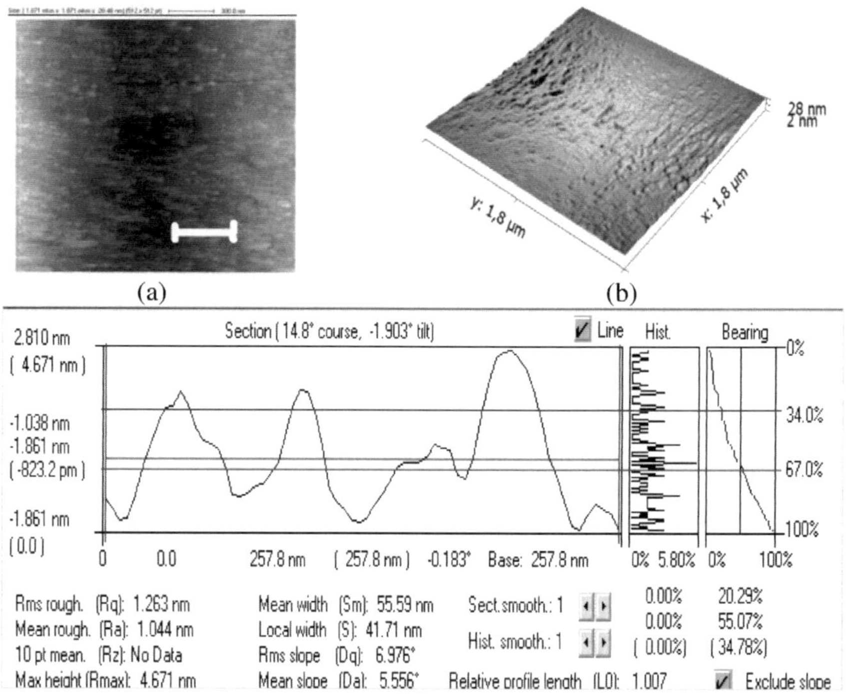

(c)

Figure 2. AFM images of the samples with Ge QDs grown on a (111) Si substrate with a buffer Si layer grown after application of 1 "cooling" pulse ((a) and (b)) and the profile of the QDs (c).The marker in (a) corresponds to 100 nm.

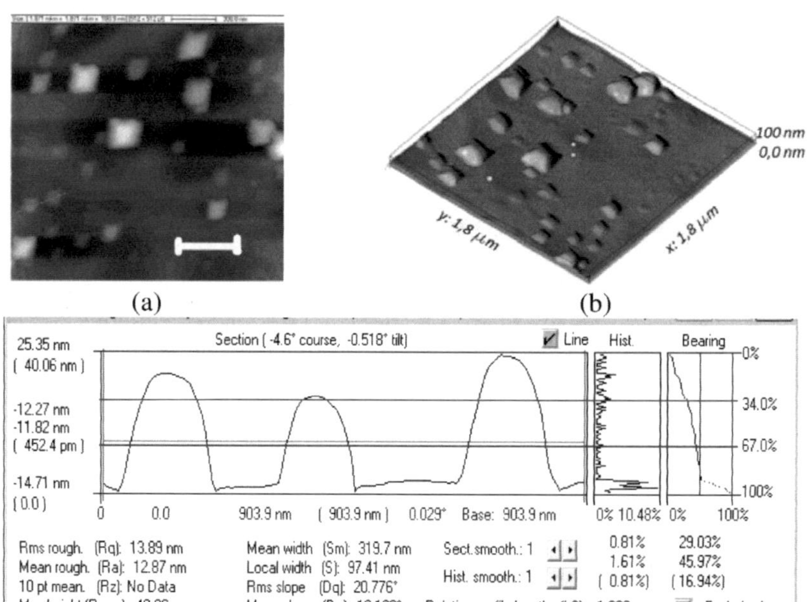

(c)

Figure 3. AFM images of the samples with Ge QDs on a (111) Si substrate with a buffer Si layer grown after application of 5 "cooling" pulses ((a) and (b)) and the profile of the QDs (c). The marker in (a) corresponds to 100 nm.

The AFM images of the substrate surface with Ge QDs grown by applying 5 cooling pulses are shown in figure 3. The QDs have a much larger size, namely, a diameter of 190 nm and a height of 70 nm. The deviation in the size values is < 35%. It has to be noted that after applying 5 cooling pulses the Ge QDs have grown in a pyramidal-like shape with a density of 2×10^{10} cm^{-1}.

As seen from the AFM images shown in figure 4 of the samples with Ge QDs grown after applying of 10 cooling pulses, the QDs have increased further in size – their diameter is 245 nm, the height is 150 nm with a deviation < 35%. The pyramidal-like Ge QDs distribution is less dense than that in the previous cases – the density of their surface distribution has decreased from 2×10^{10} cm^{-1} to 1.5×10^{10} cm^{-1}.

The data obtained from the AFM measurements are summarized in table 1.

The increase observed in the size of the Ge QDs with the increase of the number of cooling pulses is in accordance with the theoretical considerations [10,11] and the experimental data reported by other authors as well [9].

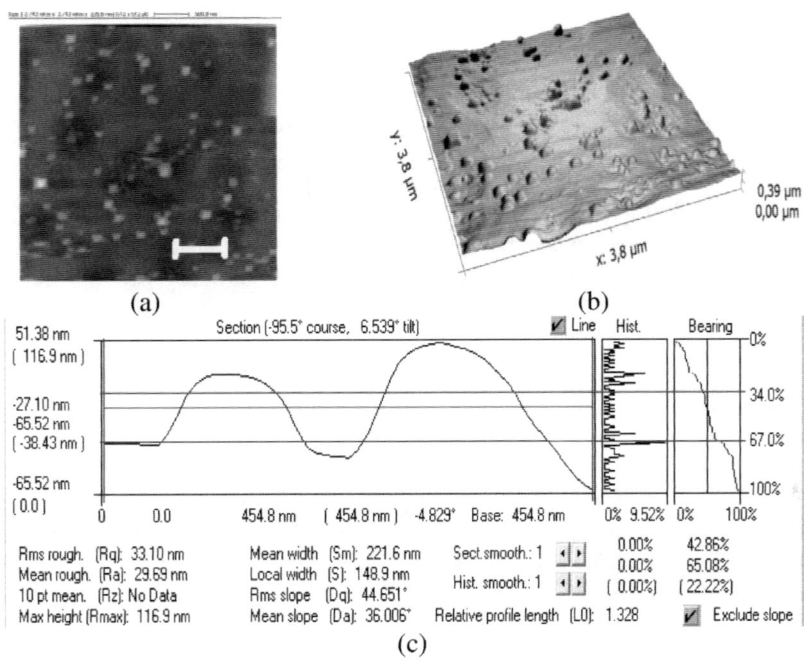

Figure 4. AFM images of the samples with Ge QDs grown on a (111) Si substrate with a buffer Si layer grown after application of 10 "cooling" pulses (a and b) and the profile of the QDs (c). The marker in (a) corresponds to 100 nm.

Table 1. Data obtained by AFM measurement of "open" Ge QDs grown on a buffer Si layer formed on a c-Si substrate with (111) crystalline orientation.

Sample No.	Number of pulses	Average diameter, nm	Average height, nm	Average deviation in the QDs diameter, nm	Density of the QDs distribution, $\times10^{10}$ cm^{-2}
1	1	25	5	20	2,0
2	3	62	10	30	2,0
3	5	196	69	35	2,0
4	10	245	150	35	1,0

4. Conclusions

Ge QDs were grown from a molten solution on a Si substrate by LPE while applying cooling pulses. The study by AFM of the structural properties of the grown structures after applying different number of cooling pulses revealed the morphology, the surface roughness, the linear dimensions of quantum dots and their density. It was observed that the surface morphology with Ge quantum dots changed with the number of cooling pulses – the QDs size and their height increased. The density of their surface distribution on the substrate decreased after the application of 10 pulses; however, a significant increase was observed of their size, while the deviation in the size remained the same as after application of 5 pulses. As the number of pulses was increased to 5, the Ge QDs shape changed from a drop-like to a pyramidal-like one. The observed increase in the size and the changes in the shape of the Ge QDs correlated with the theoretical considerations reported earlier [10,11]. Work is in progress to apply LPE to grow nano-heteroepitaxial strictures with Ge QDs on Si substrates in view of highly efficient low-cost solar cells and other optoelectronics application.

Acknowledgement

The work has been supported by the NSF of Bulgaria - grant DNTS/Ukraine - 01/0005 and by the Ministry of Education and Sciences of Russian Federation – grant №2015/702.

References

[1] Pchelyakov O P, Bolkhovityanov Y B, Dvurechensky A V, Sokolov L V, Nikiforov A I, Yakimov A I and Voitlander B 2000 *Semiconductors* (in Russian) **34** 1229
[2] Lozovoy K A, Voitsekhovskiy A V, Kokhanenko A P, Satdarov V G, Pchelyakov O P and Nikiforov A I 2014 *Opto-Electronics Rev.* **22** 171
[3] Luque A and Marti A 1997 *Phys. Rev. Lett* **78** 5014
[4] Voitekhovkii A V, Grigor'ev D V, Konhanenko A P, Pchelyakov O P, Dvurechenskii A V and Nikiforov A I 2011 *Russian Phys. J.* **53** 943
[5] Alfelov I 1998 *Semiconductors* (in Russian) **32** 3
[6] Maronchuk I E, Kulyutkina T F and Maronchuk I I 2011 Method of growth of epitaxial nano hetero-junction structures with quantum dots array *Patent Ukraine UA № 94699 Class C 30B 19/00, C 30B 29/00, H 01L 21/20. 20.09.2010-10.06.2011 bulletin №5*
[7] Maronchuk I I, Maronchuk I E, Sankovitch D, Lovchinov K and Dimova-Malinovska D 2014 *J. Phys.: Conf. Series* **558** 012049
[8] Kern W 1990 *J. Electrochemical Soc.* **137** 1887
[9] Kuhler U 1991 *Surface Sci.* **248** 321
[10] Dubrovsky V G 2009 *Theory and formation of epitaxial structures* (Moscow: MIR) (in Russian) p. 350
[11] Maronchuk I I, Bondaretz S V, Byukovskii S U, Petrash A N and Smirnov S B 2012 *Nano-heteroepitaxial structures with QDs for application in high-efficiency solar cells(deposition and characterization)* (in Russian) Scientific report № 0112U004939 (Sevastopol: SNUNEI)

Journal of Physics: Conference Series **700** (2016) 012044 doi:10.1088/1742-6596/700/1/012044

Electrochemically-deposited ZnO nanorods arrays on ITO seeding layers

M Petrov, K Lovchinov, H Nichev[1] and D Dimova-Malinovska

Central Laboratory of Solar Energy and New Energy Sources,
Bulgarian Academy of Sciences, 72 Tzarigradsko Chaussee, 1784 Sofia, Bulgaria

E-mail: nitschev@abv.bg

Abstract. In this paper, results are reported from the study of the structural and optical properties of ZnO nanostructured arrays with nanorods (NRs) grown electrochemically on ITO seeding layers with two different thicknesses deposited on a glass substrate. The properties are compared of the ZnO NRs grown on ITO seeding layers with resistivity 6.4×10^{-4} Ω cm, sheet resistance 80 Ω/□ and 8 Ω/□ and different ITO thicknesses – 80 and 800 nm, respectively. The optical spectra of reflectance, transmittance, diffused reflection and diffused transmission and the haze ratio in transmission and reflection of the samples carried out before and after ZnO NRs deposition are compared and discussed. The structures' properties are studied by scanning electron microscopy (SEM). The ZnO NRs arrays with increased values of diffused transmission and reflection compared to the seeding layer substrates can be used in thin-film solar cells for increasing the light trapping.

1. Introduction

The ZnO based nano-structured (NS) layers with nanometer-size features have highly developed surface areas per unit volume. In order to fabricate thin-film solar cells with increased efficiency, the ZnO films should possess such a size of the surface features that would ensure high light harvesting, light scattering and optical confinement [1,2]. One of the methods of fabricating ZnO layers with nanostructures is electrochemical deposition [1-4]. It offers some advantages – it is a low-cost non-vacuum method and is suitable for industrial large-scale application.

This paper reports results of the study of the optical and structural properties of electrochemically-deposited ZnO with nanorods (NRs) arrays on two different conductive ITO seeding layers with different thickness.

2. Experimental

ZnO layers were deposited by electrochemical growth in electrolyte – an aqueous solution of $ZnCl_2$ and KCl – using a three-electrode electrochemical cell and a saturated calomel electrode (SCE) as a reference electrode [1,2,4]. The concentration of the electrolyte used was $ZnCl_2$ (5×10^{-3} M), KCl (0.1 M) with pH ~ 6.5. The temperature of deposition was kept at 80 °C. Since the potential of Zn in the electrolyte is −1.05 V vs. SCE, the deposition process of ZnO was carried out at −1.0 V vs. SCE thus preventing a metal Zn deposition. Zinc peroxide (ZnO_2) was formed on the samples with bad

[1] To whom any correspondence should be addressed.

adhesion to the substrate at a redox potential higher than +0.4 V. Good quality ZnO films were obtained at a redox potential within the range +0.30 – +0.40 V vs. SCE. A spectrally-pure graphite plate electrode was used as an anode. The electrolyte was agitated by a magnetic stirrer. The duration of the process was 30 min. Highly-conductive ITO seeding layers on glass substrates with resistivity $6{,}4\times10^{-4}$ Ω cm, with two different ITO thicknesses – 80 nm and 800 nm – and sheet resistance 80 Ω/□ and 8 Ω/□, respectively, deposited by magnetron sputtering, were used as cathodes. The surface morphology of the deposited films were studied by a Philips 515 and a Jeol JSM-840A with $LaBa_6$ cathode scanning electron microscopes (SEM). The optical spectra were measured by a Shimadzu UV-3600 spectrophotometer in the range of 300 – 1200 nm employing a 60-mm integrating sphere in the cases of diffused transmission and reflection spectra measurements.

3. Results and discussion

Figure 1 shows SEM micrographs of the surface of ZnO grown on an 80-nm thick ITO seeding layer The pictures were taken at different magnifications – the markers correspond to 1 μm (figure 1 (a)) and 10 μm (figure 1 (b)). For comparison, a SEM image (marker corresponding to 200 nm) of the bare substrate – the ITO seeding layer – is shown as well (figure 1 (c)). The ZnO nanostructures have grown in the shape of hexagonal nanorods (NRs) oriented vertically with respect to the substrate (figure 1 (a)). They are distributed inhomogeneously on the surface of the bare ITO seeding layer forming large (between 30 μm and 900 μm in size) round islands with high-density grown NRs (about 200 – 300 nm in size) and areas sparsely populated with NRs between them (figure 1 (b)).

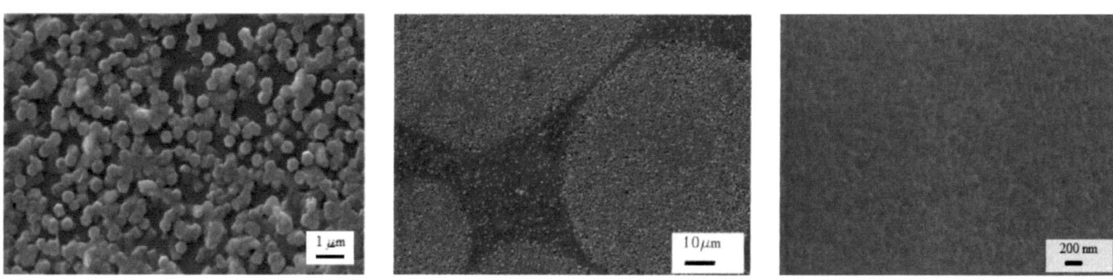

(a) (b) (c)

Figure 1. SEM images of the ZnO array grown on 80-nm thick ITO/glass bare substrate (a and b) and of the bare ITO/glass substrate. The markers correspond to 1μm (a), 10μm (b) and 200 nm (c).

SEM micrographs of the ZnO NRs array (at different magnifications) grown on the thicker ITO seeding layer (800 nm) are demonstrated in figure 2. Similarly to the previous case, an island-like ZnO NRs distribution is observed; however, the small-diameter NRs (100 – 200 nm) have grown oriented randomly with respect to the substrate surface (figure 2 (a)). Similar results have been reported for ZnO NRs grown on a 200-nm thick bare ITO substrate from an electrolyte based on $Zn(NO_3)_2$ and hexamethylenetetramine (HTM) [5]. The difference in the NRs growth could be explained by the different surface morphology of the bare ITO seeding layers as a result of their different thickness – the size of the nano-grains in the polycrystalline ITO thin films increases with the increase of their thickness. As is seen in figure 1 (c), the surface of the thinner ITO (80 nm) is very smooth with nanograins of sizes of 10 – 20 nm. However, the surface of the thicker ITO layer is very rough (figure 2 (c)) and contains large pyramidal grains sized between 100 – 200 nm with plates with well-formed shape, which causes the ZnO NRs to grow at different angles to the surface [6]. The formation of islands with a higher density of grown ZnO NRs on the surface of an ITO seeding layer could be explained by the presence of places (spots) on the surface of the seeding ITO layers with different wetting ability due to the inhomogeneous distribution of Sn that segregates on the surface of Sn-doped In_2O_3 [7, 8] and to the presence of oxygen dangling bonds on the ITO layer surface [9]. The

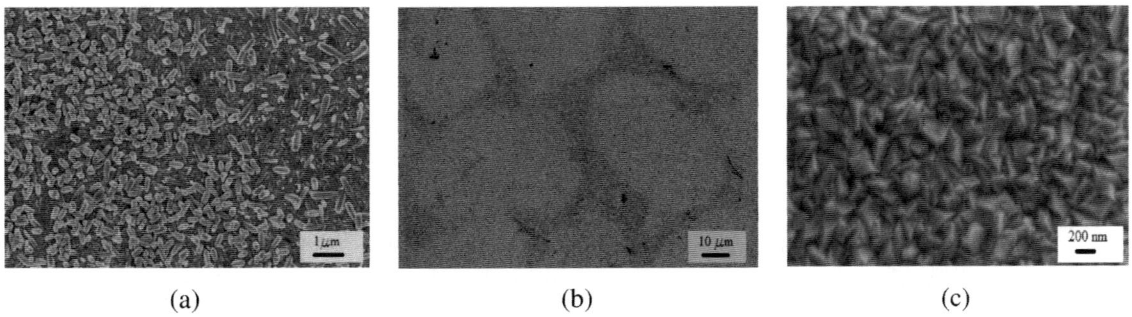

Figure 2. SEM pictures of the ZnO array grown on 800 nm thick ITO/glass bare substrate (a and b) and of the bare ITO/glass substrate. The markers correspond to 1 μm (a), 10 μm (b) and 200 nm (c).

large surface energy mismatch between the places on the bare substrate with lower wetting ability and the electrolyte makes it difficult to grow continuous ZnO films, so that the tendency to growth of islands is strong. In the places on the ITO surface with hydrophilic ability, the nucleation of the ZnO will be accelerated and the grown NRs will form islands with a denser distribution.

The optical spectra (reflectance, transmittance, diffused reflection and diffused transmission and the haze ratio in reflection and transmission) of the samples with grown ZnO NRs are shown in figures 3 and 4. For comparison, the corresponding spectra of the base ITO seeding substrates are shown as well. Note that the interference fringes in the spectra, which are due to the 800-nm thick bare ITO seeding substrate. One can observe that, upon the ZnO NRs growth, the values of reflectance and transmittance decrease, while the values of diffused reflection and diffused transmission and of the haze in reflection and of transmission increase. This tendency is much more pronounced in the case of the vertically-grown ZnO NRs on the thinner bare ITO seeding layer, where the size of the ZnO NRs (200 – 300 nm) are close to the optimal values calculated theoretically [10].

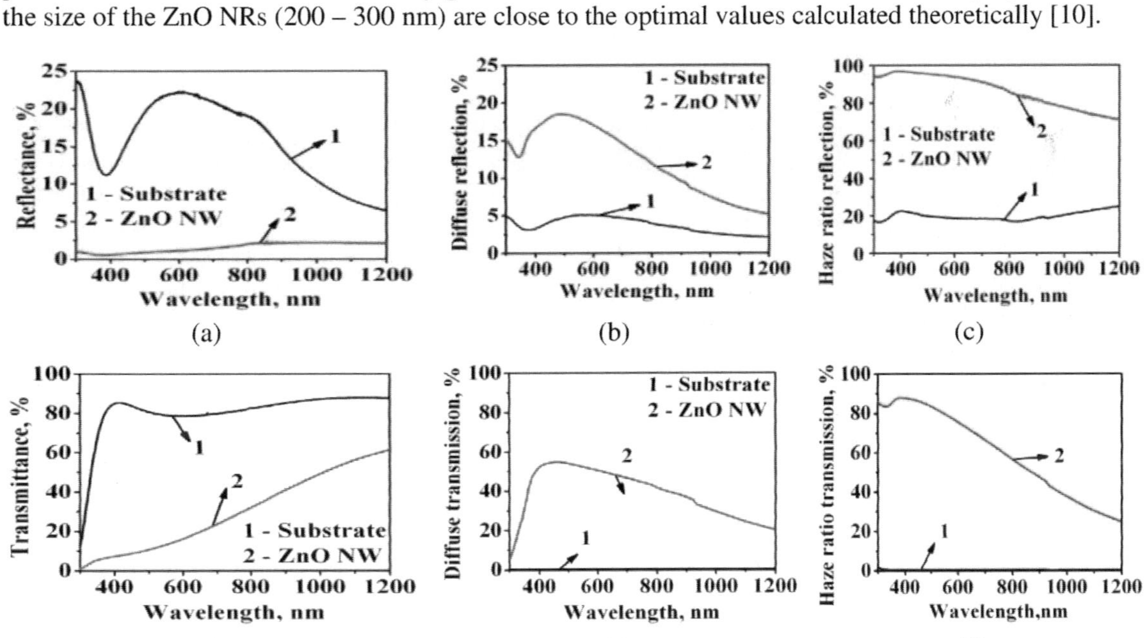

Figure 3. Spectra of reflectance (a), diffused reflection (b), haze in reflection (c), transmittance (d), diffused transmission (e) and haze in transmission (f) of samples with ZnO NRs (glass/ITO/ZnO NRs – curves 2) electrochemically deposited on 80-nm thick ITO seeding layer. For comparison, the corresponding spectra of the substrate glass/ITO (curves 1) are presented as well.

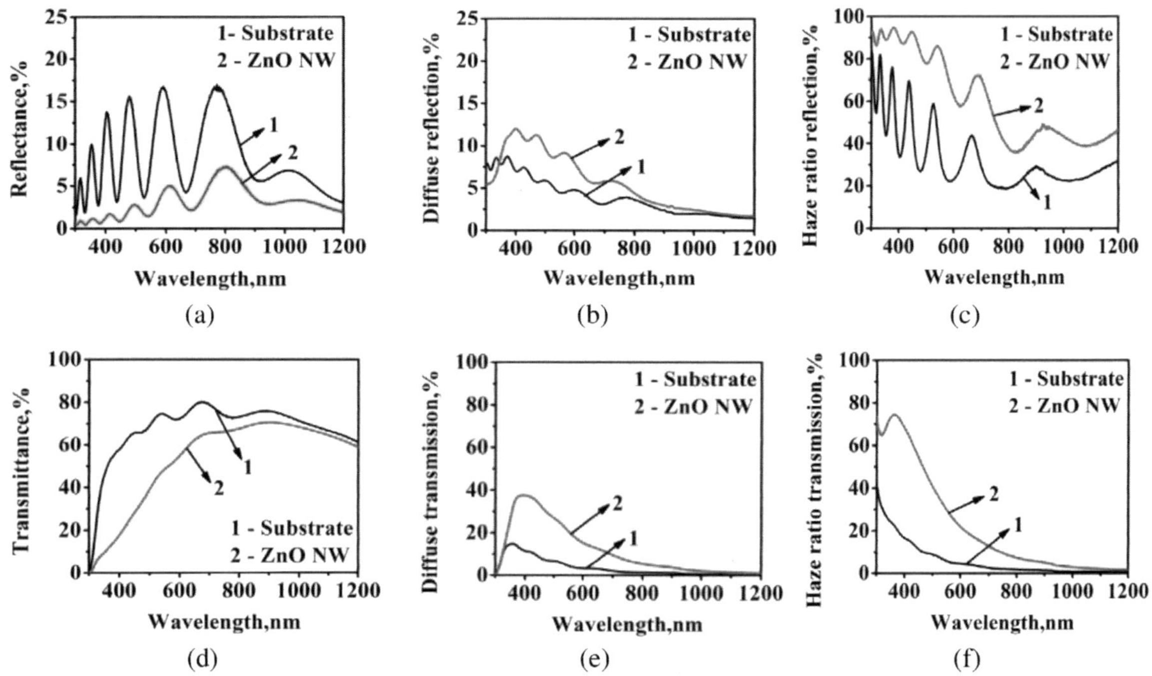

Figure 4. Spectra of reflectance (a), diffused reflection (b), haze in reflection (c), transmittance (d), diffused transmission (e) and haze in transmission (f) of samples with ZnO NRs (glass/ITO/ZnO NRs – curves 2) electrochemically deposited on 800-nm thick ITO seeding layer. For comparison, the corresponding spectra of the substrate glass/ITO (curves 1) are presented as well.

4. Conclusions

The study of the surface morphology of areas of electrochemically-grown ZnO nanorods on ITO/glass substrates demonstrated that the nanorods have grown with a different density distribution on the substrate surface. This observation is explained by the inhomogenious distribution of Sn and the presence of oxygen dangling bonds on the surface of ITO films resulting in the presence of spots with different wetting ability, which determines the different nucleation rate of ZnO NRs on the bare seeding surface. The growth of ZnO NRs on ITO/glass substrates result in an increase of the diffused transmission and reflection compared to the seeding layer substrates and can be used in thin-film solar cells for increasing their efficiency by improving the light trapping.

Acknowledgements

The work has been funded under the 7 FP of EU - project NanoPV No. 24331 and by the Bulgarian Scientific Fund – project No. DDVU 02/97.

References

[1] Petrov M, Lovchinov K, Mews M, Leendertz and Dimova-Malinovska D 2014 *Phys. Status Solidi* A **1**

[2] Zhang D, Petrov M, Dimova-Malinovska D, Dörenkämpe Mr, Bakker K, Schropp R and Soppe W 2014 *Proc. 29th EUOPVSEC* (22-26 Sept 2014 Amsterdam-Nederland) 1631

[3] Kopenkamp R, Boedecker K, Lux-Steiner M C, Poschenrieder M, Zenia F, Levy-Clement C and Wagner S 2000 *Appl. Phys. Lett.* **77** 2575

[4] Dimova-Malinovska D, Lovchinov K, Ganchev M, Angelov O, Graff J S and Ulyashin A 2013 *Phys. Stat. Sol.* A **210** 737

[5] Bubnova X, Peic A, Dimopoulos T, Abermann S, Rennhofer M, Akbarzadeh J, Peterlik H, Irista, Ertl P M, Zamini S and Bruckl H 2014 *Proc. 29th EUOPVSEC* (22-26 Sept 2014 Amsterdam-Nederland) 483

[6] Mari B, Mollar M, Mechour A, Hartiti B, Perales M and Cembrero J 2004 *Microelectr. J.* **35** 79

[7] Løvvik O M, Diplas S, Romanyuk A and Ulyashin A 2014 *JAP* **115** 083705

[8] Gassenbauer Y, Schafranek R, Klein A, Zafeiratos S, Havecker M, Knop-Gericke A and Schlogl R 2006 *Phys. Rev.* B **73** 24

[9] Cui J, Huang Q L, Veinot J C G, Yan H, Wang Q W, Hutchison G R, Richter A G, Evmenenko G, Dutta P and Marks T J 2002 *Langmuir* **18** 9958

[10] Geisendörfer S, Vehse M, Voss T, Richters J-P, Hanke B, von Meydel K and Agert C 2013 *Solar Energy Materials and Solar Cells* **111** 153

Deposition of carbon nanostructures on metal substrates at atmospheric pressure

Zh Dimitrov[1], M Nikovski and Zh Kiss'ovski

Faculty of Physics, St. Kl. Ohridski University of Sofia,
5 J. Bourchier blvd., 1164 Sofia, Bulgaria

E-mail: cozhiv@gmail.com

Abstract. The microwave-plasma-enhanced CVD of carbon nanostructures at atmospheric pressure allows shorter deposition times and reduces the complexity of the experimental set-up. In our study, the substrate temperature was varied in a wide range (300 – 700 C) using microwave plasma heating, as well as an additional heater. The distance between the substrate and the plasma flame was also varied in order to establish the conditions for an efficient deposition process, the latter being carried out at specific argon/hydrogen/methane gas mixtures. Optical measurements of the plasma flame spectrum were conducted to obtain the gas temperature and the plasma density and to analyze the existence of reactive species. The carbon nanostructures deposited on the metal samples were investigated by SEM. The relation between the morphology and the gas-discharge conditions is discussed.

1. Introduction

The carbon nanostructures (CNS) attract great interest with their unique electrical, mechanical and optical properties. CNS provide a new route to exploring the novel physics of two-dimensional materials. There are several types of CNS, the most intensively investigated ones being graphene, carbon nanotubes, nanoflakes, nanoribbons, nanowalls and vertically-standing few-layer graphene sheets [1]. Vertical few-layer graphene (FLG) sheets and carbon nanowalls (CNW) draw attention with their properties and specific morphology – intersecting vertical walls with sharp edges and a high surface-to-volume ratio. CNWs have potential applications in microelectronics as sensors, field emitters and supercapacitors [2-4]. Various methods can be used for preparation of graphene – exfoliation, chemical vapor deposition, epitaxial growth, supersonic spraying, intercalation etc. Vertical graphene structures are mainly synthesized by plasma-enhanced chemical-vapor deposition (PECVD). The deposition of FLGs and CNW is conducted usually at a low gas pressure in vacuum chambers, which results in a long period of deposition. A variety of gas discharge plasmas (DC, RF and microwave) are used for CNS deposition. CNSs are also successfully prepared at atmospheric pressure in a normal glow discharge [4] in $Ar/H_2/C_xH_y$ or N_2/C_xH_y gas mixtures. The microwave discharges are usually characterized by high-density plasma and relatively high gas temperature, which allows for efficient decomposition of hydrocarbons. Microwave plasmas are also used for PECVD of carbon nanostructures and, especially, for synthesis of carbon nanotubes [5].

[1] To whom any correspondence should be addressed.

Journal of Physics: Conference Series **700** (2016) 012045 doi:10.1088/1742-6596/700/1/012045

In this study, a novel plasma reactor was used for carbon nanostructures deposition at atmospheric gas pressure. A low-power microwave discharge in a quartz vessel [6] was used to create plasma in argon gas, with CH_4 as a precursor and additional H_2-gas to improve the deposition process. The chemical composition and diversity of radicals in the plasma jet were investigated by optical emission spectroscopy. The spectra obtained were used to determine the gas temperature, the plasma parameters and the presence of reactive species which are usually associated with the deposition process. The variation of the plasma parameters when varying the distance between the metal substrate and the tube of the plasma source are discussed and the carbon nanostructures deposited are investigated.

2. Experimental set-up and diagnostics

The plasma reactor (figure 1) consisted of a dielectric (quartz) chamber with cylindrical shape including two metal flanges (top and bottom) and a miniature microwave plasma source [6] attached to the bottom flange. The exciter of plasma surface waves had an outer diameter of 3 mm and a 1-mm thick steel capillary as inner conductor. A thin alumina ceramic tube was used as a discharge tube and as a dielectric in the coaxial exciter. This material is able to resist the high gas temperatures of the plasma.

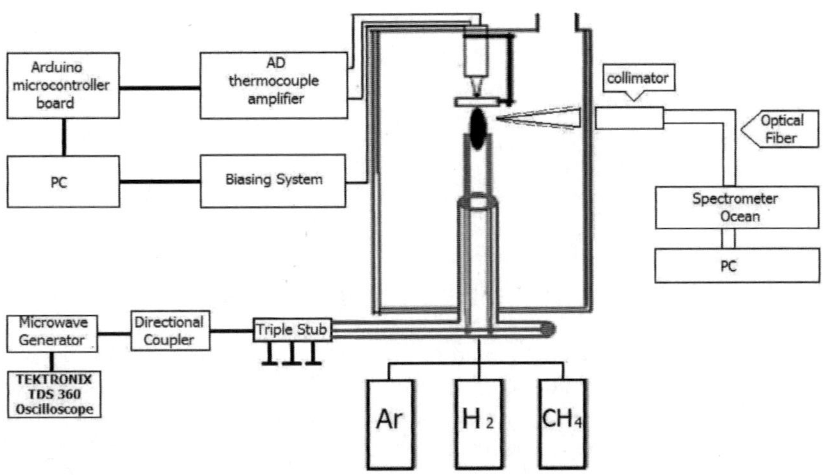

Figure 1. Experimental set-up and optical diagnostics system.

The plasma source was powered by a signal from an MPG-4M generator (0 – 100 W at a frequency of 2.45 GHz) through a coaxial line. The forward and reflected powers were measured by a Pasternack PE 2219-30 directional coupler. Precise matching to the MW source and minimum reflected power from the plasma source were achieved by a Maury 1878C triple stub. The plasma source could self-ignite in continuous and pulsed regimes at a specific threshold value of the input power. The pulses were generated at a rate in the range 70 – 700 Hz with a duration of 100 μs to 14 ms. The pulse parameters were recorded by a Tektronix TDS 360 oscilloscope. The gas mixture ($Ar/H_2/CH_4$) (187 sccm/10 sccm /3 sccm) was fed in through the capillary, the gas flow ratios being controlled by APEX-AX-MC mass-flow meters.

The substrate was positioned directly above the plasma jet. A moveable holder of the metallic substrate, made of stainless steel tube including a Teflon head and an additional heater, allowed us to vary the distance between the substrate and the plasma jet.

The temperature of the substrate was controlled by the current in the additional heater; however, the temperature was also influenced by the distance to the plasma source and by the microwave pulse parameters (frequency and duration of the microwave signal). Additional voltage (–100 V) was applied to assist the vertical growth of the carbon nanostructures. The direct current induced by the

DC voltage caused a further heating of the substrate. The temperature of the substrate was measured by an electronic system with a thermocouple [6].

The emission spectra from the excited species formed in the active region of the plasma jet were investigated by an HR Ocean Optics spectrometer (figure 2). The light from the plasma column was collected by a lens system (collimator) coupled with an optical fibre; the raw data was processed by a PC.

The gas temperature was obtained from the OH-emission band using LIFBASE software simulations compared with the experimentally acquired band. The electron temperature and density were investigated using the line-ratio method [7, 8].

3. Results

The plasma reactor was designed for deposition of carbon nanostructures on small Ni and Cu metallic plates. Before placing the plates in the reactor, they were cleaned for 5 minutes in 10% hydrofluoric acid bath in order to remove the organics and the oxides. After positioning the plate above the plasma source in the reactor, the substrates were subjected to plasma cleaning and activation in argon-hydrogen plasma at substrate temperatures below 400 °C for half an hour. During this process, the oxygen and the nitrogen should reach minimum concentrations in the reactor chamber. The experiments were carried out with a microwave signal at a rate of 100 Hz and pulse duration in the range 100 μs – 4 ms, which ensured a gradual increase of the substrate temperature and a stable plasma operation. The total gas flow was 200 sccm. The methane flow was set in the range 2 – 5 sccm during the deposition process. The analysis of the spectrum obtained by optical spectroscopy showed the presence of reactive species, as CH, C, and a Swan band, which proved that the decomposition of the methane was effective under these experimental conditions.

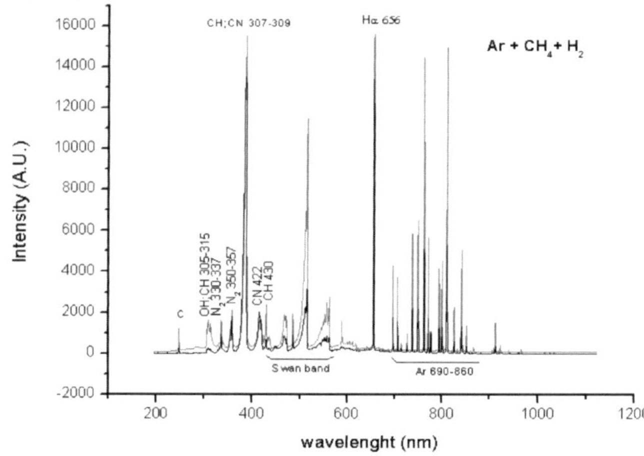

Figure 2. Optical spectrum of $Ar/CH_4/H_2$ plasma mixture.

The gas temperature in the plasma jet was investigated by OH emission band (305 nm – 315 nm) simulation using the LIFBASE software. Comparing the best fit with the experimental spectrum, the averaged gas temperatures obtained for different distances between the tube and the substrate, namely, 0.2, 0.5 and 1 cm, were 2100 K, 2000 K and 1400 K.

The simultaneous determination of the electron temperature and plasma density by the emission line ratios for low-temperature plasmas in argon gas was described in detail in [7, 8]. It is based on the collisional-radiative model (CRM) of excited particles. The model includes 21 rate balance equations for excited Ar atoms (Ar(1s5-1s1), Ar(2p10-2p1), Ar(2s3d), Ar(3p)), positive Ar^+ and Ar_2^+ ions and excited molecules Ar_2^*. Studies on the electron temperature and plasma density dependences on the line-ratios $I_{727.3}/I_{714.7}$ and $I_{738.4}/I_{727.3}$ [8] have shown that they are suitable for simultaneous determination of the plasma parameters. The isolines of the above-mentioned line-ratios had different behavior as T_e and n_e were varied. The cross-point of the isolines determined the T_e and n_e. The cross-point method was applied and initial results for different distances between the substrate and the discharge tube are shown in figure 3. The plasma parameters at a distance of 0.2 cm between the tube and substrate were close to the parameters obtained in the plasma column in the discharge tube [8, 9], while at 0.5 cm the electron temperature decreased slightly (1.5 $\pm$0.2 eV). At longer distances (1 cm), the values of the electron temperature and plasma density (1.4 $\pm$ 0.2 eV and 0.3×10^{15} cm^{-3}) were lower because the volume of the plasma column increases while the microwave energy for sustaining the discharge is constant. Substrates in the deposition process were Ni and Cu plates with an area up to 25 mm^2. A few Cu substrates were chosen for SEM imaging, which showed the presence of black layers visible by an optical microscope

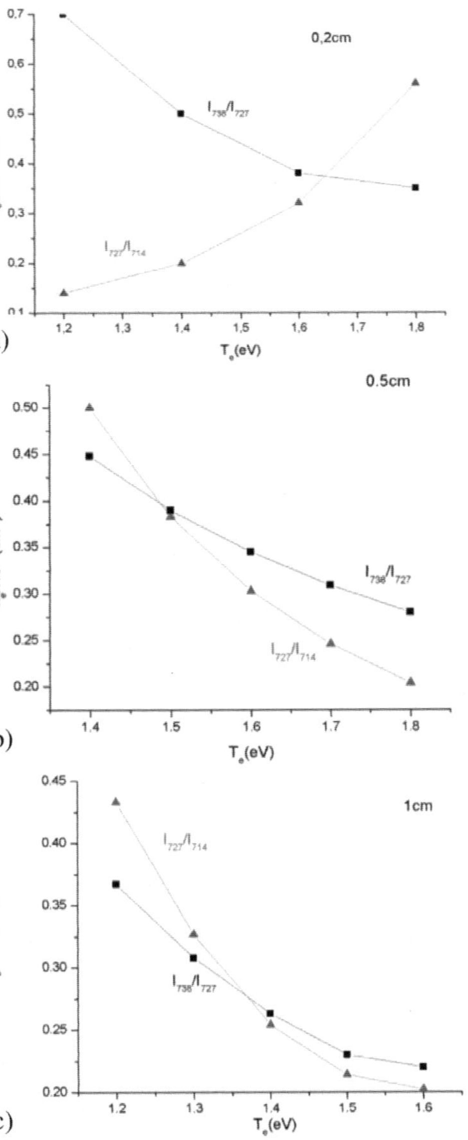

(a) (b) (c)

Figure 3. T_e, n_e dependence of the experimental line-intensity ratios $I_{727.3}/I_{714.7}$ and $I_{738.4}/I_{727.3}$ for distance between the end of the discharge tube and the substrate a) 0.2 cm; b) 0.5 cm; c) 1 cm.

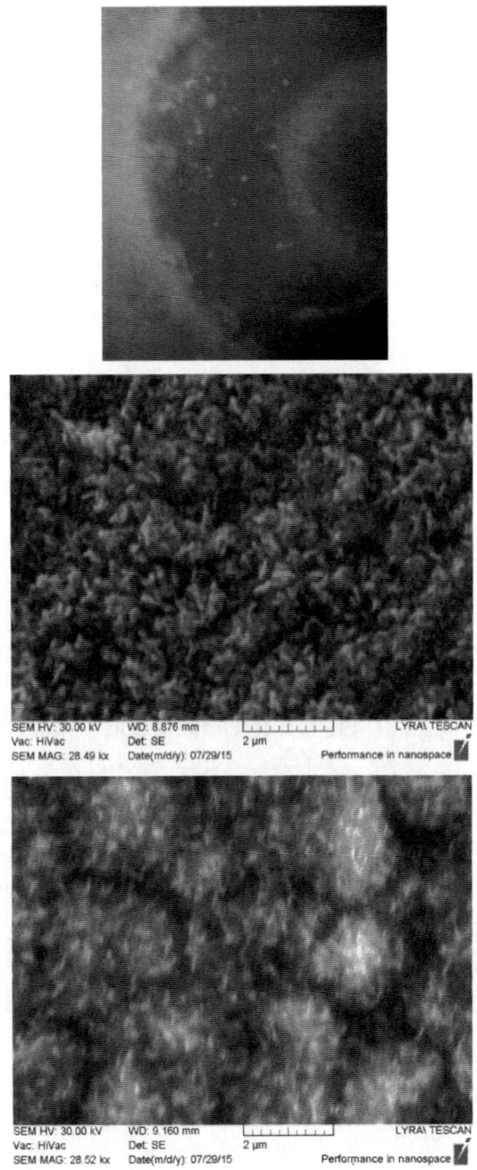

(a) (b) (c)

Figure 4. Deposited carbon structures: a) layer visible by optical microscopy on a copper substrate and b), c) SEM images of carbon nanostructures at different distances.

(figure 4 (a)). These observations showed that the black layers were in circumferences exactly around the surface directly exposed to the plasma jet. This result gives reason to assume that the most suitable conditions of deposition are in the discharge afterglow [10].

The SEM images on figure 4 (b) and 4 (c) exhibit initial forms of carbon nanowalls. The process of deposition needs optimization of the conditions. Similar structures as in figure 4 (c) have been observed in experiments (figure 5 in [11]) in a DC discharge at atmospheric pressure in the 3-min period from the beginning of the deposition process of carbon nanowalls. This is the reason for conducting experiments with varying the distance between the plasma source and substrate in order to optimize the deposition process with respect to the gas temperature and the density of the reactive species. Increasing the distance between the substrate and the discharge tube resulted in a decrease in the gas temperature. Further investigations by Raman spectroscopy and/or XPS are needed to characterize the deposited nanostructures. Future works will include additional investigations of the plasma parameters and the deposition process.

4. Conclusions

We presented experiments using a plasma reactor with a surface-wave-sustained discharge in an argon/hydrogen/methane gas mixture at atmospheric pressure and additional substrate heating for deposition of carbon nanostructures. By varying the substrate-to-discharge tube distance and analyzing the resulting changes in the plasma parameters, we demonstrated that this reactor is suitable for deposition of carbon nanostructures and that this process occurs mainly in the discharge afterglow.

Acknowledgments

This work was financially supported by the Science Fund of Sofia University, project N151/2015.

References

[1] Wu Y H, Yu T and Shen Z X 2010 *J. Appl. Phys.* **108** 071301
[2] Jiaxin Zheng, Lu Wang, Ruge Quhe, Qihang Liu, Hong Li, Dapeng Yu, Wai-Ning Mei, Junjie Shi, Zhengxiang Gao and Jing Lu 2013 *Sci. Rep.* **3** 1314
[3] Wenjing Yuan and Gaoquan Shi 2013 *J. Mater. Chem.* A **1** 10078
[4] Cai M, Outlaw R, Butler S and Miller S 2012 *Carbon* **50** 5481
[5] Jasek O, Synek P, Zajickova L, Elias M and Kudrle V 2010 *J. Electrical Eng.* **61** 311
[6] Kiss'ovski Zh, Djermanova N, Mitev D and Vachkov V 2014 *J. Phys.: Conf. Series* **514** 012007
[7] Iordanova S and Koleva I 2007 *Spectrochimica Acta* B **62** 344-56
[8] Kiss'ovski Zh, Ivanov A, Iordanova S and Koleva I 2011 *J. Phys.* D*: Appl. Phys.* **44** 205203
[9] Kiss'ovski Zh, Kolev M, Ivanov A, Lishev St and Koleva I 2009 *J. Phys.* D*: App. Phys.* **42** 182004
[10] Tatarova E, Henriques J, Luhrs C, Dias A, Philips J and Abrashev A 2013 *App. Phys. Lett.* **103** 134101
[11] Zheng Bo, Kehan Yu, Ganhua Lu, Pengxiang Wang, Shun Mao and Junhong Chen 2011 *Carbon* **49** 1849

Vacuum-deposited diphenyl-diketo-pyrrolopyrrole solar cell structures

G Georgieva[1], G Dobrikov[2], P Heinrichova[1], D Karashanova[3], D Dimov[3], M Vala[1], M Weiter[1] and I Zhivkov[1, 3,4]

[1] Materials Research Center, Faculty of Chemistry, Brno University of Technology, 118 Purkynova, 612 00 Brno, The Czech Republic
[2] Department of Microelectronics, Faculty of Electronic Engineering and Technologies, Technical University of Sofia, 8 Kl. Ohridski blvd., 1756 Sofia, Bulgaria
[3]Acad. J. Malinowski Institute of Optical Materials and Technologies, Bulgarian Academy of Sciences, Acad. G. Bonchev str., bl. 101/109, 1113 Sofia, Bulgaria

E-mail: zhivkov@fch.vutbr.cz

Abstract. Photoelectrical parameters were measured of solar cell ITO|PEDOT:PSS|composite| Al samples. The active composite film was deposited in vacuum by co-evaporation of 3,6-bis(5-(benzofuran-2-yl)thiophen-2-yl)-2,5-bis(2-ethylhexyl)pyrrolo[3,4-c]pyrrole-1,4(2H,5H)-dione (DPP(TBFu)$_2$) and fullerene (C_{60}). Additional DPP(TBFu)$_2$:C_{60} composite films were studied by spectroscopy in the ultraviolet and visible region (UV-VIS) and scanning electron microscopy (SEM). It was found that solvent annealing (SVA) of composite DPP(TBFu)$_2$:C_{60} vacuum-deposited films with tetrahydrofuran vapors improves the solar cell parameters by increasing the efficiency more than tenfold. This could be related to the more homogenized structure of the SVA composite film, as observed by SEM. The increased light absorption, as shown by UV-VIS spectroscopy, around the peak at 350 nm contributed to the better SVA solar cell performance. Photogeneration in the samples follows a monomolecular mechanism.

1. Introduction

Diphenyl-diketo-pyrrolopyrroles (DPP) are low-molecular-weight materials exhibiting outstanding luminescence and photoelectrical properties. DPPs have been the object of intensive research in view of pigment applications. They exhibit a variety of shades in the solid state and, especially, chemical, light and thermal stability [1]. Chemical tailoring of DPP allows modification of various properties, as, for example, the molar absorption coefficient and fluorescence quantum yield. Thus, the low-molecular-weight DPP derivatives have been extensively studied concerning their optical and photophysical properties [2–4]. The optical properties are related to the conjugated system which provides charge-carrier mobility across the material structure.

Potential applications of DPP derivatives have been reported as luminescent media in polymer matrices [5], solid-state dye lasers [6], OLED devices [7] and organic field-effect transistors [8]. Therefore, DPP based polymers, as well as small molecules, have found their application in the organic electronics.

[4] To whom any correspondence should be addressed.

Thin films of low-molecular-weight semiconductors are usually prepared by means of a variety of complex techniques, including physical or chemical vapor deposition, organic molecular-beam epitaxy or solution-based deposition techniques. The performance of the small molecular organic devices is highly sensitive to the films' morphology and processing conditions. Often, the solution-deposited active layers of devices (e.g. spin cast films) exhibit a high portion of microcrystallites and aggregates. The vapor deposition techniques provide high-quality crystalline films characterized by improved charge-transport properties compared with those of solution-deposited films. The vacuum technique has the advantage of a dry film deposition process excluding the usage of solvents. Moreover, this technique offers more opportunities to control such important film parameters as film thickness and uniformity. As a consequence, the films deposited exhibit more stable and reproducible properties in comparison with solution-deposited films.

This paper deals with 3,6-bis(5-(benzofuran-2-yl)thiophen-2-yl)-2,5-bis(2-ethylhexyl)pyrrolo[3,4-c]pyrrole-1,4(2H,5H)-dione, denoted hereinafter as $DPP(TBFu)_2$. Using this material as a donor for bulk-heterojunction-based solar cells leads to photovoltaic conversion efficiencies up to 4,4 % [9]. Fullerene (C_{60}) was used as and acceptor material, which facilitates the charge separation on the heterojunction. These results are strongly dependent on the film structure. For example, solvent annealing causes morphology and structure changes in bulk heterojunction devices by electron donating and electron accepting phase separation. This effect contributes to the better transport of photogenerated charge carriers from the interface to the electrodes of a solar cell device leading to higher photogeneration efficiency.

In the present work, $DPP(TBFu)_2$:C_{60} donor-acceptor composite films were prepared by vacuum co-deposition and the photoelectrical properties of ITO|PEDOT:PSS| $DPP(TBFu)_2$:C_{60}|Al solar cell samples were investigated.

2. Experimental details

The samples for photoelectrical measurements were prepared by the following procedure. Glass substrates with patterned ITO (Ossila) were preliminarily cleaned in an ultrasonic bath of 5% NaOH solution, then rinsed in water and subsequently washed twice in deionized water. Finally, ultrasonic treatment in an isopropanol bath was carried out for 10 minutes at room temperature. A PEDOT:PSS film was deposited by spin coating at 5500 rpm for 60 seconds. Subsequent annealing for 10 minutes at 150 °C in air and 10 minutes at 150 °C in nitrogen atmosphere was applied to remove the residual moisture.

Next, organic composite and electrode films were prepared in a clean room class C (M. Braun Inertgas-Systeme GmbH, Garching, Germany). $DPP(TBFu)_2$:C_{60} composite thin films were deposited by co-evaporation of $DPP(TBFu)_2$ and C_{60} on Ossila substrates through a proper deposition mask. The deposition was performed in a vacuum system from two thermally heated sources at evaporation temperatures of 180 C° and 415 °C. $DPP(TBFu)_2$ and C_{60} deposition rates of 2 and 1 Å/s, respectively, were controlled independently by a two-channel quartz crystal microbalance (Inficon SQC 310). Thus, a $DPP(TBFu)_2$:C_{60} mass ratio of 1.5:1 was achieved on a substrate rotated at 15 rpms.

Solvent vapor annealing (SVA) was carried out in nitrogen atmosphere at a temperature of 25 °C. Five ml of tetrahydrofuran (THF) was poured into a 40 mm glass Petri dish. The lid was closed for 20 min to achieve saturation vapor conditions. An as-cast film was attached on the back side of another Petri dish lid. The lids were quickly swapped on the solvent-containing Petri dish. In this way, the film was positioned about 5 mm above the solvent level in saturated THF vapors. After SVA for 20 s, the film was removed from the treatment chamber.

After that, the as-cast and the SVA samples were applied together without breaking the nitrogen atmosphere to a mask for Al deposition and put into the vacuum chamber. An Al cathode of 150 nm film thickness was thermally evaporated in a vacuum of 1×10^{-6} mbar at a deposition rate of 1 Å/s.

The solar cell samples of type ITO|PEDOT:PSS| $DPP(TBFu)_2$:C_{60}|Al were measured in nitrogen atmosphere. During the measurements, the samples were exposed to standard light produced by an LS0916 LOT Oriel class AAA solar simulator. This device provides a defined light beam according to the AM 1.5 standard. The light intensity of about 1000 mW cm^{-2} was controlled by an RR200O calibrated reference silicon cell (ReRa Solutions BV; Nijmegen, Netherlands). The current-voltage characteristics were measured by a Keithley 2601B electrometer (Keithley Instruments Inc., Cleveland, OH, USA).

Thin films for optical characterization were deposited on soda-lime-silica glass. The absorption spectra of the layers were measured in the range of 300 – 800 nm by a Varian Cary~50 UV-VIS spectrophotometer. The morphology of the as-cast and the SVA films was investigated by a Philips 515 scanning electron microscope (SEM) at an accelerating voltage of 25 kV. The samples for SEM characterization were deposited on microscope glasses. Before the SEM study, the sample surface was covered consecutively by carbon and palladium/gold films. The thickness of the layers was measured by a Decktak XT profilometer.

3. Results and discussion

3.1. Surface morphology study

Figure 1 presents SEM images of as-cast (a) and SVA in THF vapors for 20 s (b) vacuum-deposited DPP(TBFu)$_2$:C$_{60}$ composite thin films. In both cases, uniform films without visible pinholes are seen. The freshly-deposited film exhibits a relatively rough surface with a grain size of about 300 – 500 nm.

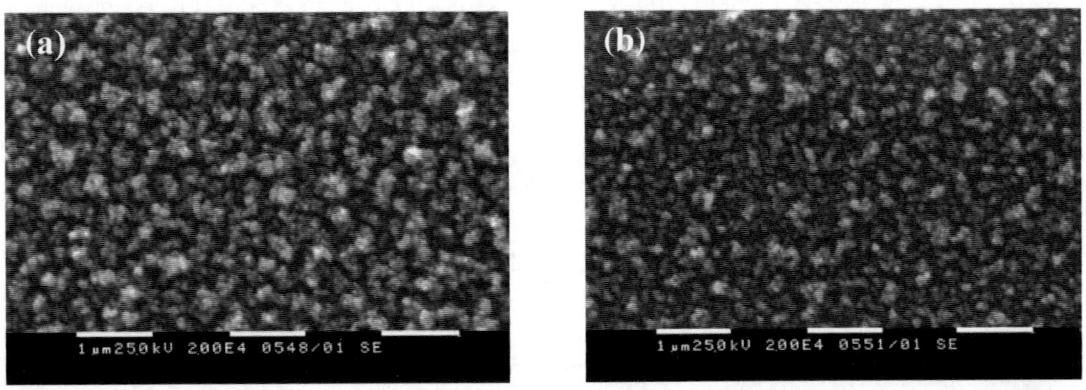

Figure 1. SEM images of vacuum-deposited DPP(TBFu)$_2$:C$_{60}$ composite thin films (a) as-cast; (b) annealed in THF vapors for 20 s.

The film treated for 20 s in THF vapors demonstrates a similar morphology, but apparently a smoother surface. The predominant grain size is about 100 – 200 nm. It could be assumed that the treatment by THF vapors partially dissolves the soluble DPP(TBFu)$_2$ clusters leading to a flatter film surface. A partial segregation of the composite could also be expected with a predominant appearance of the C$_{60}$ material on the film surface [9].

3.2. UV-VIS spectroscopy

In figure 2, optical absorption spectra of pure DPP(TBFu)$_2$ and C$_{60}$ films are presented. Curve (a) shows typical DPP(TBFu)$_2$ spectra [10] consisting of a broad doublet (500 – 750 nm) and a relatively narrow peak positioned at about 350 nm. This characteristic spectrum shows that no apparent destruction of the material is taking place. Curve (b) presents also a typical C$_{60}$ spectrum [11]. Comparing the two spectra in figure 2, it could be expected that preparing DPP(TBFu)$_2$:C$_{60}$ composite will combine the spectra in a way extending the light absorption to the UV-VIS region.

In figure 4, curve (a) is a typical spectrum of an as-cast DPP(TBFu)$_2$:C$_{60}$ composite film. As expected, increased absorption is observed, especially around the 350-nm peak. This effect should be related to the absorption by C$_{60}$.

Treatment in THF vapors for 20 s (figure 3, curve (b)) causes a decrease of the strength of the "red" positioned peak in the doublet and a remarkable increase of the peak strength around 350 nm. Thus, the impact of solvent annealing by THF on the samples' optical spectra is similar to that of thermal annealing.

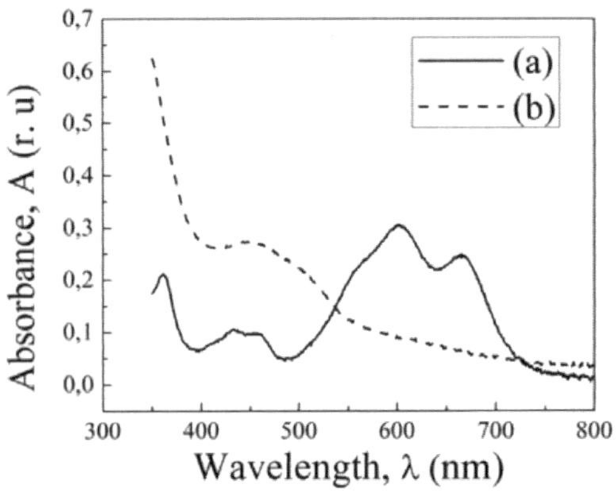

Figure 2. Optical spectra of (a) $DPP(TBFu)_2$ and (b) C_{60} vacuum deposited films.

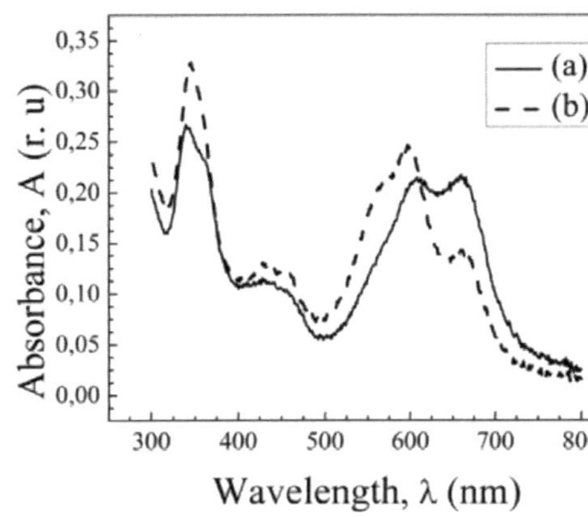

Figure 3. Optical spectra of (a) as-deposited $DPP(TBFu)_2$:C_{60} composite film and (b) after exposure to THF vapors for 20 s.

3.3. Solar cell measurements

3.3.1. Current voltage characteristics. The parameters of ITO|PEDOT:PSS|$DPP(TBFu)_2$:C_{60}|Al solar cell samples were calculated from the *I-V* characteristics measured in dark and under light exposure. The results are presented in table 1.

Table 1. Parameters* of the solar cell samples calculated from *I-V* measurements.

Sample	J_{sc} (mA/cm^2)	V_{oc} (mV)	FF (%)	η (%)
As-cast	1.0±0.03	695±2	22±0.5	0.20±0.01
SVA	6.5±0.04	730±3	50±0.7	2.2±0.05

* J_{sc} stands for current density; V_{oc}, for open-circuit voltage; FF, for fill factor; η, for efficiency.

It is clearly seen that the SVA improves the solar cell parameters. In particular, the efficiency is increased more than tenfold. This effect could be related to the more homogenized structure of the $DPP(TBFu)_2$:C_{60} SVA composite film, as observed by SEM. The increased UV-VIS absorption, especially around the peak positioned near 350 nm, also contributed to the better SVA solar cell performance.

The improvement of the solar cell parameters could also be related to a partial segregation of the $DPP(TBFu)_2$:C_{60} SVA composite film. This effect has been discussed with respect to similar measurements of $DPP(TBFu)_2$:$PC_{60}BM$ composite solar cells [12]. The SVA influence on vacuum deposited $DPP(TBFu)_2$:C_{60} composite solar cell samples seems to be similar. This phenomenon needs more extensive investigations.

3.3.2. Dependence on the light intensity. The dependence of the photocurrent density on the light intensity is plotted in a log-log scale in figure 4. It is clearly seen that the graph follows a linear dependence. The linear regression performed shows a slope (γ) of 1.0 ± 0.017. The dependence $J=I^{\gamma}$, where γ is close to 1 indicates a typical monomolecular mechanism of charge carrier generation under light exposure.

Figure 5 is a plot of the open-circuit voltage (V_{oc}) as a function of the light intensity. The graph follows a nonlinear dependence. This phenomenon could not be connected with the predominant influence of the shunt or serial resistance. This nonlinearity is more probably connected to the nonlinear changes in the electric field in the sample as a result of the changed conditions for charge carrier generation and recombination under the influence of the light.

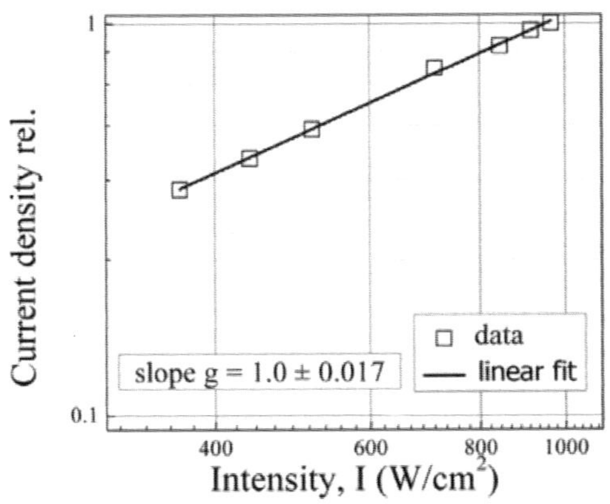

Figure 4. Dependence of the photocurrent density on the light intensity measured in ITO|PEDOT:PSS|DPP(TBFu)$_2$:C_{60}|Al samples.

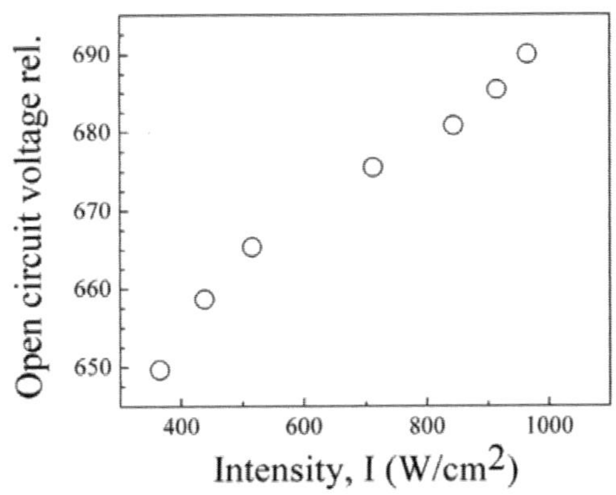

Figure 5. Dependence of the open-circuit voltage on the light intensity in ITO|PEDOT:PSS|DPP(TBFu)$_2$:C_{60}|Al samples.

4. Conclusions

Solvent annealing by THF vapors of vacuum-deposited DPP(TBFu)$_2$:C_{60} composite films improves the solar cell parameters of ITO|PEDOT:PSS|DPP(TBFu)$_2$:C_{60}|Al samples. In particular, the efficiency is increased more than tenfold. This effect could be related to the more homogenized structure of the DPP(TBFu)$_2$:C_{60} SVA composite film, as observed by SEM.

The increased light absorption, as shown by UV-VIS spectroscopy, around the peak positioned near 350 nm also contributed to the better SVA solar cell performance.

The photogeneration in the samples follows a monomolecular mechanism.

Acknowledgement

This work was supported by the Czech Science Foundation under project No. 15-05095S and research infrastructure project MŠMT No. LO1211.

The authors would also like to acknowledge the project BGBG161PO003-1.2.04-0034-C0001 with the Ministry of Economy of Bulgaria for supplying the preparation technique for SEM samples.

D. Dimov and G. Dobrikov would like to thank Erasmus and CEEPUS programs, respectively, for the financial support of their mobility abroad.

References

[1] Zhang G, Liu K, Fan H, Li Ya, Zhan X, Li Yo and Yang M 2009 *Synthetic Metals* **159** 1991
[2] Fukuda M, Kodama K, Yamamoto H and Mito K 2002 *Dyes Pigments* **53** 67
[3] Mizuguchi J and Rochat A C 1988 *J. Imag. Sci.* **32** 135
[4] Mizuguchi J 2000 *J. Phys. Chem.* A **104** 1817
[5] Smet M, Metten B and Dehaen W 2001 *Tetrahedron Lett.* **42** 6527
[6] Fukuda M, Kodama K, Yamamoto H and Mito K 2004 *Dyes Pigments* **63** 115
[7] Yamamoto H and Dan N 2004 Fluorescent diketopyrrolopyrroles *Patent WO 2004/090046*
[8] Yanagisawa H, Mizuguchi J, Aramakil S and Sakai Y 2008 *Jpn. J. Appl. Phys.* **47**/6 4728
[9] Walker B, Tamayo A, Dang X, Zalar P, Seo J, Garcia A, Tantiwiwat M and Nguyen T 2009 *Adv. Funct. Mater.* **19** 3063
[10] Liu J, Walker B, Tamayo A, Zhang Y and Nguyen T 2011 *Adv. Energy Mater.* **1** 610
[11] Fravventura M, Hwang J, Suijkerbuijk J, Erk P, Siebbeles L and Savenije T 2012 *J. Phys. Chem. Lett.* **3** 2367
[12] Kovalenko A, Stoyanova D, Pospisil J, Zhivkov I, Fekete L, Karashanova D, Kratochvílová I, Vala M and Weiter M 2015 *Int. J. Photoenergy* Article ID 238981, I:10.1155/2015/238981

Morphological and optical investigation of Sol-Gel ZnO films

T Ivanova[1,3], A Harizanova[1] and A Petrova[2]

[1]Central Laboratory of Solar Energy and New Energy Sources,
Bulgarian Academy of Sciences, 72 Tzarigradsko Chaussee, 1784 Sofia, Bulgaria
[2] Institute for Space Research and Technology, Bulgarian Academy of Sciences,
Acad G. Bonchev Str., Bl. 1, 1113 Sofia, Bulgaria

E-mail: tativan@phys.bas.bg

Abstract. This paper presents morphological and optical studies of the properties of spin-coated ZnO films depending on the annealing temperatures. The films microstructure was explored using a scanning nano-hardness measuring device of the NanoScan family, based on the principles of atomic force microscopy, in a constant frequency mode. The surface study revealed that the root-mean-square (RMS) surface roughness of 985.64×985.64 nm ZnO films becomes greater with the increase of the annealing temperature, but the film surface remains uniform and smooth. The results were confirmed by XRD analysis, which demonstrated that the crystallite size grew from 25 to 36 nm with the thermal treatments. The ZnO films possessed high transmittance in the visible spectral range and the optical band gaps in ZnO films varied from 3.25 eV to 3.52 eV. The optical and morphological properties of the ZnO films formed on Si and quartz substrates are very good. The sol-gel approach proposed for deposition of nanostructured ZnO films is promising for applications in optoelectronic devices or solar cells.

1. Introduction

Zinc oxide has been widely studied as one of the binary II–VI semiconductor compounds with a hexagonal wurtzite structure possessing an optical band gap of 3,37 eV and a large exciton energy [1]. Other interesting properties of ZnO include its optical, acoustical and electrical properties, high photosensitivity, high optical transparency in the visible region and stability, which can be applied in the fields of electronics, optoelectronics and sensors [2, 3]. Some potential applications are as piezoelectric transducers, optical waveguides, acousto-optic media, surface acoustic wave devices, conductive gas sensors, transparent conductive electrodes, solar cell windows, and varistors [4].

Currently, many methods are used to prepare ZnO films, such as molecular beam epitaxy (MBE), chemical vapor deposition, electrochemical deposition, pulsed laser deposition (PLD), reactive evaporation, magnetron sputtering, spray pyrolysis and atomic layer deposition [5, 6]. Among these deposition methods, the sol-gel technology has proved to be a low-cost and flexible technique, offering advantages such as simple deposition equipment, easy composition adjustment and incorporation of different dopants, fabrication of uniform films over substrates of large area [7].

In this work, a sol-gel technology was applied to achieve uniform and transparent ZnO thin films. Two types of sol solutions differing in the solvent used were developed. The structural and optical

[3] To whom any correspondence should be addressed.

properties were compared of ZnO films prepared from the two sol solutions by spin coating. The films morphology was studied by a scanning nano-hardness measuring device depending on the different thermal treatments. Their vibrational and optical properties were also investigated by using FTIR and UV-VIS spectroscopy.

2. Experimental

Two types of sol solutions were developed using zinc acetate dehydrate with concentration of 0,4 M as zinc precursor.

Solution 1) The zinc acetate was dissolved in isopropanol. The complexing agent and stabilizer used was trietanolamine (TEA). The molar ratio TEA/Zn was fixed to 1.

Solution 2) The solvent used was absolute ethanol and the corresponding complexing agent was trietanolamine (TEA). The molar ratio TEA/Zn was fixed to 1.

The films were deposited by spin coating at 4000 rpm on glass and Si substrates. The films were subjected to annealing at the temperatures of 400, 500 and 600 °C. The thickness of the films obtained from the two solutions was close, which made their properties comparable. The thickness values were in the range of 130 to 100 nm depending on the annealing temperatures.

The films were investigated by a NanoScan scanning nano-hardness measuring device, which is based on the principles of scanning force microscopy. The NanoScan data [8] can yield complex information on the surface roughness, chemical composition and local surface hardness. The FTIR measurements were performed in the spectral region 350-4000 cm^{-1} by an IRPrestige-21 Shimadzu FTIR Spectrophotometer. The optical measurements were conducted by using a UV-VIS-NIR Shimadzu 3600 spectrophotometer.

3. Results and discussions

Previous XRD study [9] have shown that sol-gel ZnO films (Solution 2) start to crystallize at the relatively low temperature of 400 °C. The degree of crystallization increases with the annealing temperature. The XRD diffraction pattern corresponds to the hexagonal wurtzite structure (JCPDS 01-07-8070) with no preferential orientation observed. The crystallite sizes change from 22 nm (400 °C annealing) to 35 nm (600 °C annealing). The XRD study of ZnO films prepared from solution 1 (600 °C) reveals a polycrystalline structure with a wurtzite phase. It exhibits a lower degree of crystallization (lower intensity and broader XRD peaks) compared to the corresponding ZnO films deposited from solution 2. The estimated crystallite size was 31 nm, revealing smaller crystallites compared to ZnO film (solution 2).

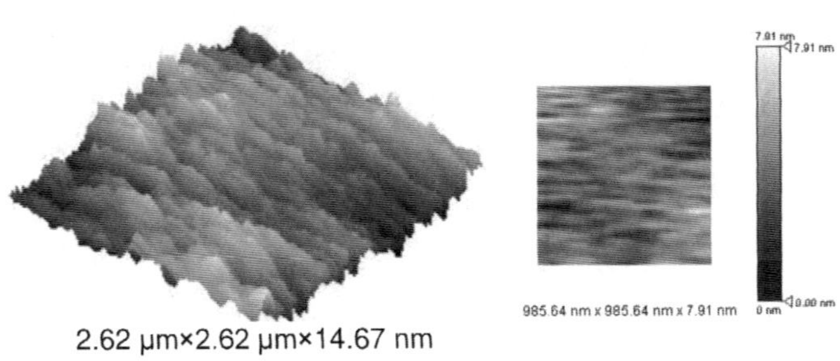

Figure 1. Three-dimensional and plane view images of sol-gel ZnO films obtained from solution 1 and annealed at 600°C.

Our morphological study was applied to ZnO films obtained from solution 1 and treated at 600 °C (figure 1), and ZnO films (solution 2) annealed at 400, 500 and 600°C (figure 2). Table 1 presents the roughness parameters determined for the ZnO films. The RMS parameter is the standard deviation of the surface heights values within a given area, where R_a is the average roughness (the arithmetic mean of the absolute values of the height of the surface profile $Z(x)$) and R_z is the arithmetic mean of the highest peaks added to the deepest valleys over the evaluation length measured. The maximum roughness is defined as the distance of the highest to the lowest point within the image. This parameter is very sensitive to single peaks and valleys in the surface, whereas the R_a value is only slightly affected by a single surface feature. These parameters are derived from the NanoScan software.

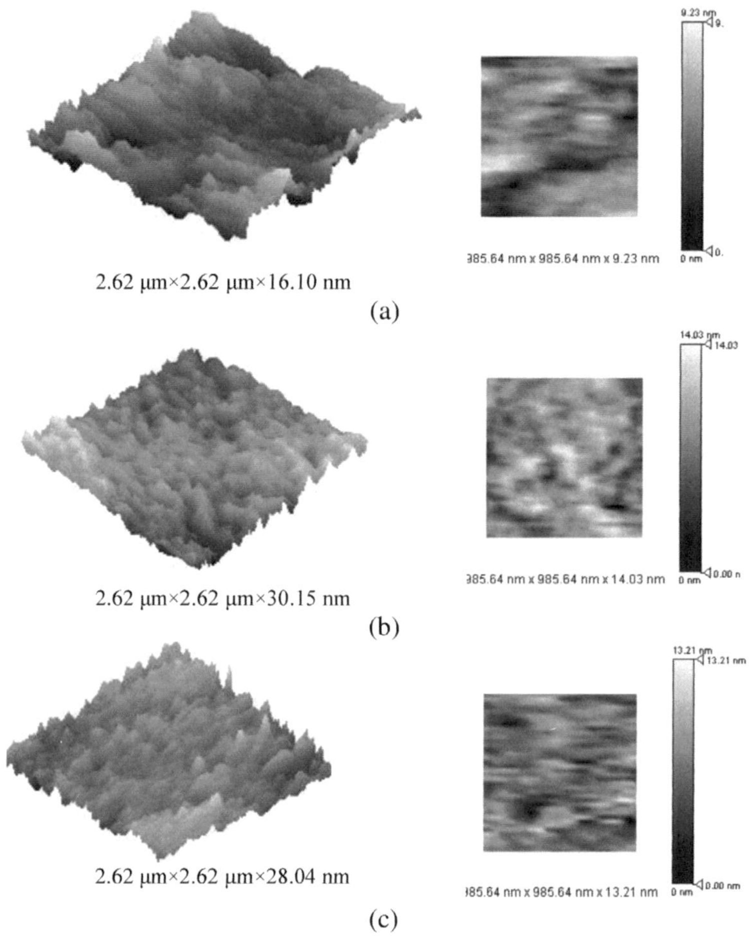

Figure 2. Surface images of ZnO films (solution 2), annealed at 400 (a), 500 (b) and 600°C (c).

The AFM micrographs reveal that the sol-gel ZnO films possess a smooth and uniform surface. The roughness is influenced by the thermal treatment. The RMS values rise with the increase of the annealing temperature. This is in agreement with the XRD results, which indicate that the crystallites grow larger with the annealing. The surface morphology of the films, obtained from the two solutions, is very similar, but the ZnO films obtained from solution 1 show a lower roughness.

Table 1. AFM parameters for sol-gel ZnO films.

Solution	Thermal treatment	RMS [nm]	R_a [nm]	R_z [nm]	Max h [nm]
1	600°C	3.48	2.77	14.78	33.12
2	400°C	4.56	3.52	17.80	37.75
2	500°C	6.32	4.96	20.95	44.27
2	600°C	4.58	3.38	19.83	43.00

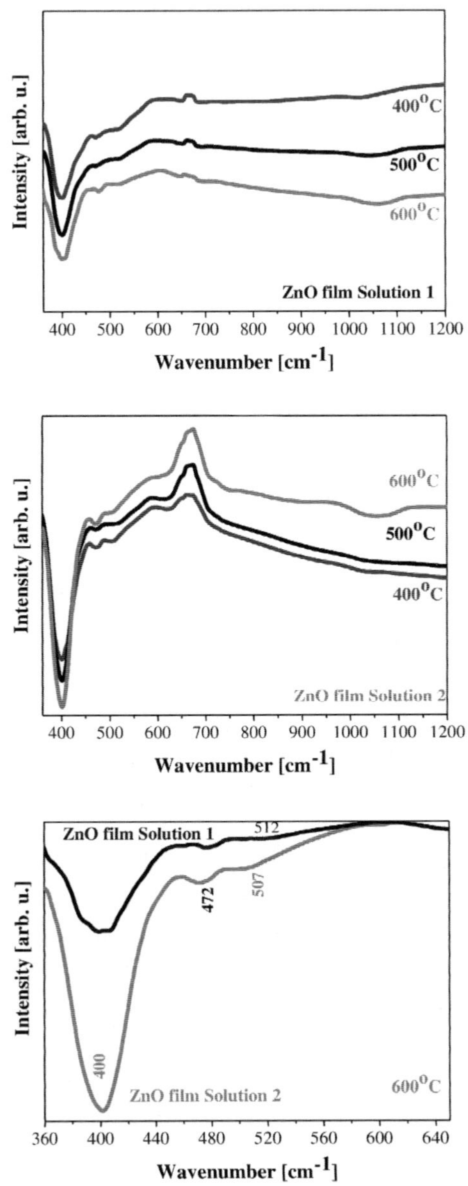

Figure 3. FTIR spectra of the sol-gel ZnO films annealed at different temperatures and a comparison of ZnO films prepared from the two solutions and annealed at 600°C.

The vibrational properties of the ZnO films were studied by FTIR spectroscopy in the spectral range 350-4000 cm^{-1}. The absorption bands positions, shape and number depend on the films' chemical composition, crystalline structure and morphology [10]. The FTIR spectra of the films are given in figure 3. The spectra are narrower in the spectral region of 350-1200 cm^{-1}, where metal – oxygen vibrations are expected. The weak broad bands around 1058 cm^{-1} can be assigned to Si-O-Si bonds originating from the Si substrate. The absorption bands observed for the films obtained from the two solutions are at close positions; they differ only in the intensity of the main strong line.

The absorption band of ZnO films (Solution 2) is considerably more intense compared to the other ZnO films. This agrees with the XRD study showing that the ZnO films from solution 1 present a lower crystallinity. The weak broad peaks at 510 cm^{-1} (ZnO, Solution 1) and near 502 cm^{-1} (solution 2) are related to crystalline wurtzite ZnO. The strong main bands of ZnO around 404 cm^{-1} are attributed to Zn-O bonding in wurtzite ZnO. The IR active modes (408 and 513 cm^{-1}) are theoretically confirmed for the wurtzite phase [11]. The weaker bands at 425, 470 cm^{-1} are assigned to Zn-O stretching vibrations [12]; they appear in the FTIR spectra of ZnO (solution 2). The FTIR analysis reveals that the ZnO films possess different vibrational characteristics depending on the sol solution indicating that the solvent influences considerably the vibrational properties of the resulting sol-gel ZnO films.

The optical properties of the ZnO films were determined by spectrophotometric measurements (see figure 4). The transmittance spectra reveal very different optical behavior depending on the sol solution used. The ZnO films (solution 1) show an almost negligible change in the optical transparency, which reaches 91 % at 550 nm. On the other hand, the ZnO films deposited from solution 2 reveal a clear tendency of decreasing transmittance with the increase of the annealing temperature. The ZnO film treated at 400 °C shows a transparency of 91 % at the wavelength of 550 nm, while the films annealed at 500 and 600 °C reveal transmittance values of 86 and 80% (λ = 550 nm), respectively. This reduction of the transmittance in the visible range can be explained by the increase in the film crystallinity and bigger crystallites, causing scattering of the incident light. The same tendency of film transmittance decrease with the annealing temperature has been reported by other authors [13]. The comparison of the ZnO films prepared using the two solutions (figure 5a) shows that the transparency of ZnO (solution 1) is considerably higher, which is in agreement with the AFM investigation that reveal a smoother surface and lower RMS (table 1).

The small shoulders below the absorption edge indicate excitonic absorption. They differ for the two sets of sol-gel ZnO films. The ZnO films (solution 1) exhibit excitonic peaks at 345-350 nm, while the peaks of the ZnO films from solution 2 are located at 345 nm. However, these shoulders show a substantial blue shift relative to bulk ZnO (~373 nm), which could be due to the confinement effects in films [14]. The presence of excitonic absorption is an indication that the sol-gel ZnO films possess a high optical quality.

Using the data of the transmittance measurements one can estimate the optical band gap considering a direct gap semiconductor or to determine it by the second derivative of the transmittance spectrum [15, 16]. As can be seen from figure 5b, the optical band gaps of the ZnO films derived from solution 1 are larger and increase slightly with the annealing temperature. On the other hand, the optical band gap of ZnO films (solution 2) tends to decrease after a high temperature annealing. Other authors [17] have observed a similar reduction of the optical band gap as the annealing temperature is increased, which they assigned to the improved crystallinity of the films and the lower number of defects in them. It is known that the energy band gap of the ZnO film could be affected by the residual strain defects and grain size confinement [18]. The values are smaller than the optical band gap 3.37 eV of bulk ZnO.

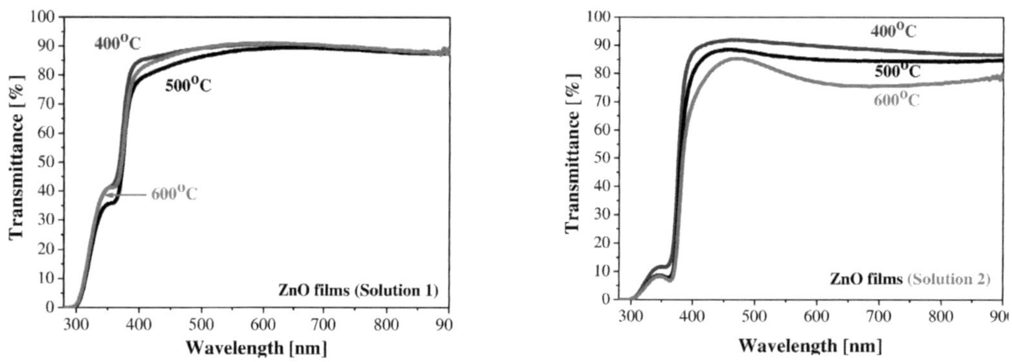

Figure 4. UV-VIS transmittance spectra of ZnO films annealed at different temperatures.

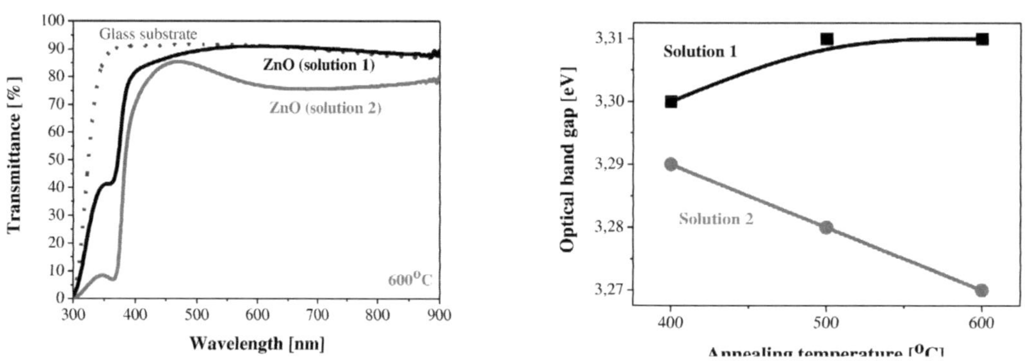

(a) (b)

Figure 5. Comparison of ZnO films obtained from the two solutions and annealed at 600°C (a). The spectrum of bare glass is also given. Optical band gaps of the sol-gel ZnO films (b).

4. Conclusions

This works presents an easy to perform sol-gel process for depositing ZnO films, which proved to be a very flexible approach to modifying the structural, vibrational and optical properties by only changing the solvent used in the sol solution. The surface morphologies of ZnO films are very smooth with a tendency to become rougher with increasing the annealing temperatures up to 600°C. The FTIR investigation shows the appearance of absorption features indicative of hexagonal wurtzite ZnO. The optical study reveals that the sol-gel ZnO films are very transparent (up to 90 %) in the visible spectral range with an optical band gap ranging from 3.27 to 3.31 eV, characteristic for ZnO.

References

[1] Heo J H, Ryu H and Lee W-J 2013 *J. Industrial Engin. Chem.* **19** 1638
[2] Ilican S, Caglar Y and Caglar M 2008 *J. Optoelectr. Adv. Mater.* **10** 2578
[3] Znaidi L 2010 *Mater. Sci. Engin.* B **174**/1-3 18
[4] Zhong J B, Li J Zh, Xiao Zh H, Hu W, Zhou X B and Zheng X W 2013 *Mater. Lett.* **91** 301
[5] Gomes M A, Valerio M E G, Rey J F Q and Macedo Z S 2013 *Mater. Chem. Phys.* **142**/1 325
[6] Yuan N Y, Wang S Y, Wang X Q, Chen G G and Ding J N 2013 *J. Crystal Growth* **366** 43
[7] Cui L, Wang G-G, Zhang H-Y, Sun R, Kuang X-P and Han J-C 2013 *Ceram. Inter.* **39** 3261
[8] http://www.nanoscan.info
[9] Ivanova T, Harizanova A, Koutzarova T and Vertruyen B 2010 *Mater. Lett.* **64**/10 1147
[10] Yang Z, Ye Z, Xu Z and Zhao B 2009 *Physica* E **42** 116
[11] Yuvaraj D and Rao K N 2008 *Vacuum* **82** 1274
[12] Music S, Dragovic D and Popovic S 2007 *J. Alloys Compounds* **429** 242
[13] Xue S W, Zu X T, Deng H X, Xiang X, Zhang L and Deng H 2008 *J. Alloys Compounds* **448** 21
[14] Liu Y L, Liu Y C, Liu Y, Shen D, Lu Y, Zhang J Y and Fan X W 2002 *Physica* B **322** 31
[15] Wang M, Kim E, Kim S, Yoo I, Shin E, Hahn S and Park Ch 2008 *Thin Solid Films* **516** 1124
[16] Xu L, Zheng G, Miao J and Xian F 2012 *Appl. Surf. Sci* **258** 7761
[17] Lv J, Gong W, Huang K, Zhu J, Song X and Sun Z 2011 *Superlattices Microstruct.* **50** 98
[18] Wahab H, Salama A, El-Saeid A, Nur O, Willander M and Battisha I 2013 *Results Phys.* **3** 46

Optical characterization of Sol-Gel ZnO:Al thin films

T Ivanova[1], A Harizanova[1,4], T Koutzarova[2] and B Vertruyen[3]

[1]Central Laboratory of Solar Energy and New Energy Sources,
Bulgarian Academy of Sciences, 72 Tzarigradsko Chaussee, 1784 Sofia, Bulgaria
[2]Emil Djakov Institute of Electronics, Bulgarian Academy of Sciences,
72 Tzarigradsko Chaussee, 1784 Sofia, Bulgaria
[3]LCIS/SUPRATECS, Institute of Chemistry B6, University of Liege,
Sart-Tilman, B-4000 Liege, Belgium

E-mail: tonyhari@phys.bas.bg

Abstract. This paper presents a sol-gel technological process for preparing thin films of ZnO and ZnO:Al. The effect of annealing treatments (500, 600, 700 and 800 ^{o}C) on their properties was studied. The structural evolution with the temperature was investigated by using X-Ray diffraction (XRD). Fourier Transform Infrared (FTIR) and UV-VIS spectrophotometry were applied to characterizing the films' vibrational and optical properties. The ZnO and ZnO:Al films possessed a polycrystalline structure. The films studied are highly transparent in the visible spectral range. The optical band gap values and the haze parameter were also determined.

1. Introduction

Nanostructured metal oxide films have been extensively studied due to their numerous applications, especially in optoelectronics. Among these materials, ZnO has prompted extensive scientific research with respect to its interesting properties, such as a wide band gap (3.37 eV) with a large exciton binding energy (60 meV), a high chemical stability, a low dielectric constant, a high thermal conductivity, antibacterial and UV protection properties [1]. On the other hand, the ZnO thin films exhibit a high transmittance in the infrared and visible spectral regions and a high refractive index, which extends their application range [2]. ZnO films are used as transparent conductors, chemical and biological gas sensors, ultraviolet light emitters, thin film transistors, solar cell windows [3-4].

Doping and mixing ZnO with different metals leads to changing and enhancing some of their electronic and optical properties. Recently, Al, P, N, Mg, Sb, Mn, B etc. [5] have been studied extensively as promising dopants for ZnO using various technological approaches. Aluminium-doped zinc oxide coatings manifest high transparency and low resistivity [6]. The doping concentration influences the optical and structural properties. The sol-gel technology offers the possibility to produce high-quality homogeneous thin films over a large area, excellent control of the stoichiometry, composition modification (mixing on a molecular level), a cost-effective process, and inexpensive equipment. [7].

[4] To whom any correspondence should be addressed.

This paper presents a sol-gel method for depositing ZnO and ZnO:Al thin films. The effect of the doping concentrations of Al on the optical and structural properties is investigated. The influence of the annealing temperature is also studied. The measurement techniques used are XRD, FTIR and UV-VIS spectrophotometry.

2. Experimental

Two sol solutions were prepared for depositing ZnO and Al_2O_3 films. They were synthesized as follows:

Solution 1) Zinc acetate was dissolved in absolute ethanol (0.4 M concentration). The complexing agent and stabilizer was monoethanolamine (MEA/Zn molar ratio = 1);

Solution 2) Alumina-sec-butoxide was dissolved in isopropyl alcohol (0.2 M concentration). Acetylacetone was used as chelating agent. The full details are given in [8].

The two sol solutions were found to be stable and to retain their film forming properties for more than three months. The mixing of the two sols was made in molar ratios (Al sol/Zn sol) of 0.01, 0.03, 0.05 and 0.1; the films thus prepared are referred to with these names throughout the text. The films were deposited by spin coating at 4000 rpm on Si and quartz substrates; the samples underwent five layer depositions. The preheating temperature was 300 °C/10 min. After repeating the coating procedures five times, the films were annealed at 500, 600, 700 and 800 °C in air for one hour.

The XRD spectra of the sol-gel films were recorded by a Bruker D8 XRD diffractometer at the grazing angle of 2°, a step time of 8 s at a step of 0.1°. The FTIR measurements were performed on an IRPrestige-21 Shimadzu FTIR Spectrophotometer. The optical measurements were conducted by using a UV-VIS-NIR Shimadzu 3600 spectrophotometer in direct, diffuse and total transmittance modes.

3. Results and discussions

The XRD spectra of ZnO and ZnO:Al (0.01) films are presented in figure 1 after thermal treatments at different temperatures. The five strongest lines are assigned to a wurtzite phase (JCPDS 01-07-8070). The corresponding lines for ZnO:Al (0.01) films appear with lower intensities suggesting that the crystallization has decayed. This means that the Al additive affects the films' crystalline structure.

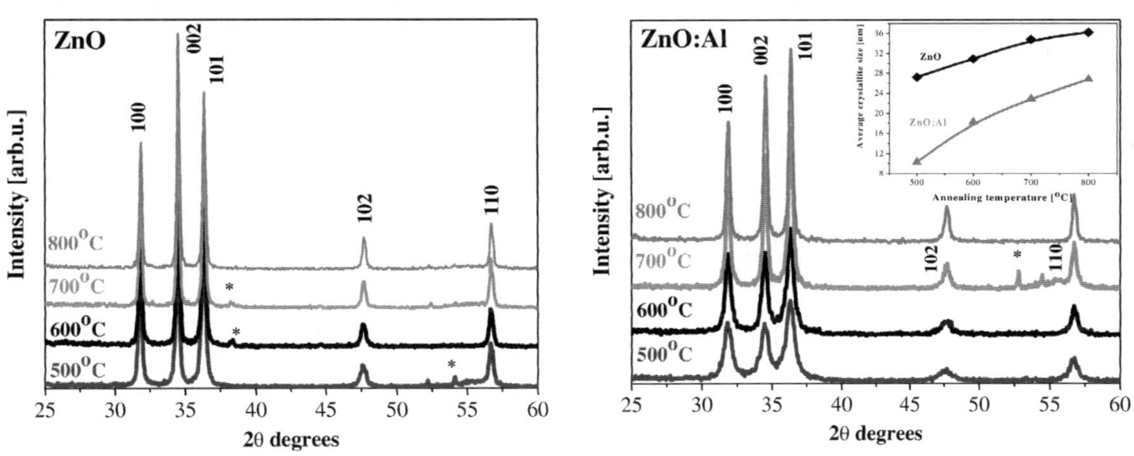

Figure 1. XRD spectra of ZnO and ZnO:Al (0.01) films. The asterisks show lines due to ZnO_2.

The ZnO film annealed at 500 °C shows five intense peaks of wurtzite ZnO and a small line at $2\theta = 54.1$, attributed to the ZnO_2 cubic phase (200) (JCPDS 13-0311). The ZnO:Al film (500 °C) manifests XRD lines corresponding to wurtzite only. The ZnO film (600 °C) crystallizes in a wurtzite. phase with a weak reflection at 38.4, associated with cubic ZnO_2. The thermal treatment at 700 °C leads to the detection of two additional weak lines in the spectrum of ZnO:Al at 2θ =54.4 and 55.6, assigned to cubic ZnO_2. The highest annealing temperature results in good polycrystalline films with a wurtzite ZnO crystal phase. The average crystallite sizes are estimated according to Scherrer's formula (for peaks: 100, 002 and 101) and the results obtained are shown in the inset (figure 1) and table 1. The Al component influences the crystallization of the films, namely, the crystallite size significantly diminishes compared to that of ZnO films. The average crystallite size of ZnO is 32 nm, while the value for ZnO:Al is 20 nm. Table 1 presents also the calculated lattice parameters and the dislocation density of the films. Dislocations are imperfections in a crystal associated with misregistry of the lattice in one part of the crystal with respect to another part. The dislocation density of the films is given by the Williamson and Smallman's relation ([9]): $\delta = n/d^2$ (1), where n is a factor equal to unity at the minimum dislocation density and d is the grain size. The dislocation densities decrease as the annealing temperature is raised, which indicates a lower concentration of lattice imperfections.

Table 1. Average crystallite size (d), dislocation density (δ) and lattice parameters.

Material	T, (°C)	d, nm	$\delta\times10^{-4}$ (1/nm^2)	a (Å)	c (Å)
ZnO	500	27.2	13.5	3.243	5.197
	600	30.9	10.5	3.243	5.193
	700	34.8	8.3	3.242	5.189
	800	36.2	7.6	3.241	5.185
ZnO:Al	500	10.3	94.3	3.244	5.192
	600	18.3	29.9	3.242	5.200
	700	22.9	19.1	3.242	5.196
	800	26.9	13.9	3.241	5.187
JCPDS 01-07-8070 card, wurtzite ZnO				3.2489	5.2049

The lattice parameters a, c of the two materials decrease slightly with the annealing temperature. The XRD study reveals that the degree of crystallization worsens with the addition of Al, resulting in crystallites of a smaller size. The finding that there is no indication of mixed Zn/Al oxide phases can imply that the Al atoms substitute for the zinc in the respective sites or are incorporated in the lattice [10].

FTIR spectroscopy was used as a supplement for revealing the structure of the films studied. FTIR analysis reveals that both the metal additive and the annealing procedures affect the vibrational properties of the films. The absorption bands due to metal-oxygen bonds are expected in the spectral range 350-1200 cm^{-1} (figure 2). The line around 1100 cm^{-1} observed in all spectra is due to Si-O bonding. For the ZnO:Al films, weak peaks at 665 and 681 cm^{-1} are observed, which might be assigned either to Zn-O [11] or to O-Al-O bonding [12]. However, a line at 667 cm^{-1} is also observed in the ZnO spectra. The main IR lines are located in the spectral range 390-425 cm^{-1} (see the insets).

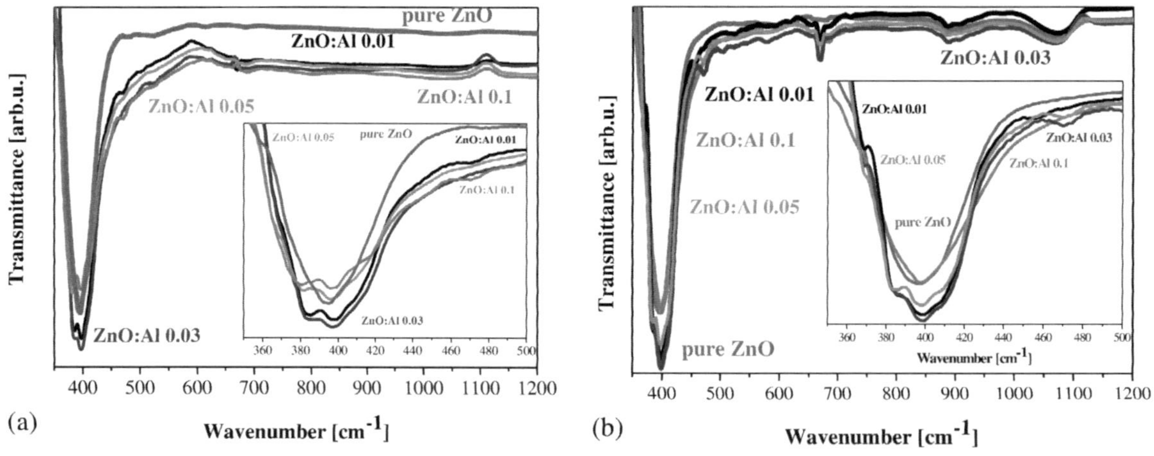

Figure 2. FTIR spectra of sol-gel ZnO and ZnO:Al films annealed at 600 (a) and 800 °C (b).

The Al component influences the shapes of the absorption bands. The bands are broader and with different intensities compared to the ZnO spectra. The IR spectra may be influenced by the crystallite's size, shape and degree of crystallinity and are often used to corroborate the XRD study. Upon addition of Al, the main absorption band shifts from 394 to 400 cm^{-1}. Clear shoulders of the main bands appear at 415 cm^{-1}. They are related to Zn-O stretching vibrations. It must be noted that there are no clearly identified absorption bands that are connected with vibrations of the Al-O bonds. This conclusion is in agreement with the XRD study, where the only detected crystalline phase is wurtzite ZnO.

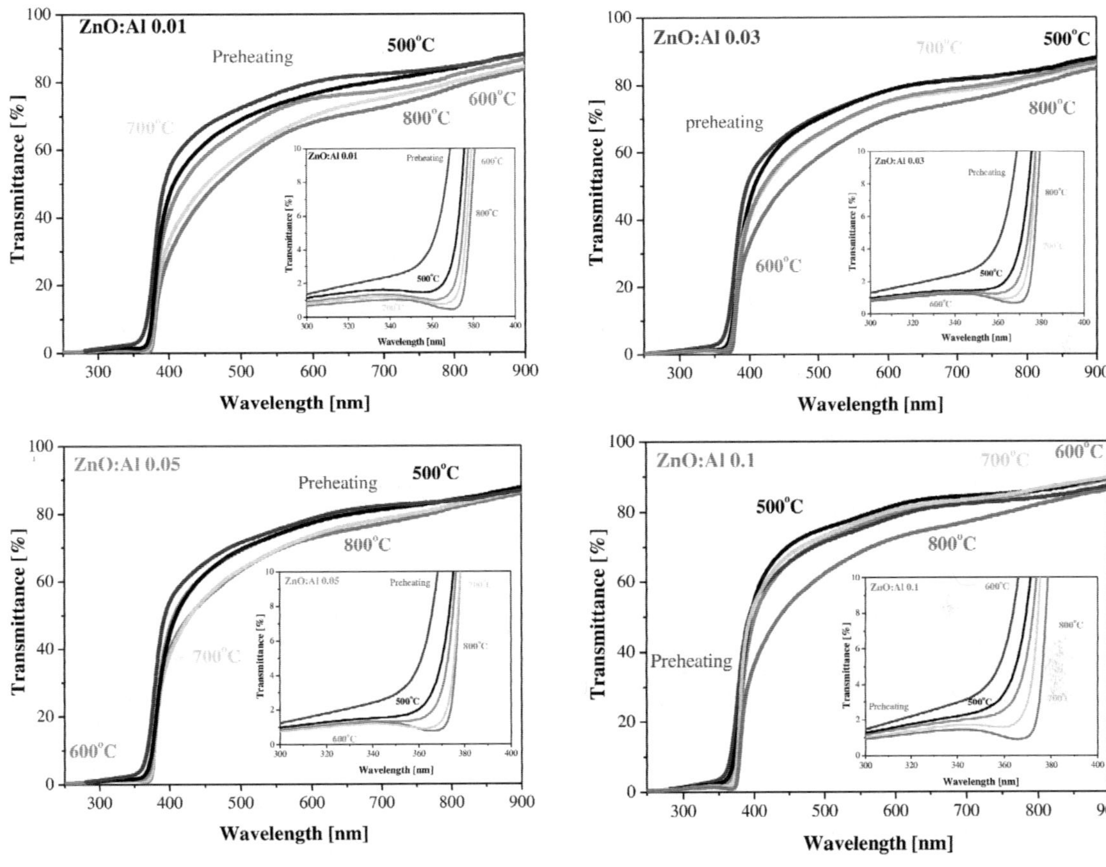

Figure 3. Transmittance spectra of sol-gel ZnO:Al films annealed at different temperatures.

The optical properties of the sol-gel ZnO and ZnO:Al films were also investigated; the transmittance spectra are presented in figure 3. The ZnO films possess a lower transparency compared to ZnO:Al films; their transmittance decreases as the annealing temperature is raised. The ZnO:Al films demonstrate the same tendency of decreasing the transmittance after higher temperature annealing.

Our previous study [13] showed that as a higher amount of the Al component was introduced to the sol solution, the films produced exhibited a significantly higher transparency in the visible spectral range. This effect is more clearly pronounced for ZnO:Al 0.05 and ZnO:Al 0.1, as can be seen in figure 4.

The specific bands in the optical spectra below the absorption edge are due to the excitonic absorption of ZnO (insets in figures 3 and 4). The bulk ZnO excitonic feature is located at 373 nm [14]. The appearance of an excitonic peak is a sign of the degree of ZnO crystallinity. Respectively,

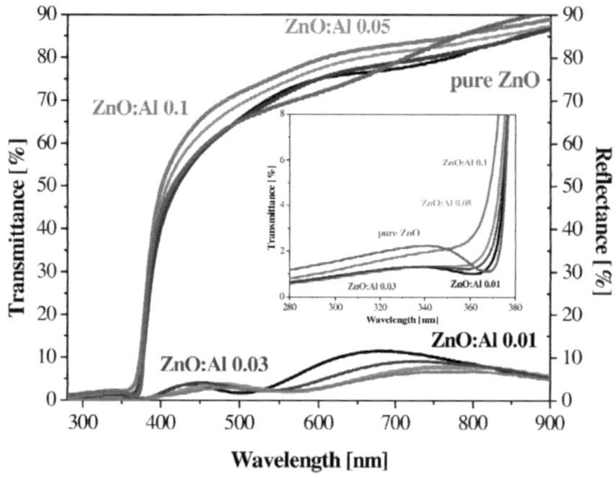

Figure 4. Comparison of transmittance and reflectance spectra of ZnO and ZnO:Al films annealed at 600°C.

the excitonic absorptions prove that the ZnO:Al films are less crystalline compared to the ZnO thin films, as it has been found by the XRD analysis.

The excitonic absorptions of ZnO are found to be in the range of 331 to 347 nm. The ZnO:Al films show excitonic absorptions only after annealing at 700 and 800°C. This is due to the smaller degree of crystallization of the ZnO:Al films annealed at lower temperatures, as has been proved by the XRD study.

Table 2 shows the values of the optical band gap E_g of ZnO; there is no clear trend of an increase or a decrease of the values observed after annealing at different temperatures. The optical band gap of ZnO:Al films decreases with the increase of the annealing temperature. The results are close to those for ZnO films. It has been found earlier [15] that increasing the Al component results in a widening of E_g due to the lower degree of crystallization and/or formation of Zn-Al–O phases. The optical band gap values of the films studied by us agree with the data reported in [16]. On the other hand, the band gap values of all sol-gel ZnO and ZnO:Al films

Table 2. Optical band gap values [eV] determined from spectrophotometrical data.

	ZnO	ZnO:Al 0.01	ZnO:Al 0.03	ZnO:Al 0.05	ZnO:Al 0.1
Preheating temperature	3.30	3.27	3.28	3.28	3.29
500°C	3.28	3.27	3.25	3.25	3.28
600°C	3.27	3.27	3.26	3.26	3.27
700°C	3.28	3.26	3.27	3.27	3.27
800°C	3.28	3.26	3.26	3.27	3.26

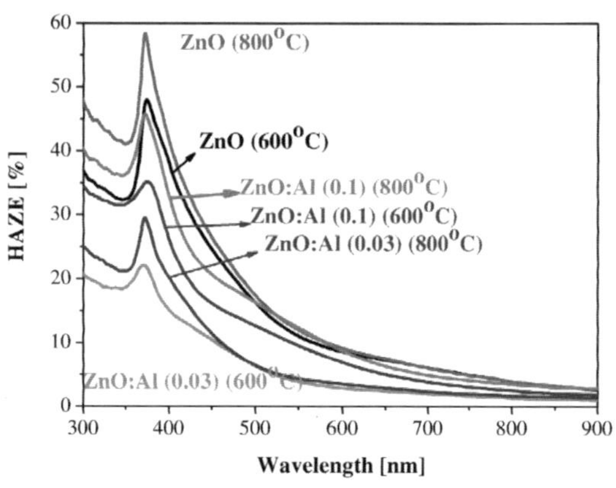

Figure 5. Haze of ZnO, ZnO: Al (0.03) and ZnO: Al (0.1) films annealed at 600 and 800 °C.

are smaller than the optical band gap of single crystalline ZnO (3.37 eV). It has also been reported that the lower values of the optical band gap could be due to the presence of growth stress, thermal expansion and mismatch stress.

Another interesting optical property is the haze value. The haze parameter serves as an indicator for the diffuse scattering of light by thin films. The haze parameter for transmitted light is defined as the ratio between the diffuse part of the transmitted light to the total transmitted light. Total and diffuse transmittance spectra were measured using an integral sphere attachment.

Figure 5 shows the haze parameter. It can be seen that the ZnO films exhibit higher values of haze compared to ZnO:Al. It is interesting to note that the films with a higher content of Al reveal values higher than those of ZnO:Al (0.1). When the Al component amount is increased in the sol solution, the spectra of the direct and total transmittance are very similar, indicating that the films are very smooth and there is no light scattering.

4. Conclusions

In this work, a sol-gel technology was successfully applied to depositing transparent ZnO and ZnO:Al films. The XRD analysis reveals the existence of wurtzite ZnO and a decrease of the film crystallinity upon Al doping. The crystallite sizes of ZnO and ZnO:Al films differ, as those of the doped films are smaller. No Al containing phases are detected by XRD and FTIR analysis. The optical transmittance is improved for the ZnO:Al films in comparison with pure ZnO films. The optical study also reveals that the optical band gap is not significantly changed by Al addition and thermal treatments. The technological approach proposed for ZnO:Al film deposition is very promising.

References

[1] Poongodi C, Kumar R M and Jayavel R 2015 *Ceramics Int.* **41** 4169
[2] Tiwari N, Shieh H P D and Liu P T 2015 *Mater. Lett.* **151** 53
[3] Chu J, Peng X, Sajjad M, Yang B and Feng P X 2012 *Thin Solid Films* **520** 3493
[4] Lim D C, Shim W H, Kim K-D, Seo H O, Lim J-H, Jeong Y, Kim Y D and Lee K H 2011 *Solar Energy Mater. Solar Cells* **95** 3036
[5] Özgür U, Hofstetter D and Morkoç H 2010 *Proc. IEEE* **98** 1255
[6] Marotti R E, Bojorge C D, Broitman E, Cánepa H R, Badán J A, Dalchiele E A and Gellman A J 2008 *Thin Solid Films* **517** 1077
[7] Aydemir S and Karakaya S 2015 *J. Magnetism Magnetic Mater.* **373** 33
[8] Vitanov P, Harizanova A, Ivanova T and Dimitrova T 2009 *Thin Solid Films* **517** 6327
[9] Ergin B, Ketenci E and Atay F 2009 *Int. J. Hydrogen Energy* **34** 5249
[10] Gong L, Ye Zh, Lu J, Zhu L, Huang J, Gu X and Zhao B 2010 *Vacuum* **84** 947
[11] Li C P, Jang B H, Wang X C, Li M J, Su L and Li X W 2011 *Appl. Surf. Sci.* **257** 5998
[12] Adamczyk A and Długoń E 2012 *Spectroch. Acta* A **89** 11
[13] Ivanova T, Harizanova A. Koutzarova T and Vertruyen B 2015 *Superlattices Microstr.* **85** 101
[14] Kuo S-Y, Chen W-C and Cheng C P 2006 *Superlattices Microstr.* **39** 162
[15] You Q, Cai H, Hu Zh, Liang P, Prucnal S, Zhou Sh, Sun J, Xu N and Wu J 2015 *J. Alloys Comp.* **644** 528
[16] Wang M, Lee K, Hahn S, Kim E, Kim S, Shin E and Park C 2007 *Mater. Lett.* **61** 1118
[17] Yen W T, Lin Y C and Ke JH 2010 *Appl. Surf. Sci.* **257** 960

Journal of Physics: Conference Series **700** (2016) 012049 doi:10.1088/1742-6596/700/1/012049

Influence of the surface pre-treatment of aluminum on the processes of formation of cerium oxides protective films

R Andreeva[1], **E Stoyanova**[1], **A Tsanev**[2] **and D Stoychev**[1,3]

[1]Rostislaw Kaischew Institute of Physical Chemistry, Bulgarian Academy of Sciences, Acad. G. Bonchev Str., Bl. 11, 1113 Sofia, Bulgaria
[2]Institute of General and Inorganic Chemistry, Bulgarian Academy of Sciences, Acad. G. Bonchev Str., Bl. 11, 1113 Sofia, Bulgaria

E-mail: stoychev@ipc.bas.bg

Abstract. It is known that there is special interest in the contemporary investigations on conversion treatment of aluminum aimed at promoting its corrosion stability, which is focused on electrolytes on the basis of salts of metals belonging to the group of rare-earth elements. Their application is especially attractive, as it enables a successful substitution of the presently applied highly efficient, but at the same time toxic Cr^{6+}-containing electrolytes. The present paper presents a study on the influence of the preliminary alkaline activation and acidic de-oxidation of the aluminum surface on the processes of immersion formation of protective cerium oxides films on Al 1050. The results obtained show that their deposition from simple electrolytes (containing only salts of Ce^{3+} ions) on the Al surface, treated only in alkaline solution, occurs at a higher rate, which leads to preparing thicker oxide films having a better protective ability. In the cases when the formation of oxide films is realized in a complex electrolyte (containing salts of Ce^{3+} and Cu^{2+} ions), better results are obtained with respect to the morphology and protective action of cerium oxides film on samples that have been consecutively activated in alkaline solution and deoxidized in acidic solution.

Electrochemical investigations were carried out in a model corrosion medium (0.1 M NaCl); it was shown that the cerium protective films, deposited by immersion, have a cathodic character with regard to the aluminum support and inhibit the occurrence of the depolarizing corrosion process – the reaction of oxygen reduction.

1. Introduction

Cerium-based conversion coatings have been recognized as an environmentally friendly alternative to chromate conversion coatings for the corrosion protection of aluminum and aluminum alloys, such as the Al 1050 and Al 2024 alloys. Despite providing excellent corrosion protection, chromate conversion coatings contain hexavalent chromium, a known toxin and carcinogen. Federal mandates have required implementation of strict environmental controls to reduce workplace exposure to Cr^{6+}, motivating the development of environmentally benign alternatives [1]. Rare earths conversion coatings, principally cerium-based conversion coatings, have been shown to improve the corrosion resistance of aluminum alloys [2,3]. The deposition of cerium-based conversion coatings occurs via a

[3] To whom any correspondence should be addressed.

precipitation reaction driven by an increase in the near surface pH that is formed by electrochemical reactions between the ions of coating solution and the alloy substrate. The cerium oxides conversion coatings are commonly considered environmentally friendly corrosion inhibitors for aluminum alloys. Nevertheless, the effectiveness of the cerium-based conversion coatings on the corrosion protection has been highly dependent on surface pre-treatment, deposition parameters, and post-treatment [4,5]. Recent results demonstrated a surface pre-treatment for deposition of cerium-based conversion coatings on aluminum alloys that consisted of acid then alkaline cleaning [6,7]. The corrosion performance of cerium-based conversion coatings can also be improved by using a subsequent weak acidic phosphate bath treatment after cerium oxides deposition. This additional bath treatment results in the formation of insoluble $CePO_4$ species that can provide long-term protection [8]. Although several authors have reported an increase in the corrosion resistance of certain aluminum alloys by using cerium-based conversion coatings, there is not a reliable common method of assessment. To date, no systematic studies have been reported that optimize the corrosion resistance of the cerium oxides coatings in terms of process parameters such as thickness, microstructure, composition, and bath temperatures, among others.

The purpose of the present study was to investigate the influence of the alkaline activation and acid deoxidation of Al 1050 on the formation and corrosion resistance of Ce-based conversion films on aluminum.

2. Experimental

Layers of cerium oxide (chemical conversion treatment) were deposited on substrates of "technical degree of purity" Al 1050 (0.40% Fe, 0.25% Si, 0.05% Mn, 0.05% Cu, 0.07% Zn, 0.05% Mg), widely applied as a construction material. The samples studied of dimensions 1×1 cm were cut out of Al sheets (thickness 0.1 cm). Their pre-treatment involved degreasing in an organic solvent, degreasing (chemical cleaning) in aqueous solution of NaOH (1.5 mol dm^{-3}) at 60°C (τ = 1 min) or chemical cleaning in NaOH and consecutive etching and surface activation in aqueous solution of HNO_3 (50 wt %) (acid deoxidation) at room temperature (τ = 30 sec). After each one of these operations, the obligatory standard rinsing of the samples was accomplished with distilled water. The Ce-containing protective layers were deposited on the so pre-treated samples using two different kinds of electrolytes containing: 1) $CeCl_3$ - 5×10^{-1}mol dm^{-3}; 2) $CeCl_3$ - 5×10^{-1} mol dm^{-3}+ $CuCl_2$ - 1×10^{-5} mol dm^{-3}. The studies were carried out at temperatures of deposition 25 $^{\circ}$ and 60 $^{\circ}$C, while the time interval of the deposition was 120 minutes. The morphology, structure and chemical composition of the conversion films, as well as the distribution of the elements on the aluminum surface prior to and after the deposition of the protective layers, were observed by the electron microscope JEOL JSM 6390 (under the conditions of secondary electron image - SEI) and by X-ray photoelectron spectroscopy (XPS). The XPS measurements were carried out on an AXIS Supra electron spectrometer (Kratos Analitycal Ltd.) using monochromatic AlKα radiation. The depth profile was performed by layer-by-layer sputtering (the energy of the Ar gun was 500 eV). The corrosion behavior of the samples was tested in 0.1 M NaCl model medium at 25°C. Platinum electrode was used as the counter electrode, while the reference electrode was saturated calomel electrode (SCE → +0.240 V vs. SHE). All the potentials in this study are compared to SCE. The polarization curves were obtained by means of a Gamry Interface 1000 potentiostat/galvanostat. The curves were recorded at a sweeping rate of the potential 10 mV s^{-1} in the range of potentials from –2300 up to + 1800 mV.

3. Results

The cathodic and anodic polarization curves (E-lgi), illustrating the electrochemical behavior of aluminum samples coated with a cerium oxide film from electrolyte containing also $CeCl_3$ (degreased in advance only in a hot solution of NaOH), as well as samples treated consecutively in NaOH and HNO_3 are represented in figure 1. It is seen that the conversion films, formed on samples activated only in NaOH, possess a better protective ability with regard to the cathodic depolarization reaction of the corrosion process in chloride solution – compared with the films formed on samples after activation

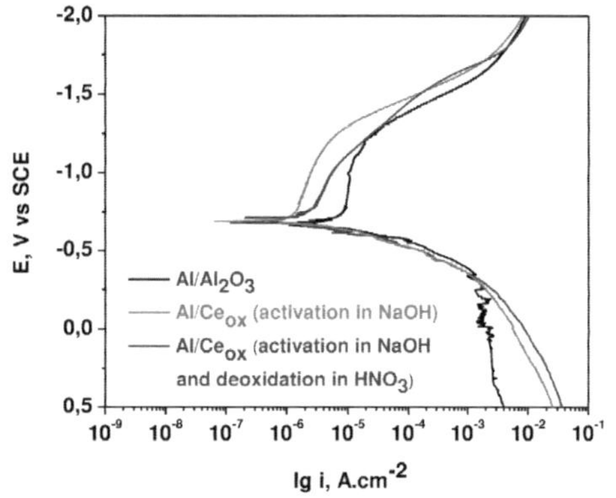

Figure 1. Polarization curves of tested systems in 0.1 M NaCl at 25 °C.

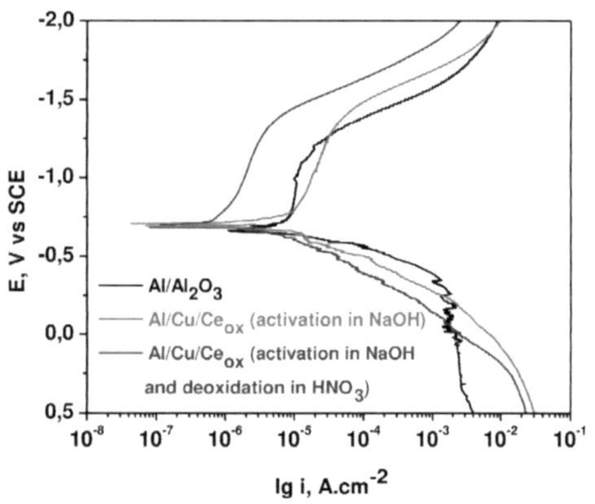

Figure 2. Polarization curves of tested systems in 0.1 M NaCl at 25 °C.

in solution of NaOH and deoxygenation in HNO_3. A considerable inhibition of the reaction rate of oxygen reduction is observed, which is connected with a substantial decrease in the cathodic currents (table 1). The calculated degree of protection $z=(i_{corr(Al)}-i_{corr(CL/Al)})/i_{corr\ (Al)})$ % on the basis of the values of the corrosion currents determined in the model potentiodynamic investigations, is reaching a value of 93.75 %. This value is a proof that the formed oxide layer is a more efficient barrier for the corrosion process in the case when the samples have been pretreated only in NaOH.

The analogous polarization curves, obtained with samples coated with cerium oxide film from electrolyte containing $CeCl_3$ and $CuCl_2$, are represented in figure 2. In this case, better results with respect to the protective ability of the ceria films are observed when the samples have been treated consecutively in NaOH and in HNO_3. As it was noted above, the conversion treatment is leading to an insignificant decrease in the corrosion current observed with the samples which have been treated in advance in NaOH only. At the same time, the additional acidic deoxygenation results in the formation of conversion films acting as an efficient cathodic inhibitor. In such a case, the degree of protection against corrosion reaches the value of 95 %, which is evidence of the importance of the preliminary acidic treatment of the Al 1050 samples during the formation of conversion films on their surface from electrolyte containing $CeCl_3$ and $CuCl_2$.

Figures 3 and 4 illustrate SEM micrographs of the formed oxide films, showing the differences in the morphology of the deposited layers depending on the preliminary treatment procedures and on the composition of the electrolytes for chemical conversion treatment. The morphology of the coating, formed in electrolyte containing $CeCl_3$ on substrates which have been activated in advance only in NaOH, is more homogeneously distributed on the aluminum surface (figure 3a). This result is in correspondence with the data from the potentiodynamic investigations, showing lower corrosion currents, respectively better protection ability, in comparison with the coating which is being formed on substrates consecutively treated in NaOH and in HNO_3 (figure 3b).

Table 1. Pre-treatment solutions, operation conditions and electrochemical parameters obtained from the potentiodynamic curves.

Pre-treatment	**Electrolyte**	$C_{comp.}$, M	**Time,** min	i_{corr}, Acm^{-2}	*z*, %
Activation in NaOH	$CeCl_3$	0.5	120	5×10^{-7}	93.75
Activation in NaOH and HNO_3	$CeCl_3$	0.5	120	9×10^{-7}	88.75
Activation in NaOH	$CeCl_3$ $CuCl_2$	0.5 1.10^{-5}	120	1×10^{-6}	87.50
Activation in NaOH and HNO_3	$CeCl_3$ $CuCl_2$	0.5 1.10^{-5}	120	4×10^{-7}	95.00

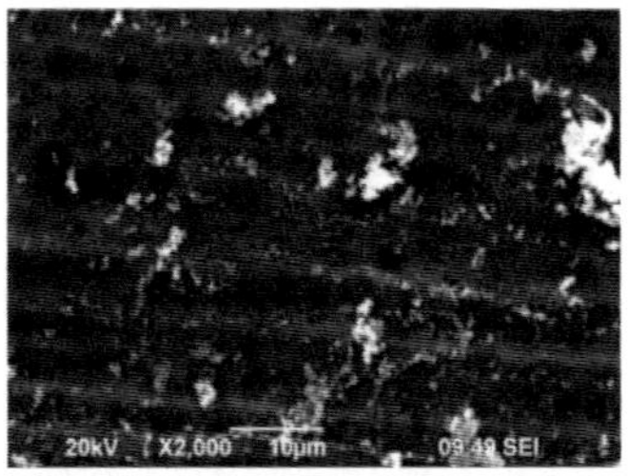

Figure 3a. SEM micrograph on the surface of Al/Ce_{ox} (pre-treatment in NaOH).

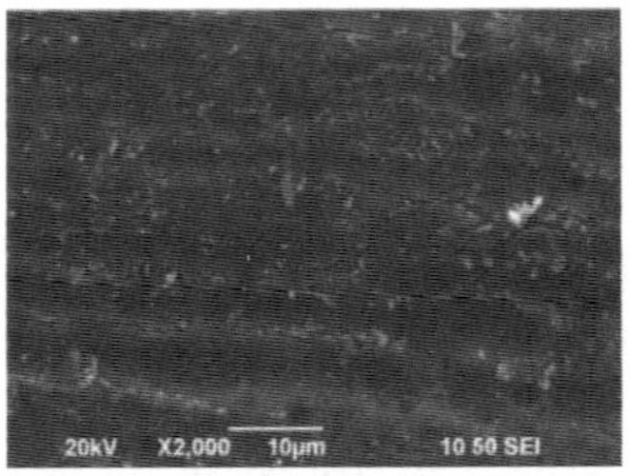

Figure 3b. SEM micrograph on the surface of Al/Ce_{ox} (pre-treatment in NaOH and HNO_3).

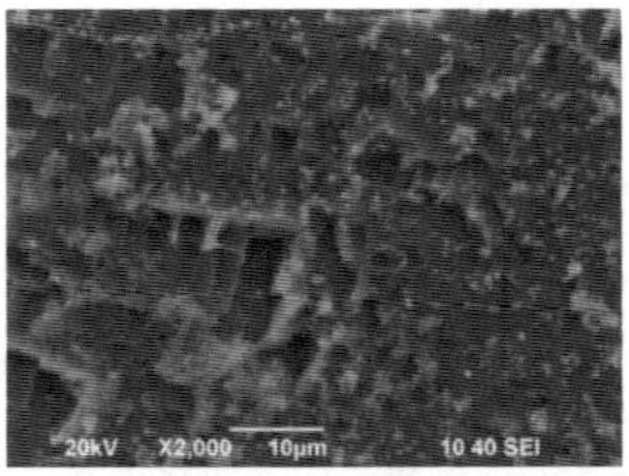

Figure 4a. SEM micrograph on the surface of Al/Ce_{ox} (pre-treatment in NaOH).

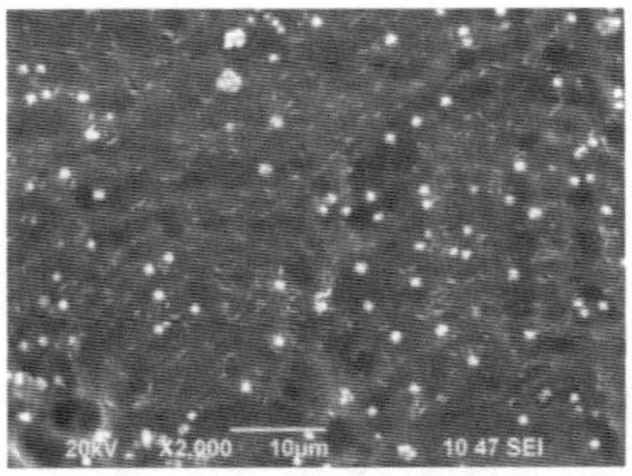

Figure 4b. SEM micrograph on the surface of Al/Ce_{ox} (pre-treatment in NaOH and HNO_3).

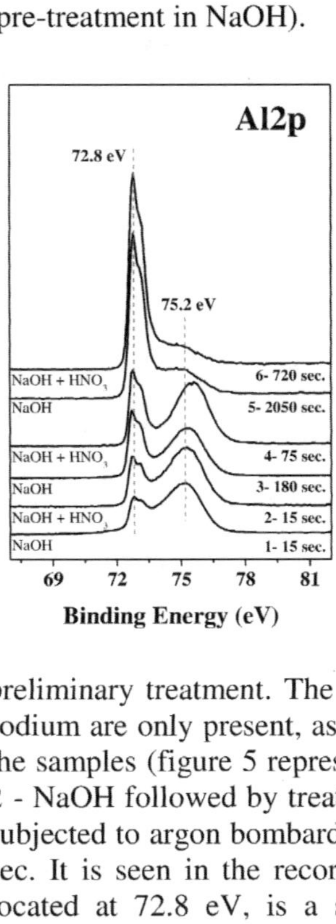

Figure 5. XPS - Al2p - spectra, recorded from surfaces located in different depths of Al - samples:
1. Al, pre - treated in NaOH, at etching time intervals 15 sec. (Al/NaOH/15sec.);
2. Al/NaOH+HNO_3/15sec.;
3. Al/NaOH/180 sec.;
4. Al/NaOH+HNO_3/75 sec;
5. Al/NaOH/2050 sec.;
6. Al/NaOH+HNO_3/720 sec.

One has the impression that in the case of the samples treated consecutively in NaOH and HNO_3 (figures 4a and 4b), the homogeneity of the distribution of the cerium oxide layer, being formed in a complex electrolyte (containing $CeCl_3$ and $CuCl_2$), displays a substantially more uniform distribution in comparison with the layers formed in electrolyte containing only $CeCl_3$. This effect is obviously associated with the availability of higher concentrations of copper/copper oxides on the surface of the sample treated by immersion in the solution containing cerium and copper ions, designed for conversion treatment [9]. Such kind of distribution is leading to a considerable increase in the overpotential of the cathodic reaction of oxygen reduction (figure 2). The activation of the aluminum surface (in NaOH and HNO_3) as a result of the increase in the number of the cathodic sections during the contact (electroless) deposition of Cu on the aluminum substrate leading to an increase in the rate of deposition of Ce oxides/hydroxides on these sections (figure 4b).

The surfaces of the substrates of Al, treated in solutions of NaOH and NaOH and HNO_3, was studied by means of XPS with the aim to establish the composition and the valence states of the elements on the surface after the preliminary treatment. The XPS analysis of the surface shows that peaks of aluminum, oxygen and sodium are only present, as well as the peak of C1s with a lower intensity. The spectra of Al2p of all the samples (figure 5 represents only the spectra of Al samples treated in solutions of: 1 - NaOH and 2 - NaOH followed by treatment in solution of HNO_3.) The Al substrates pretreated in this way were subjected to argon bombardment (etching) in the course of: 15 sec, 75 sec, 180 sec, 720 sec and 2050 sec. It is seen in the recorded spectra that they are composed of two components. The first peak, located at 72.8 eV, is a superposition of two peaks having a spin-orbital splitting of 0.44 eV, characteristic of $Al2p_{3/2}$ and $Al2p_{1/2}$ of metallic state Al^0 [10]. The second peak at 75.2 eV is wider than the first one. During its deconvolution for the sample treated in solution of NaOH (figure 5,

15 sec etching), it becomes clear that it is built up of two peaks (figure 5 with deconvolution of Al). By separating the two peaks we can determine the ratio between the oxide and the hydroxide on a given surface. Accounting for the energy of the respective peaks and their areas, as well as comparing them with the deconvoluted spectrum of O1s for the same sample (figure 6 – deconvolution of O1s)), we can state that the first peak at 73.9 eV is corresponding to Al^{3+} in Al_2O_3. Its respective peak in the spectrum of O1s is the one located at 530.6 eV [11].

The second peak in the spectrum of Al2p, located at 75.2 eV, is characteristic of Al^{3+} in the composition of AlOOH. Its corresponding peak in the spectrum of O1s is the one at 531.9 eV [12]. The small peak in the spectrum of Al2p at 71.7 eV is owing to the peak of Na2s, to the availability of Na^+ - ions in the solution, in which the sample was treated. The oxygen peak at 533.3 eV is attributed to the availability of water molecules adsorbed on the surface.

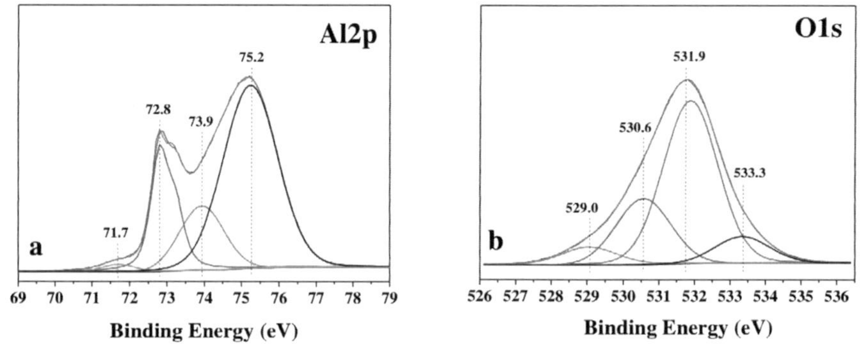

Figure 6. Deconvolution of Al2p (a) and O1s (b) spectra on the surface of Al-sample etched for 15 sec. The sample has been treated in solution of NaOH only.

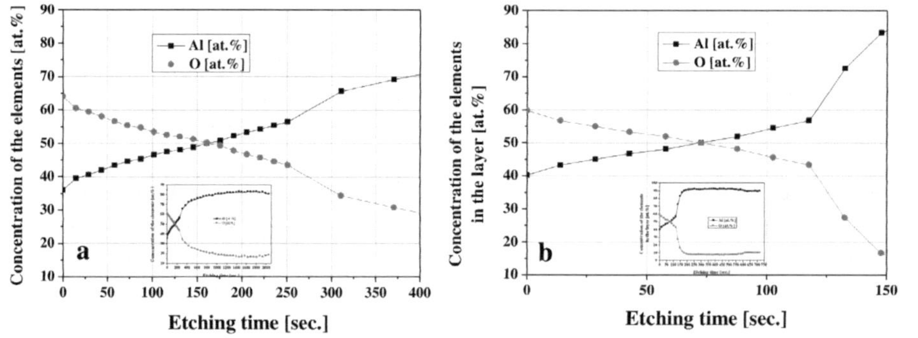

Figure 7. Depth profiles reflecting the change in the ratio between the elements Al and O in the layers obtained on the sample treated only in solution of NaOH (a) and the sample treated in solutions of NaOH and HNO_3 (b).

Accounting for the number of peaks and the magnitude of the areas of the deconvoluted peaks of Al2p and O1s, one can conclude that the surface consists mainly of AlOOH, and some smaller amounts of Al_2O_3, at a ratio of AlOOH/Al_2O_3 of approximately 3.6 : 1, whereupon this ratio between the oxide and the hydroxide is preserved in the depth of the layer.

A layer-by-layer XPS analysis was carried out in order to elucidate the influence of the pre-treatment on the composition and on the valence states of the elements in the so treated samples.

It becomes clear from figures 7a and 7b that upon increasing the time interval of etching (through argon bombardment) of the Al surface, the quantity of oxygen (respectively the oxide and the hydroxide in the layer) is decreasing, while that of aluminum is increasing. Then an intersection point is reached, after which the increase in the concentration of Al, respectively the decrease in the O concentration take place in an abrupt way. This sharp change gives us reason to assume that this intercept point could serve as an indicator for the fact that the interphase boundary is reached between the layer formed during the pre-treatment and the aluminum substrate. In the case of the sample treated in solution of NaOH, this point is reached at the etching time interval of 160 sec, while in the case of the sample activated and deoxygenated in solutions of NaOH and HNO_3, this intercept point is reached after the etching time interval of 72 sec. Comparing the time intervals for reaching the interphase boundary observed with the two samples, which are proportional to the thickness of the layer, we can draw the conclusion that the thickness of the oxide/hydroxide layer for the sample treated with NaOH is much bigger (about 2-3 times) in comparison with that of the layer treated additionally in HNO_3.

The chemical composition on the surface of the samples with deposited conversion ceria layers was investigated by XPS as well, for both types of pre-treatments. It is as follows: Ce – 31.1÷39.6 at.%; O –56.6÷65.8 at.%; Al – 0.3 ÷ 3.8 at.% (and Cu – about 2.3÷3.8 at. %).

The results obtained on the basis of these analyses show that the preliminary treatment procedures of the Al substrates determine the formation of surface oxide layers differing in their thickness and in their composition. These layers, as it was shown by the above electrochemical investigations, possess a higher activity acquired during the immersion process of deposition of cerium oxide layers. This activity obviously influences the rate of formation and the homogeneity of the ceria layers. Moreover, they can contribute to the appearance of a more efficient barrier against the corrosion process.

4. Conclusions

The effect was investigated of alkaline activation and acid base deoxidation on the process of the deposition of cerium oxide conversion coatings on Al 1050 substrates. The results obtained prove that the conversion deposition of ceria on an alkaline-activated Al surface from electrolyte containing only $CeCl_3$ offers a high rate of deposition and a better option for cathodic protection.

In the case when the conversion coatings are formed in electrolyte containing $CeCl_3$ and $CuCl_2$ on an Al substrate pretreated by alkaline cleaning and acid deoxidation, the rate of the process of conversion deposition is much higher and the coatings exhibit an even higher cathodic protection. Moreover, the acidic pre-treatment decreases the amount of Al oxide/hydroxides on the surface of the Al 1050.

Acknowledgements

The work was supported by the Bulgarian National Science Fund under grant T02/22.

References

[1] Zhao J, Xia L, Sehgal A, Lu D, McCreery R L and Frankel G S 2001 *Surf. Coat.Technol.* **140** 51
[2] Hinton B R, Arnott D R and Ryan N E 1984 *Met. Forum* **7** 211
[3] Wang C, Jiang F and Wang F 2004 *Corrosion* **60**/3 237
[4] Cole G 2003 *Mater. Sci. Forum* **419-422** 43-50
[5] Tomlinson C E U.S. Patent 005952049A
[6] Pinc W, Geng S, O'Keefe M, Fahrenholtz W and O'Keefe T 2009 *Appl. Surf. Sci.* **255** 4061–5
[7] Joshi S, Fahrenholtz W G and O'Keefe M J 2011 *Appl. Surf. Sci.* **257** 1859
[8] Heller D K, Fahrenholtz W G and O"Keefe M J 2009 *J. Electrochem. Soc.* **156**/1 C400 - 6
[9] Andreeva R, Stoyanova E, Tsanev A, Stefanov P and Stoychev D 2014 *TIMF* **92** 203-11
[10] Hinnen C, Imbert D, Siffre J M and Marcus P 1994 *Appl. Surf. Sci.* **78** 219-31
[11] Liao H M, Sodhi R N S and Coyle T W 1993 *J. Vac. Sci. Technol.* A **11**/5 2681-6
[12] Barr T L 1978 *J. Phys. Chem.* **82**/16 1801-10

Transmission photocathodes based on stainless steel mesh and quartz glass coated with N-doped DLC thin films prepared by reactive magnetron sputtering

N I Balalykin[1], **J Huran**[2], **M A Nozdrin**[1,3], **A A Feshchenko**[1], **A P Kobzev**[1] **and J Arbet**[2]

[1]Joint Institute for Nuclear Research,
6 Joliot-Curie Str., 141980 Dubna, Moscow Region, Russian Federation
[2]Institute of Electrical Engineering, Slovak Academy of Sciences,
9 Dúbravská cesta, 84104 Bratislava, Slovakia

E-mail: nozdrin@jinr.ru

Abstract. The influence was investigated of N-doped diamond-like carbon (DLC) films properties on the quantum efficiency of a prepared transmission photocathode. N-doped DLC thin films were deposited on a silicon substrate, a stainless steel mesh and quartz glass (coated with 5 nm thick Cr adhesion film) by reactive magnetron sputtering using a carbon target and gas mixture Ar, 90%N_2+10%H_2. The elements' concentration in the films was determined by RBS and ERD. The quantum efficiency was calculated from the measured laser energy and the measured cathode charge. For the study of the vectorial photoelectric effect, the quartz type photocathode was irradiated by intensive laser pulses to form pin-holes in the DLC film. The quantum efficiency (QE), calculated at a laser energy of 0.4 mJ, rose as the nitrogen concentration in the DLC films was increased and rose dramatically after the micron-size perforation in the quartz type photocathodes.

1. Introduction

Electron emission from carbon materials has been based on two effects: field enhancement from conducting nanostructures and barrier lowering due to the negative electron affinity of surfaces [1]. The photoemission properties of different samples were rationalized by considering the electron emission process located at the a-C/diamond/vacuum triple border and the quantum efficiency governed by the ratio of amorphous sp^2-C to crystalline sp^3-C [2]. In [3], the authors investigated the electron field emission properties of sulphur-assisted nanocrystalline carbon (n-C:S) thin films grown on molybdenum substrates by the hot-filament CVD technique using a methane-hydrogen (CH_4/H_2) and hydrogen sulphide-hydrogen (H_2S/H_2) gas mixtures. In [4], the results were reported and discussed of the external quantum efficiency (QE) measurements, in the range 150–210 nm, of poly-, nano- and single-crystalline diamond (PCD, NCD and SCD) film photocathodes (PCs). The process of diamond-like carbon deposition by plasma-assisted chemical vapor deposition (PECVD) using deuterated hydrocarbons as precursor gases was investigated in [5]. Diamond-like carbon thin films were deposited from a pure graphite target by DC magnetron sputtering. Experimental

[3] To whom any correspondence should be addressed.

parameters, i.e., substrate temperature and negative bias voltage, were varied to finely tune the chemical bonding property (sp^2/sp^3) of the as-deposited DLC films [6]. DLC films deposited under a mixture of acetylene/argon plasma over a large area and three dimensional deposition in electron cyclotron resonance (ECR) systems were reported in [7]. The possibility of application of ion implantation for the modification of the physicochemical properties of carbon coatings was presented in [8]. Diamond-like carbon thin films were deposited on p-type Si (100) substrates by the RF hollow-cathode method under different RF power and pressures using ethane as the precursor gas [9]. The photo-gun is a key element of injectors for producing high-charge electron bunches. Research continues on various designs of cathodes. A promising, but technically challenging, DC gun with a hollow cathode was proposed by JINR [10]. In the development of this concept, the use of transmission photocathodes in the form of a grid was the most promising direction. The vectorial photoelectric effect has long been known. It is necessary to notice that research of vectorial photoeffect demands that the photocathode surface be smooth. In the presence of a roughness, the polarized light will have various polarizations in relation to the plane of incidence on variously oriented surface elements of the rough photocathode.

In this work, we investigated the influence of N-doped diamond-like carbon (DLC) films properties on the quantum efficiency of a prepared transmission photocathode. The N-doped DLC films were prepared by reactive magnetron sputtering with Ar as an inert gas and $90\%N_2$+$10\%H_2$ mixture as a reactive gas. The structural properties of the DLC films were investigated by RBS and ERD measurements. The properties of the prepared transmission photocathodes were studied by measurement of the quantum efficiency. The vectorial photoelectric effect of the quartz type photocathode was studied.

2. Experimental

N-doped DLC thin films were deposited on a silicon substrate, a stainless steel mesh (samples MS), and quartz glass coated with a 5-nm-thick Cr adhesion film (samples MQ), by reactive magnetron sputtering. The mesh grid consisted of stainless steel wire with a diameter of 0.03 mm, the cell size being 0.04 mm. The quartz glass substrates were two-side polished, with dimensions 10×10×0.5 mm. Prior to deposition, standard cleaning was performed to remove impurities from the silicon surface, stainless steel mesh and quartz glass. The magnetron target was a high-purity graphite disk with a diameter of three inches. High-purity argon was used as an inert gas and the gas mixture 90% nitrogen + 10% hydrogen, as a reactive gas. The flow rates of $90\%N_2$+$10\%H_2$ and Ar gases were 2 sccm and 30 sccm for sample MS2 (MQ2), 4 sccm and 30 sccm for sample MS3 (MQ3), and 8 sccm and 30 sccm for sample MS4 (MQ4). The DLC film for sample MS1(MQ1) was sputtered without a reactive gas. The DLC film was deposited at a working pressure of 0.6 Pa and the magnetron input power was a combination of RF power (200 W at 13.56 MHz) and DC power (U_{DC} = 550 V, I_{DC} = 150 mA). During the magnetron sputtering, the substrate holder was biased at U_b = -150 V. The substrate holder temperature during deposition was 100 °C. The film thickness for the silicon substrate and the stainless steel mesh was about 200-250 nm, and for quartz glass, about 25 nm. The DLC films were deposited on both sides of the mesh under the same deposition conditions. The DLC films on the Si substrate were used for structural characterization. The elements' concentrations in the films were analyzed using Rutherford backscattering spectrometry (RBS) and elastic recoil detection (ERD) simultaneously [11]. A lift-off technique was used for Al mesh preparation on DLC/quartz structures. Fig. 1 shows the technological steps of preparation and shaping of Al mesh on quartz-type transmission photocathodes. For the study of the vectorial photoelectric effect, the quartz-type photocathode was irradiated by intensive laser pulses (three-four pulses with laser power density of about ~5 MW/cm^2) to create micron-size perforation in the Cr and DLC films by laser ablation. The transmission photocathode quantum efficiency testing was performed at JINR Dubna. The 15-ns UV laser pulses (quadrupled Nd:YAG laser, 266 nm) with a laser spot size of 5 mm were used to illuminate the backside of the photocathode. To draw the electrons from the DLC-film-coated stainless-steel mesh and the quartz-glass photocathode, the cathode was biased by a negative voltage of roughly 20 kV. The bunch charge was measured by using a Faraday cup (FC) connected to the ground through a measuring capacitor with a matched cable; the voltage on the charging capacitor (U_c) was monitored by a 500 MHz oscilloscope.

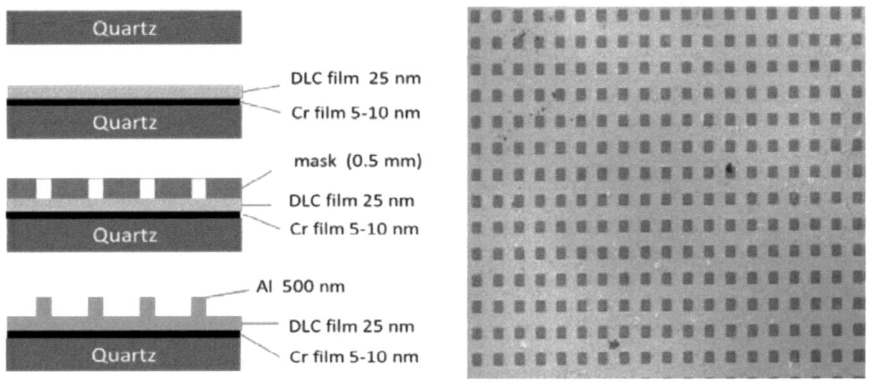

Figure 1. Transmission photocathode technology based on quartz glass coated with a DLC film.

3. Results and discussion

The measured and simulated RBS and ERD spectra for the MS1 to MS4 samples are shown in figure 2. The RBS and ERD analyses indicated that the films contain carbon, nitrogen, hydrogen, and a small amount of oxygen. In figure 2a, the RBS spectra show a leading edge of about ~530 ch corresponding to the Si substrate. The slight difference of the Si edge position between the samples can be attributed to the differences in the films' thickness. The higher counts corresponding to C are observed in the profile of the bulk Si at around ~320 ch. Higher counts corresponding to N are observed in the profile of the bulk Si at around ~370 ch. In figure 2b, the leading edge of H occurs at approximately 530 ch. The concentrations of elements for all samples were: sample MS1(carbon 90 at.%, nitrogen 1-2 at.%, hydrogen 7 at.%, oxygen 1-2 at.%), sample MS2 (70, 15, 14, 1-2), sample MS3 (64, 21, 14, 1-2), and MS4 (61, 23, 15, 1-2), respectively. We assume that the concentration of hydrogen and nitrogen in the MS1 film was mainly caused by the vacuum chamber atmosphere. Adding the gas mixture 90%N_2+10%H_2 to the working gas results in reducing the carbon concentration in the films from 90 at.% to 61 at.%. The hydrogen concentration in samples MS2-4 was practically the same. We also assume that the elements' concentrations in the films for MQ samples were the same as in the MS samples due to the simultaneous preparation in one vacuum cycle. The DLC film sputtering on a quartz substrate takes place at the end of the DLC film sputtering on a stainless steel mesh and silicon to obtain a thickness of about 25 nm. Figure 3a) shows the bunch charge as a function of the laser

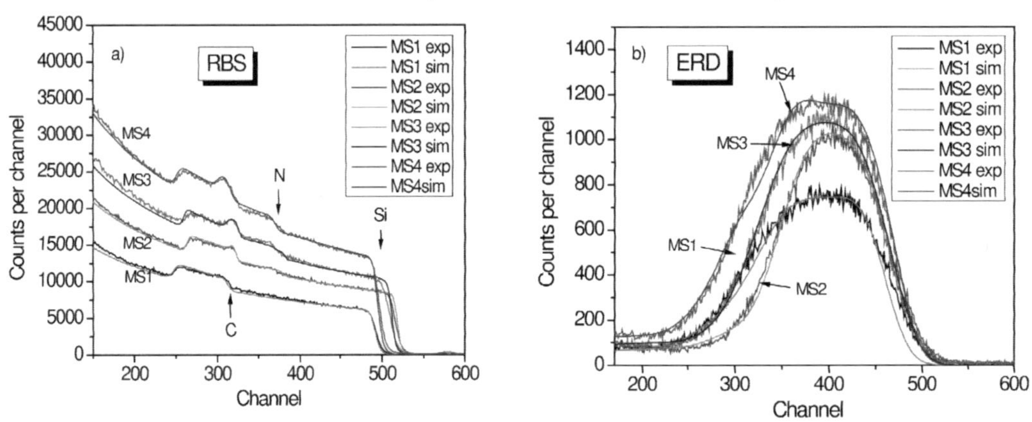

Figure 2. RBS a) and ERD b) experimental and simulated spectroscopy results for samples MS1-4.

energy for MQ prepared photocathodes after micron-sized perforation. The bunch charge as a function of the laser energy for MS prepared photocathodes was about three times lower. The bunch charge increased as the laser energy increased. The increase in the laser energy means an increase in the number of photons causing photoemission, which leads to an increase in the number of photoelectrons produced at the cathode surface. The bunch charge intensity trend of the MQ photocathodes reported in figure 3a evidences a dependence of the photoemission properties on the graphitic component inside the DLC films and the nitrogen concentration. Incorporated nitrogen mainly exists as N-sp^{2}C and N-sp^{3}C bonds and small amounts exist as N-H bonds. The N-N bonds may be related to encapsulated nitrogen [12]. Usually, we measure the extracted charge Q as a function of the laser energy EL. The QE was calculated from a linear fit to the data points at low charge densities, where space charge effects are negligible and the relation $Q(EL)$ is linear. The quantum efficiency was calculated using the formula

$$QE = N_e/N_\upsilon, \quad (1)$$

where N_e is the number of electrons and N_υ is the number of photons. Figure 3b shows the quantum efficiency as a function of the type of the prepared transmission photocathode. The quantum efficiency for MS type transmission photocathodes were: MS1 – 0.9×10^{-4} %, MS2 × 1.0×10^{-4} %, MS3 – 1.5×10^{-4} % and MS4 – 1.8×10^{-4}%. The quantum efficiency for MQ type transmission photocathodes after micron-size perforation were: MQ1 – 4.4×10^{-4} %, MQ2 – 5.0×10^{-4} %, MQ3-$7.2\text{x}10^{-4}$% and MQ4 – 1.8×10^{-4}%. On the basis of the results presented, it can be concluded that the QE of the MQ photocathode is higher than the QE of the MS photocathode. The emission mechanism in DLCs has been difficult to understand. The experimental evidence suggests that the main barrier to emission from a DLC is large and at the front surface. Lowering the barrier requires some types of heterogeneity [13]. In our case, the results obtained show that lowering the barrier can be achieved by nitrogen doping of the DLC film. The quantum efficiency (QE) calculated at a laser energy of 0.4 mJ rose as the nitrogen concentration in the DLC films was increased, and dramatically rose after micron-size perforation of the quartz type photocathodes. The vectorial photoelectric effect was observed in all MQ and MS cathodes investigated, but was much more efficient in the MQ cathodes. In the case of MS cathodes, the effect is suppressed, as the net-substrate geometry does not allow sending p polarized laser light at an optimal angle, so that the highest possible amount of charge for a given laser energy cannot be achieved. These results are not only useful in the optimization of a photocathode, but also allow a deeper understanding of the photoemission physics.

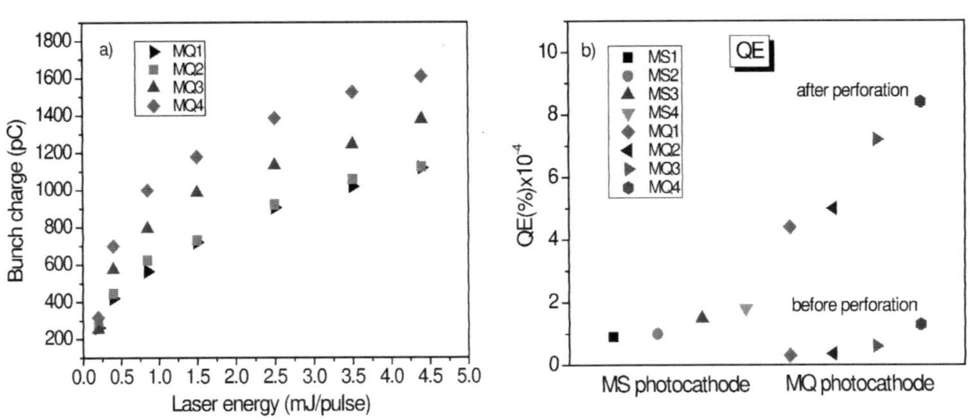

Figure 3. Bunch charge versus laser energy (a) for MQ transmission photocathodes and quantum efficiencies (b) for MS photocathodes and MQ photocathodes before and after micron-size perforation.

4. Conclusions

We investigated the influence of N-doped diamond-like carbon (DLC) films properties on the quantum efficiency of prepared transmission photocathode. N-doped DLC thin films were deposited on a silicon substrate, a stainless steel mesh and quartz glass (coated with a 5-nm-thick Cr adhesion film) by reactive magnetron sputtering. The RBS and ERD analyses showed that the films contain carbon, nitrogen, hydrogen and a small amount of oxygen. The nitrogen concentration rose with the higher flow of reactive gas and reduced the carbon concentration. A microelectronic technology was used for preparation of quartz-glass-type transmission photocathodes. The quantum efficiency (QE) calculated at a laser energy of 0.4 mJ rose with the increase in the nitrogen concentration in the DLC films for all transmission photocathodes and dramatically rose after micron-size perforation of the quartz-type photocathodes. The best results for the QE (%) was 8.4×10^{-4} for a quartz-type transmission photocathode. The results indicate that the surface properties are critical for the vectorial photoelectric effect.

Acknowledgment

This research has been executed in the framework of the Topical Plan for JINR Research and International Cooperation (Project 02-0-1067-2013/2017) and supported by the Slovak Research and Development Agency under contract APVV-0443-12.

References

[1] Nemanich R J, Bilbro G L, Bryan E N, Koeck F A, Smith J R and Tang Y 2010 *Proc. 3rd Int. Conf. Nanoelectronics (IEEE010)* ed Paul K Chu (2010 Hong Kong) p 56
[2] Sessa V, Orlanducci S, Fiori A, Terranova M L, Tazzioli F, Vicario C, Boscolo I, Cialdi S and Rossi M 2005 *Proc. SPIE* **5838** p 216
[3] Gupta S, Weiner B R, Weiss B L and Morell G 2001 *Proc. Symp Mater. Res. Soc.* (2001 San Francisco) **675** W6.9.1
[4] Nitti M A, Colasuonno M, Nappi E, Valentini A, Fanizza E, Bénédic F, Cicala G, Milani E and Prestopino G 2008 *Nucl. Instrum. Methods* A **595** 131
[5] Maia da Costa M E H and Freire Jr F L 2010 *Surf. Coat. Technol.* **204** 1993
[6] Lu Z G and Chung C Y 2008 *Diamond & Related Materials* **17** 1871
[7] Guo C T and Ueng H Y 2014 *Vacuum* **107** 304
[8] Batory D, Gorzedowski J, Rajchel B, Szymanski W and Kolodziejczyk L 2014 *Vacuum* **110** 78
[9] Pang X, Peng H, Yang H, Gao, Wu X and Volinsky A A 2013 *Thin Solid Films* **534** 226
[10] Dudarev A, Budagov U A, Balalykin N, Kobets V, Lyablin M V, Sabirov B M, Shirkov G, Syresin E and Trubnikov G V 2013 *Proc. IPAC 2013* (2013 Shanghai) p 1631
[11] Kobzev A P, Huran J, Maczka D and Turek M 2009 *Vacuum* **83** S124
[12] Liu Y, Yu H, Quan X, Chen S, Zhao H and Zhang Y 2014 *Sci. Rep.* **4** 6843
[13] Robertson J 2003 *IEICE Transaction on Electronics* **E86-C (5)** 787

Transformation of electromagnetically induced transparency into absorption in a thermal potassium optical cell with spin preserving coating

S Gozzini[1], A Lucchesini[1], C Marinelli[2], L Marmugi[4], S Gateva[3], S Tsvetkov[3,4] and S Cartaleva[3]

[1]Istituto Nazionale di Ottica CNR – UOS Pisa, Via Moruzzi 1, 56124 Pisa, Italy
[2]Department of Physical, Geological and Environmental Science, University of Siena, via Roma 56, 53100 Siena, Italy
[3]Emil Djakov Institute of Electronics, Bulgarian Academy of Sciences, 72 Tzarigradsko Chaussee , 1784 Sofia, Bulgaria
[4]INO and Department of Physics and Astronomy, University College London, Gower Street, London WC1E 6BT, United Kingdom

E-mail: stocvet@ie.bas.bg

Abstract. We report a new experimental approach where an order of magnitude enhancement of the electromagnetically induced absorption (EIA) resonance contrast, thus making it similar to that of the EIT resonance contrast is observed under the same conditions. The EIA signal results from the interaction of a weak probe beam with a ground state that has been driven by the pump (counter-propagating) beam. Probe absorption spectra are presented where the laser frequency is slowly detuned over the D_1 line of ^{39}K vapor contained in a cell with a PDMS antirelaxation coating. In addition to the frequency detuning, a magnetic field orthogonal to the laser beams is scanned around zero value at a higher rate. With both laser beams linearly polarized, an EIT resonance is observed. However, changing the pump beam polarization from linear to circular reverses the resonance signal from EIT to EIA.

1. Introduction

The phenomenon of electromagnetically induced transparency (EIT) has many applications in laser physics, precision laser spectroscopy, quantum information, all-optical magnetometers, miniaturized atomic clocks, precision measurements of fundamental symmetry, etc. In the fluorescence signal from a vapor cell, the EIT resonance is usually observed as a narrow dip. Consequently, the EIT signal is also called "dark resonance". One of the most significant and useful features of the EIT resonance is its width, which can be much smaller than the natural line width. Perhaps these resonances are the narrowest ones that can be observed in a cell filled with 'hot' atomic vapor (i.e. at room temperature).

The electromagnetically induced absorption (EIA) resonance, a sub-natural-width resonance ("bright resonance"), has applications in an enhancement of the group velocity of the light ("fast" light), four-wave mixing, magnetic field mapping of cold atomic samples. However, the scope of EIA applications is rather small due to some problems that are yet to be solved. So far, the EIA has been

[4] To whom any correspondence should be addressed.

Journal of Physics: Conference Series **700** (2016) 012051 doi:10.1088/1742-6596/700/1/012051

predominantly observed in closed "bright" hf transitions. The most effective methods for sufficient narrowing of the resonance width are based on the usage of a buffer gas or an optical cell with an anti-relaxation coating on the walls. These methods, although demonstrating perfect results with EIT signals, happen to be useless for EIA ones in conventional schemes of resonance observation. The collision process between atoms and buffer gas or cell walls rapidly destroys the EIA signal, or even changes its sign. For a long time, EIA resonances with a good contrast-to-width ratio could not be achieved [1,2].

In this paper, we suggest a new approach to observing EIA signals with a potential for significant (up to an order of magnitude) enhanced resonance contrast. The scheme is based on open hyperfine (hf) transitions of resonance lines in alkali atoms. The scheme proposed allows one to use a cell with anti-relaxation coating of the walls or a buffer gas, in order to improve the resonance properties. In this case, the buffer gas does not lead to destroying the EIA resonance [3] as in regular schemes for EIA observation. In the new scheme, the EIA signal results from the interaction of a low intensity probe beam with the ground atomic state of K (D_1 line) that has been driven in Hanle configuration by a pump laser beam. The pump and probe beams are taken from the same mono-mode laser source and arranged as for the case of saturated absorption (SA) observation [4]. The sign of the coherent resonance (i.e. EIT or EIA) can be controlled by the polarization (linear or circular) of both laser beams.

2. Experimental results and discussion

2.1. D_1 line of potassium

The optical cells used in the experiment contained atomic vapor with natural abundance of the two stable isotopes of potassium with mass numbers 39 and 41. The potassium-39 isotope atoms are predominant (93.3%). In figure 1 we show the level diagram of the D_1 line (770.1 nm) of potassium-39. The ground level frequency splitting is smaller than the Doppler width of the optical transitions (0.77 GHz at 300 K). The spectrum of potassium is much less studied than those of Na, Rb, and Cs. Several features distinguish K from the other alkali atoms; in particular, the hf ground-state splitting is smaller than the Doppler width. Thanks to this fact, the hf transitions starting from a single ground level are completely overlapped. Moreover, the hf components from both the ground state levels cannot be spectrally resolved. As a result, the crossover (CO) resonances that involve both ground state levels, being negligible for Cs and Rb and suppressed for Na, are usually dominant for K [5].

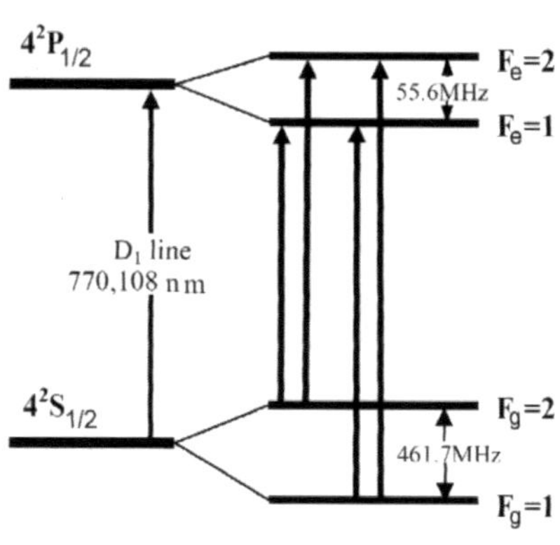

Figure 1. Energy level diagram and hyperfine transitions of the D_1 line of potassium

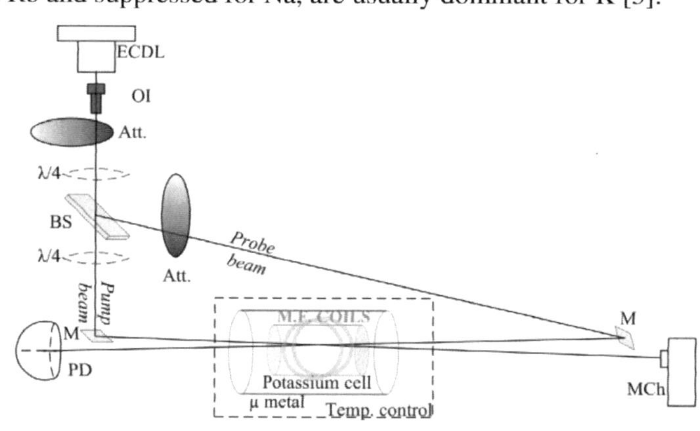

Figure 2. Experimental setup. ECDL – extended cavity diode laser; OI – optical isolator; PD – photodiode detector, M – mirror, BS – beam splitter, Att – attenuator, MCh – mono-chromator.

2.2. Experimental setup

A sketch of the experimental setup is shown in figure 2. The laser beam produced by an extended cavity diode laser (ECDL, Toptica DL 100) is tuned to the D_1 line of potassium (770.1 nm). It is divided in a pump and a probe beams that enter into a PDMS coated optical cell from two opposite directions and overlap inside it. The laser intensities are controlled by two adjustable attenuators in order to obtain the needed power ratio between the pump and probe beams. The polarization of the probe beam is set either linear or close to circular by a $\lambda/4$ plate placed in both beam arms. The pump beam polarization is linear or close to circular, controlled by a second $\lambda/4$ plate placed in its beam arm. A heater controls the temperature of the measurement cell. The PDMS coated cell is protected against stray laboratory magnetic fields by a μ-metal shield. The cell is situated between two coils in a Helmholtz configuration providing a magnetic field variable around zero value and orthogonal to the laser beam propagating direction. The laser frequency is scanned across the D_1 line spectrum, and the transmission of the probe beam is detected by a photodiode.

2.3. Saturated absorption resonances on the D_1 line of K

Transmitted spectra of the probe beam with the SA resonances are shown in figure 3, in case of uncoated optical cell. Three well separated groups of SA resonances can be clearly distinguished.

(i) The first group is formed on the ground state level $F_g = 2$ set of hf transitions and is situated at the lower frequency part of the transmission profile of the D_1 line. Here, both hf transitions start from the same hf level of the ground state. This group consists of three resonances of reduced absorption. The conventional crossover resonance situated in the middle of the group combines transitions from the same ground level to different excited hf levels. Here the dip in absorption shows that at each resonance the probe beam sees the depleted atomic population of the ground level that is caused by the counter propagating pump beam. The behavior of the crossover resonance produced by the $F_g = 2$ transitions shows that the resonance peak is influenced by the pump beam polarization, being of higher amplitude in case of elliptically polarized pump beam.

(ii) The second group of SA resonances is due to the $F_g = 1$ set of transitions and is situated on the higher frequency wing of the absorption profile. Here also, both hf transitions start from the same ground level.

(iii) Between the two groups, a third group of sub-Doppler-width resonances is formed and is identified as a particular CO group, which involves the hf transitions starting from both ground-state levels. The contrast of the resonances belonging to the last group is much larger than that of the previous two groups. Moreover, they are with opposite signs compared to those of the two previous groups. Here, the enhanced absorption resonances combine transitions sharing the same excited hf level, as well as four separate hf transitions that do not share any of the excited or ground hf levels. In this group, the probe beam measures a very strong accumulation of atoms on the ground levels caused

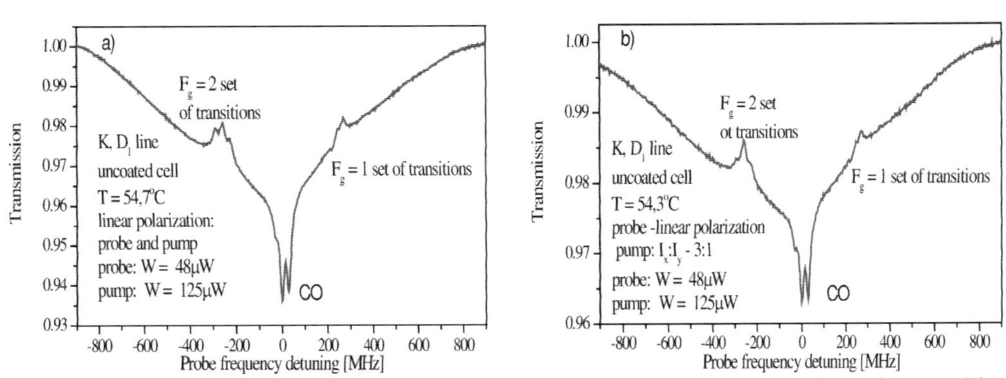

Figure 3. Transmission of the probe beam with SA resonances on the D_1 line of K. In this work, the polarization of the probe beam is kept linear, while that of the pump beam is changed: (a) the pump beam is linearly polarized; (b) the pump beam is with elliptical polarization.

by the pump beam, due to the velocity selective hf optical pumping. Such an accumulation is the reason of the saturated resonance sign reversal from reduced to enhanced absorption.

Further discussion of the processes involved shows that the area of the last group of transitions is the most promising for formation of the EIA resonances with a good contrast in a coated cell.

2.4. EIA resonance observation in optical cell with anti-relaxation coating

The EIA resonances in Hanle configuration have been first observed in uncoated optical cells, irradiating Rb vapor by linearly [6] or circularly [7] polarized mono-mode laser light. In the case of a buffered cell containing alkali atoms, the EIA resonance transforms to an EIT one due to the depolarization of excited state resulting from Rb atom collisions with the buffered gas atoms [1].

While in [3] a new method has been theoretically proposed for observing a high-contrast and narrow-width EIA resonances using a buffered gas cell and a single hf transition of alkali atoms in Hanle configuration under counter propagating pump and probe light waves, here we decided to use an anti-relaxation coated optical cell that shows better efficiency for excitation by mono-mode laser light. Moreover, instead of using a single hf transition, in our case the EIA resonance formation takes place in the CO area (figure 3), where all 4 hf transitions contribute to the EIA resonance contrast.

The first experimental result observed by the new experimental approach is illustrated in figure 4. A coated optical cell containing pure K vapor is irradiated by a counter propagating: pump beam and a probe beam of significantly lower intensity. The absorption profile of the probe beam is measured under conditions of slow laser frequency scan and much faster variations around zero value of a magnetic field orthogonal to both laser beams. At low probe beam intensity that is used in our experiment, no Hanle resonance is observed experimentally when the pump beam is switched off. With the pump beam (of a significantly higher intensity) involvement, a Hanle resonance with a very good contrast is observed in the probe beam absorption. In the case of both beam being linearly polarized, EIT resonances are observed with contrast that is more than 70% of the maximal absorption

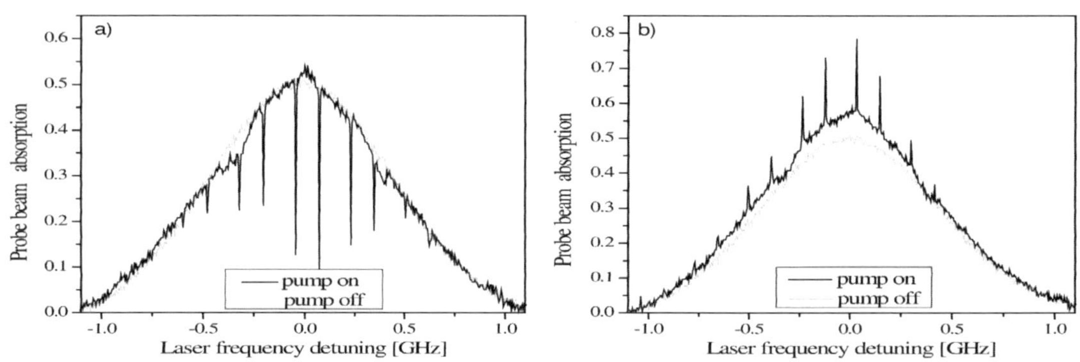

Figure 4. EIT (a) and EIA (b) resonances observed by means of scan of the laser frequency over the absorption profile and magnetic field varying around zero value. (a) pump and probe beams are of linear polarization; W_{pump} = 2.92mW, W_{probe} = 0.98mW; (b) pump beam is almost circularly polarized (1:1.4), probe beam is linearly polarized; W_{pump} = 2.91mW, W_{probe} = 0.57mW.

of the D_1 line (figure 4a). When the pump beam polarization is changed to circular, the Hanle resonance sign is reversed and an EIA resonance is demonstrated with a dramatically increased contrast, more than 40% of the atomic absorption (figure 4b). Thus, the new experimental method proposed shows extremely good efficiency of EIA resonance preparation. The resonance width observed is 150 mG. It is enhanced by the low-quality magnetic shielding and not completely homogeneous magnetic field applied.

2.5. Discussion of processes involved in EIA resonance observation

Figure 5 illustrates the main hf transitions and processes leading to the EIA resonance formation. When the polarization of the pump beam is circular, the hf transition $F_g = 1 \rightarrow F_e = 1$ leads to atomic accumulation on the magnetic sublevel m = 1 of the $F_g = 1$ level. Such an accumulation is provided by the fluorescence from the $F_e = 1$ Zeeman sublevels. Note that the m = 1 sublevel does not interact with the pump beam, i.e. its population is significant. Both the $F_g = 2 \rightarrow F_e = 1,2$ transitions produce strong

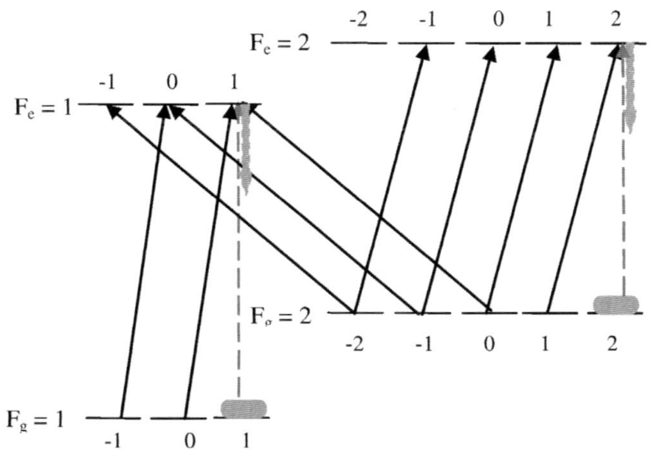

Figure 5. Magnetic sublevels of the potassium D_1 line hf levels and hf transitions involved in the EIA resonance formation. The circularly polarized pump beam is involved in σ+, $F_g = 1 \rightarrow F_e = 1$ and $F_g = 2 \rightarrow F_e = 1,2$ transitions (black solid arrows) between magnetic sublevels. The linearly polarized probe beam transitions are only shown that are responsible for the EIA resonance observation (red dash arrows).

atomic accumulation on Zeeman sublevel m = 2 of the $F_g = 2$ that also does not interact with the pump light. It is worth noting that the population of the ground-state magnetic sublevels not interacting with the pump beam is the highest in the case of magnetic field $B = 0$, i.e. at the maximum of the Hanle EIA resonance profile that is measured by the lower intensity probe beam. Further on, with the enhancement of the magnetic field value B in both directions with respect to $B = 0$, the atoms accumulated on m = 1,2 ground Zeeman sublevels evolve on the other Zeeman sublevels and the probe beam sees a reduction of atomic population, i.e. the low-absorption wings of the EIA profile.

3. Conclusions

A new approach is proposed and demonstrated that provides formation of an EIA resonance with a contrast enhancement by one order of magnitude. Now the EIA and EIT resonances can be of comparable contrast. The new method is experimentally demonstrated in K vapor contained in a coated cell, thus realizing advanced combination of a narrow resonance with its high contrast. The amplitude and the sign (EIT or EIA) of coherent resonances can be easily controlled by the polarization of the involved laser beams. Further work is planned related to the magnetic shielding improvement and providing a homogeneous, orthogonal to the laser beam magnetic field, aimed at a significant resonance narrowing.

This result is expected to contribute significantly to expanding the EIA resonance applications in numerous new important fields.

Acknowledgements

This work was supported by Marie Curie International Research Staff Exchange Scheme Fellowship within the 7th European Community Framework Program and by the Italian National Research Council and the Bulgarian Academy of Sciences (Program CNR/BAS 2013-2015).

References

[1] Andreeva C, Cartaleva S, Dancheva Y, Biancalana V, Burchianti A, Marinelli C, Mariotti E, Moi L and Nasyrov K 2002 *Phys.Rev.* A **66** 012502
[2] Ho-Jung Kim and Han Seb Moon 2011 *Opt. Express* **19**/1 168

[3] Brazhnikov D, Taichenachev A, Tumaikin A and Yudin V 2014 *Laser Phys. Lett.* **11** 125702
[4] Gozzini S, Lucchesini A, Marinelli C, Marmugi L, Gateva S, Tsvetkov S and Cartaleva S 2015 *Proc. SPIE* **9447** 944708
[5] Bloch D, Ducloy M, Senkov N, Velichansky V and Yudin V 1996 *Las. Phys.* **6** 670–8
[6] Dancheva Y, Alzetta G, Cartaleva S, Taslakov M, Andreeva Ch 2000 *Opt. Commun.* **178** 103
[7] Renzoni F, Cartaleva S, Alzetta G and Arimondo E 2001 *Phys.Rev.* A **63** 065401

The impact of active layer nanomorphology on the efficiency of organic solar cells based on a squaraine dye electron donor

D Stoyanova[1], S Kitova[1,4], J Dikova[1], M Kandinska[2], A Vasilev[2], I Zhivkov[1,3] and A Kovalenko[3]

[1]Acad. J. Malinowski Institute for Optical Materials and Technologies, Bulgarian Academy of Sciences, Acad. G. Bonchev str., bl. 109, 1113 Sofia, Bulgaria
[2]Faculty of Chemistry and Pharmacy, St. Kliment Ohridski University of Sofia, 1 James Bourchier blvd., 1164 Sofia, Bulgaria
[3]Brno University of Technology, Faculty of Chemistry, Materials Research Center, Purkyňova 118, 612 00 Brno, Czech Republic

E-mail: skitova@iomt.bas.bg

Abstract. The possibilities were studied of improving the photovoltaic performance of solution processed BHJ solar cells by solvent vapor annealing (SVA) of the active layers, based on a squaraine dye Sq1 as a donor and the fullerene derivative PCBM as an acceptor. For this purpose, the optical properties were determined of as-deposited and of annealed with tetrahydrofuran (THF) vapors for different duration Sq1/PCMB layers, as well as the efficiency of cells built on their basis. A considerable change was established in the absorption spectra of treated for only a few seconds films and a twofold increase of the power conversion efficiency after 6 sec SVA. The results obtained are explained in terms of solvent vapor induced phase separation and formation of squaraine dye small aggregates in the blend nanostructure. The assumption made was confirmed by morphological investigation of as-deposited and of annealed Sq1/PCBM blended layers. On this basis, the impact of the active layer nanomorphology on the efficiency of solar cells based on squariane dye as electron donor is discussed.

1. Introduction

During the past two decades, the organic solar cells (OSCs) have drawn considerable interest due to their high potential of low-cost, light-weight and simple fabrication of roll-to-roll production on flexible substrates [1]. In the most successful implementation, a soluble n-type fullerene derivative, (6,6)-phenyl C_{61} butyric acid methylester ($P_{61}CBM$) as acceptor, is blended with the donor from conjugated *p*-type polymers or small molecular weight organic semiconductors to form bulk-heterojunction (BHJ) cells. Although high power-conversion efficiencies (PCEs) of 10.8 % have been achieved [2], the polymeric solar cells still suffer from some shortcomings, such as batch-to-batch variation, indefinite molecular weight, polydispersity and impurity. In contrast, small molecular weight materials intrinsically do not have such drawbacks. Additionally, they have a well-defined molecular structure with a definite molecular weight, giving the advantages of high purity, tunable

[4] To whom any correspondence should be addressed.

electronic structures and better device reproducibility [3]. Even though the highest PCE of 9.96 %, obtained very recently in the solution-processed small molecule OSCs [4], in the most cases PCE is still lower than that of the polymer devices. It is considered that this is mainly due to the difficulty of controlling the nanomorphology of the active layers, where donor and acceptor components self-assemble to form bicontinuous interpenetrating networks [5]. The latter should allow the excitons to diffuse to the donor-acceptor interface and charge carriers to transport to the respective electrodes after separation. Therefore, the challenge of solution processed BHJ small molecule OSCs is to enhance the molecular order in the active layers for a better charge carriers' mobility, and at the same time, to restrict the domain size due to the limited exciton diffusion length in these materials. Different strategies for resolving the problem have been reported, among which the solvent vapor annealing (SVA) seems very promising [6].

Recently, we studied the potentiality of a symmetrical n-hexyl substituted squaraine dye, marked as Sq1, for using as electron donating component in the active layer of solution processed BHJ organic solar cells [7]. The dye was in-house synthesized by optimizing a method proposed earlier in the literature [8]. Optical modelling was performed to predict and improve the photovoltaic performance of a BHJ device with a blended Sq1/$PC_{61}MB$ active layer; encouraging results were thus obtained. However, the initial experiments with constructed real cells showed a moderate PCE, not exceeding 0.7-0.8 %. Obviously, the improvement of the active layer morphology is a must for achieving an increased power conversion efficiency.

In this work, studies are presented on the influence of solvent vapor annealing treatment on the photovoltaic performance of solution processed BHJ organic solar cells based on Sq1 dye as an electron donor and PCBM as an acceptor. For the purpose, the absorption spectra of as-deposited and annealed Sq1/PCMB blended layers were investigated and the efficiencies of the solar cells were determined. The morphological changes induced by the solvent vapors were followed by TEM observations.

2. Experimental

The squaraine dye Sq1, synthesized earlier [7], is 2-(5-(((2,2-diphenylhydrazono) methyl)-1-hexyl-1H-pyrrol-2-yl)-4-(5-(2,2-diphenyl-hydrazono)-ylidenemethyl)-1-hexyl-2H-pyrrol-1-ium)-3-oxocyclobut-1-enolate. The soluble $PC_{61}BM$ acceptor, with a purity of more than 99 %, was supplied by Sigma-Aldrich. The blended active layers were obtained by spin casting of 20 mg/ml chloroform solution of Sq1 and $PC_{61}MB$ in a weight ratio 1:2, previously determined as optimal.

The annealing of the blended Sq1/PCBM films was performed by exposing the samples to vapors of anhydrous tetrahydrofuran (THF $\geq$ 99.9 %, Sigma-Aldrich). The choice of THF was done on the basis of literature data according to the requirements for used solvents to have high vapor pressure and medium dissolving ability of the respective donor component [9]. The SVA processing of Sq1/PCBM blend films was carried out in nitrogen atmosphere at room temperature 21°C.

The absorption spectra of as-cast and solvent-annealed Sq1/PCBM blend films were recorded using a Varian Cary 50 Probe Spectrophotometer (UV-VIS-NIR). The samples studied were spin-coated over glass substrates from a 20 mg/ml chloroform solution. The non-annealed film served as a reference.

For studying the effect of the SVA treatment on the photovoltaic performance, BHJ solar cells were built with the structure ITO/PEDOT:PSS(~60nm)/Sq1:PCBM(~100nm)/Al(~100nm), developed in the previous work [7]. The depositions of the photoactive and Al layers, as well as the encapsulation procedure, were fulfilled in a nitrogen filled glove box. The Al cathode was thermally evaporated on the top through a shadow mask giving an active device area of 4.76 mm^2. The thicknesses of the individual layers in the cell's structure were controlled by using a Bruker's DektakXT Stylus Profiler.

The current density-voltage (*J-V*) characteristics were obtained by a Keithley 2601B System Source Meter. The photocurrent was measured under AM 1.5G illumination at 98.8 mW cm^{-2} under LOT – QuantumDesign solar simulator, as the light intensity was preliminary determined by a Si reference solar cell. From the *J-V* curves, the basic characteristics J_{sc}, V_{oc} and *FF* were determined, whose product gives the maximum power of the electrical energy P_{max} generated. Here J_{sc} is the short circuit current, V_{oc} is the open circuit voltage and FF is the fill factor [10]. The power conversion efficiency PCE was determined as the ratio between the maximum power of the electrical energy generated and the power of the incident sun light energy: PCE = $P_{max}/P_{(solar\ in)} = J_{sc}V_{oc}FF/P_{(solar\ in)}$.

The investigation of the SVA impact on the morphology of the Sq1/PCBM blend films was performed by an HR STEM JEOL JEM 2100 high-resolution scanning transmission electron microscope at an accelerating voltage of 200 kV. All inspected films were spin-coated from 20 mg/ml chroloform blend solution on KBr lamellas.

3. Results and discussion

Figure 1 shows the molecular structure of the Sq1 dye and the fullerene derivative $PC_{61}BM$ used in the study.

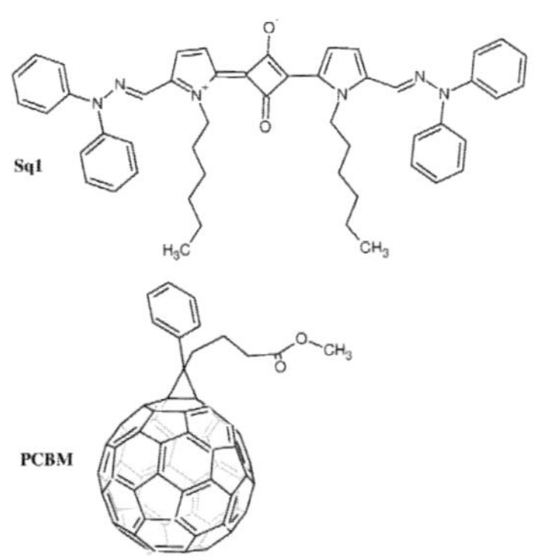

Figure 1. Molecular structures of Sq1 and $PC_{61}BM$.

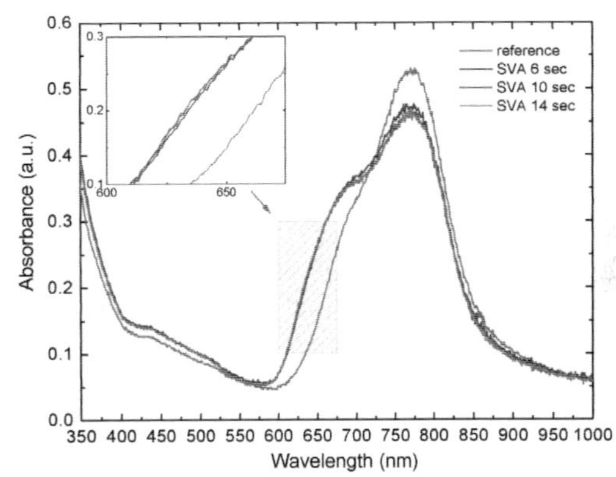

Figure 2. Absorption spectra of solvent-annealed and non-annealed 100 nm Sq1/ $PC_{61}BM$ layers.

Figure 2 presents the absorption spectra of as-cast and annealed for different durations by THF vapors Sq1/PCBM films. As seen, the spectrum of the reference virgin sample shows absorption of light across a broad range of wavelengths – from UV to NIR, with a maximum at 775 nm. The absorption in the range 350 – 600 nm is due to the presence of PCBM in the blended film, whilst the main absorption peak is caused by electron excitations in the squaraine dye molecules. After solvent vapor annealing with THF, the absorption of PCBM and Sq1 changes notably. The PCBM absorption of light is enhanced, while the maximum of the main peak is reduced considerably after only 6 seconds of treatement by THF vapors. Besides, the appearance of a shoulder-type peak is seen at ~ 690 nm in the visible part of the solar spectrum. Such an absorption band splitting into two components after SVA could be due to a good packing of the squaraine molecules in the presence of THF into a manner that favors aggregation of the chromophore molecules, when they are in close contact to each other. Similar appearance of blue-shifted absorption shoulder in relation to the band of monomeric state is explained in the literature as due to the formation of small aggregates, like dimers, trimmers and tetramers [11]. Besides, it is seen from the figure that the SVA duration affects also the magnitude of the absorbance, which slightly decreases from 6 to 10 and 14 seconds treatment. The absorption bands owing to the PCBM molecules in that case appear rather unaffected. In general, the results from the optical measurements give us a reason to suppose that some rearrangements in the active layer nanomorphology occur under SVA with THF vapors, which undoubtedly will affect the photovoltaic performance of the cells studied.

Figure 3 shows the architecture and J-V characteristics of BHJ cells with untreated and Sq1/PCBM active layers treated for different times by THF vapors. As seen from the figure, the photovoltaic response increases as a result of the SVA processing, the response being maximal for the 6 sec treatment time. A further extension of the annealing duration reduces the J_{sc} values.

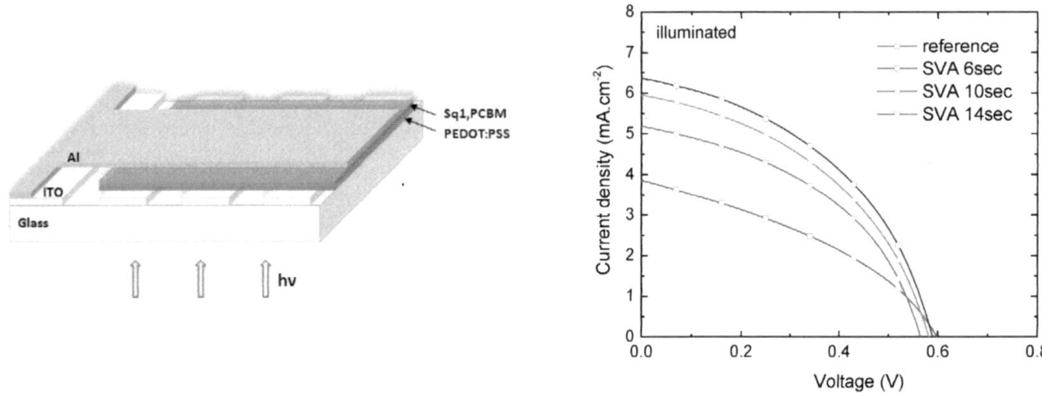

Figure 3. Architecture and *J-V* response of the BHJ solar cells for different solvent vapor annealing duration of the Sq1/PCBM active layers.

Most probably, the structural transformations in the active layers, induced by THF vapors for the 6 sec treatment, are the most favorable with respect to photocurrent generation. This statement is confirmed by the basic electrical parameters calculated from *J-V* curves and summarized in table 1. From these data, some general performance trends can be clearly discerned: i*)* the solvent treatment with THF vapors for only 6 seconds is enough to double the PCE; ii*)* the improvement of the fabricated cell's performance is associated mainly with the increase of J_{sc} as well as of *FF* and R_{sh}; iii) The values of V_{oc} are almost unaffected by the solvent annealing. Hence, the tendency observed of the electrical characteristic's changing under the action of THF vapors is similar to the literature data for other small molecule donor/acceptor blended active layers [6].

Table 1. Electrical characteristics of BHJ cells with Sq1/PCBM) active layers annealed in THF vapors for different duration.

Treatment duration (s)	*J*sc (mA cm^{-2})	*V*oc (V)	*FF* (%)	PCE (%)	*R*shunt (Ohm cm^2)	*R*series (Ohm cm^2)
0	3.66	0.59	36.93	0.79	6670.49	34.34
6	6.22	0.59	43.10	1.57	8147.78	14.61
10	6.08	0.57	41.92	1.46	7871.89	15.07
14	5.03	0.55	42.60	1.18	8128.73	15.29

Figure 4 presents TEM bright field images and the corresponding SAED patterns of Sq1/PCBM blended films as-deposited (a) and treated for 6 sec (b) and 14 sec (c) by THF vapors. A slightly inhomogeneous structure of the virgin film is clearly seen (a). After only 6 sec SVA treatment, the film nanomorphology is changed visibly, most probably due to the occurrence of phase separation (b). A further increase of the treatment duration results in an enhanced contrast of the TEM images and a better developed phase separation. Simultaneously, the diffraction patterns, shown in the lower frame of figure 4, evidence the amorphous structure of all inspected specimens. As seen, the diffraction peak intensity is increased visibly after 6 seconds SVA treatment with THF, which is probably due to obtaining a more ordered network. The latter can be regarded as optimal, having in mind that the best photovoltaic performance of the cells studied is observed for this time of solvent vapor treatment.

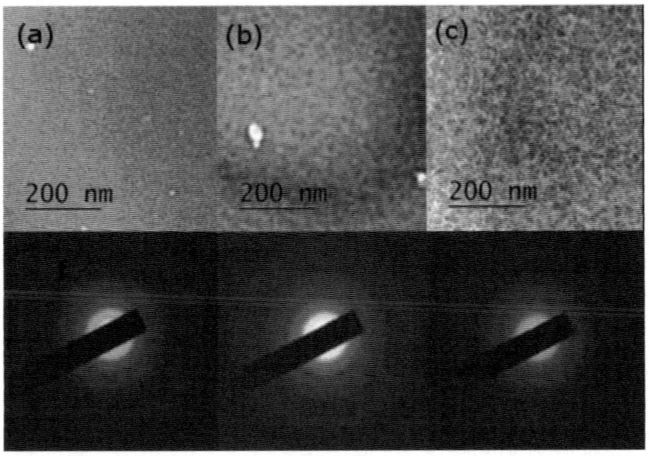

Figure 4. TEM bright field images (up) of Sq1/PCBM blend films with weight ratio of 1:2 before (a) and after SVA by THF vapors for 6 seconds (b) and for 14 seconds (c), with the corresponding SAED patterns (down).

4. Conclusions

The results of the present study demonstrate the possibilities of improving the photovoltaic performance of solution-processed BHJ solar cells by solvent vapor annealing of Sq1/PCBM blended active layers. It is established that only a few seconds processing of Sq1/PCBM films by THF vapors results in a double increase of the solar energy conversion efficiency. Simultaneously, a modification of the active layer's nanomorphology after annealing for different duration is observed, which is most probably due to assembling of the squaraine dye monomers into small aggregates. The latter statement stems from the observed notable changes in the absorption spectra of solvent vapor treated samples, including the appearance of a shoulder-type peak, explained reasonably in the literature by aggregation of the chromophore molecules. Thus, the very important role is evidenced of a controlled active layer's morphology for producing highly efficient small molecule OSCs.

Acknowledgments

This work was supported by the Czech Science Foundation by project No. 15-05095S and research infrastructure project MŠMT No. LO1211. The Bulgarian Science Fund project "SunStore" (DFNI E 02/11) also supported the investigations performed.

References

[1] Dennler G, Scharber M and Brabec C 2009 *Adv. Mater.* **21** 1323–38
[2] Chen J, Cui C, Li Y, Zhou L, Ou Q, Li C, Li Y and Tang J 2015 *Adv Mater.* **27** 1035–41
[3] Xue J, Uchida S, Rand B and Forrest S R 2004 *Appl. Phys. Lett.* **85** 5757–9
[4] Kan B, Zhang Q, Li M, Wan X, Ni W, Long G, Wang Y, Yang X, Feng H and Chen Y 2014 *J. Am. Chem. Soc.* **136** 15529–32
[5] Peet J, Kim J, Coates N, Ma W, Moses D, Heeger A and Bazan G 2007 *Nat. Mater.* **6** 497-502
[6] Wei G, Wang S, Sun K, Thompson M and Forrest S 2011 *Adv. Energy Mater.* **1** 184–7
[7] Kitova S, Stoyanova D, Dikova J, Kandinska M, Vasilev and Angelova S 2014 *J. Phys.: Conf. Series* **558** 012052
[8] Bagnis D, Beverina L, Huang H, Silvestri F, Yao Y, Yan H, Pagani G A, Marks T and Facchetti A 2010 *J. Am. Chem. Soc.* **132** 4074–5
[9] Reichardt C 2003 *Solvents and Solvent Effects in Organic Chemistry* (Weinheim: WILEY-VCH Verlag GmbH & Co. KGaA)
[10] Chaar El 2011 *Renewable and Sustainable Energy Reviews* **115** 2165–75
[11] Dimitriev O 2005 *J. Mol. Liq.* **120** 131–3

Journal of Physics: Conference Series **700** (2016) 012053 doi:10.1088/1742-6596/700/1/012053

Performance and stability of two types of bulk heterojunction polymer solar cells with sputtered back contacts

M Sendova-Vassileva[1,3], G Popkirov[1], P Vitanov[1], G Grancharov[2], V Gancheva[2], H Dikov[1], E Lazarova[1] and R Gergova[1]

[1]Central Laboratory of Solar Energy and New Energy Sources, Bulgarian Academy of Sciences, 72 Tzarigradsko Chausse, 1784 Sofia, Bulgaria
[2]Institute of Polymers, Bulgarian Academy of Sciences, 1113 Sofia, Bulgaria

E-mail: marushka@phys.bas.bg

Abstract. Two types of polymer solar cells with active layers of P3HT:PCBM and PCDTBT:PCBM in a standard configuration were deposited by spin-coating the polymer layers and sputtering the Ag back contacts. After encapsulation, they were subjected to post-deposition annealing and were studied by measuring their current-voltage characteristics and quantum efficiency spectra. Their stability was examined by exposing them to solar radiation in open-circuit conditions for several hours every day for a total of 80 hours. Their characteristics were measured regularly throughout this period. The performance of the two types of cells are compared and discussed.

1. Introduction

Polymer-based solar cells have been the subject of intensive research in recent years. The most thoroughly studied bulk heterojunction (BHJ) system has been P3HT:PCBM. It is comparatively stable under ambient conditions. The usual architecture of polymer solar cells includes ITO and PEDOT:PSS as a transparent front contact and an anode, as well as evaporated Al as a back contact. In our previous studies, sputtered contacts of both Al and Ag were employed [1]. It was observed that post-production annealing could improve their performance considerably. Another advantage of the sputtered Ag contacts that was observed was the fact that samples with such contacts did not degrade with storage time in the dark as much as those with evaporated Ag and Al ones.

The polymer poly[N-9'-heptadecanyl-2,7-carbazole-alt-5,5-(4',7'-di-2-thienyl-2',1',3'-benzothiadiazole] (PCDTBT) was introduced mainly due to its lower HOMO and the potential to obtain higher open-circuit voltages [2]. In this study, we combined it in a BHJ with PCBM and used sputtered Ag back contacts and post-deposition annealing. This system was compared to the one with a P3HT:PCBM active layer with regard to their performance through current-voltage measurements, quantum efficiency spectra and stability studies under solar irradiation.

2. Experimental

The samples studied were deposited on glass substrates covered with structured ITO from Ossila Ltd. The glass plates were cleaned using warm water, acetone and isopropanol and treated with UV-ozone

[3] To whom any correspondence should be addressed.

before spin-coating PEDOT:PSS from a water solution, followed by annealing at 130 °C in an Ar ambient on a hot plate for 30 min. This was followed by the deposition of the active layer. For the P3HT:PCBM samples, a 1:1 weight ratio of the two constituents in 12 mg/ml chlorobenzene solution was employed. For the PCDTBT:PCBM samples, the weight ratio was 1:4 in a 20 mg/ml solution of dichlorobenzene. All spin-coating depositions were carried out in air. After deposition, the samples were kept in solvent vapor for 30 min in order to achieve a better active layer structure. The Ag back contacts were deposited by magnetron sputtering. The samples thus obtained, representing a matrix of six solar cells each with an area of 4.5 mm^2, were encapsulated using a cover glass and a light-cured resin from Ossila Ltd.

The encapsulated samples were subjected to post-deposition annealing on a hot plate in ambient atmosphere. The current-voltage characteristics were measured using a fully automated computerized system employing a halogen lamp for illumination. The quantum efficiency spectra were taken by a computer-controlled equipment including a halogen lamp, a Digikrom 240 monochromator, a chopper, a lock-in amplifier and a calibrated Hamamatsu Si photodiode used as a reference.

The stability experiments were implemented by exposing the samples to sunlight for five hours every day under open-circuit conditions during the month of August in Sofia and measuring their current-voltage characteristics before and after each exposure. Seven cells of each kind were studied. The whole experiment lasted for 80 hours. The temperature of the samples during the irradiation experiments was 42 °C.

3. Results and discussion

Figure 1 shows typical current-voltage characteristics of the solar cells with a P3HT:PCBM active layer before and after post-deposition annealing, as well as after degradation in the sun for 80 hours. The post-deposition annealing for this type of samples was performed at 150 °C. It has been shown in previous studies [1] that this is the optimal temperature for this kind of active layers.

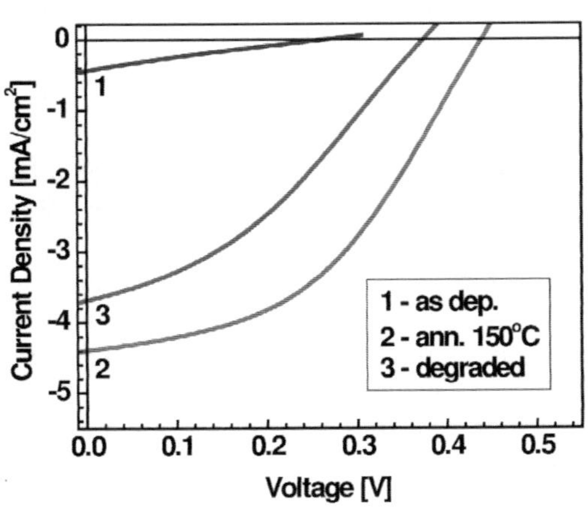

Figure 1. *J-V* characteristics of a solar cell with a P3HT:PCBM active layer – as-deposited, after annealing and after 80 hours degradation in the sun.

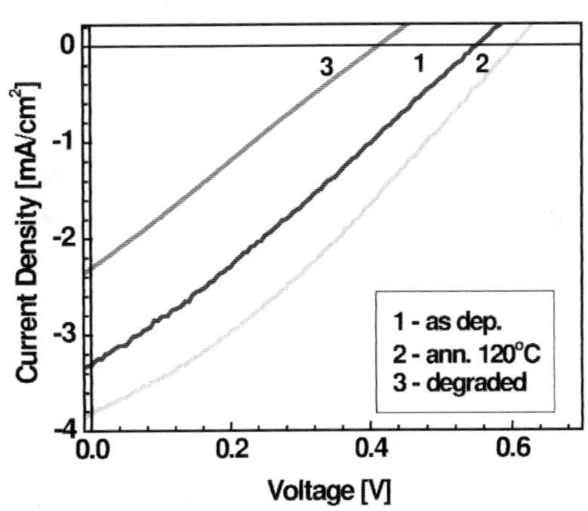

Figure 2. *J-V* characteristics of a solar cell with a PCDTBT:PCBM active layer – as-deposited, after annealing and after 80 hours degradation in the sun.

Figure 2 illustrates the current-voltage characteristics of a typical solar cell with a PCDTBT:PCBM active layer, again, as-deposited, after annealing and after degradation under the same conditions as above. The annealing this time was performed at 60 °C, 90 °C and 120 °C (the characteristic after annealing at the highest of these temperatures is only shown in the figure). Although there is data in the literature that PCDTBT should not be annealed at temperatures higher than 90°C [3], our data showed an improvement of the cell parameters after annealing at 120 °C.

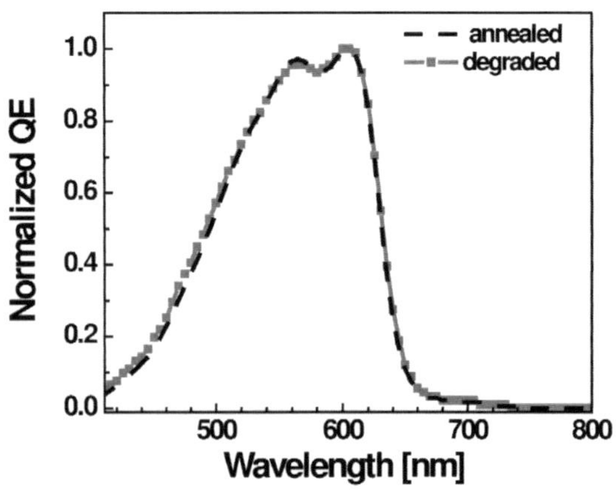

Figure 3. Normalized quantum efficiency curves of polymer solar cells with a P3HT:PCBM active layer – after annealing at 150 °C and following 80 hours of degradation under real conditions.

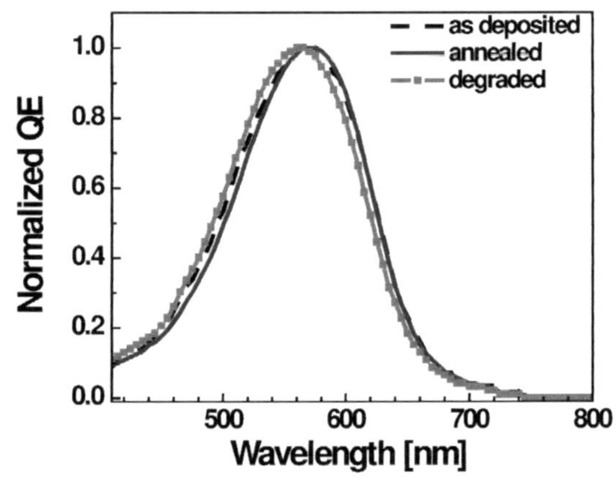

Figure 4. Normalized quantum efficiency curves of polymer solar cells with a PCDTBT:PCBM active layer – as-deposited, after annealing at 90 °C and after 80 hours degradation in real conditions.

When comparing the two figures, the following differences are noticeable. The characteristics of the cells with a P3HT:PCBM active layer, which we will denote as A-cells (as opposed to B-cells for the ones with a PCDTBT:PCBM active layer), are very poor in the as-deposited state, but improve significantly after post-deposition annealing. On the other hand, the as-deposited B-cells show better characteristics, but do not improve as much after annealing. The B-cells have a greater open-circuit voltage (U_{oc}), which is to be expected bearing in mind the literature data [2]. However, their short-circuit current density (J_{sc}) is smaller and they have a much higher series resistance, which leads to a lower fill factor as well.

Figure 3 and figure 4 display the normalized quantum efficiency spectra of the two types of solar cells before and after degradation. Both types of solar cells do not show any effective absorption of photons with wavelengths longer than 700 nm, which is a shortcoming of these active layers as they do not utilize an important part of the solar spectrum. As shown in our previous work [4], degradation under concentrated UV radiation leads to a change in the shape of the quantum efficiency spectrum of P3HT:PCBM solar cells. We interpreted this as evidence for deterioration of the interchain order in the active layer after UV irradiation. Keeping the cells under real conditions in the sun, however, does not lead to such effects in the quantum efficiency spectrum, as seen in figure 3. As evident from figure 4, annealing and degradation in the sun do not lead to a noticeable change in the shape of the quantum efficiency of solar cells with a PCDTBT:PCBM active layer.

Figure 5 presents the evolution of the solar cell parameters, fill factor (FF), short-circuit current (I_{sc}), open-circuit voltage (U_{oc}) normalized to their initial values, of two different A-cells. Analogous data for the B-cells is presented in figure 6. When comparing the degradation of the solar cell parameters, a number of peculiarities can be observed. As a whole, the B-cells have degraded to a greater extent at the end of the experiment. However, all their parameters undergo a sharp initial drop in the first few hours of solar irradiation, after which the degradation is very slow, especially for one of the samples. In the case of the A-cells, the degradation is more gradual throughout the experiment. A similar behavior has been observed for these two types of active layers in reference [5], where the samples had evaporated Al back contacts instead and the degradation experiments were carried out indoors using a solar simulator. In our experiments, the initial sharp drop is larger, takes place for a shorter time and is observed for J_{sc} as well.

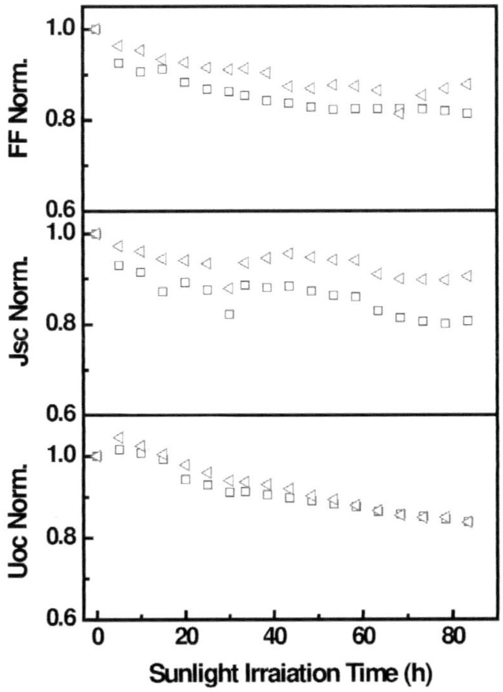

Figure 5. Evolution of the *J-V* parameters with the time of solar irradiation of two different samples with a P3HT:PCBM active layer.

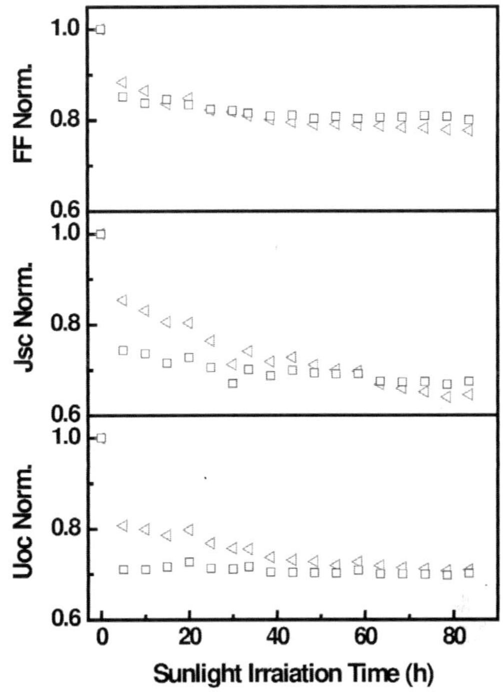

Figure 6. Evolution of the *J-V* parameters with the time of solar irradiation of two different samples with a PCDTBT:PCBM active layer.

4. Conclusions

It is shown that polymer solar cells with two different active layers, namely, P3HT:PCBM and PCDTBT:PCBM, prepared in air and having sputtered Ag back contacts demonstrate the following differences: The P3HT:PCBM solar cells show a higher J_{sc} and fill factor (*FF*), while the PCDTBT:PCBM ones have a higher U_{oc}, but their series resistance is very high. The PCDTBT:PCBM cells do not improve as much as the P3HT:PCBM ones after annealing. Some of these issues are thought to be related to the sputtered back contacts, which will be the subject of further studies.

The degradation experiments under real conditions show a sharp drop in the characteristics of the PCDTBT:PCBM solar cells in the first hours of irradiation, after which their parameters stabilize, while those of the P3HT:PCBM samples keep on falling slowly and steadily. If the initial drop is taken into account, the PCDTBT:PCBM cells degrade to a greater extent.

Acknowledgements

The work was supported financial by the Bulgarian National Scientific Fund under Contract DFNI E02/5.

References

[1] Sendova-Vassileva M, Popkirov G, Vitanov P, Dikov Ch, Gancheva V, Tsocheva D and Mokreva P 2012 *J. Phys.: Conf. Series* **398** 012049

[2] Park S H, Roy A, Beaupré S, Cho S, Coates N, Moon J S, Moses D, Leclerc M, Lee K and Heeger A J 2009 *Nat. Photonics* **3** 297

[3] Kingsley J W, Marchisio P P, Yi H, Iraqi A, Kinane C J, Landridge S, Thompson R L, Cadby A J, Pearson A J, Lidzey D G, Jones R A L and Parnell A J 2014 *Sci. Rep.* **4** 5286

[4] Sendova-Vassileva M, Popkirov G and Vitanov P 2014 *J. Phys.: Conf. Series* **558** 012048

[5] Peters C H, Sachs-Quintana I T, Kastrop J P, Beaupré S, Leclerc M and MacGehee M 2011 *Adv. Energy Mater.* **1** 491

Optical and electrical properties of nanolaminate dielectric structures

Hr Dikov[1,3], P Vitanov, T Ivanova and V Stavrov[2]

[1,]Central Laboratory of Solar Energy and New Energy Sources,
Bulgarian Academy of Sciences, 72 Tzarigradsko Chaussee, 1784 Sofia, Bulgaria
[2]AMG Technology Ltd., Microelectronica Ind. Zone, 2140 Botevgrad, Bulgaria

E-mail: dikov@phys.bas.bg

Abstract. The aim of this study was formation of a multilayered transparent conductive nanolaminate structure with optimized conductivity vs. transparency parameters. The nanolayered structure comprised one or two thin layers of dielectric materials. The overall electrical conductivity was modified by tuning the size of the planar metal granules. Magnetron sputtering system with three different targets was used for layers deposition. The advantages are: the good reproducibility of a low-temperature process allows for effective process control and, optionally, selective formation of conductive areas in a dielectric structure. Our studies revealed that the relation between the sheet resistance and the maximum transparency in the visible spectral range depends on the size of the metal granules and the film thickness of the dielectric coating. The technology provides transparent conductive coatings with well-controlled optical and electrical properties.

1. Introduction

The wide application of transparent and conductive oxide (TCO) films for optoelectronics and photovoltaics has promoted the development of low-cost materials and technological processes. Most of the TCO films are binary or ternary compounds containing one or two metallic elements. The combination of conductivity and transparency is impossible to achieve in intrinsic stoichiometric oxides. The successful approach is to create electron degeneracy in wide band gap (greater than 3 eV) oxides by introducing in a controlled way non-stoichiometry and/or appropriate dopants [1]. These conditions are conveniently achieved in oxides of tin, indium, zinc and their alloys in a thin-film form. There is an inherent limitation in the metal oxide conductivity that can be obtained by increasing the carrier concentration. When the densities become greater than 2×10^{21} cm^{-3}, the TCO exhibits plasma frequencies that shift from absorbing infrared wavelengths to visible light, reducing the transparency in the visible region [2]. The requirement for transparency and the fundamental scattering mechanism establish an absolute limit to the TCO resistivity of about $4\times10^{-5}\Omega$ cm, or, to obtain conductivity in such materials, they should be doped with an appropriate element. An alternative approach recently proposed consists of TCO/metal/TCO stacks. In order to improve the transparency and durability of a TCO film, a thin semitransparent metal layer is embedded between two TCO films. This multilayer structure has many advantages compared to a single-layer TCO coating. Good performance has been

[3] To whom any correspondence should be addressed.

achieved with other TCO/metal/TCO structures, such as ZnO/Ag/ZnO [3,4], Al:ZnO/Ag/Al:ZnO [5, 6], $InZnSnO_x$/Ag/$InZnSnO_x$ [7], and $InZnO_x$/Ag/$InZnO_x$ [8], all with sheet resistance below 10 Ω/sq and maximum transmittance of about 85% with less than 100 nm total thickness. Thus, the substitution of ITO by alternative TCOs becomes easier with TCO/metal/TCO structures, because the metal interlayer allows one to decrease the overall resistivity even though the TCO electrical quality is not optimal [9].

The aim of this study is formation of an optically transparent and conductive nanolaminate dielectric structure. The idea is based on using the electronic conductivity in granular (discontinuous) types of materials [10]. The granules are metallic particles of sizes ranging usually from a few to hundreds of nanometers, embedded into an insulating matrix [10]. The nanolayer structure is formed using one or two different dielectric materials. The electrical conductivity can be modified by varying the size of the planar metal granules [11]. The advantages of this method are reproducibility of the deposition process, effective control of the layers and compatibility with methods for low-temperature selective formation of conductive areas into the dielectric structure. Our studies revealed that the relation between the sheet resistance and the maximum transparency in the visible spectral range depends on the dimensions of the metal granules and the film thickness of the dielectric layers. Another option for engineering the optical and electrical properties of nanolaminate structures is to use different metal and dielectric layers.

2. Experimental

Magnetron sputtering system with three different targets was used for the deposition of the layers. This system has been developed as low-temperature high-speed sputtering equipment to be used in experiments and is suitable for formation of thin-films of metals and insulating materials. The substrate holder has a size of 200 mm and three targets (75 mm) placed at eccentric positions. They are cooled by water (3 *l /min*). The targets and the substrate holder are vertically arranged at a distance of 8 cm. The substrate holder is spinning at a rate of 80 rpm/min. This configuration allows us to prepare thin films with a homogeneous thickness. The three targets give the possibility to deposit successively three different layers without opening the chamber.

The most important part of the deposition process of a nanolaminate structure with specified optical and electrical properties is the deposition of a discontinuous metal layer. The size and density of the metal granules depend on the duration of the sputtering, the RF power, the type of reactive gas (Ar or He) and the substrate holder rotation rate. Evaluation and optimization of the deposition process was performed by AFM observation of the structures deposited.

Figure 1 presents AFM images of nanogranular coatings produced by Ag sputtering on p-type Si wafers with thermally grown SiO_2 (40 nm) under the following deposition conditions: reactive gas Ar, RF power 200 W and two deposition times. The metal nanoparticles with a maximum height

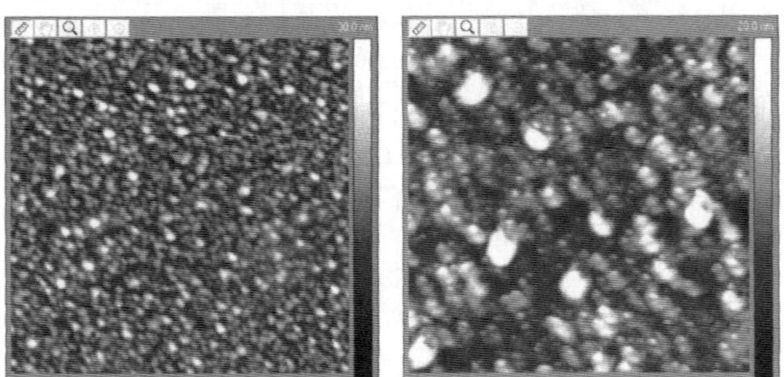

(a) (b)

Figure 1. AFM image of the Ag layers prepared at: a) sputtering time 10 s; b) sputtering time 20 s.

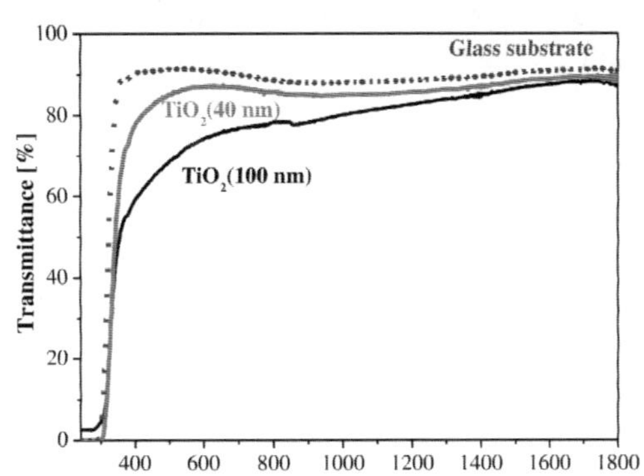

Figure 2. Transmittance spectra of sputtered TiO_2 films with thickness 40 nm and 100 nm.

approaching 20 nm (figure 1a) form a discontinuous layer and are randomly distributed. When the deposition time is raised, the granules become bigger and are located closer to one another. Indication for the density of the nanogranular layer may be obtained indirectly from the transmittance spectra. The optical properties of the nanolaminate structure dielectric/metal/dielectric (oxide/metal/oxide - OMO) were studied by means of a UV 3300 UV-VIS-NIR Shimadzu spectrophotometer.

TiO_2 films were deposited at an RF power of 300 W in Ar at a deposition rate of 10 nm/min. The film thickness was determined by a laser ellipsometer. Figure 2 shows the optical spectra of 40 nm and 100 nm thick layers.

The sheet resistance was measured using the four-point probe method, the van der Pauw method and I-V measurements of the resistors.

3. Results and discussions

The nanolaminate structures studied were TiO_2/metal (Ag)/TiO_2. The investigations revealed that the relation between the sheet resistance and the maximum transparency in the visible spectral range depends on the size of the metal granules, the film thickness of the dielectric coating, and the type of the metal used for the granular structure formation.

3.1. Dependence on the sputtering time of Ag granules

The nanolaminate structures TiO_2/metal (Ag)/TiO_2 were investigated by keeping the thickness of the TiO_2 layers at 50 nm, but varying the Ag sputtering time (4 s, 7 s and 25 s), which determines the dimensions and density of the metal granules. Table 1 presents the sheet resistance values for nanolaminate structures with different Ag layers.

For the structures investigated, the sheet resistance decreased as the optical transparency diminished [3, 6, 9]. The optical transmittance (see figure 3) and the sheet resistance are functions of the metal nanoparticles density in the middle layers. Similar results are reported for ZnO/Ag/ ZnO [12] and ZnO/Au/ ZnO [13] structures. The reflection of the multilayer system increased with the increase in the Ag layer thickness. The increase in the reflection in the near infrared region is due to the interaction of free electrons with the incident radiation.

Table 1. Sheet resistance measurements of nanolaminate structures with different Ag layers.

Structure	Sheet resistance [Ω/sq]
TiO_2/4'' Ag/TiO_2	75
TiO_2/7'' Ag/TiO_2	22
TiO_2/25'' Ag/TiO_2	6.5

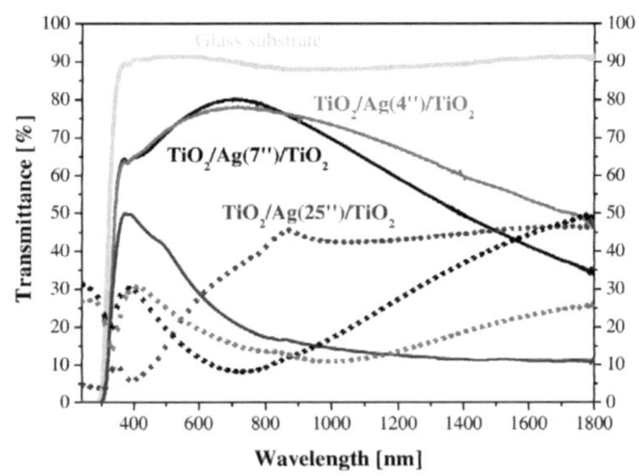

Figure 3. Optical spectra of TiO_2/Ag/TiO_2 with Ag layers of different thickness. The reflectance is shown as dotted lines.

3.2. Dependence on the metal granules type

In order to study the effect of the nanogranules of different metals, the dielectric layer used was TiO_2 deposited under the same conditions, but Ag was replaced by Ni or Cr. Table 2 presents the measured sheet resistance values of these new structures.

Table 2. Sheet resistance measurements of nanolaminate structures with different metal layers.

Structure	Sheet resistance [Ω/sq]
TiO_2/5'' Cr/TiO_2	680
TiO_2/5'' Ni/TiO_2	680
TiO_2/10'' Ni/TiO_2	310

The optical spectra shown in figure 4 reveal a significant difference in the transmittance of the two nanolaminate structures $TiO_2/Cr/TiO_2$ and $TiO_2/Ni/TiO_2$ (Ni and Cr layers being deposited at a sputtering time of 5''), but it must emphasized that the sheet resistances of both structures are equal.

3.3. Engineering the optical properties of nanolaminate OMO structures

The studies, related to the research on the influence of the sputtering deposition process parameters on the optical and electrical properties of oxide/metal NPs/oxide (OMO), were aimed at finding the optimal conditions for the formation of transparent and conductive coatings. The $TiO_2/Ag/TiO_2$ nanolaminate structures produced possessed a very good reproducibility, high transparency and sheet resistance of 7 Ω/sq. The sputtering time was 370 sec. The procedure included a variation of the film thickness of the top dielectric layer of TiO_2 that causes variation of the transmittance spectra. In figure 5, the optical spectra of three nanolaminate structures are presented; they have the same values of the sheet resistance, but the thickness of the second TiO_2 layer was varied. The structures were as follows:

TiO_2 (d = 28 nm)/Ag/TiO_2 (d = 28 nm)
TiO_2 (d = 28 nm)/Ag/TiO_2 (d =56 nm)
TiO_2 (d = 28 nm)/Ag/TiO_2 (d =86 nm).

It can be seen that there is a shift of the transparency maximum from 480 nm to 640 nm and, respectively, the structures are slightly colored from bluish to yellow. This result can be used for spectral selectivity where it is needed.

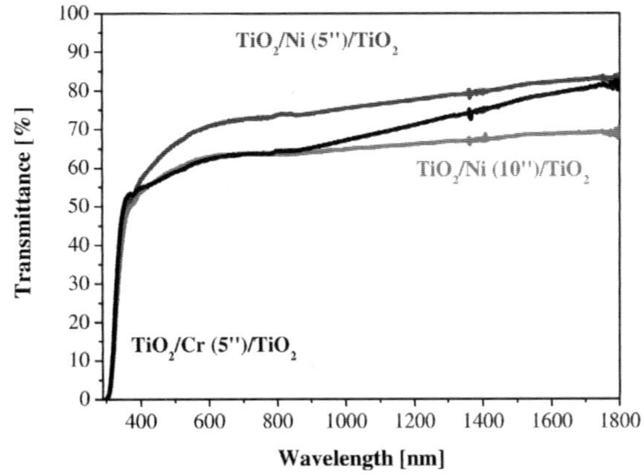

Figure 4. Transmittance spectra of nanolaminate structures with different metals.

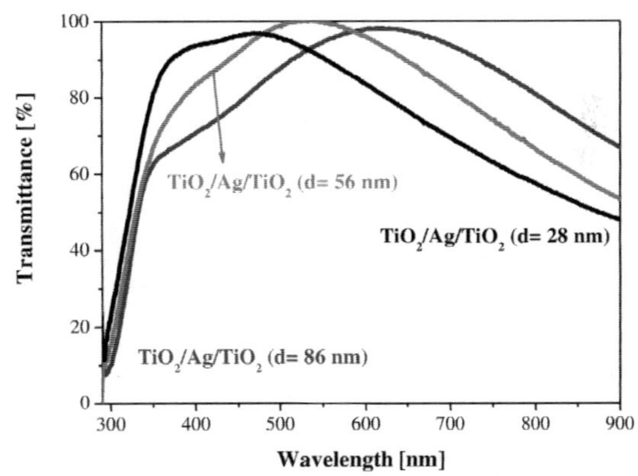

Figure 5. Transmittance spectra of $TiO_2/Ag/TiO_2$ nanolaminate structures with different thickness of the second dielectric layers. The spectra are measured against the glass substrate.

The performance of a nanolaminate structure as a transparent conductive coating is described by the figure of merit (F_{TC}), as defined by Haacke [14] $F_{TC} = \frac{T^{10}}{R_{sh}}$, where T is the transmittance at 550 nm and R_{sh} is the sheet resistance. Table 3 summarizes the figure of merit values for similar dielectric structures prepared with different metals.

It can be concluded that the highest figure of merit, F_{TC}, achieved for a $TiO_2/Ag/TiO_2$ structure reached a value, which is competitive with the ones reported in literature [15].

3.4. Application of the electrical conductivity of OMO structures

Figure 6 presents the temperature dependence of the specific resistivity of the nanolaminate structures. The electrical resistivity of the conductive area at room temperature is $\rho = 9{,}9\times10^{-5}$ Ω.cm. Usually, the TCOs exhibit a semiconductor ehaviour, namely, a negative temperature coefficient of receptivity (TCR) [15]. The nanolaminate structure with metal granulate demonstrates a metallic conductivity, because it has a positive TCR. Therefore, the carriers are delocalized due to thermal activation and the conductivity is dominated by phonon scattering. Similar observations of high conductivities, but with a positive TCR have been reported in the cases of TiO_2: Nb [16] and ZnO: Ga [17] layers. The

Journal of Physics: Conference Series **700** (2016) 012054 doi:10.1088/1742-6596/700/1/012054

deposition method for obtaining OMO nanolaminate structures allows for a selective formation of conductive buses or areas without using an etching process. Figure 7 shows the structure of the conductive area on an oxidized silicon surface, which functions as a resistor.

Table 3. Figure of merit for nanolaminate structures with different metal layers.

Nanolaminate structure	T (λ=550 nm)	Sheet resistance [Ω/sq]	F_{TC} [10^{-3} Ω^{-1}]
TiO_2/Ag/TiO_2	0,97	7	105,3
TiO_2/Cu/TiO_2	0,69	16	1,5
TiO_2/Al/TiO_2, annealed 400°C	0,87	40	6,3

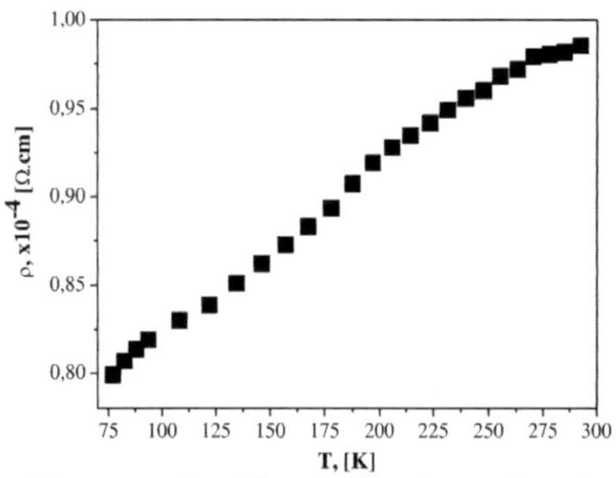

Figure 6. Plot of the electrical resistivity vs temperature.

Figure 7. Resistor formed by an OMO structure on an oxidized Si substrate.

4. Conclusions

The oxide/metal/oxide (OMO) structure is a good alternative to TCO films based on SnO_2:F, ZnO:Al, ZnO:Ga, Sn:In_2O_3, Zn_2SnO_4 etc. The technological process proposed is magnetron sputtering deposition, which is a new approach to the formation of a low-resistivity transparent electrode at room temperature without using high substrate temperature or a post-annealing process. Therefore, it is compatible with polycarbonate or flexible substrates as well. The advantages of the proposed technology are reproducibility of the parameters, effective methods for process control and a selective formation of conductive areas in a dielectric structure.

The nanolaminate structures TiO_2/Ag/TiO_2 were studied and optimized in terms of a high electrical conductivity and transparency.

References

[1] Chopra K I, Major S and Pandy D K 1983 *Thin Solid Films* **102** 1
[2] Knickerbocker S A and Kulkani A K 1996 *J. Vac. Sci. Technol.* A **14** 757
[3] Sahu D R, Lin S Y and Huang J L 2006 *Appl. Surf. Sci.* **252** 7509
[4] Vedraine S, Hajj A, Torchio P and Lucas B 2013 *Org. Electr.* **14** 1122
[5] Guillen C and Herrero J 2010 *Phys. Status Solidi* A **207** 1563
[6] Crupi I, Boscarino S, Strano V, Mirabella S, Simone F and Terrasi A 2012 *Thin Solid Films* **520** 4432
[7] Choi K H, Nam H J, Jeong J A, Cho S W, Kim H K, Kang J W, Kim D G and Cho W J 2008 *Appl. Phys. Lett.* **92** 223302
[8] Jeong J A, Park Y S and Kim H K 2010 *J. Appl. Phys.* **107** 023111
[9] Guillen C and Herrero J 2011 *Thin Solid Films* **520** 1

[10] Beloborodov I, Lopatin A, Vinokur V and Efetov K 2007 *Rev. Mod. Phys.* **79** 469
[11] Vitanov P and Dikov Ch 2015 *BG patent application N=111911/ 27.01.2015*
[12] Yu X, Zhang D, Wang P, Murakami R, Ding B and Song X 2012 *Int. J. Mod. Phys.* **6** 557
[13] Kim D 2012 *Optics Comm.* **285** 1212
[14] Haacke G 1975 *J. Appl. Phys.* **47** 4086
[15] Girtan M 2012 *Solar Energy Mater. Solar Cells* **100** 153
[16] Furubayashi Y, Hitosugi T, Yamamato Y, Inaba K, Kinodo G, Hirose Y, Shimada and Hasegawa T 2005 *Appl. Phys. Lett.* **86** 252101
[17] Bhosle V, Tiwari A and Narayan J 2006 *Appl. Phys. Lett.* **88** 032106